EINSTEIN

May 21, 2000

To Mariah Stelle:

Congratulations on being selected to receive the Science Achievement Award 2001

Best wishes for continued success.

Bella Jacobson
Science Dept. Chair

Also by Denis Brian

Murderers and Other Friendly People:
The Public and Private Worlds of Interviewers

The Enchanted Voyager: The Life of J. B. Rhine

The True Gen: An Intimate Portrait of
Ernest Hemingway by Those Who Knew Him

Fair Game: What Biographers Don't Tell You

Genius Talk: Conversations with
Nobel Scientists and Other Luminaries

EINSTEIN

A Life

Denis Brian

John Wiley & Sons, Inc.

New York • Chichester • Weinheim • Brisbane • Singapore • Toronto

Copyright © 1996 by Denis Brian
Published by John Wiley & Sons, Inc.
All rights reserved. Published simultaneously in Canada.

Library of Congress Cataloging-in-Publication Data:
Brian, Denis.
    Einstein : a life / Denis Brian.
        p.   cm.
    Includes bibliographical references and index.
    ISBN 0-471-11459-6 (cloth : acid free paper)
    ISBN 0-471-19362-3 (paper : alk. paper)
    1. Einstein, Albert, 1879–1955.   2. Physicists—United States—
Biography. I. Title.
QC16.E5B737   1995
530′.092—dc20
   [B]   95-12075

Printed in the United States of America.

10 9 8 7

To Martine, Danielle, Alex, and Emma,

with love

Contents

Preface   ix
Acknowledgments   xiii

1	Childhood and Youth	1
2	First Romance	9
3	To Zurich and the Polytechnic	15
4	Marriage Plans	25
5	Seeking a Position	30
6	The Schoolteacher	35
7	Expectant Father	39
8	Private Lessons	44
9	The Patent Office	51
10	The Olympia Academy	54
11	The Special Theory of Relativity	60
12	"The Happiest Thought of My Life"	69
13	To Prague and Back	79
14	The War to End All Wars	89
15	In the Spotlight	100
16	Danger Signals	107
17	Einstein Discovers America	118
18	The Nobel Prize	137
19	The Uncertainty Principle	155
20	The Perfect Patient	165
21	The Unified Field Theory	173
22	On the International Lecture Circuit	191
23	Einstein in California	208

24	Weighing Options	225
25	Einstein the Refugee	241
26	A New Life in Princeton	256
27	Settling In	274
28	Family Matters	285
29	Politics at Home and Abroad	299
30	World War II and the Threat of Fission	311
31	The Race for the Bomb	326
32	Einstein Goes to War	336
33	The Atomic Bomb	342
34	Toward a Jewish State	347
35	The Birth of Israel	358
36	The FBI Targets Einstein	374
37	The Communist Witch-Hunt	382
38	Conversations and Controversies	393
39	Einstein's Mercy Plea for the Rosenbergs	404
40	The Oppenheimer Affair	415
41	The Last Interview	421
42	Einstein's Legacy	430
Appendix	Einstein's Brain	437
	Notes	441
	Bibliography	481
	Index	491

Preface

After Albert Einstein's death in 1955, his close friend and executor, Otto Nathan, frustrated those researching Einstein's life by steering them away from information that he considered too personal or revealing. He also threatened into submission others who independently discovered new material and wished to make it public. Even the site where Einstein's ashes were scattered was kept secret except from a handful of friends, although his brain has been preserved in the so far vain hope it might provide clues to his genius.

As keeper of the flame, Nathan sought to sustain the shining image of his friend as a secular saint. He was helped in this well-meant subterfuge by Einstein's loyal secretary, Helen Dukas.

Dukas, Einstein's secretary-housekeeper, outlasted both of his wives. She lived with him for twenty-seven years and was with him at the end. Over the years she saved almost every scrap of his mail and copies of his compassionate, witty, and sometimes acid replies. She even collected incoming anonymous insults, threats, and accusations, and eagerly grabbed scribbled notes he had tossed in the wastebasket—much of which went into the archives. But controversial material was withheld from researchers while Nathan and Dukas were in joint control of it.

"Nathan and Dukas felt, and perhaps they were right, that Einstein (and his memory) needed protection," explains Einstein expert Dr. Robert Schulmann of Boston University. "Einstein himself characterized Dukas as his Cerberus." According to Schulmann, "Dukas knew a lot more about Einstein than anyone." But when pressed to discuss him, she invariably replied: "There's nothing more I can say. It's all in the books."

It wasn't, of course. Although in recent years some four hundred books about Einstein and his work have appeared, they tell only part of the truth. As Walter Moore points out, quoting Montaigne, in his illuminating biography of Einstein's friend and colleague Erwin Schrödinger, "Our life is divided betwixt folly and prudence; whoever will write of it only what is revered and canonical will leave about one-half behind." My aim is to balance the equation by retrieving as much of the missing half as I can, to reveal some of the long-guarded secrets. My intent is not to diminish but to enlarge the man.

I have been lucky in my timing. Legal action freed the Einstein Archives from Otto Nathan's clutches a few years before his death in 1987. Now, clues to Einstein's thinking and previously concealed facts about his private life can be found in the tens of thousands of papers in the archives—the originals in Jerusalem, and copies in Princeton and Boston. Still hidden, however, are some letters between Albert and Elsa, his second wife, and between Albert and his sons, and other material judged too sensitive for contemporary eyes. They are likely to be kept locked away until well into the twenty-first century.

Despite those restrictions, previously untapped sources—especially intimates willing to talk freely now that all the principals are dead—have provided a less sanitized and more authentic picture of Einstein than ever before.

He was loved and hated to excess. One admirer called him the greatest Jew since Jesus, and another the greatest since Moses. Some even spoke of him as "a Jewish saint." Critics damned him as a self-promoting fraud who stole others' ideas. Others saw him as a humanitarian in the same league as Mahatma Gandhi and Albert Schweitzer, yet Einstein's liberal views, and his support of world government and of left-wing causes, made both the Nazi authorities and the FBI suspect him as an enemy of the state. Two wives testified to his failure as a husband, and some believed he was a misogynist—though certainly not his adoring friend, Dorothy Commins, who recalled how, gazing into his eyes, she had thought, "Here is the noblest of creatures." Such discrepancies might explain his remark, "There is far too great a disproportion between what one is and what others think one is."

His scientific speculations baffled the general public, but not knowing what Einstein was talking about did nothing to quench their thirst for any information about him.

One aspect of Einstein's life that has been hidden is his way with women. One who could have told but didn't was Helen Dukas. She helped Banesh Hoffmann write his informed but adoring biography, *Albert Einstein: Creator and Rebel*. To commemorate Einstein's 100th birthday, the same team published *Albert Einstein: The Human Side*, which gave no hint of his love life, the tempestuous nature of his first marriage, and the less-than-perfect nature of his second marriage, concentrating instead on the mostly amusing or delightful letters to and from him.

On the subject of women in Einstein's life, Ronald Clark's standard biography does concede that "Einstein's pleasure in the company of women lasted all his life, but there is little more to it than that." There is much more, but Clark and other writers before and since were unable to reveal Einstein's love affairs, because the information was suppressed by the Nathan–Dukas protection agency. That was true until 1982, when all of Einstein's papers were handed over to the Hebrew University as he had wished. Abraham Pais's *'Subtle is the Lord . . .': The Science and the Life of Albert Einstein*, published that same year, and free from Nathan's restrictions, was the first Einstein biography to hint that there was a human side to his subject, a side that had been kept under wraps.

Eleven years later Roger Highfield and Paul Carter, in their *The Private Lives of Albert Einstein*, did some energetic unwrapping. In the London *Sunday Times Books*, reviewer John Carey applauded the freshness of their research, during which they questioned Einstein's adopted granddaughter, Evelyn, and tracked down his former maid. Carey concluded, however, that the biography was a "dredging-job," in which "Einstein can do nothing right [and] despised women, it appears, while expecting them to wait on him . . . Maybe he beat Mileva [his first wife] . . . Maybe he died of syphilis [though] there is no medical evidence to support it."

Fortunately, the material available in the archives allows us to piece together the previously missing parts of Einstein's personal life. In the archives are personal letters to and from two women he loved, and irrefutable evidence for the existence of his illegitimate daughter. For the first time it is possible to tell almost everything—often from the vivid viewpoints of protagonists.

Einstein's life, I discovered, was full of triumphs and tragic ironies. The scientist whose mind took him to the further reaches of space had a schizophrenic son who couldn't find his way across the street. The pacifist who literally wouldn't hurt a fly felt compelled to urge the making of a devastating bomb. The humanist who showed compassion and concern for the children of strangers neglected his own sons and kept the existence of his first, illegitimate child a secret. The lover of solitude was invariably surrounded by women, hounded by the press, and mobbed by crowds. And the dedicated democrat was constantly accused of being a Communist or Communist dupe.

This account will surprise those fixed on him as a secular saint devoted only to good works and to solving cosmic mysteries. It will surprise because it explores what previous biographers left out or covered superficially—his private life, the earthbound Einstein. It reveals him as far more compelling, complicated, and controversial—still in all his glory, but with his halo slightly askew. He is no less a genius for that, of course, and to many perhaps even more endearing.

Denis Brian

Rockport, Massachusetts
October 1995

Acknowledgments

No one helped me more in bringing Einstein to life than Thomas Bucky, his friend for over twenty years, who regarded Einstein as a second father. Others who recollected their encounters and relationships with Einstein include I. F. Stone and his son, Christopher; Elizabeth Roboz Einstein (the widow of Einstein's son Hans Albert); Evelyn Einstein; Dean Hill Jr.; Ashley Montagu; Otto Nathan; Mansfield Williams; James Blackwood; Eric Rogers; Peter Burr; Mrs. Joseph Copp; Jane Leonard Swing Chapman; Mrs. Dorothy Joralem; Mrs. John Wheeler; Theodore von Laue; Freeman Dyson; Dr. Eva Short; Sidney Hook; John Oakes; Senator Alan Cranston; David Sinclair (Upton's son); Mrs. John Kieran; Clair Gilbert; Henry Abrams; Mark Abrams; Gillett Griffin; Dorothy Commins; and Alice Kahler.

Einstein's scientific colleagues — excitement still in their voices after some forty years — told of working with him on his last great quest: to show that nature's fundamental forces, electromagnetism and gravitation, are different aspects of the same universal force. Many considered this a tragic waste of time on an impossible quest — but, said his former assistant John Kemeny, "He was absolutely convinced there would be a unified theory." The hope is still alive. Other scientists provided insights into his intellectual development, his "coming-from-left-field" thoughts, day-to-day conversations, and working methods. They include Banesh Hoffmann, John Kemeny, John Wheeler, S. M. Ulam, Valentin Bargmann, Abraham Pais, Peter Bergmann, I. Bernard Cohen, John Stachel, Victor Weisskopf, I. I. Rabi, Eugene Wigner, George Wald, Murray Gell-Mann, Walter Moore, Karl von Meyenn, Linus Pauling, and Michio Kaku — several of them, like Einstein, Nobel Prize winners.

Eugene Wigner described how he and Leo Szilard got Einstein's signature on a letter urging FDR to support research for an atom bomb; Alice Kahler sent me a letter from Helen Dukas about the uproar in the Einstein household when they got news of the atomic attack on Hiroshima; I. Bernard Cohen recalled his last interview with Einstein — the last ever — just two weeks before he died. Thomas Harvey, the surgeon who removed Einstein's brain, revealed its fate. And Dr. Lucy Rorke recalled what she discovered from examining sections of the brain.

I have concentrated on what I believe to be the most reliable sources for this biography, including those already mentioned, the unpublished memoir by Einstein's sister, Maja, letters to and from him, and the memories of those willing to record their encounters with him.

My wife, Martine, was a tremendous help at every stage of this book — during our frequent discussions about a man who fascinated us both, in shaping the manuscript, and in researching Einstein material in Switzerland, Princeton University, Boston University, the National Archives, and the American Institute of Physics.

Hana Umlauf Lane, my editor, through phone calls, faxes, and Federal Express, has enhanced the book with valuable advice and probing questions. My thanks, too, for the exceptional assistance of managing editor Marcia Samuels.

I also owe thanks particularly to Dr. Robert Schulmann of the Einstein Papers Project at Boston University, as well as to the Hebrew University, the Einstein Archives, the Niels Bohr Library, the Library of Congress, the National Archives, the American Jewish Archives, Charles Niles of the Boston University Library Special Collections, the Lilly Library, Caltech, the Smithsonian, the American Institute of Physics, the FBI, the Columbia University Oral History Collection, Zurich Polytechnic (ETH), Stadt Ulm Stadtarchiv, and the Institute for Advanced Study.

Special thanks to Richard Alan Schwartz, who kindly gave me copies of Einstein's FBI file.

The following public libraries in Massachusetts were of great help: Rockport, Gloucester, Newburyport, Beverly, Salem, Marblehead, and Ipswich. So were the following libraries in Florida: the University of Florida in Miami; the Florida Atlantic in Boca Raton; the Four Arts of Palm Beach; Palm Beach County; West Palm Beach; and Fort Lauderdale.

EINSTEIN

CHAPTER 1

Childhood and Youth

1879 to 1895

From birth to age 16

Albert Einstein was born on March 14, 1879, in Ulm, Germany, with a swollen, misshapen head and a grossly overweight body, causing his grandmother, Jette Koch, to wail, "Much too fat! Much too fat!" His alarmed parents, Hermann and Pauline, consulted the doctor, who assured them that time would heal the deformities, and he was right. Within months everything had become normal, except for the back of Albert's skull. That always remained unusually angular.

But for his parents the worrying was not over, as Albert's speech was late in developing. What in fact happened is in dispute. Einstein maintained he made no attempt to talk until he was past three, and his parents feared that he was mentally retarded. His explanation was that he consciously skipped baby babbling, waiting until he could speak in complete sentences. He stuck to this account throughout his life, responding to an inquiry by his biographer, Carl Seelig, in 1954, as follows: "My parents were worried because I started to talk comparatively late, and they consulted a doctor because of it. I cannot tell you how old I was at the time, but certainly not younger than three."

His version of the facts is contradicted in a letter written by his doting maternal grandmother, Jette. After she and her husband, Julius, visited the Einsteins when Albert was just two years, three months old, she wrote: "He was so good and dear and we talk again and again of his droll ideas." How could he have conveyed droll ideas without speaking?

Albert's sister, Maja, supports their grandmother. Presumably using her parents as the sources, Maja reports that before her birth on November 18, 1881, when Albert was still four months shy of three, he had been promised a new baby to play with. Evidently expecting a toy, he greeted her appearance with a disgruntled "Where are the wheels?" Not bad for a "backward" two-year-old!

Albert was certainly a late and reluctant talker, but not nearly as late as he

1

recalled. He clearly hoarded his words, doling them out at rare intervals to a favored few: the child equivalent of an introspective adult who shuns small talk.

The Einsteins had many relatives in southern Germany, but Albert never knew his paternal grandparents. Hermann's father, Abraham Einstein, reputedly a decent, intelligent man, and his unremarkable wife, Hindel, from Buchau on the Federsee, both died while Albert was an infant. His maternal grandfather, Julius Derzbacher, an enterprising and hardworking small-town baker from Cannstatt, adopted the family name of Koch and made a fortune in the grain trade. He and his brother were partners, and they and their families shared a house. As Maja remembered it, "Their wives [also] shared the cooking, each taking charge of it . . . in weekly turns. If such an arrangement is rather rare, and not only in Germany, theirs was all the more remarkable because it lasted for decades without any friction."

While Albert was an infant, his generally imperturbable father was preoccupied with the family's foundering electrochemical business. Hermann's brother and partner, Jakob, suggested a way out — that they jump ship before it sank. Why stay in a backwater? Ulm's only distinction then was a cathedral boasting the country's biggest organ. Less than a hundred miles away Munich beckoned, the fast-growing political and intellectual capital of southern Germany. The dynamic, ambitious younger brother painted a not altogether fanciful picture of the vibrant metropolis with its sophisticated population crying out to be converted from gaslight to electricity. Hermann was persuaded.

So the Einsteins moved from their modest apartment in Ulm to a large, welcoming house, which they shared with Jakob, in a shady, tree-filled garden on Munich's outskirts. The brothers added plumbing to their electrochemical enterprise and planned to use their profits to mass-produce and market a dynamo Jakob had recently invented.

Despite their early concern, the Einsteins hardly coddled their firstborn, giving him unusually early training in self-reliance. We have his sister's word for it that before Albert was four his parents encouraged him to roam the neighborhood and even to cross the local streets on his own. The first few times they watched surreptitiously to ensure that he looked both ways before crossing. After that he went solo. True, the traffic was mostly horse-drawn carriages, but a child of four at large in the streets was certainly at risk.

Munich did provide the Einstein brothers with more business opportunities than Ulm, but it also teemed with fierce competitors. Once again the brothers ran into trouble. Hermann faced it with characteristic calm. He appeared to regard a financial emergency as an immutable law of nature and was confident that things would work out in the end, or that his wife's wealthy father, Julius, would fling him a lifeline.

Even though the business was shaky, Pauline decided it was time to imbue Albert with her passion for music: she bought a fiddle and hired a teacher. Albert resisted, throwing a chair and a tantrum—which sent his teacher scurrying for the nearest exit. She never came back. Pauline persisted and hired a replacement.

Albert was still liable to express his discontent with the nearest weapon at hand, but the new teacher was made of sterner stuff than the first, and the lessons continued under duress.

On the rare occasions when Albert mixed with children his age, he was quiet and withdrawn—the onlooker. Relatives thought of him as a dear little fellow who never joined in the other children's squabbles, except to separate the combatants. His younger sister knew the other Albert, the little hellion with a wild temper, and she bore the brunt of his ferocity.

Maja escaped serious and frequent injury because she could detect the onset of his rages—his face turned yellow—and would run for cover. His color change was not a foolproof warning signal, however. Once she barely missed getting a concussion from a bowling ball Albert aimed at her head. The next time his face turned from pink to yellow, either her luck ran out or she wasn't watching. He closed in for the attack and smashed Maja over the head with a garden hoe. Years later, when her brother was a dedicated pacifist and literally wouldn't swat a fly, Maja quipped, "A sound skull is needed to be the sister of a thinker."

The "thinker" suddenly emerged from the junior Jekyll and Hyde when Albert was five and ill in bed. To keep him amused, his father gave him a magnetic compass. Instead of throwing it at his sister, Albert shook it and turned it, hoping to catch it unawares. He was both intrigued and puzzled by the compass. What invisible force, he wondered, always made the needle point north? It seemed like a mechanical homing pigeon with a built-in direction finder. He grappled with the enigma for a long time, trying to discover the answer for himself.

Two years later when he began elementary school, his teachers, far from rating him a thinker, revived early fears that he was mentally retarded. Perhaps his mother was partly to blame for having had him tutored at home until he was seven. She had proudly accepted the tutor's inflated assessment of Albert as a whiz kid. By prolonging his isolation from other children she had helped to create a misfit, the odd boy out, which he was inclined to be in any case.

Classmates regarded Albert as a freak because he showed no interest in sports. Teachers thought him dull-witted because of his failure to learn by rote and his strange behavior. He never gave a snappy answer to a question like other students, but always hesitated. And after he had answered, Albert silently moved his lips, repeating the words.

This was obviously his way of coping with the required rote learning. Teachers punished wrong answers with painful whacks on the knuckles. To avoid the pain and humiliation, Albert played for time until he could conjure up the proper reply. After giving it, he silently checked himself to make sure he'd got it right, and perhaps to ensure he wouldn't forget it.

Those school days were a taste of things to come. He ignored whatever bored him, making no attempt to master it; but if something caught his interest, he embraced it with the purposeful concentration of a watchmaker. Once, for example, his sister Maja watched him slowly and carefully build a house of cards.

She had seen others do it and tried it herself, but these houses never reached more than four stories before collapsing. Her brother stuck at it until his house of cards grew to an astonishing fourteen stories.

After two years in elementary school, Albert had shown a talent both for math and for Latin; he liked the latter because it was logical. He was hopeless at everything else and was subjected to angry complaints and painful knuckle raps, which he bore with a faint grin.

One bright spot in the week was Thursday. That was the day when, following a traditional charitable practice of European Jews, the Einsteins hosted a poor student for lunch. At this time, the beneficiary of this practice was Max Talmud (later Talmey), a medical student, who regularly (and for years) joined the family for lunch. As a result, what Albert missed in school he more than made up for at the Thursday lunches. Sensing the youngster's intellectual hunger, Talmud fed him tidbits of the latest scientific breakthroughs, recommended groundbreaking scientific authors, and discussed mathematics and philosophy as if he and Albert were contemporaries.

Albert flourished in this atmosphere and showed his prowess when Uncle Jakob turned up for the lunchtime seminars armed with tricky math problems. When he solved them, Albert yelled triumphantly like a soccer player scoring an unlikely goal. Jakob had a way with words and got Albert interested in algebra by describing it as "a merry science in which we go hunting for a little animal whose name we don't know. So we call it X. When we bag the game we give it the right name."

At twelve, now in high school, Albert had his first encounter with geometry. In later life he was to describe his discovery of Euclid as one of the great delights of his life and refer to a book on Euclidean geometry as "holy." (He had vivid recall of such early events. It was personal details, he said, that eluded him.)

Meanwhile, God waited in the wings. The Almighty was at best a shadowy figure in the Einstein household, where both parents were agnostics. They occasionally discussed Jewish traditions with their relatives. But if the subject of organized religions came up, Albert's father dismissed them all with derisive impartiality as ancient superstitions.

The only Jew in his overwhelmingly Catholic class, he felt neither uncomfortable nor singled out. For example, he didn't take it personally when, in a show-and-tell attempt to enliven religious instruction, the teacher held up a big nail and compared it with those used to crucify Christ. Albert simply thought it was a botched attempt to arouse sympathy. Discussing the event as an adult with his friend and biographer Philipp Frank, Einstein speculated that such a vivid portrayal of brutality was more likely to arouse latent sadism than pity for victims of cruelty. The crucifixion aside, he enjoyed learning about Christianity. It reminded him of Jewish traditions occasionally mentioned at home and gave him a comforting sense of living in a caring, harmonious world.

The state required that Albert be instructed in his faith. Even though both parents were without religious convictions, he had to be taught Judaism because of his Jewish heritage. This was easier said than done, as no one in his school or home was equipped for the task. Eventually a distant relative came to the rescue and was spectacularly successful.

To say God stepped out of the shadows understates the case. The Almighty dazzled Albert. All aglow, the twelve-year-old abandoned his devotion to math and science for the wisdom of Solomon and the ethics of his religious forefathers, all in the worshipful service of the Master of the Universe. Religious ecstasy converted him into an extrovert who composed songs in praise of God, which he belted out as he walked to and from his high school, the Luitpold Gymnasium. His bemused, indulgent parents took it in their stride, though they resisted vigorously when he tried to shame them into giving up pork.

A year later, faith lost out to reason. Albert was lured back from his impassioned proselytizing to the fascination of science and philosophy. With his parents, Uncle Jakob, and friend Max Talmud rooting for him, Albert tried out his intellectual wings. At thirteen, he began to study higher mathematics on his own, along with the intricate suppositions of Immanuel Kant—tough going at any age.

Kant—the tiny, fragile eighteenth-century philosopher—proposed some bizarre ideas, such as that all planets have been or will be inhabited. Moreover, he peppered his discourse with snappy asides like a stand-up comedian—panning pompous philosophers, for example, as those who live on the higher towers of speculation "where there is usually a great deal of wind." His comments on time and space — that they are not products of experience but concoctions of our minds which clothe our sense perceptions — reinforced Einstein's already aroused interest in the subject.

At his most daring, Kant suggested that God might not exist. This idea enraged God-fearing Germans, especially clergymen, prompting some to name their dogs Kant. Ironically, having his name shouted all over the country provided great free publicity for the eccentric philosopher. Kant soon recanted, however, conceding that God was at least a possibility. Einstein reversed the process, first believing as a child, then doubting in a personal God for the rest of his life. He agreed with Kant about many things, however, including that the way to end world wars was through a world government.

Though Einstein abandoned his uncritical religious fervor, feeling he had been deceived into believing lies, he never lost his admiration for the ethical and aesthetic aspects of some Christian and Jewish teachings. Attaching himself to no sect, repulsed by the rigid rules and compulsory behavior dictated by most organized religions, he was still considered by those who knew him to have been deeply religious. They cite his almost childlike wonder at the splendors of the universe and his belief in its ultimate harmony, his concern for the fate of others, and his active commitment to social justice. They also mention his frequent allusions to a cosmic intelligence.

Philosophy and science were not the only enthusiasms to replace Albert's brief, enraptured encounter with organized religion. Herr Reuss, his literature teacher at the Gymnasium, introduced him to the works of such authors as Schiller, Shakespeare, and Goethe—all of whom became enduring favorites of his. (At home, Albert's father sometimes read those same writers aloud to the assembled family after dinner.) Reuss was one of the few teachers Albert liked, especially because he never demanded rote learning.

Against the odds, his years of enforced, mechanical violin lessons had made Albert a music addict. In time his violin became a near-permanent attachment, like his pipe. No one rejoiced more than Maja, who spent many evenings listening to her mother and brother play duets, mostly Mozart and Beethoven sonatas. After mastering the violin, Albert took up the piano, becoming so adept, Maja recalled, that he "constantly searched for new harmonies and transitions of his own invention." He also used music as a study aid. More than once Maja witnessed him solve a problem after a session on the violin or piano. He would play, then suddenly stop and cry out, "There, now I've got it!"

If only he could have solved his father's business problems with a few bars of music. Tough competition and Hermann's free-and-easy management style had doomed the Munich enterprise. The brothers hadn't cleared enough even to market Jakob's newly invented dynamo.

As the optimistic Hermann had anticipated, however, his wife's relatives came to his aid. The wealthy Kochs in Genoa offered to back him and Jakob in another venture, also installing electric lighting. There was one string attached. It would mean a second move, this time out of the country, from Germany to Italy. There the investors could keep a close watch on their investment and curb Hermann's overgenerous impulses toward others in trouble. The offer was accepted, although it meant leaving Albert alone in Munich.

Now fifteen, he would soon be subject to the draft. The law required all fit young German males to complete military service before they could leave the country, so Albert had to stay behind to finish high school and then to tote a rifle. The rest of the family, Uncle Jakob included, traveled south across the border for a new life in Milan, Italy, near the new factory in Pavia.

Reluctantly, Albert moved from his comfortable home with its large, tree-filled garden to a lonely and solitary existence in a boardinghouse. There were no friends to share his after-school hours, and his only visitor was a distant relative who looked in on him occasionally. He wrote regularly to his family but gave no hint of a growing depression. School had become a dread prospect. He went in trepidation to face teachers who barely tolerated him and schoolmates who kept their distance. They treated as a curious outcast this healthy male who never kicked a ball, would sooner read than run, and was less than eager to serve in the army.

Greek was his biggest ordeal. His failure to grasp it had so exasperated the teacher that he conducted the class as if Albert weren't there. Sitting at the back of the room, Albert didn't always catch what was said and probably wouldn't have understood it anyway. Bored and uneasy, aware of the teacher's animosity, he just sat there, smiling slightly. Whether the smile was the grin-and-bear-it variety or derisive is uncertain, but it turned the teacher's tongue from Greek to plain German, in which he told Albert that he would never amount to anything, that he was wasting everyone's time, and that he should leave the school immediately. When Albert protested that he had done nothing wrong, the teacher complained,

"But you sit there in the back row, smiling. And that undermines the respect a teacher needs for his class."

In fact, Albert longed to fulfill the Greek teacher's wishes. During a visit to the family doctor for a minor illness he discussed the situation, unable to hide his desperation. Albert Einstein was never the calm, imperturbable Buddha-like character of popular myth. At fifteen, he was both high-strung and emotional, and felt abandoned. The doctor recognized this and gave him a to-whom-it-may-concern letter, warning that unless Albert was allowed to recuperate with his family he might suffer a complete breakdown.

To strengthen his hand, Albert showed the doctor's letter to a math teacher—on the face of it a bad choice, because the teacher was one of the few who would have liked Albert to stay. Nevertheless, he too saw that Albert was sick with worry, and sympathized with his plight. He wrote a note saying that Einstein was so proficient at math there was little more he could teach him.

Albert took these possible passports to the school principal and they did the trick. He was free from school and the country. Had he waited until he was sixteen he would have first been required to complete his military service before joining his family abroad.

If he had in fact been heading for a breakdown, the family reunion in Italy was the cure. Maja had never seen him in such high spirits. He soon had a coterie of friends. Yet he had arrived at a bad time. His father was struggling to get the new business off the ground and hardly welcomed the added burden of a school dropout. They constantly argued over Albert's future. For now he seemed happy to drift aimlessly, stirring himself only to visit Milan's museums and art galleries or to explore the countryside with his newfound friends.

When pressed, Albert said that he might eventually consider teaching philosophy, Kant especially. This was too much for his harassed father, who ridiculed the "philosophical nonsense" and pushed for a practical job such as electrical engineering. They were doubtless too emotionally upset to appreciate the irony that Hermann was advocating a trade rapidly taking him down to financial ruin for the third time.

Albert put up a strenuous fight, saying that even the thought of a practical job was unbearable. He clung to his dream of teaching philosophy, although it was impossible without a degree, and as a high-school dropout he had spoiled any chance of attending a university. Without further education only practical jobs were open to him, and the best of those required technical college training. He needed a college diploma even to teach in high school.

Hermann finally prevailed, persuading Albert to bear the unbearable and to apply to a technical college for a course in electrical engineering. Albert set his sights high—on the Zurich Polytechnic, a Swiss technical college with an international reputation. Its other big attraction was that he had merely to pass the entrance exam; he wouldn't have to finish high school. Meanwhile, finally moved by his father's plight, he agreed to help out at the family's failing business.

Maja was astounded by the change in her brother in just six months. The nervous, withdrawn dreamer had become an amiable, outgoing young man with a tart sense of humor. Was it the Italian air? The warmhearted people? His escape from purgatory?

What hadn't changed was his laserlike ability to focus on whatever held his interest. On social occasions, when the living room was filled with conversation and music, Albert brought pen, ink, and notebook to the party rather than shutting himself up in his room to study. Maja was amused to see him squeeze himself onto the sofa, balance the inkwell on the armrest, then set to work apparently oblivious of all distractions.

His favorite uncle, Caesar Koch, proudly displayed an essay Albert had mailed to him describing a proposed experiment to determine if electricity, magnetism, and the ether were connected. In those days, the ether was thought to be a rarefied and invisible element permeating all space. Einstein's essay showed original thinking, even though experiments later demonstrated there was no such thing as the ether.

When not studying for the college entrance exam, Albert kept his word and helped out at the family business. There his Uncle Jakob and an assistant engineer had been stymied for hours over calculations needed to solve a technical problem. Albert offered to try his hand. He had the answer in fifteen minutes. Jakob enthusiastically seconded Caesar in predicting Albert's great future.

These lofty predictions only compounded the shock when Albert flunked the entrance exam to Zurich Polytechnic. He had been floored by French, chemistry, and biology, subjects he had neglected through lack of interest.

Despite this traumatic failure, his uncles were not alone in detecting Albert's unusual talents. Heinrich Weber, the Polytechnic's professor of physics, was so impressed with Albert's high scores in math and science that he invited him to audit his lectures. And Albin Herzog, the college principal, noting that at sixteen Albert was two years younger than most taking the exam, took that into consideration and promised to admit him the following year. Albert wouldn't even have to take the exam again. He just had to get a high-school diploma from any school of his choice. Failure had turned into near triumph.

CHAPTER 2

First Romance

1895 to 1897

16 to 18 years old

T he Einsteins focused on a high school in Aarau, a small valley town twenty-five miles west of Zurich, within walking distance of the Swiss-German border. Its mostly Protestant population spoke German, so Albert would not have the burden of studying another language. The school's appeal was further enhanced by an offer from one of its teachers, Jost Winteler, to take Albert as a lodger. And although Winteler and wife Pauline had four sons and three daughters, Albert was assured of his own bedroom. He doubtless would have accepted a broom closet rather than languish alone in another dreary boardinghouse. The offer was accepted, and Albert's wealthy Koch relatives agreed to pay his school fees.

After only a few months in Italy, Albert had thought the Italians the most civilized people he'd known. Now, on the way to northern Switzerland, where many people were of German origin, his old fears returned that once again he would be at the mercy of rigid, unsympathetic teachers.

Was he any better off than an escaped convict reluctantly returning to captivity to endure a year's hard labor and strict discipline? And for what? To prepare for something he scorned—a "practical" job!

But as his train approached his destination he had a vivid, euphoric change of mood. Riding through glistening meadows, forests, and vineyards, breathing the sweet air, his fears evaporated. This glowing first impression never faded, and he later described Aarau as "an unforgettable oasis in that European oasis, Switzerland."

The Wintelers lived up to the scenery. The children treated Albert as one of the family, and the parents welcomed him with such affection that he was soon calling them "Mommy" and "Daddy." Jost Winteler, who taught history and classical literature, gave a taste of his teaching style at family meals, sustaining spirited discussions and encouraging everyone to take part in an unrestricted give-and-take about anything sparking his or her interest. He aimed to fascinate rather than force-feed. If the other teachers were even a pale reflection of Winteler, Aarau would be paradise compared to Munich.

But paradise seemed to have a trapdoor. Albert's first syllabus listed singing, gymnastics, and military instruction. He loathed the idea of singing or exercising in public and had been allergic to the military since childhood, when the sight of soldiers on the march made him cry. He was deeply relieved to find that the first two subjects were optional and that the class in how to handle guns and march in unison was required only of Swiss nationals. Being a young man without a country had its advantages.

Albert reveled in the relaxed atmosphere in which teachers freely discussed controversial topics with students, even politics—unthinkable in German schools—and encouraged them to devise and conduct their own chemistry experiments, short of blowing up the place. Naturally, he concentrated on subjects he had flunked at Zurich: chemistry, French, and biology. And while most of his class were raising their voices in song, he took advanced violin lessons.

Some weekends Jost Winteler led two of his older children and a handful of students into the country to stalk rare birds. On the lookout for a sudden flash of added color among the bluebells, buttercups, and wild red rhododendrons, they walked in a silence broken only by muted voices, cowbells, or birdsong. At times, Albert and Winteler's daughter Marie went along. A sensitive, affectionate eighteen-year-old, she was following in her father's footsteps in more ways than one, training to be a teacher. If she and Albert spent less time bird-watching than in watching each other, it was because these field trips were their few chances to be almost alone together. In Italy he had been infatuated by a flock of young women who found him charming, but had kept his distance and his head. Now, at sixteen, he believed he was really in love.

Even so, he didn't neglect the younger Winteler children, endearing himself to them by his prowess in making kites out of paper and scraps of wood and by showing them how to keep these homemade toys airborne.

Perhaps he spent too much time at play and in musical interludes with Marie—she on the piano, he on the violin—because his school report in December 1895 revealed the usual peaks and valleys, high in math and physics, low in French and almost everything else. Jost Winteler was worried, but Albert's father wrote to reassure him that he was used to his son "getting not-so-good grades along with very good ones."

Albert seemed to share his father's confidence. Instead of cramming over the next few months, he spent much of his free time with the school orchestra and on bird-watching trips. He even took an active part in a local May Day celebration, when a classmate, Emil Ott, saw Albert acting very much out of character.

Among that day's traditional events was a mock sword battle between high-school students and young men from the town. According to Einstein biographer Carl Seelig, the students were dressed to kill—or, more likely, to impress female onlookers—wearing black fencing jackets and black top hats with white skulls embroidered on the fronts of both. Because the student group was a man short, Albert agreed to fill the gap—proof, remarked Ott, that despite Einstein's already strong pacifist views, he was no killjoy. But before the battle could commence, a heavy downpour cleared the field. The fight was called off. Somehow Albert didn't get the message, and while everyone else ran for cover he was left standing alone, rain dripping from the black top hat, still grasping his sword. Recalling the scene years later, Ott felt sure that it was the first and last time Albert Einstein was ever seen with a weapon in his hand.

On the other hand, Albert was almost welded to his "dear child," as Marie called his violin, even taking it with him on visits home. Home was now Pavia, where, after a rocky year in Milan, the Einsteins had moved into a less expensive house to keep their business afloat. In these new surroundings, he rarely missed a chance to make music with the local young women.

In his letters to Marie during these visits he showed sensitivity for her feelings, saying he much preferred her as a partner, and disparaging the Pavia women, with whom he had only days before been infatuated, as fussy, overdressed, and stuffy perfectionists. They irritated him, he wrote, by insisting that the music they played be both fast and flawless — implying that he failed on both counts — and he escaped from them as soon as he could. He also complained of Pavia's rundown, dingy streets and buildings; the only good word he had to say was for the city's children.

Unlike the Pavia women, Aarau classmate Hans Byland, the double bassist in the school orchestra, had nothing but praise for Einstein the Musician. "What fire there was in his playing!" Byland recalled, especially when they played Mozart sonatas together in the school refectory and the first notes of Einstein's violin seemed to make the room expand. In such moments Albert was so dramatically transformed that Byland thought of him as a split personality hiding his deep emotions—brought out by music—with a prickly exterior. Once asked if he counted the beats, he replied with a laugh, "Heaven, no! It's in my blood!"

As soon as the music was over, however, Einstein reverted to his protective cover. According to Byland, "his attitude towards the world [then was] that of a laughing philosopher and his witty mockery lashed any conceit or pose"; he was a youngster who "loathed any display of sentimentality, and kept a cool head even in a slightly hysterical atmosphere." Byland also noted Albert's luck in lodging with the Wintelers: "Fate decreed that this precise thinker should pitch his tent with the romantically inclined Winteler family where he felt completely happy."

Free from the transforming influence of music, Einstein became, in Byland's admiring view, a charismatic young man who "strode energetically up and down in a rapid, almost crazy tempo of a restless spirit . . . Nothing escaped his bright brown eyes. Whoever approached him came under the spell of his superior personality. A sarcastic curl of his rather full mouth with the protruding lower lip did not encourage Philistines to fraternize with him." Byland also saw him as a much-traveled, sophisticated, wise, and well-informed conversationalist. Perhaps compensating for years of submissive silence, Einstein now frequently voiced his thoughts, however provocative.

Once he and Byland were on a field trip to the Jura Mountains led by geology professor Fritz Muhlberg. Although Albert found the subject less than enthralling, he thought the man interesting and original. Others resented Muhlberg's curt manner but merely grumbled about it. Einstein responded to Muhlberg in kind. When the professor asked in his usual gruff way, "Now, Einstein, how does the strata run from here? From below upwards, or vice versa?" Einstein replied, "It's pretty much the same to me whichever way they run, Professor." To Byland, this impudent rebuff demonstrated Einstein's integrity and "courageous love of truth," and "gave his whole personality a certain cachet which, in the long run, was bound to impress even his opponents."

Ott treasured his memory of Albert, the good sport, in lonely splendor, soaked to the skin and clutching a sword. Byland rhapsodized over his dynamic personality and described his mastery of the violin in ecstatic terms as if each time he played was an epiphany. Both chuckled over his put-down of Professor Muhlberg. But neither noted his budding genius.

Yet it was during this time that his extracurricular activities included teaching himself calculus and speculating on the possibility of splitting the atom. He had also begun to think about light, wondering especially what things might look like if someone went along for the ride with a light wave, keeping pace with it as it traveled through space. This visionary flight of fancy seemed less like scientific speculation than like an illustration for a Jules Verne adventure story. But when he got the answer, after ten years of thought on the subject, it became one aspect of his relativity theory.

Listeners have been deadlocked over their views of Einstein as a violinist. Some, like Byland, spoke of him as an inspired virtuoso, while others panned him as a clumsy amateur. It seems that he was good but not remarkable. Still, he stood out from nine other students taking a music exam at school in the spring of 1895. The examiner praised him for his "sparkling" rendition of an adagio from a Beethoven sonata, played with "deep understanding."

He had reason to sparkle. He and Marie Winteler were in love, and his letters to her show how the sardonic, intimidating Einstein relished by classmate Byland could quickly dissolve into a mushy, lovesick romantic.

At home with his family for spring break, Albert wrote to Marie as his "beloved sweetheart," thanking her for her "charming little letter, which made me endlessly happy . . . Only now do I realize how indispensable my dear little sunshine has become to my happiness. My mother has also taken you to her heart, even though she does not know you; I only let her read two of your charming little letters." He told Marie that his mother laughed at him because he had lost all interest in girls he once found enchanting. "You mean more to my soul," he concluded, "than the whole world did before."

When Albert graduated from Aarau with the required diploma and moved to Zurich to study at the Polytechnic, still with the financial support of relatives, Marie left home to teach first-graders in a school some miles away. Separation intensified their passion, and they exchanged eager letters full of longing. Then, without warning, Albert suggested that they stop writing. Marie was distraught. Was it, she wondered, because he suspected she loved someone else? If so, he was mistaken; she would love him forever. Hadn't they each sworn eternal love? A week later Albert sent a consoling reply, and their exchange of love letters continued.

His heart wasn't in it, though. Mileva Maric, a young Serbian woman in his college physics class, had caught his interest, and his "eternal love" for Marie was fading fast. For Einstein eternity had lasted less than a year. But until he could find a way to tell Marie, he kept writing as if still in love.

"My dear, dear sweetheart!" Marie replied on November 30, 1896. "Finally, finally I felt happy, happy, something only your dear letters can bring about . . . How I sometimes yearn to stroke a little your dear tired brow when I picture to myself how you now sit in your little room tired and pensive, the way you often did back at home, you know. In such moments I would like to fly to my sweetheart and tell him how much I love him." She longed to visit him in Zurich "to see where my darling dreams away his days. I am looking forward to that so much. Then I will arrange everything the way I like it, and you will enjoy your little study room twice as much."

She never made the trip. And instead of meeting her during the spring break of 1897, Albert went home to Pavia. There he and his sister spent a carefree time, playing duets that were often interrupted by their helpless laughter and teasing their mother until she rushed from the room.

Unable to tell Marie the truth, Albert simply stopped writing to her. Distressed by the halt in the flow of letters from him, Marie wrote to his mother to complain. Pauline Einstein was fond of Marie, and not knowing of Albert's change of heart, she sent a chatty reply, reporting the music and happy laughter in the next room and attributing Albert's failure to write to the family weakness — laziness.

If Albert ever told Marie that their romance was over, the letter is not in the archives. When he did write to Aarau it was to her mother, declining an invitation to stay with the Wintelers. Then, although he referred elliptically to the breakup for which he blamed himself, he never once mentioned Marie by name.

Their breakup caused Marie to have a physical and emotional breakdown. Einstein was sympathetic but adamant. For him it was over. His college classmate Mileva Maric was to become the woman in his life.

Even though he took great care to avoid meeting Marie, he always remained on affectionate terms with the rest of the family. He loved the father, Jost Winteler, and admired him both as an outstanding teacher-scholar and as a liberal of great integrity. Jost became his role model. Under his influence, Albert decided to defy his father and to study with teaching in mind rather than for a "practical" job.

Emboldened by his golden days at Aarau, Einstein approached the Zurich college professors with breezy informality. The autocratic bureaucrats were not amused and eventually made him pay for it. Still, a handful of kindred spirits among his fellow students gleefully responded to his caustic wit and audacity. One, Marcel Grossmann, showed remarkable perception in recognizing Einstein's incipient genius. All he had to go by were a few conversations with Einstein over iced coffee in the Metropole café. Despite the faculty's almost unanimous opinion that Einstein was out of his depth at college, Grossmann told his own parents that in his opinion this eighteen-year-old classmate would be a very great man.

CHAPTER 3

To Zurich and
the Polytechnic

1897 to 1900

18 to 21 years old

B ecause of Einstein's arresting looks and personality, his college friends were puzzled when he gravitated to Mileva Maric, a somewhat shapeless woman of awkward gait caused by a congenital dislocation of the hip. She was the only female in his class, four years his senior, and showed little sense of humor, while even the tamest joke set off his explosive laughter.

It was assumed that they had linked up for moral support as "outsiders" in the largely insular Swiss student body. Mileva came from Bacska, a region of what was then southern Hungary. Bacska was north of Serbia, but had a very large Serbian population. Mileva was an ethnic Serb, as were both of her parents (today the area is part of Serbia). Albert was a transplant from Germany via Italy. Having renounced his German nationality, he was now a man without a country.

Carl Seelig, an early well-informed Einstein biographer, took a sympathetic view of Mileva, concluding that "her dreamy, ponderous nature often curdled her life and studies. Her contemporaries found Mileva a gloomy, laconic, and distrustful character. Whoever got to know her better began to appreciate her Slav open-mindedness and the simple modesty with which she often followed the liveliest debates from the background." The background might partly explain her gloom. She was also burdened with the growing certainty that her younger sister, Zorka, was mentally unbalanced. Nevertheless, in Einstein's presence she showed a vitality and an enthusiasm not evident to others.

For her to be studying physics at all testified to her talent and steely persistence. In her country, where her father was a middle-rank civil servant with a few acres of land, physics was a strictly male preserve. Mileva broke the barrier by persuading a high-school teacher in Zagreb to let her attend an otherwise all-male physics class. That became the end of the road for her in Zagreb, where she might as well have expected to join a dueling club as to continue with advanced physics.

Her hope lay across the frontier in more liberal Switzerland, where higher learning was open to outstanding women. Mileva's record qualified her to join the select group at the prestigious Polytechnic in Zurich, intellectual capital of the German-speaking Swiss. Once again she was the only woman in the class.

Although Einstein had broken up with Marie Winteler early in 1897, his relationship with Mileva later that same year was still casual, even cautious. It seemed likely to stay that way, because she intended to leave Zurich to spend the fall semester auditing physics and math classes at Heidelberg University, some 150 miles to the north. There was still no hint of romance between them when he saw her off on the train and suggested that, if she ever got bored in Heidelberg, she might drop him a line.

She did write, in October, when she was stuck in her room because of a thick fog, to say that she intended to rejoin him in Zurich when she resumed studying at the Polytechnic in the spring. He replied four months later, apologizing for the delay and encouraging her to hurry back to catch up with the courses she had missed while away.

Separation from Mileva sharpened Einstein's romantic interest, and when he went home to Milan for the spring break he carried her photo with him. As his mother stared at the picture, Albert murmured that Mileva was a clever creature. His mother and sister teased him, as they had during his infatuation for Marie, and he told Mileva of this in an exuberant letter containing a playful triple pun, which, unfortunately, defies English translation.

When he and Mileva spent vacations at their respective homes, her usual complaint was his infrequent letters. He compensated when he did write, giving colorful details of his laughter-and-music-filled home life and expressing his excitement at living during a revolution in physics, which he meant to join.

Some who had been exposed to his absentmindedness scoffed when they heard of his high ambition. After one weekend with friends, for example, he left his suitcase behind, and the host predicted that Albert would never amount to anything because of his rotten memory. Both landladies he had in Zurich during his student years suffered from it. Surprisingly, neither tried to evict him, even though they were frequently groggy from being awakened in the early hours because he'd forgotten his keys.

Albert never forgot his beloved "fiddle," however, and took every opportunity to play it, as his second landlady's daughter, Susanne Markwalder, recalled. She was at home one warm summer day when windows in the neighborhood were wide open. Through one came the sound of Mozart sonatas. Enchanted, Albert immediately wanted to know who was playing and where. Told it was a piano teacher who lived in an attic, he grabbed his violin and hurried out, ignoring protests that he wasn't wearing a collar or tie—almost a crime in those days. The slamming of the front gate was quickly followed by the sound of Albert's violin joining the piano in a spirited Mozart duet. Shortly afterward a little old woman called at Einstein's boardinghouse, tentatively asking for the name of the extraordinary young man who had burst into her attic crying, "Keep playing!" She left reassured that the excited, half-dressed intruder was a harmless student with a passion for Mozart.

Einstein's indifferent or bizarre approach to clothes amused the mother of his friend and classmate Jakob Ehrat. Once when visiting their home he wore something odd around his throat. Asked what it was, he explained that because he had a cold he'd grabbed a runner off a piece of furniture to use as a scarf.

Lunchtimes at Einstein's lodging were enlivened by tourists, foreign students, and Anita Augsburg, a resident actress and suffragette addicted to cigarettes and men's clothes. Albert didn't see much of this group, however, because he usually lunched in his room. His favorite meal was a slice of apple or plum tart from a local bakery, followed by a big cigar.

When Susanne Markwalder, an elementary-school teacher, brought girlfriends home to tea, he often joined them and invariably walked them home. Albert treated Susanne with easygoing affection, once comparing her to a donkey stuck on a mountain after she missed her cue while accompanying him on the piano.

On another occasion, a widow and her two daughters invited themselves to a rehearsal. As Albert and Susanne played, all three visitors sat on a nearby sofa knitting, clashing their needles and sighing loudly when they—frequently—dropped stitches. When Albert very deliberately closed the sheet music and replaced his violin in its case, the widow asked why they had stopped playing. "Because," he replied, "we wouldn't dream of interrupting your work." Einstein later found a unique way of apologizing to the knitting trio for his sarcasm: he took his violin into the garden, where he and an Italian tenor serenaded all the women in the boardinghouse.

Susanne first met Mileva when she came to the house to study with Albert, and got the impression of a modest and unassuming woman. She thought Mileva was hardly a match for the "irresistible" Albert, whom she and her mother knew as a thoroughly decent, shy, but impulsive and outspoken young man.

His offbeat attitude and penchant for punning eased embarrassing situations. When someone referring to Mileva's obvious limp remarked that he wouldn't have the courage to marry a woman who wasn't fit, Einstein calmly mentioned Mileva's lovely voice as if it more than compensated. During one social evening at Professor Alfred Stern's home, after a fellow guest had fiercely argued some point, Einstein pointed to his violin and said they could now play "what you obviously like—Handel." In German, *Handel* is also a pun on the word *händel*, which means bickering (without the umlaut it means commerce). *Händelsuchen* means to pick a quarrel.

Not that he was always blameless in these situations, and he knew it. He exasperated his autocratic professors because he regarded most of them as irrational or ignorant, and he showed it. His independent, disdainful manner irritated them even more than it had his insecure high-school teacher of Greek. He infuriated physics instructor Jean Pernet, who saw Einstein dump the official instructions on how to conduct an experiment into the wastebasket without a second glance. Pernet complained to an assistant, who daringly replied that Einstein's methods were interesting and his solutions always right. Pernet disagreed. He confronted Einstein. "You're enthusiastic," he conceded, "but hopeless at physics. For your own good you should switch to something else, medicine maybe, literature or law." Math professor Hermann Minkowski didn't see even the enthusiasm in his classes, and called Einstein a lazy dog.

His casual study habits also irritated Heinrich Weber, who had expected great things of him and was irked because Einstein called him "Herr Weber" instead of the more respectful "Herr Professor." For his part Albert was disappointed in Weber for excluding from his history of physics the stunning ideas of James Maxwell.

Maxwell died the year Einstein was born, 1879. Like Einstein, he was considered a poor student as a child. Later, at Cambridge University, Maxwell tried to create more study time by going to bed in the afternoon, then waking at 9 P.M. and working through much of the night. Maxwell's mistake was to include an exercise period in the small hours, between two and two-thirty, when he ran up and down the stairs and along the corridors of his quarters. He abruptly ended the experiment after sleep-starved fellow students bombarded him with shoes. Such mild eccentricity appealed to Einstein.

He cut Weber's classes to read Maxwell's theory that light and electricity were different aspects of the same phenomenon, and that electromagnetic action moves through space in waves similar to light waves and at the same velocity. Maxwell based his research on Faraday's pioneering work, but his equations were uniquely his own. Maxwell's equations, as they are known, soon became a mathematical key to unlocking many of the mysteries of electricity and led to radio, radar, and television.

Criticized for neglecting his lab work, Einstein attempted to catch up, only to have his efforts backfire. Determined to do things his way, he ignored the printed instructions provided by Pernet and caused an explosion, severely injuring his right hand. Yet when he proposed a safer and more ambitious experiment, to measure the earth's movement against the ether (then thought to exist), Professor Weber turned him down—just one of several hints that Weber had switched from well-wisher to antagonist.

Weber's failure to update his lectures to include the breakthroughs of the past quarter century is like a physics teacher's surveying the history of physics in 1941 and excluding Einstein. But Weber redeemed himself by encouraging students to read for themselves where he left off. Einstein hardly needed telling: "Day and night he buried himself in books, from which he learned the art of erecting a mathematical framework on which to build up the structure of physics." Yet apart from attending a few lectures by Hermann Minkowski and Carl Geiser, he was a no-show at other math classes.

If he needed an excuse to neglect math, it was that he was captivated by physics, and was already developing ideas destined to rank him with Galileo and Newton. With physics he could go intuitively to the heart of the matter, and see what needed to be done. Wherever he was—on the veranda smoking his pipe and watching the sun set, sipping iced coffee in a Viennese café, at a picnic in the nearby forest—he let his imagination loose on the revolutionary theories of his contemporaries. Not that he didn't find time for music or his favorite poet, Heine, a witty and lyrical former law student; but never at the expense of his main interest. Even when sailing on Lake Zurich with his landlady's daughter, he took notes or read until the wind caught the sails. When the wind died he returned to his books, absorbed in the works of Maxwell, Hertz, Kirchhoff, or Helmholtz.

Hermann Helmholtz, professor of physics at the University of Berlin, wanted to purge all metaphysical speculation from science. He suggested rejecting every theory not verifiable by observation or experiment, and advised fellow scientists to ask themselves two vital questions: How do our ideas correspond to reality? and What in our sense perception and thought is true? He practiced what he predicated. His paper *On the Conservation of Energy* is the basis of the science of thermodynamics. A man of astonishing versatility, he was also a biologist, physician, mathematician, and philosopher. Helmholtz did pioneering work in physiology, acoustics, chemistry, magnetism, and optics; during the latter he invented the ophthalmoscope to measure the interior of the eye. His final research, of special interest to Einstein, involved the relation of matter to the ether, but he died in 1894 before he could put his ideas to the test.

Helmholtz's assistant, Heinrich Hertz, had confirmed Maxwell's electromagnetic theory in 1886, and during his experiments had produced electromagnetic (radio) waves. He showed that they traveled at the velocity of light, and like light could be reflected, refracted, and polarized. Hertz also died in 1894.

Gustav Kirchhoff, another scientist who demanded a more rigorous approach, was professor of physics at the University of Berlin at the same time as Helmholtz. He said that Newton's theories failed to explain the proliferating physical phenomena discovered in the last years of the nineteenth century. Einstein read about Kirchhoff's work in electricity, thermodynamics, and spectroscopy, which he would soon study in greater detail.

Michele Besso, a friend studying engineering, recommended Ernst Mach's *Science of Mechanics*, which ridiculed the concept of absolute space and absolute motion and suggested that Newton's laws should be reexamined and rewritten. Einstein loved this no-nonsense attitude, and Besso believed it was Mach's influence that led Einstein into thinking of "observables—and to become profoundly skeptical of concepts like absolute space and absolute time."

When Mach was a child in Austria, his teachers, like Einstein's and Maxwell's, thought him backward and difficult. Yet he grew into an intellectual giant, rivaling Helmholtz for versatility and agreeing with many of his views. Mach's *Principle of Economy*, for instance, states that scientists must use the simplest means of arriving at their results and exclude everything not perceived by the senses. Einstein admired Mach for his independence, incorruptibility, and ability to see the world as if through the eyes of a curious child. The charming, unassuming man dazzled psychologist William James, who had never had such a strong impression of pure intellectual genius: Mach seemed to know everything about everything.

With what amounted to clairvoyance, Einstein's classmate Marcel Grossmann judged him to be in the same league as Mach. For his part, Einstein saw Grossmann as an ideal and popular student and himself as a daydreamer, aloof, discontented, and unpopular.

He was certainly not unpopular with classmate Jakob Ehrat, who almost idolized him for his uncompromising honesty and complete lack of pettiness. As fellow Jews, Einstein and Ehrat speculated about the cause of worldwide anti-Semitism. Einstein concluded that it was obviously not because Jews were worse than others, but just because they were different.

Friedrich Adler, son of psychiatrist Viktor Adler, founder of the Austrian Social Democratic party, was another friend. Einstein called him the purest and most idealistic spirit he ever knew. Friedrich in turn sympathized with Einstein as a brilliant misfit, whom professors treated with such contempt that they locked him out of the college library. He believed that Einstein's problem was an inability to ingratiate himself with those in authority. Not that Friedrich was the man to emulate; his own method of dealing with those in power was deadly force. He eventually assassinated Austria's prime minister in the belief that he was helping the cause of peace, and was sentenced to death. Indirectly, Einstein helped to save his life. (Adler's death sentence was commuted when he was judged insane, partly because he had criticized Einstein's theory of relativity, which by then was accepted by many physicists.)

But that was years ahead. These days Albert couldn't even help his own parents solve their financial problems. He shared his feelings of frustration and guilt with his sister, Maja, saying that their parents would be better off if he were dead. At the same time, he consoled himself with the thought that he had never been extravagant, and had deprived himself of entertainment and other distractions — except, of course, for his studies.

In this mood, he found consolation in the home of history professor Alfred Stern, who, with his wife and daughter, always made him welcome. Thanking Stern for those therapeutic visits, Einstein acknowledged his kindness and ability to cheer him up, although he claimed to be an essentially cheerful individual and that his bad moods were usually caused by indigestion or something of the sort. This was quite at odds with how he would soon describe himself to a friend, Julia Niggli.

Most friends, however, would echo his self-assessment. They cherished him for his wit and ebullience, and for entertaining them with his spontaneity, his passion for Mozart, and his mockery of the pompous and pretentious.

During the summer of 1899, Albert joined his mother and sister for a vacation at the quiet and charming Paradise Hotel in Mettmenstetten, Switzerland. Anna Schmid, the seventeen-year-old sister-in-law of the hotel's owner, found Einstein in a playful mood when she asked him to sign her autograph book. He exceeded her expectations by writing:

Little girl, small and fine,
What should I inscribe for you here?

I could think of many things
Including a kiss
On your tiny little mouth.
If you're angry about it
Do not start to cry.
The best punishment is—
To give me one, too.
This little greeting
Is in remembrance of your rascally little friend,
Albert Einstein.

This carefree, flirtatious doggerel belied his mood. He didn't understand why, but his sister and mother seemed to have become conventional and narrow-minded. Instead of showing affection, they behaved like estranged friends who stayed together through habit. He missed the family fun when all three of them had collapsed with helpless laughter. Puzzled and disappointed, he consoled himself by thinking of Mileva and her "divine composure." He had not yet realized that Mileva was the cause of the changed atmosphere. The more his mother learned about Mileva the less she liked her, and Mrs. Einstein did not hide her feelings.

Nevertheless, Albert was not too disconsolate nor too absorbed in scientific speculations to write a breezy letter to Julia Niggli, a young woman with whom he had played duets at the Wintelers'. In it he gives a revealing and unflattering self-analysis, and possibly an insight into how he treated Marie Winteler during their brief romance.

Julia had confided to Einstein that she was in love, and with the wisdom of his twenty years he cautioned her not to expect permanent happiness with someone else, even the man she was in love with. He knew exactly how the man would turn out, Einstein said, because he had a similar personality. She must expect her lover to be sullen one day, high-spirited on the next, cold and depressed on the third. Should she find this not too discouraging, he added that he'd forgotten to mention that the man would also be unfaithful, ungrateful, and selfish.

If a killjoy in his advice to the lovelorn, Einstein willingly helped in other areas. Julia had been looking for work, and Einstein mentioned that his Aunt Julie Koch in Genoa wanted a governess for her seven-year-old daughter. He described the girl as slightly spoiled but basically good and intelligent. He also warned Julia what to expect if she landed the job, praising his aunt for two qualities, intelligence and honesty, but damning her for many more: being poorly educated, vain, domineering, tactless, and insensitive—and wearing the pants in the house. A recent visit to the vacation hotel by that same aggravating aunt had sharpened Einstein's view of her as an arrogant monster.

He was also irritated by the boring chatter of his mother's acquaintances, who had flocked to the hotel. He usually slipped away from them on some pretext, often to walk in the nearby mountains, but to his regret couldn't avoid their company at mealtimes.

Mileva would gladly have changed places with him. A local diphtheria epidemic made it dangerous to venture into town, so she was trapped at the family's summer home in Kać, Hungary, during a heat wave, where the only relief was the shade of flowering cherry trees in the garden. She told Albert she envied his ability to escape to cool mountain slopes.

As his mother expected Albert to spend his vacations with her, he searched for ways to avoid the visiting bores without hurting her feelings. He recalled promising Julia Niggli that the next time they were guests of the Wintelers he would bring his violin to play duets. However, that hardly seemed a compelling excuse for getting away. More persuasive was an agreement he had made to discuss scientific topics at the Wintelers' with a Professor Haab. For this he got his mother's blessing to go to Aarau.

Before leaving, he wrote to Mileva in September 1899, telling her not to worry: he would not be visiting the Wintelers at Aarau very often, because he risked running into Marie. He admitted that he had been in love with Marie four years earlier, but felt invulnerable now in his "high castle Peace of Mind." Yet he said he was very reluctant to meet Marie—as she would drive him crazy again and he feared that "like fire." This was a striking simile—but did it mean that Marie Winteler had the power to drive him crazy, into her arms, or both? His explanation seems less illumination than smokescreen. Only three years before, his "dear little sunshine" had been indispensable to his happiness, and now he avoided contact with her as if his sanity were at stake.

When he reached the Wintelers' home in Aarau, Julia Niggli was waiting for him. She was much more eager to discuss her love life than to make music, although he managed to sit her at the piano for a few duets. He also managed to have several discussions with the scientist Haab, who was another houseguest, and so was able to justify the visit to his mother and Mileva. Apparently he escaped an encounter with Marie. On his return to the Paradise Hotel after the Aarau trip, Einstein wrote to thank Pauline Winteler for her continued affection despite the great distress he had caused. The subject was still too sensitive for him to mention Marie by name.

He then buried himself in his studies, learning the laws of thermoelectricity (electricity produced by heat) and devising a simple way to find out if an electrically charged body has a different heat from an uncharged one.

Returning to college for the 1889 winter semester, Albert again stopped at Aarau, which was on the way, this time to drop off his sister. She was to live with the Wintelers, as he had done, while studying to be a teacher. Albert promised Mileva to make the Aarau visit brief so as to avoid meeting Marie. He kept his word, continuing his journey and arriving in Zurich the same day.

There he persisted in deciding for himself which mandatory courses to attend, often missing classes for a more engrossing pursuit. Although he believed that we will never know the "real nature of things," he shared Spinoza's view that the fundamental secrets of nature were to some extent accessible. He spent his time trying to unravel those secrets instead of attending math lectures—a risky business, because he faced expulsion whenever the math professor gave an exam.

Fortunately, his classmate Marcel Grossmann, who had so quickly recognized his potential, always came to his rescue. Grossmann faithfully attended every math lecture and took clear, detailed notes. Whenever an exam loomed, he lent the notes to Einstein. With this help Albert was able to stay in college and concentrate on the work that led to his astonishing discoveries.

Einstein was not the only student to antagonize teachers. He once overheard a woman student, Margarete von Uexküll, argue with her physics professor about his "impossible" requirements for writing up lab experiments. After the professor stormed out of the room, Einstein commiserated with Margarete and offered to write a report for her, using her notes. When she hesitated, he guaranteed that it would placate the angry professor. She was glad she took the chance. The outcome was as he had predicted, with the professor remarking that despite his "impossible" demands, she had somehow produced satisfactory results. It also showed that Einstein could give his professors what they wanted—if he chose to.

During his fourth and last year at Zurich Polytechnic, Einstein was still a man without a country. He had been waiting several months for an answer to his application for Swiss citizenship. On first applying he was disconcerted by personal questions such as "Do you lead a respectable life?" and "Were either of your grandfathers syphilitic?" Afterward, a detective named Hedinger had investigated him and his relatives, confirming his claims to syphilis-free grandfathers, and describing Albert as "a very eager, industrious and extremely solid man" and a "teetotaler." That was one hurdle cleared. Now it was a matter of waiting for the detective to finish his inquiries.

Of more immediate concern to him and Mileva was the imminent final exam in the summer of 1900. Failing it would mean they could never teach in college.

Three days before the all-important exam, Professor Weber jeopardized Einstein's chances. He made him rewrite an entire article because his first submission was on nonregulation paper. This took up much valuable pre-exam study time. Such a mean-spirited enforcement of a petty rule shows how much Einstein had antagonized Weber.

Five students took the exam—four men, including Albert's helpful friend, Marcel Grossmann, and Mileva. Louis Kollros got the highest total mark of the group—60. Marcel Grossmann came second with 57.5, closely followed by Jakob Ehrat with 56.5. Einstein was fourth and last of the men with 54. Mileva had the lowest score of 44. She equaled Albert's 10 in experimental physics, and got 9 to his 10 in theoretical physics, 4 to his 5 in astronomy, and 16 to his 18 for the diploma thesis. She did poorly in theory of functions, however, getting 5 to his 11.

Mileva was naturally upset to be the only one to fail the exam. Albert encouraged her to try again the following summer, and she said she would.

Meanwhile, there was a wedding to look forward to—she and Albert had decided to get married. When he told his sister, she advised him to keep it a secret from their mother.

A few days later, Mileva's close friend Helene Kaufler, a history student at

Zurich University, returned from a visit to Einstein's parents in Milan. Both Albert and Mileva were anxious to know how they might respond to his wedding plans, so he asked Helene for her opinion, urging her to be frank. Helene gave the good news first: she very much liked Albert's mother and admired his father's looks. Then came the bad news: Mrs. Einstein had openly ridiculed Mileva.

Trying to divert the blow, Albert pretended to be jealous of his handsome father. But Mileva took it hard, especially as Albert was about to join his family in Italy for the summer vacation, when his mother would have a chance to poison his mind against her. She already feared that Pauline Einstein would be an implacable obstacle to their marriage.

CHAPTER 4

Marriage Plans

1900 to 1901

21 to 22 years old

Soon after Albert dropped the bombshell of his intended marriage, he described its impact to Mileva, now at home with her parents and eagerly awaiting his letter. Like a war correspondent covering the latest raid, he humanized the drama, highlighting both the pain and the comic relief. To give it to her straight, he explained, he was writing the account in bed, away from prying, censorious eyes.

On the very first day of the vacation, he said, his sister again cautioned him not to tell their mother how far "the affair" had gone because Pauline was still bitterly opposed to the relationship. Albert ignored this advice. That evening, alone with his mother in her hotel room discussing the final exam results, he mentioned that Mileva had failed. Pauline then casually asked, "What's to become of her?" Albert, in his customary direct manner, just as casually replied, "My wife." There was nothing casual about Pauline's cry of dismay. She threw herself on the bed, buried her head in the pillow, and sobbed like a distraught child. Stunned, Albert stood there mute and helpless. After crying herself out, Pauline turned from tears to a tirade, angrily accusing him of destroying his future for a woman who "cannot gain entrance to a good family." She ignored his attempts to calm her hysterical outburst, which reached a crescendo with a prophetic, "If she gets with child you'll be in a pretty mess!"

Albert played the injured innocent, protesting at her "outrageous" suggestion that they had been living in sin. This protestation was sharply at odds with his impassioned message to Mileva a few days later: that he couldn't wait to hug her and live with her again. Yet he kept up the pretense of outrage and was about to walk out when Pauline's friend, Mrs. Bar, walked in. Albert portrayed what followed as a welcome comic interlude that lightened the acrimonious atmosphere. He described the encounter to Mileva with Dickensian zest: "Mrs. Bar is a small, lively little woman full of life, a sort of hen of the nicest kind." When she walked in, "we immediately started to talk with the greatest eagerness about the weather, the new spa guests, ill-behaved children, etc."

The next day Pauline was less frantic. She admitted to being terribly afraid that Albert and Mileva had been sleeping together, but now accepted his word that they hadn't. She also seemed confident that if they didn't rush into bed or marriage a mutually acceptable solution might be found. Albert believed that for his mother the acceptable solution would be Mileva's total disappearance, and that her biggest nightmare was having to acknowledge that they were in love and meant to live together permanently. The prospect horrified her, he wrote, because she considered Mileva an unsuitable wife for him, being too old, too unfeminine, and too unhealthy.

By relaying such insulting comments verbatim Albert put Mileva through an emotional wringer. Yet she apparently welcomed his brutally honest reporting, in turn feeling free to confide in him without pulling any punches.

Although reassured that there was no fear of Mileva's imminent pregnancy, Pauline soon resumed her tirades, saying Albert needed a wife, not another "book" like himself, and that when he was thirty Mileva would be "an old hag" (in fact she would be thirty-three). It was enough to drive him crazy, he said. When his mother saw that her insults simply infuriated him without changing his mind, she switched to sulking. Maja couldn't wait to get away from the place, she told Albert, and was deliriously happy at the prospect of her imminent return to Aarau. He had no such prospective paradise, and the weather, being as grim as his mood, prevented an escape to the mountains.

In his desperation, he "fled to Kirchhoff"—not a place, but a book by the scientist Gustav Kirchhoff. Einstein's "escapist" literature quickly transported him out of this world; it also taught him the danger of saying "never." In 1844 Auguste Comte predicted that man would never know the composition of stars and planets, that it would remain an eternal mystery. Just two years after Comte died in 1857, Kirchhoff proved him wrong by helping to develop the spectroscope.

Albert had never given up his boyhood fantasy of riding beside or on a beam of light and noting his observations. While that was not possible, reading the pioneering work of scientists such as Kirchhoff on the nature of light and electricity served as absorbing if more down-to-earth substitutes.

He assured Mileva that his studies were no substitute for her presence, that he yearned for her kisses and pitied a group of Catholic priests in the hotel for having no Mileva in their lives. He approved of only two guests, a young married couple he knew from Zurich. Perhaps he enjoyed their blissful happiness in anticipation of his own. Albert dismissed all other females at the hotel, except for his mother, sister, and Mrs. Bar, as soft, lazy, disgruntled, and overdressed.

When the weather broke, he was off to the mountains with his sister for a rare, sparkling day of fresh air and freedom. They marveled at the white-and-yellow edelweiss, its flowers scattered over the slopes like fallen stars, and, far below, fields of snow. When the rains came again, he returned to the bliss of reading Kirchhoff.

When his mother began handing Mileva's letters to him without a murmur,

Albert believed she was reconciled to the inevitable. To keep her in this passive mood, he bowed to her wishes and gave a one-man concert to hotel guests. His lively playing helped drown out the fierce rainstorm that had assured a captive audience. There was nothing restrained about their enthusiastic cries for more.

Basking in her son's success and popularity, Pauline continued to avoid mentioning Mileva. But it was only a brief reprieve. Soon after, his father took over.

Too preoccupied with his business to join the family on vacation, Hermann still found time to mail Albert his fatherly advice. The gist of his sermons, as Albert called them, was that only a man of means can afford the luxury of a wife. Albert, being both penniless and unemployed, was in no position to marry anyone. Einstein responded by ridiculing his father's view, with the provocative assertion that it defined a wife as being more mercenary and less principled than a prostitute.

After the vacation, the trio returned to Milan. Despite his mother's outbursts and his father's sermons, Albert had not wavered in his determination to marry Mileva. Consequently, Pauline called off the truce she had kept during the last days at the hotel. Now she angrily predicted disaster. Mileva, she said, was unhealthy and had bewitched him.

When rational arguments and fearful predictions failed to shake him, his parents resorted to a duet of despair—"mourning me," Albert explained to Mileva, "as if I were dead." He sent her blow-by-blow accounts of his parents' sighs, tears, and altercations, with passionate declarations of his undying love for her, using almost the same phrases he had written to Marie Winteler.

He might have placated his father by at least getting a job, but he was not willing to take just anything; when his college friend Jakob Ehrat recommended him for a stopgap position in life insurance, he rejected it as idiotic drudgery.

He escaped from his restless, embattled mood only by studying. Then he became engrossed in what became a lifelong search for the probably unattainable—objective reality. He wanted to discover what the world is really like, from its smallest constituent to the cosmos as a whole, and if nature's laws apply uniformly throughout the universe.

Albert won a small personal victory when his parents resigned themselves to his spending Christmas with Mileva, and he eagerly told her they'd soon be together. He now believed his parents had grudgingly given up their opposition, and accounted for this victory to Mileva as due to his having more stubbornness in his little finger than his parents had in their entire bodies.

Despite their quarrels, he agreed to learn about the family business in order to be able to take over in an emergency. So father and son planned a combined business and pleasure trip, first to survey the nearby power plants installed by the Einsteins, which supplied entire villages with lighting, then on to Venice.

Early in October, he and Mileva were reunited in Zurich, where he also met her younger sister, Zorka. In person she struck him as cheerful but obstinate, and if she showed signs of her eventual mental illness, he didn't mention it in any of his letters.

Since his graduation, his wealthy aunt, Julie Koch, no longer supported him, so to survive, the young lovers gave occasional private lessons. Meanwhile, they daydreamed of a glorious future as a team, discovering nature's greatest secrets and earning enough to buy bicycles for trips to the country. Albert's parents had not abandoned their efforts to scuttle his marriage plans, however, and now they persuaded him to leave Mileva to spend Christmas with them.

Describing her anguish to confidante Helene Savic (formerly Kaufler), Mileva said she would give her the details later, but because of Helene's recent marriage did not want "filth" to spoil her friend's beautiful days. It seems likely that the Einsteins had advised Albert that any appearance of impropriety in Zurich would jeopardize his chances of becoming a Swiss citizen—and consequently of getting work in Switzerland. At twenty-one he could have married without his parents' consent, or he might have lived openly with Mileva, but with a detective from the Naturalization Department still on his track, it would have been risky.

Still, he and Mileva were buoyed by the news that his paper on capillarity (the theory of liquids) had been accepted for publication by a prestigious scholarly journal, *Annalen der Physik*. Since the paper had been inspired by Ludwig Boltzmann's work on the second law of thermodynamics, Einstein sent a copy to the great Austrian physicist, hoping for a comment. There was still another reason to rejoice. Professor Alfred Kleiner of the University of Zurich had approved Albert's proposed subject for his Ph.D. thesis—the kinetic theory of gases.

With his father's approval, Albert took the next step in his move to become Swiss, filling in the necessary forms containing the inevitable questions: Yes, he had been a permanent resident of the country for five years; yes, he was supplying an affidavit attesting to his good character. But nothing was taken on trust by the cautious Swiss. Now he had to await a painstaking investigation.

On December 14 he faced an eight-man committee considering his wish to become a Swiss citizen, and confirmed the Naturalization Department detective's initial report that Albert Einstein was a teetotaler living on a modest income from teaching. He told them he had saved 800 francs to pay the naturalization fee (from money his aunt had provided for his support at the Polytechnic). The committee accepted his application, but it still had to be approved by the Great Municipal Council.

That same day, December 14, Max Planck was establishing the birth of modern physics by reporting his recent astounding discovery to the Berlin Physical Society. Planck had succeeded Kirchhoff (whose book had provided Albert with his escapist vacation reading) as associate professor at the University of Berlin—a position Einstein would eventually hold.

Like Einstein, Planck had been investigating light's mysterious properties. He had clinched his new, disturbing theory while strolling with his son in a Berlin park. The theory was disturbing, he told the assembly, because it shattered the comforting illusion that the physical world was fundamentally harmonious. It showed that instead of flowing in a steady, uninterrupted stream, light, heat, and other forms of radiation moved in separate pieces of energy, which he called "quanta" (Latin for "how much"). Consequently, he said, many previous reassuring ideas about matter, energy, and cause-and-effect were bound to be false. Instead, he said, we lived in an uncertain, more frightening universe. With this speech Planck fathered atomic-particle physics and a revolutionary new view of the physical world.

Einstein first read Planck's paper a few months afterward with mixed feelings. His later, more informed, reaction echoed Planck's—the quantum theory greatly disturbed him. Yet he made it a lifelong preoccupation and would eventually expand it, applying quantum theory to all radiant energy in the universe, including electromagnetic waves, gamma rays, and X rays. He could "never make his peace with it. The quantum theory was his demon," said Abraham Pais, an Einstein colleague and biographer.

Christmas with his parents passed without major confrontations. In the new year, back in Zurich with Mileva, Albert began converting some of his notes for a prospective Ph.D. thesis, which was titled *Eine neue Bestimmung der Molekïdimension* ("On a New Determination of Molecular Dimensions"). Meanwhile, Mileva studied for the exam she intended to retake.

With their future still uncertain, and hardly able to feed themselves on their occasional income, Albert swallowed his pride and principles and applied for a practical job in Vienna. Yet despite their problems, Mileva assured her friend Helene, she and her "darling" were very much in love—and in high spirits. They had taken advantage of the recent snowfalls to go sledding on the slopes of Zurichberg Mountain. "As you can see, we still have our innocent passions," she wrote. "Albert felt in seventh heaven whenever we went downhill 'like the devil.' "

CHAPTER 5

Seeking a Position

1901

21 to 22 years old

When the Zurich city fathers read the dismal final report on Einstein, the odds of obtaining Swiss citizenship seemed heavily against him. By now even his occasional income from private lessons had dried up, putting him in a risky category—an unemployed student with neither job prospects nor money of his own. The report revealed that his father owned no property and barely survived on a modest income. If granted citizenship, Einstein seemed sure to become a burden on the state.

Even worse, the Swiss officials deciding Einstein's fate also "mistrusted the unworldly, dreamy young scholar of German descent. They could not be sure that he was not engaged in dangerous practices." They continued to mistrust him until they questioned him. Then they got an impression of him so glaringly different from the Einstein known to friends and teachers that he might have been putting on an act. If he was, it worked: "They observed how harmless and how innocent of the world the young man was. They laughed at him, teased him about his ignorance of the world, and finally [on February 21, 1901] honored him by recognizing his right to Swiss citizenship."

Three weeks later, the day before his twenty-second birthday, he stripped for a medical examination for the military service required of all able-bodied adult Swiss males. The doctor gave him a welcome birthday present, excusing him from the army because of flat feet and varicose veins. The medical report describes him as 5 feet 7½ inches tall, with a 34½-inch chest, flat, sweaty feet, and varicose veins. Told of this report, his longtime physician friend Thomas Bucky, whose father was Einstein's doctor, confirmed the height and chest measurements but adamantly denied that Einstein had either flat feet or varicose veins.

Whatever his physical condition, he should have been euphoric. He was in love, no longer a man without a country, and free to pursue his engrossing studies. Yet none of the positive aspects of his life compensated for the humiliation he suffered from his former professors at Zurich Polytechnic.

All his fellow graduates had been appointed assistant professors at the college. He alone had been singled out for rejection. Even Professor Heinrich Weber—who had implied, despite their differences, that he would hire Einstein after graduation—had changed his mind and now wanted nothing to do with him.

When Einstein had first applied to enter the college, Weber had been an enthusiastic supporter. Over the years, however, the professor had become infuriated with what he saw as Einstein's arrogance, once snapping, "You're a clever fellow, Einstein, but you have one fault. You won't let anyone tell you a thing." Weber's disenchantment made him irrational. Although he needed two assistants, he avoided hiring Einstein, a fellow physicist. Instead, he looked far afield and took on two mechanical engineers.

Einstein had talked himself out of a job; his outspoken, sardonic manner, which delighted friends, annoyed most professors, especially Weber. Used to the respect due to the guardians of scientific absolutes, they resented this young heretic, with his rigorous mind and don't-give-a-damn manner, who clearly regarded them as expendable. Einstein later defined his professor-baiting views as follows: "What we learn up to age twenty is taken for primordial truth accepted once and for all and inviolate, what we meet after that is pure speculation without form and weight."

Even the few enlightened professors who might have agreed with him in principle resented his threat to their authority, which blighted his prospects of working for any of them. His only good friend on the staff, Alfred Stern, was a historian, so he was in no position to recommend or employ him. And math professor Hermann Minkowski—who later made use of Einstein's work—would hardly consider hiring someone he had derisively dubbed a lazy dog.

The application for the "practical" job in Vienna had come to nothing, so Albert mailed job applications to scientists throughout Germany and anxiously awaited replies. Not yet fully aware of the intensity of Weber's animosity toward him, he gave his former professor's name as a reference. Meanwhile, Albert and Mileva survived by the occasional tutoring job and food parcels from home.

Mileva suspected that Albert's outspoken, tell-it-like-it-is attitude was only part of the trouble, confiding to a friend: "My sweetheart has a very wicked tongue and is a Jew into the bargain." Being a Jew in that time and place was certainly no passport to success, although there was probably less overt anti-Semitism in Switzerland than in Germany. Professor Minkowski was Jewish, and several of Einstein's Jewish friends had been hired by Zurich Polytechnic. But he had another strike against him; as a recently naturalized citizen he was treated by many natives as a second-class citizen, or "paper Swiss."

As soon as he was aware of this, Einstein sought opportunity further afield. In March he sent a copy of his paper in the *Annalen der Physik* to Wilhelm Ostwald, a physical chemist at Leipzig University who was to win a Nobel Prize in 1909. Mentioning that his work was inspired by Ostwald's own research, Albert invited comments and asked Ostwald if he could use a mathematical physicist familiar with absolute measurements. He concluded his appeal by writing, "I am without means and only such a position would give me the possibility of further education." Getting no reply, Einstein sent a follow-up letter, using the excuse that perhaps he had forgotten to include his return address in the first letter. Ostwald again failed to respond. The next month Einstein applied for a job in Leyden, the Netherlands, with Heike Kamerlingh Onnes, a Dutch physicist (also to become a Nobel laureate, in 1913)—again without success.

In the spring of 1901, Albert was so broke that he returned home to Milan. There at least he was well fed. He continued job hunting, but after several disappointments he struck Germany off his list and concentrated on Italy, a country he loved and one virtually free of anti-Semitism. Finally, convinced that Weber was compromising his chances, Albert dropped him as a reference, citing instead former teachers in Aarau and even some in Munich.

He felt like a stranger in his own home, his sense of alienation doubtless fed by his separation from Mileva, humiliation by Zurich's academic establishment, and failure to get a job. He was also still torn between loyalty to Mileva and sympathy for his parents, especially his father. Hermann's health and business were failing, and he and Pauline were being constantly harassed by Albert's rich Uncle Rudolf, a disgruntled investor in the company. Despite these distractions, Albert applied himself to his scientific speculations with passionate intensity, "burning with desire," he told Mileva, "to . . . make a gigantic step in the exploration of the nature of latent heat."

A planned visit from Michele Besso, Einstein's student friend, was to be a welcome distraction. In addition to their similar interests, both men had close ties to the Winteler family. Albert, of course, had lived as a member of the family and had had a romantic relationship with the youngest daughter, Marie; Albert's sister, Maja, would eventually become the wife of the Wintelers' son, Paul; and Besso had married their eldest daughter, Anna. Besso had just become a father, and he and his young family were coming from Trieste to visit the Einsteins.

Einstein was drawn to eccentric misfits if they amused him or challenged his thinking. Besso did both. Although Einstein called him a schlemiel, the Italian electrical engineer became a lifelong friend and confidant.

Besso called Einstein an eagle and himself a sparrow that, taken under the eagle's wing, fluttered higher than otherwise possible. This modest image had some truth to it. Besso was so absentminded that he made Albert's occasional lapses appear insignificant.

Knowing the Bessos were en route, Mileva begged Albert not to talk about her, afraid that if he did, Michele or Anna might gossip to the Wintelers in Aarau and cause more trouble. Albert assured her that there was nothing to fear, that no one would dare nor want to hurt her.

During the Bessos' visit Albert avoided talk of love and marriage. Instead, he spent almost four hours with Michele discussing the fundamental separation of matter and luminiferous ether (supposed to be an invisible, jellylike sea in space through which radiation was transmitted) and trying to define absolute rest as a key to the movements of sun, stars, and planets. They then went on to discuss something that had not yet been achieved — a method of measuring and counting molecules; from that, to surface phenomena; and finally, the effect of heat on gases. Besso held Einstein's enthralled attention and responded to his flood of ideas like a naturalist rather than an electrician in a concluding prediction: "If they are roses, they will bloom."

One blighted plant was the luminiferous ether—eventually shown not to exist. And in 1905, four years after their conversation, Einstein destroyed the notion of absolute rest.

Despite his dim view of Besso as a lazy, uncreative wimp, Einstein relished his "extraordinarily fine mind whose working, though disorderly, I watch with great delight." Soon after their absorbing marathon conversation, Albert informed Mileva that although Besso tended to concentrate on petty details and so miss the overall picture, he was very interested in "our investigations" (it's not clear if "our" referred to Mileva or to Besso as coinvestigator).

The absorbing conversation with Besso was the only bright interlude in Einstein's life for several weeks. For a while he hoped that Besso's influential Italian uncle, a Professor Jung, might be his savior. Jung had sent a paper by Albert and Mileva to two leading physicist friends, Professor Battelli in Pisa and Professor Augusto Righi in Bologna. This too came to nothing—and the few scientists who replied to Albert's many job applications turned him down. Einstein maintained a lighthearted front for Mileva's benefit, despite an awful feeling, which he expressed to others, that he was a pariah abandoned by everyone and without a future.

Hermann demonstrated his loving concern for Albert by secretly writing to Wilhelm Ostwald, describing Albert's qualifications, talent, and great love of science, and asking Ostwald to send a few words of encouragement. Not even this plea brought an answer from Ostwald. Apparently it was filed and forgotten with other job applications, to be unearthed years later from the Nobel Prize–winner's archives.* It is doubtful that Einstein ever learned of this moving, secret attempt to help him, but if he did it was probably many years after his father's death.

Yet encouragement was near at hand. The day Hermann wrote to Ostwald, Albert heard from Marcel Grossmann, the former Polytechnic classmate who had lent him math notes and spotted his potential genius. Sympathizing with his friend's predicament, Marcel had persuaded his father to recommend Einstein to Friedrich Haller, director of the Swiss Patent Office in Bern.

Albert immediately replied to Marcel, moved by his letter and deeply grateful to him. He said he'd be delighted if he got the job. Einstein later acknowledged that Grossmann's help, in a way, saved his life; not that he would have actually died without it, but that he would have been intellectually stunted.

Still, there was no guarantee he would get a job in the Patent Office. The

---

*Wilhelm Ostwald redeemed himself by being the first to propose Einstein for the Nobel Prize in 1910, and again in 1912 and 1913. Einstein eventually received the award in 1921.

announcement of a vacancy there was many months away, and he would have to compete for the job with other applicants.

Then his luck changed. He was offered temporary work, teaching math at a progressive technical school, from May to July 1901, in the northern Swiss town of Winterthur. He would be filling in for the head teacher, Professor Gasser, while he was away on military duty. Albert hadn't even applied for the job, and was curious to know what well-wisher had recommended him, certain that it was not one of his teachers. He soon found out: it was two college friends, Ehrat and a fellow named Amberg, who had made this friendly gesture, which he greatly appreciated.

En route to his job, Albert took the first morning train north from Milan, stopping off at Lake Como to spend the weekend with Mileva. She had traveled south from Zurich to meet him on the station platform and had been waiting since 5 A.M., anxious not to lose a moment of their time together. They spent the morning in Como, then took a steamboat to Cadenabbia, where they visited Villa Carlotta and its fabulous flower gardens.

The next day they rode in a horse-drawn sledge through snow flurries up to Splugen Pass in the Alps. Their driver, perched on a small plank at the back, delighted Mileva by mistaking them for honeymooners. The journey lasted several hours, and the higher they went the deeper the snow became until she could see nothing but snow, which, as Mileva told her friend Helene, gave her the shivers and the chance to hold "my sweetheart firmly in my arms under the coats and shawls."

They walked part of the way down, laughing as they caused several small avalanches. And then it was time for Albert to catch his train for the journey to his new job — and a new life.

"I am beside myself with joy," he told his friend, Professor Alfred Stern.

CHAPTER 6

The Schoolteacher

Summer 1901

—

22 years old

A lbert told Mileva that the technical school at Winterthur was even better than he had expected and that his room on the edge of town was spacious, bright, and airy, with parquet floors, beautiful carpets and pictures, and a comfortable couch. When he glanced through the large window, he felt at one with nature, part of a huge, colorful flower garden.

A former Aarau classmate, Hans Wohlwend, had an adjoining room and was training for Volkert, an import-export business based in India. He and Albert joined a local amateur orchestra. After an evening of playing, Albert usually returned to the comfort of his couch and a simple supper of bread and sausage.

At first Einstein found his students hard to handle. His intelligence and wry humor eventually won them over, however. On one notable occasion, when a rambunctious youngster kept scraping his stool, Einstein asked, as if only mildly interested, "Is it you or is it the stool making such a noise?"

He mostly reveled in his work at Winterthur, surprised to find how much he enjoyed teaching mathematics. Despite a heavy morning workload of up to six classes, he still had the resilience to spend the rest of the day studying Boltzmann's work on the kinetic theory of gases. Einstein then wrote a paper on the subject, providing "the keystone in the chain of proof that he had started."

Those engrossing self-imposed tasks completed, he picked up Arthur Schopenhauer's *Aphorisms on the Wisdom of Life*. The opening line—"What a man *is* contributes much more to his happiness than what he *has*, or how he is regarded by others"—must have been some consolation in his present situation. He later acknowledged the therapeutic value of his studies: "I hold with Schopenhauer that [the strongest motive leading men to art and science] is the desire to escape the rawness and monotony of everyday life, so as to take refuge in a world crowded with images of his own creation."

A personal questionnaire Einstein once completed at a friend's request attests to the fact that he was happiest when alone, but he was not, like Schopenhauer, a pessimistic, mother-hating misanthrope who claimed that his only friend was his pet poodle. Einstein was hardly a misanthrope. Apart from Mileva, Einstein had at this time at least five close friends in whom he confided, and he was soon to make two more. Throughout his adult life he would surround himself with women and marry twice. His main focus was science, however, and his intimate friendships were invariably with those who shared his obsession for solving cosmic mysteries—or at least for discussing them. This is reflected in his letters to Marcel Grossmann, for example. In one, he makes a cursory mention of the beautiful weather and his music making, then continues:

> As regards science, I have got a few wonderful ideas in my head . . . I am certain now that my theory of the attractive forces . . . can be extended to gases . . . Then the decision about the question of the close relation of molecular forces with the Newtonian forces acting at a distance will come a big step nearer . . . It is a wonderful feeling to recognize the unifying features of a complex of phenomena which present themselves as quite unconnected to the direct experience of the senses.

Nowhere in his letter does Albert mention Mileva, who was anxious for him to meet her parents. They also had serious objections to the relationship, but Mileva was confident that when her parents saw them together their misgivings would evaporate. Albert shared her optimism, and most Sundays he made the seventeen-mile train journey south to join Mileva in Zurich.

At Winterthur, Albert met a Professor Weber who showed a friendly interest in his work—a marked contrast to his bête noire of the same name in Zurich. Encouraged, Albert showed Weber his and Mileva's joint paper, presumably the one Besso's father had sent to the Italian physicists. Albert's subsequent correspondence with Mileva was cautiously comforting; he envisioned their happy, productive future together as a team, their lives enhanced by the obstacles they had overcome.

Pauline's fear of Mileva becoming pregnant was justified, judging by Albert's comment in a letter he wrote to Mileva in late May: "How delightful it was last time when I was allowed to press your dear little person to me in the way nature created it." His next letter to her revealed the natural consequence: "How is the boy? How are our little son and your doctoral thesis?" His mention of "boy" and "son" was wishful thinking, because the baby eventually turned out to be a girl.

Mileva could hardly have conceived at a worse time: Hermann's tottering business had finally collapsed. No longer able to pay for Maja's education, he called on Albert to help. With a pregnant mistress and a temporary job about to end, Albert was expected to take over the support of his sister, who he believed was spreading the rumor that he led a life of debauchery in Zurich. News of Mileva's pregnancy would surely confirm it to the Wintelers, not to mention Albert's and Mileva's parents.

What caused Maja's disaffection? One reason, he suspected, was that she had accepted the Wintelers' account of his heartless breakup with Marie and so empathized with the heartsick young woman. Albert also attributed her animosity to her turbulent adolescence (she was then nineteen).

Again Albert demonstrated his remarkable ability to absorb himself in work when under stress. During this trying time, one of his letters to Mileva begins with a rave review of a paper he had just read on cathode rays and ultraviolet light, by Philipp Lenard, while another reports a letter he had written to Paul Drude, a leading German physicist, pointing out two flaws in his electron theory.

Even though his focus was on science, Einstein did not shirk his responsibilities — as he was soon to show. Having once defined the two mainsprings of life as love and hunger, he found himself with the fruit of one and facing the imminent prospect of the other. Now he tried to calm Mileva's fears by confidently predicting that she had only to be patient and they would ultimately reach their goal, working as a team and sharing a joyful, unhurried, peaceful, idyllic life. As for the immediate future, he intended, of course, to stand by her and to care for their "boy."

Aware that the Patent Office job was still a distant and iffy prospect, he decided that to support her and their child, as well as his sister, he must temporarily abandon his dream of a scientific career. With that decision made, he inquired about a job in an insurance company, prepared to face what he had recently scorned as idiotic drudgery. But, dreading the prospect of getting that job, he also applied for a more acceptable teaching position in Frauenfeld, naming Jost Winteler as a reference, and anxiously awaited a reply.

Meanwhile, Einstein had mailed the letter to Paul Drude, who replied that an eminent colleague agreed with his electron theory, and that was good enough for him, implying, Who is this nobody Einstein to dare to criticize me? Einstein concluded, not for the first time, that authority was the greatest enemy of truth. He resolved then to discontinue his correspondence with the Webers and Drudes of the world, and instead to go public and criticize them mercilessly in journals as rotten scientists. His experiences with Drude and others made him believe that he was becoming even more like Schopenhauer, a misanthrope in the making.

In July, with only a few days left of his Winterthur employment, he again reassured Mileva that he would postpone his scientific goals and swallow his pride to work at any job that paid enough for them to get married. When the Winterthur job ended, he traveled south to spend his summer vacation with his mother at Mettmenstetten. As Hermann was virtually bankrupt, they were doubtless the guests of Pauline's wealthy relatives.

At the Paradise Hotel, despite his intention to put a scientific career on hold, Albert couldn't resist working on the "thermodynamics of liquid surfaces" in a vain effort to link gravitation and molecular forces. There were compensations, however; the weather was marvelous, and the view from his writing table revealed fresh delights each time he glanced up. He also believed his mother's opposition to Mileva had ended. He quickly concluded an affectionate letter to Mileva as his mother walked in the room to join him for coffee.

He had been overly optimistic. She still hated the idea of his marrying Mileva and vigorously restated her objections. He kept Mileva informed of this renewed opposition. How ironic, she replied, that in contrast to their parents, people in Zurich envied them for their relationship.

The news from the Paradise Hotel wasn't all gloomy, however. Maja's ambivalent attitude had changed for the better, and she and Albert were friends again.

But Albert still had to resist and refute his mother's constant and at times hysterical warnings that Mileva would wreck his career and ruin his life. Meanwhile, Mileva had asked him to write a brief, cordial note to her father to prepare him for the news that as their baby was due in about five months he would soon be a grandfather.

That summer of 1901 Mileva got bad news: she had failed in her second attempt at a college diploma, with exactly the same average mark of 4. The five others who took the exam all passed with averages of 5 or more. The results meant that even a teaching career was closed to her.

If Mileva was devastated, neither she nor Albert mentioned it in their letters. He still encouraged her to think of them working as a scientific team and she, now several months pregnant, focused on marriage and motherhood.

CHAPTER 7

Expectant Father

August 1901 to February 1902

22 years old

Although desperate enough to consider entombing himself in an insurance office, Einstein clung to the hope of rescue from this fate. He had two chances: the teaching job at Frauenfeld, and another recently advertised in the *Swiss Teachers Journal*, for which he had promptly applied. The latter was in Schaffhausen, hometown of Conrad Habicht, a former fellow student at Zurich Polytechnic, who had added his recommendation on the off chance it might help.

By early September, one of his hopes had been eliminated. The Frauenfeld post had gone to Marcel Grossmann. Einstein could hardly begrudge it to someone who had helped him in college, tapped him as a future genius, tried to get him a Patent Office job, and was better at math.

When a few days later he wrote to congratulate Grossmann, Einstein was out of danger and breathing freely, having landed the other job, at Schaffhausen. It seemed a cinch. He would have a miniclass of two high-school students—one a nineteen-year-old Englishman, Louis Cahen—who had failed to graduate because of pathetic math scores. Einstein was to bring them up to scratch and steer them through the final exam. In return, starting in a few weeks, he was promised a year's free board and lodging in the home of the school's owner, Dr. Jakob Nuesch (who was also head teacher), and a modest minimum wage of 150 francs (about $30) a month. (By contrast, he had to pay a fee of 230 francs when submitting his Ph.D. thesis.)

The salary was hardly a bonanza for an expectant father with marriage in mind, impoverished parents, a sister in urgent need of support, and nothing in the bank. Still, it was a year's reprieve from the prospect of purgatory in an office job and from constant battles with his parents. As he told Grossmann, "Although such a position is not ideal for an independent nature . . . I believe it will leave me some time for my favorite studies so that at least I shall not become rusty."

Einstein made his way to Schaffhausen—as far north as one could go in Switzerland without crossing the German border. He found it to be an ancient fortress town in hill country, some thirty miles from Mileva in Zurich.

In October, barely a month later, Mileva, now seven months pregnant, began planning to leave for her family's home in Novi Sad, where she was to have her baby. Fearful of her condition becoming public knowledge and news of it reaching Albert's parents, she went into hiding, telling no one but Albert her new address.

Mileva pleaded with him from her hiding place in early November not to tell his sister where she was, afraid Maja might inadvertently cause trouble. She also thanked him for the books he had sent, saying that one of them had made her laugh a lot, while another, by Max Planck, she had found interesting. She promised to tell him what she thought of the third book, by psychiatrist Auguste Forel, as soon as she finished it. Meanwhile, Einstein spent his free time putting the finishing touches to his doctoral thesis, a research paper on thermodynamics, then submitted it to Professor Kleiner at the University of Zurich.

When Mileva wrote again on November 13, she mentioned "Lieserl" for the first time in her letters. This was the name they'd chosen should their child, as she hoped, be a girl. She also suggested keeping in touch with their friend Helene Savic (formerly Kaufler), and said that they would now have to treat her "nicely," not only because she was so kind, but because she could help them in something important.

One plausible reason why they might have needed Helene's help would have been to find a home for their baby until they could safely support it. If Albert was known to be living in Bern with his mistress and their offspring, it would jeopardize—if not scuttle—his chance of the government job at the Patent Office. Moreover, he couldn't afford to marry Mileva and "legitimize" their baby until he had a steady income—just as his father had warned. Who, then, might help them out of this dilemma? Not Albert's parents, who were financially ruined and still so opposed to the relationship that Mileva's pregnancy had been kept from them. As for her parents, they had their hands full with Mileva's mentally unstable younger sister, Zorka. The married Helene Savic, on the other hand, who had recently become a mother, was a good candidate. Even if she were unable to raise the child herself, she might persuade her parents, other relatives, or friends who lived near Mileva's home in Serbia to help out.

Albert and Mileva kept in touch almost daily. Once, when Albert hadn't heard from her for three days, he wrote to say he feared that her letters had been lost or destroyed by the mailman. Apart from that worry, he told her, he was cheerfully ensconced in a cozy but colorless room in which he and the lampshade were the sole ornaments. He took a daily stroll around the town, and the one evening he wasn't stuck in Woldemar Voigt's book on theoretical physics he went to a chamber music recital.

In mid-December, Albert admitted that although they talked and wrote about having a girl named "Lieserl," he really hoped for a son and thought of him as "Hanserl."

To his great relief, Mileva's missing letters soon turned up and made him "unspeakably happy," because they brought news that her parents were calmer about the intended marriage and more inclined to trust him. He assured her that their confidence was justified and promised to marry her as soon as feasible. He was still far from the ideal suitor in her mother's eyes, though, judging by his remark that he was looking forward to the thrashing her mother promised to give him.

He seemed in an upbeat mood, expressing high hopes for his doctoral thesis and daring shortsighted Professor Alfred Kleiner at Zurich's Physics Institute to reject it. If Kleiner had the gall to do so, Albert promised to make a fool of him by publishing the rejection letter together with the thesis.

Kleiner was not the only authority figure Albert was eager to take on. For three months he had been living and eating at the home of the school's owner and head teacher, Dr. Jakob Nuesch. Albert could isolate himself in his cosy but colorless room much of the time, but his contract called for him to have meals with Nuesch, his wife, and their four children, and he loathed the lot of them. He also suspected Nuesch of gouging him over the room-and-board arrangement.

After a few heated arguments, Albert shamed the "snotty" profiteer into handing over money for his meals. From then on, instead of eating with Nuesch and his obnoxious family, he went to a nearby inn, where he accepted the invitation of two young pharmacists to join them. His plan worked out as he had hoped. By skimping on food he was able to save enough for a trip to see Mileva when he could get the time off. Sharing with her his triumph over the money-grubbing landlord, he crowed: "Long live impudence! It's my guardian angel in this world."

Things had dragged on so long that he had given up hope for the Patent Office job. He was therefore especially buoyed by a tip from Marcel Grossmann that the job would soon be advertised and was his for the asking. Optimistically predicting to Mileva that in two months their struggles would be over, Albert said that it made him dizzy with joy to contemplate their idyllic future together as eternal students who didn't give a damn about the rest of the world. Of course, he added, he would never forget Marcel's efforts to get him the job, and promised always to help gifted youngsters if he ever got the chance—a promise he kept.

Albert already sounded like a concerned parent by expressing the odd superstition that cows' milk might make their child stupid, then dismissed the thought by reasoning that since Mileva would do the feeding, the milk would be both safe and nourishing. With the approaching prospect of fatherhood and a steady, not uninteresting job, Einstein's communications to Mileva became more carefree; on December 17, he even described his life as screamingly funny. But certainly not at lunchtime. The two young pharmacists soon bored him, and he spent much of the time maneuvering to avoid their conversation, toying with his knife and fork between courses or staring through the window.

Though he won the first round with Nuesch in their battle over money, confrontations reerupted because of Einstein's easygoing style of teaching. Nuesch had been influenced by his colleagues across the border in Germany, where the unholy trinity of rote learning, rigid discipline, and kowtowing to authority was an article of faith. Einstein could neither kowtow, discipline, nor teach by rote. Despite his dire financial needs and imminent parental responsibilities, he refused to compromise. So did Nuesch. Incensed by Einstein's earlier triumph and ongoing inflexibility, he stormed away from many of their encounters red with rage.

The day Einstein applied for the advertised job of technical expert, second class, at the Patent Office was also Mileva's birthday. He had forgotten it, and when he remembered the next morning, he wrote her a letter in which he apologized briefly before announcing the good news: although his qualifications did not fit him for the job, Marcel Grossmann and his father had promoted him so effectively that the director cut the red tape and established a new position for which he did qualify. Sure of the outcome, Marcel congratulated Albert in advance. Almost overwhelmed by the sudden bright prospects, Albert told Mileva that he wanted to hug and kiss her and say how much he loved her, and that he longed for the time when he could call her his in front of the world.

In a confident, exuberant mood, he took the train to Zurich, impatient to hear Kleiner's reaction to his Ph.D. thesis, which he had sent to the professor several weeks before. To his dismay, Kleiner had not yet read it. Nevertheless, when Einstein described another paper, his recently completed "electrodynamics of moving bodies," Kleiner was impressed and urged him to publish the theory and a description of the experiment he had devised to test it. The professor also gave Einstein permission to list him as a reference. This encouragement changed Einstein's opinion of Kleiner from shortsighted to not as stupid as he first thought.

After his meeting with Kleiner, Einstein chatted over coffee with his friends Marcel Grossmann and Jakob Ehrat. He was heartened to discover that despite all of the distractions and obstacles he had faced, he was the first of their group to complete a doctoral thesis. Aware of Mileva's sensitivities—how could he find time for these friends when he hadn't been able to meet her shortly before?— Albert later told her that he seemed separated from everyone but her by an invisible barrier, and emphasized her preeminence in his life.

Not if his mother could help it. Pauline saw the looming marriage as a nightmare she felt compelled to prevent. Her mood explains why, in desperation, frustrated by her failure to influence Albert, she wrote a poisonous letter to Mileva's parents. Speaking for herself and her husband, she blamed the wicked "older" woman—meaning Mileva—for leading her son astray.

Expecting her baby in six weeks, Mileva reached her home in Novi Sad on the heels of Pauline's acrimonious letter. She confided to her friend Helene:

> I wouldn't have thought it possible that there could exist such heartless and outright wicked people! [They wrote] a letter to my parents in which they reviled me to such an extent that this really was a shame . . . In spite of all the bad things, I cannot help but love him very much . . . especially when I see he loves me just as much . . . He is now in Schaffhausen, where he is employed as a tutor. You can imagine that he does not feel good in such a state of dependency. Yet, it is not likely that he will soon get a secure position; you know that my

sweetheart has a very wicked tongue and is a Jew into the bargain. Pray for us, dear little Helene, that things no longer go so terribly wrong for us!

If Helene did pray, her prayers fell short of Schaffhausen, because Albert was in for another setback. Sensing that his days there were numbered, he planned to beat Nuesch to the punch by leaving for Bern and taking his student Louis Cahen with him. Cahen expected his parents to approve the move. His fees would then revert to Einstein, enabling him to survive until he started the Patent Office job. But the plan collapsed when Cahen's father went mad and his mother became too distraught to make any decisions.

As Einstein had anticipated, the authoritarian Jakob Nuesch fired him after four stormy months for being what he termed a bad influence. Einstein had heard the same complaint as a Munich schoolboy, but Nuesch used more inflammatory language, claiming that in Einstein's easygoing manner he detected the start of a widespread revolution. Once more Einstein had shown his inability to cope with an authority figure. On this occasion, he had driven one to hysterical hyperbole. What widespread revolution could Einstein launch as a twenty-two-year-old math tutor with a cadre of two students?

Albert moved to Bern with no immediate prospects and little money. Once there, he sent a jaunty note to Conrad Habicht, who had helped him get the job with Nuesch, saying that he'd made a spectacular exit from the job—though in what way it was spectacular he didn't explain.

That same day, February 4, 1902, Einstein learned that he was the father of a daughter. Mileva was too weak to write, so her father wrote for her. Einstein immediately replied that he had been scared out of his wits when he got her father's letter "because I had already expected some trouble . . . At once I felt like being a tutor [at Nuesch's school] for two more years if this could make you healthy and happy. But you see, it has already turned out to be Lieserl, as you wished." He wanted to know about the baby's eyes, who she looked like, how she was being fed, and if she was completely bald. He said he loved her very much without even knowing her and asked Mileva to send him a photograph or drawing of their daughter as soon as she was completely well again.

After his declarations of love, he gave Mileva a glowing picture of what she had to look forward to in Bern, an exquisite, ancient city with long arcades on both sides of the main street enabling one to walk in a rainstorm from one end of the city to the other and not get wet. While looking for a place to stay, he had noticed that all the homes were spotlessly clean. The room he finally chose was both extremely large and beautiful, with a very comfortable sofa, like the one he had in Winterthur. To put Mileva in the picture, he drew a detailed bird's-eye-view of the place, with its six upholstered chairs and three wardrobes. He also told her that until he started at the Patent Office, he hoped to support himself by giving at least two private lessons a day, and that he had already advertised his services in a local newspaper. This letter to Mileva was unique; the excited young father had not written a word about science.

CHAPTER 8

Private Lessons

Spring 1902

———

22 to 23 years old

When Einstein advertised his services in a Bern newspaper, he ended with a sweetener. His notice read:

> *Private lessons in*
> MATHEMATICS AND PHYSICS
> for students and pupils
> given most thoroughly by
> ALBERT EINSTEIN, holder of the fed.
> polyt. teacher's diploma
> GERECHTIGKEITSGASSE 32, 1ST FLOOR.
> Trial lessons free.

First to bite was a poor Romanian Jew, Maurice Solovine, who climbed the stairs and crossed the dark corridor to Einstein's spacious apartment. His bell ringing was greeted by a thunderous "Come in!" and after a few moments of formality, the two young men, seated in upholstered chairs reeking of tobacco smoke, began a rapt conversation. They were two of a kind, driven by a burning desire to probe the elusive secrets of the universe.

Solovine had first sought answers from the great philosophers but had yawned through the lectures. Hegel, for one, was so contradictory that his views were quoted to support opposing factions and in time would be cited to justify fascism, communism, democracy, and both supporters and critics of Christianity. Solovine had therefore turned from philosophy to physics and math as more reliable guides. One question especially intrigued him: Did the recently discovered radioactive properties of radium threaten the principle of the conservation of energy?

Einstein had also been disenchanted with philosophy but continued to dip into Schopenhauer and Nietzsche for fun, using their limpid and lively words as stimulants, like music or poetry, to set him off on speculations of his own. He found one Schopenhauer aphorism especially comforting: "Man can do what he wills, but he cannot will what he wills." To Einstein this meant that free will in the philosophical sense was a myth, and "mercifully mitigates the sense of responsibility which so easily becomes paralyzing, and it prevents us from taking ourselves and other people too seriously; it conduces to a view of life in which humor, above all, has its due place."

Solovine's first and lasting impression of Einstein, aside from the extraordinary radiance of his large eyes, was of "a great liberal, a very enlightened spirit." They shared a scorn for material possessions—both despised people whose goal was fame or wealth—and a fervent belief in social justice for all.

After two hours of animated talk, no fees had been mentioned. Even when Einstein walked his potential student to the street and chatted on the sidewalk for half an hour, they did not broach the subject. Instead they agreed to meet next day and pick up where they had left off.

On the third day Einstein told Solovine that rather than tutoring him, he'd prefer to discuss physics informally any time he chose to turn up. Solovine took him at his word, often returned, and was always welcome.

His initial liking for Einstein strengthened into affection. Solovine found him to be remarkably lucid in their discussions and able to make abstract thought coherent by using everyday experiences to illustrate what he meant.

A few weeks after Solovine's arrival, Conrad Habicht stopped by. Habicht had recommended Einstein for the job from which he'd just been fired and was in Bern to complete his training as a math teacher. After hearing a colorful account of Albert's battles with the terrible Nuesch, Habicht agreed to join him and Solovine in their weekly search for ultimate reality.

Two was a small catch—Einstein had hoped for at least ten students—but their contributions to the meetings, which were to continue for years, had an enormous impact on his future work. Moreover, Solovine and Habicht paid Einstein 2 francs apiece for each session in his apartment. It hardly kept him in pipe tobacco, but at this point any income was welcome. Einstein had withdrawn his Ph.D. thesis for the sake of the refund of the 230-franc submission fee; and perhaps because Kleiner had persuaded him that it contained too much harsh criticism of Drude, Planck, and other science greats. This money, together with his savings from his previous job, would be enough to cover ten weeks' rent. He was a teetotaler, and food was always a modest part of his expenses, as he could count on occasional food parcels from his relatives and Mileva. He would scrape by somehow.

The trio chose to study as equals rather than as teacher and student, and to read and discuss great books. Their eclectic reading list included works by Karl Pearson, Plato, Ernst Mach, John Stuart Mill, Racine, and David Hume. They kicked off with Pearson, a contemporary English scientist who used statistics in his research on heredity and evolution. From him they moved on to Mach, the staunch enemy of "metaphysical obscurities," still a great Einstein favorite.

Einstein and Solovine rekindled their interest in some philosophers. How, for example, could they resist the provocative, open-minded suggestion of John

Stuart Mill that two and two might equal five on another planet? At first, though, they focused on Mill's more earthbound analysis of inductive proof (generalizing from observed facts) in his *System of Logic*, published in 1884. Mill believed nature to be uniform and predictable so that an isolated incident can be expected to recur in similar circumstances. In a discussion with friends, Einstein emphatically disagreed: "There is no inductive method which could lead to the fundamental concepts of physics. Failure to understand this fact constituted the basic philosophical error of so many investigators of the nineteenth century."

Einstein came to regard philosophy as a tempting illusion, once saying to a student, "Is not philosophy as if written in honey? It looks wonderful when one contemplates it, but when one looks again it is all gone. Only mush remains." He admitted that there were exceptions, praising several British philosophers, among them David Hume, a plainspoken Scot he characterized as "representative of the English Enlightenment." (To Einstein—and many Europeans—the Scots and the English were—and are—indistinguishable.) He especially liked Hume's direct, no-nonsense prose, his insistence that experience and mathematics are science's only legitimate tools, and his warning that nothing can be verified as the absolute truth because all human ideas and impressions are subjective.

This was brought home to Einstein one Saturday. He had been strolling through Bern streets when he was stopped by Hans Frosch, a classmate at Aarau and now a medical student en route to a lecture on forensic pathology. He joined Frosch at the lecture and was spellbound, particularly when the professor in charge brought two live specimens into the lecture hall as evidence to support his theories about mentally deranged criminals. One was a megalomaniac swindler, the other a female pyromaniac who set fire to buildings when drunk.

Fascinated, Einstein planned to attend every week. In an enthusiastic letter to Mileva, now slowly recovering from childbirth, he said that the lecture had reminded him of the book they had both recently read by Auguste Forel. In it the Swiss psychiatrist had cited cases of pathological swindlers he had hypnotized in an attempt to understand and reform them.

Mileva remembered the book vividly because it had left her feeling disgusted with hypnosis as "a violent attack on human consciousness" and agreeing with the critic who had called it "immoral." Before taking physics at the Polytechnic, Mileva had studied medicine for a year, during which she had learned the importance of suggestion in human behavior. That was a far cry from Forel's experiments, several of which she thought questionable. She dismissed him as better-informed and more self-assured than most quacks, but a quack nevertheless, who deceived stupid people. Her overall view of hypnosis was that if it existed, it was merely suggestion or autosuggestion. (Which, of course, does not invalidate its therapeutic effects. In fact, that is how it is defined today by even its strongest advocates.)

If Mileva was warning Albert not to be taken in by hypnosis, it was hardly necessary. It was the criminal mind, rather than hypnosis, that had piqued his interest, and he noted that in his friend Frosch he had a shrewd guide. Frosch was so remarkably intelligent and respected, said Einstein, that whenever the professor made a smart remark he turned to Frosch for approval.

Some four weeks after the birth of their daughter, Mileva felt vulnerable and neglected, especially as Albert seemed to be flourishing without her and absorbed in exciting discussions with his friends. He assured her there was no cause to be jealous of Habicht and Frosch, that he missed her terribly but hid his feelings because it wasn't manly to show them, and that even studying lost half its appeal when she wasn't with him.

By chance, he met a former Polytechnic student who was employed at the Patent Office working under the direction of Friedrich Haller. The man called the work dull and the atmosphere depressing. When Einstein mentioned that he had applied for a job there with Haller's encouragement, the man said that the job was in the bag, because the authorities invariably rubber-stamped the director's recommendations. He added that as he had applied for the lowest rank, Einstein was unlikely to have any competition.

Dismissing this chance acquaintance as a chronic complainer, Albert felt sure that he would enjoy the work and be eternally grateful to Haller. He was also delighted to know he might be the sole applicant for the job. As for the sneering comment about the lowest rank, Einstein was far from humiliated, remarking to Mileva, "We two don't give a damn about height!"

What Mileva gave a damn about was Albert's mother and her continued fierce opposition to their marriage plans. Ever since her pregnancy became visible, Mileva had been scared Albert's sister might find out and tell her mother. In fact, Maja suspected only that Albert and Mileva were secretly engaged, and on a brief visit home from Aarau told her mother so. The mere possibility so infuriated Pauline that she forbade any discussion of the subject at home or among the Wintelers, where there was still considerable emotional fallout from Albert's breakup with Marie.

Evidently Pauline Einstein did not know she was a grandmother even as late as six weeks after Lieserl's birth, when she confided her misery to Pauline Winteler: "We strongly oppose the liaison of Albert and Miss [Mileva] Maric . . . We don't want to have anything to do with her . . . and there is constant friction with Albert because of it . . . This Miss Maric is causing me the bitterest hours of my life. If it were in my power, I would make every possible effort to banish her from our horizon. I really dislike her. But I have lost all my influence with Albert. You can imagine how unhappy this makes me." How she reacted when she learned of Lieserl's existence, if she ever did, is not known. And what happened to her granddaughter—the how, why, where and when—was a closely guarded secret and remains a mystery still.

On first hearing of Lieserl's birth, Albert had responded with a barrage of excited questions, wanting to know everything about her, from her eyes to her appetite. Was she like either of them? He asked for a sketch of her, saying he already loved his daughter without even a glimpse of her. This letter was then followed by silence. When he wrote to Mileva a week later, and again three days after that, he did not even mention Lieserl. Not one word, as if she no longer existed. If that was so, of course, he would be expected to have tried to console Mileva. He didn't. Nor did he ask about her health. A few subsequent letters give only tantalizing hints. They indicate that the child for whom Albert had shown so much concern and loving anticipation was put up for adoption. She survived a childhood illness at about eighteen months. And then, again, silence.

The brief paper trail resumes some thirty years later, when Einstein was famous and living in Princeton, New Jersey. In 1935, a British friend sent him an urgent warning that a woman in Europe was trying to persuade people in "high circles" that she was Einstein's daughter. His unusual response—hiring a detective to find out if the claimant was genuine—indicates that he believed his daughter might still be alive.

The Einstein expert most likely to come close to the truth about Lieserl is Dr. Robert Schulmann of Boston University, now director of the Einstein Papers Project, which is charged with the official publication of Einstein's papers. He is also an editor of *The Collected Papers*. In the 1980s, while in Zurich hunting for new material to add to the tens of thousands of papers in the archives, Schulmann learned that Einstein's early love letters were in California. Einstein's son Hans Albert and his first wife had both tried to publish them in the late 1950s, but had been frustrated by Otto Nathan. The letters then came into the possession of Hans Albert's second wife, Dr. Elizabeth Roboz Einstein. When arbitration ruled Nathan out of the picture in the 1980s, the letters were made available to the Einstein Papers Project. Lieserl's existence, therefore, was not publicly revealed until eighty-five years after her birth—when Nathan could no longer hold things up—in *The Early Years, 1879–1902*, the first volume of *The Collected Papers of Albert Einstein*, published by Princeton University Press in 1987, with John Stachel as editor and Robert Schulmann as associate editor.

The author questioned Schulmann about Lieserl:

D.B. Isn't it puzzling that none of the Einstein–Mileva letters refer to their daughter's death?

SCHULMANN. If I'm right, the child survived and died after Einstein died.

D.B. Einstein died in 1955. You think the girl survived and lived as an adult?

SCHULMANN. That is correct.

D.B. So there was a daughter of Einstein's unknown to the world.

SCHULMANN. And possibly unknown to herself. But I'm not sure about that.

D.B. Did it seem strange to you that Einstein wasn't with Mileva when their daughter was born?

SCHULMANN. No. The child was born in what was then Hungary and is now Serbia.

D.B. So the daughter may never have traveled to Switzerland, where Einstein was living at the time.

SCHULMANN. That is correct.

D.B. What happened to his first love, Marie Winteler?

SCHULMANN. She married another man whose name was Albert who was ten years younger than herself. It was an unhappy marriage. She had two children with that man and lived out her life in rather impecunious circumstances, giving piano lessons and waiting tables and writing a little poetry, but primarily by giving piano lessons in Zurich. I think she always retained a crush on Einstein.

D.B. What surprises me is that he wrote very similar love letters to Marie and Mileva only a few years apart, as if equally in love with them. Don't you agree?

SCHULMANN. Yes. I think he definitely liked women, but his priority was always science. He liked to be in the company of women and he liked women sexually, but I would hazard a guess that it did not deflect him one bit from his first priority, physics. That is, of course, my interpretation.

D.B. It seems so unlike Einstein—the way he appears to have abandoned his daughter, Lieserl.

SCHULMANN. I disagree. But I can't spell it out. It's just a gut feeling. I think he was much more opportunistic than one would be willing to see. I think one sees him too much as the noble sage of Princeton and one must be willing to assume at least that he was more opportunistic. How one fears the word opportunistic, but in its neutral sense one has to take that factor into account. That isn't done enough. If you do that, certain directions he went become easier to understand.

D.B. But judging his mother and father, for example, from their letters and milieu, if they knew of the baby and heard that Albert and Mileva were going to get rid of her somehow, don't you think they would have been horrified?

SCHULMANN. I would say the parents' reaction would have been, "Look, you got the Serbian girl into trouble," as his mother predicted in that scene in 1901, when she threw herself on the bed. And then she would say, "Look, now you've got it. And you told me you weren't having an affair with her!" I think that because of bourgeois sensibilities they would immediately push it on the woman. "She tempted my son. She's ugly. She's older than he is. She tempted my boy. Get rid of the kid!" That would be my feeling. And this is nothing to do with anti-Semitism or philo-Semitism.

D.B. That's right, because his mother approved of his friendship with Marie Winteler, the Protestant.

SCHULMANN. Yes. Because that was a respectable Swiss family, so to speak. While Serbians [Mileva was Serbian] were thought of as some kind of bandits from the borders of civilization. I evince the German sensibility that I can attest to.

D.B. They might have regarded Mileva as some regarded gypsies?

SCHULMANN. Absolutely. Which is, of course, unfair. Because Mileva's father was a civil servant. Some people have called him a Serbian peasant. That's ridiculous.

D.B. Socially they were more or less on a level.

SCHULMANN. And probably Mileva's father had done better than Albert's, because he was a bureaucrat and Albert's father had been in the entrepreneurial sector and hadn't done well. But certainly they're of a similar social class.

D.B. Can you reconcile Einstein writing that he was already in love with his daughter without having seen her, that he looked forward to having her join them in Switzerland as soon as he gets work—and then not taking her, unless she was too ill to travel or dying?

SCHULMANN. No. There you have to give too much weight to the words he writes and I'm not prepared to do that. He was of good intentions but the roads to hell are paved with good intentions. I'm not convinced that alone can bear the burden of conjecture. While he wants his child as I assume most fathers do, and that therefore there must have been something really wrong with her or he would have taken her, I don't accept that line of reasoning.

D.B. But Mileva wasn't a pushover and she lived with him for many years after that and had two more of his children. So it wasn't a case of her saying or thinking, "You're a monster not to have kept our daughter!"

SCHULMANN. I would argue, and I grant you this is conjecture, that the marriage was strained already, that it was poisoned. That's a pure guess. We don't know what role the illegitimate daughter played in the emotional landscape. But I can't imagine that Mileva gave up her daughter without emotion.

CHAPTER 9

The Patent Office

April to October 1902

23 years old

By the spring of 1902 Albert's frugal existence had become life-threatening. His capital almost exhausted, he had moved to a cheap apartment at 49 Kramgasse, a block from Bern's famous clock tower. A friend, Friedrich Adler, said Albert came close to starving. But he disdained to ask wealthy relatives for a handout.

Meanwhile, Mileva languished at home in Novi Sad. Much of their correspondence during this critical time has disappeared, almost certainly to conceal their agonizing over Lieserl. In later years, after they were married, visitors to their home probably came close to the truth when they sensed Mileva was brooding over some secret sorrow.

Albert continued to exchange news and ideas by mail with Besso and Grossmann. He also batted around his theories with Habicht and Solovine during their sometimes daily meetings—theories he would soon clarify, crystallize, and publish. He did not, however, share his personal problems with them.

Max Talmey, the former medical student who had enlivened Albert's boyhood lunchtimes, was traveling around Europe. In Italy he called on the Einsteins, now back in Milan, and found Hermann Einstein in poor health. Still, Hermann's heart trouble hardly explained the excessively gloomy atmosphere or the couple's reluctance to discuss their son. Talmey thought he hit on the reason when shortly afterward he made a surprise visit to Albert in Bern. Evidently, from Albert's appearance and his small, almost squalid apartment, he was down on his luck. Talmey assumed that the parents' taciturnity was a reflection of their disappointment over Albert's failure to live up to his promise, knowing nothing of the more likely cause—their futile efforts to end Albert's affair with Mileva.

Relief from one of the strains upon them, however — worry about Albert's financial situation — was at last close at hand. After waiting through the winter, spring, and early summer, Albert was told to begin work at the Patent Office on June 23, 1902. He was overjoyed.

Six days a week he walked from his apartment to the Patent Office a few blocks away. There, perched atop stools for eight hours a day, he and twelve other examiners began to sort out the feasible from the fatally flawed. Easiest to handle were the simple inventions—toasters, mechanical vegetable peelers, and mousetraps. Equally easy were the inevitable and automatically rejected perpetual-motion machines.

Many of the most ingenious and intelligent inventors came a cropper when describing their inventions by writing in an obscure or confusing manner. Rather than irritate the examiners, this obstacle delighted them. They took it as a game-like challenge and felt as if they were breaking secret codes. Einstein became adept at this exercise, zeroing in on the fundamental aspects of a promising invention, unscrambling the cryptic or convoluted language, and converting gobbledygook into lucid prose. He called it "cobbling." Then he checked to see if the model matched the creator's claims before sending it on to a higher authority for a final yes or no decision.

Instead of bemoaning his lowly status and salary, Einstein appreciated the perks of his job: a constant and diverse flow of the creative thoughts of others that stimulated his own ideas; friendly workmates; and a congenial atmosphere. The job also left Albert with eight free hours a day and uninterrupted Sundays to pursue his own interests. He stretched those hours of freedom by stealing moments at the office to work on his own ideas. At the director's approach, he would hurriedly stuff his notes into his desk. As there is no account of Albert being caught, he had either quick reflexes or an indulgent director.

Einstein came to think of the Patent Office as "a worldly monastery" in which its director, Friedrich Haller, with his "splendid character" and "good brain," could pass for the father superior. The only sour note was Haller's coarse language, which at first disconcerted him—a surprising reaction from one who enjoyed earthy humor and was known for his "wicked tongue." Some wit described Haller as a benevolent taskmaster running the place "with a whip in one hand and a bun in the other."

Albert acknowledged that his father had been right to push him to get a practical job free from academic pressures, and that examining and describing many varied inventions was better for a man of his temperament than a university appointment. He proved his point in 1905, some three years later, by the astonishing volume and quality of his scientific production while he was still working at the Patent Office.

Patent Office examiners spoke and wrote in German and were also expected to have a fair knowledge of French or Italian. Albert had picked up conversational Italian during the happy months with his family in Milan and Pavia. According to a new friend, Lucien Chavan, Albert was also fluent in French, which was enhanced by his "compelling and vibrant" cellolike voice and slightly foreign accent. The thirty-one-year-old Chavan, an electrical engineer in the Patent Office building, had learned about Albert's discussion group and had become its fourth member.

Impatient for their next meeting, Solovine would waylay Albert at noon as he left his office for lunch. Then they continued "the discussion of the previous evening: 'You said ... but don't you think?' Or: 'I'd like to add to what I said yesterday ...'" At times just a sentence or two in a book would spark a long, engrossing discussion.

In Henri Poincaré's groundbreaking 1902 book *La Science et L'hypothese,* for example, the mathematician wrote: "There is no absolute time; to say two durations are equal is an assertion which has by itself no meaning and which can acquire one only by convention ... Not only have we no direct intuition of the equality of the two durations, but we have not even direct intuition of the simultaneity of two events occurring in different places." This idea, said Solovine, "held us spellbound for weeks." The concept foreshadowed Einstein's original space-time hypothesis.

By the fall of 1902 Albert had been at the Patent Office four months and hadn't seen Mileva for a year. She was apparently still weak from a difficult childbirth. He had never seen their daughter, Lieserl, who, if still alive, would have been nine months old. She is last mentioned in her parents' available letters as suffering from scarlet fever in September 1903.

Early in October Albert traveled south to Milan, where his fifty-five-year-old father was dying from heart disease. The past quarrels were forgiven, and Hermann gave his approval for Albert to marry Mileva. Pauline Einstein was in no mood to hold out and presumably gave her approval, too.

Albert wanted to stay beside his father till the end, but Hermann insisted on being alone in his room during his last moments. Although it was his father's wish, ever after Albert could not recall Hermann's lonely death without feeling guilty. As Helen Dukas told Banesh Hoffmann, "Many years later, he still recalled vividly his shattering sense of loss. Indeed on one occasion he wrote that his father's death was the deepest shock he had ever experienced."

Maja's big regret was that "sad fate did not permit [her father] even to suspect that two years later his son would lay the foundation of his future greatness and fame."

CHAPTER 10

The Olympia Academy

January 1903 to September 1904

23 to 25 years old

Grief-stricken, Albert blamed himself for failing to dissuade his father from risky business ventures doomed to failure, believing that the stress had surely hastened his death. "Dazed" and "overwhelmed by a feeling of desolation," Albert repeatedly asked himself "why his father should have died rather than he."

That winter, soon after the funeral, the Einstein family left their Italian home for good. Pauline moved to Hechingen, Germany, to live with a sister, Fanny Koch; she later became housekeeper to a widower. Maja resumed her schooling at Aarau. Albert took the train back to Bern.

Mileva joined him there after their long separation. They were married in a civil ceremony on January 6, 1903. He was twenty-three and she was twenty-seven. It was a simple affair, which they celebrated by treating their witnesses, Habicht and Solovine, to supper at a nearby restaurant. From there, unable to afford a honeymoon, the couple went directly to their apartment at 49 Kramgasse (now an Einstein museum). Mileva had to wait outside shivering in the cold while Albert searched for the key, which he had either forgotten or mislaid. Eventually he roused the sleeping landlord to let them in.

Matrimony did not cramp his style. On the contrary, he now began a period of intense productivity. He completed a fourth research paper on thermodynamics, and sent it off for publication in *Annalen der Physik*, confident, as he informed Besso, that it was both clear and simple. Unfortunately, it was also flawed, depending on "an erroneous assumption."

Over the next few months he immersed himself in his work, both in and out of the office. He and like-minded friends frequently met in each other's homes, and on weekends continued their animated conversations while hiking or mountain climbing.

To mock stuffy academicians, they began calling themselves "The Olympia Academy." And though these Bern-based Olympians rarely dined on anything more succulent than sausage, cheese, and fruit, and drank nothing more potent than coffee, their boisterous talk and laughter made their neighbors suspect otherwise. "Our means were frugal," Solovine recalled, "but our joy was boundless . . . These words of Epicurus applied to us: 'What a beautiful thing joyous poverty is!' " Their conversations ranged beyond science and philosophy; they also took on Sophocles, Dickens, and Cervantes. Almost fifty years later Einstein treasured these idyllic times "when we ran our happy 'Academy' which after all was less childish than those respectable ones which I got to know later." He remembered their meetings as "a delight."

The meetings were less than delightful to Mileva. She never went walking with the group, and when "the Olympians" met in the apartment she rarely if ever joined in, though she listened intently. Some speculate that the talk was over her head; that twice having failed the final college exam, she had abandoned her dream of teaming up with Albert as a carefree student of scientific exploration. Supporting this view was his remark that after their marriage she seemed almost bored with his attempts to share ideas with her. A friend of Albert's described Mileva as gloomy and taciturn, implying perhaps that she resented not being an active member of the Academy. Solovine, more sympathetic than most, saw her as intelligent and reserved, suggesting that she simply preferred the onlooker's role. Because it was such a closely guarded secret, no one in their circle, except of course Einstein himself, knew the reason for her gloomy manner: that she was deeply depressed, mourning the loss—or abandonment—of her daughter, Lieserl.

Biographer Peter Michelmore came close to the truth when, several years after Albert's death in 1955, he questioned Hans Albert, the Einsteins' eldest son and probably the best available source. Afterward, Michelmore wrote:

> Friends had noticed a change in Mileva's attitude and thought the romance might be doomed. Something had happened between the two, but Mileva would only say that it was "intensely personal." Whatever it was, she brooded about it and Albert seemed to be in some way responsible. Friends encouraged Mileva to talk about her problem and to get it out in the open. [Because] she insisted that it was too personal and kept it a secret all her life, a vital detail in the story of Albert Einstein has been shrouded in mystery. Mileva married Albert despite the incident. She knew her love for the man was strong enough to survive. She did not think of the shadow her "experience" would cast over their lives together.

Was the "experience" that of putting their daughter up for adoption, which was the child's most likely fate? Was it a mutual decision, or had Albert pressured Mileva into it? And had the motive been to protect his career, to avoid scandalizing society, to placate his parents—or all three?

Answers to these questions may be in letters still hidden from public scrutiny. One thing is certain, though; however distressed Mileva may have been at having to abandon Lieserl, and however much she may have blamed Albert, she still had married him.

According to biographer Philipp Frank, Mileva was "a free-thinker with progressive ideas." So was Albert. But, in Frank's view, they were temperamentally unsuited. Mileva "did not possess to any great degree the ability to get into intimate and pleasant contact with her environment. Einstein's very different personality, as manifested in the naturalness of his bearing and the interesting character of his conversation, often made her uneasy. There was something blunt and stern about her character. For Einstein, life with her was not always a source of peace and happiness."

Einstein, however, found life with his friends to be idyllic, despite neighbors' complaints about the noise that sometimes drove the trio into the deserted streets to continue their arguments. Einstein relished these occasions, remarking, "One can really quarrel with brothers and close friends."

One day Einstein ended their arguments with a violin solo and a quip, saying he could earn more as a street musician than at the Patent Office. Solovine agreed and offered, if he ever made the career change, to join him on the guitar.

Because Einstein had never tasted caviar, his friends schemed to surprise him with it as a treat for his twenty-fourth birthday. They were holding a meeting of the Olympia Academy that same night. While the three of them discussed Galileo, Solovine made the switch, replacing the usual simple snack with caviar. Then he and Habicht watched furtively as Albert ate the lot, talking all the time but without so much as a murmur of appreciation. He had been completely absorbed in the subject—Galileo's inertia principle. When they laughed and explained why, Albert sat for a moment in stunned silence, then said apologetically that delicacies were wasted on him.

Some days later they again dipped into their meager savings for caviar. This time they repeatedly chanted, "Now we are eating caviar!" to Beethoven's Symphony in F. Albert admitted it was delicious, but added, "You have to be an epicure like Solovine to make so much fuss about it."

One late-night excursion took them through a wood, where they were intoxicated by the heady mixture of pine scent and their own fermenting ideas, and then onward and upward to the summit of Mount Gurten. There the shimmering stars seemed close enough to touch—with a slight jump—and they turned their talk to astronomy. They discussed cosmology until the stars faded and the rising sun painted the Alps pink. If their enthusiasm hadn't inoculated them against a need to sleep, coffee in a mountainside café on the descent did the trick.

On another occasion they rose at dawn on a Saturday for an eighteen-mile hike to Thun, northeast of Bern. Again the scenery set the agenda, as the sight of the Alps peaking to over 14,000 feet spurred conversation about the earth's history and how mountains are made. After lunch in Thun, they returned by evening train to Bern—talking all the way.

Chavan had been impressed by Einstein's cellolike voice. Others noticed an air of serenity, even of childlike innocence, in his manner of speaking—a startling contrast to his frequent bursts of lusty laughter. If aroused in conversation he became as animated as his friends, but normally he expressed himself in an even tone, with long, silent lapses as though lost in another world. It was a world, he later explained, full of symbols: "When I have no special problem to occupy my mind I love to reconstruct proofs of mathematical and physical theorems that have long been known to me. There is no goal in this, merely an opportunity to indulge in the pleasant occupation of thinking." And, of course, it was an escape from everyday worries.

In June 1903 he and Mileva had saved enough to enjoy a delayed honeymoon in Lausanne. They had been looking forward to having another child, so Albert was pleased shortly after their return when Mileva told him she was pregnant.

Other members of the Olympia Academy had been impatient to resume their meetings, which had been interrupted by the honeymoon. They were so fired up by the discussions that they rarely missed one for any reason. But they were all music lovers, and when Solovine learned that a visiting Czech orchestra was playing the night a meeting was to be held at his home, he suggested going to the concert instead. Einstein talked him out of it—or so Einstein believed. On the day of the performance, however, Solovine was offered a cheap ticket and couldn't resist. Not having time to warn the others, he hoped to compensate for his absence by leaving four extra eggs with the usual sausage snack, and covered the dish with a conciliatory note: "Hard-boiled eggs and a greeting to very dear friends."

The "bribe" didn't work, and they agreed to punish Solovine for his absence. Knowing that he detested tobacco smoke, they shut the window and, as they talked far into the night, filled the room with smoke from Habicht's cheap cigars and Albert's pipe. Then they piled everything movable—including a table, chairs, plates, cups and saucers, cutlery, books, a teapot, and a sugar bowl—onto Solovine's bed. Finally, before leaving they pinned a message to the wall: "Thick smoke and a greeting to a very dear friend." When Solovine returned he saw the mess through an eye-stinging haze. Holding his breath, he flung open the window, cleared cigar stubs and stale tobacco from various dishes, and moved the mountain of litter from his bed. Nevertheless the pillow and curtains still reeked, and he couldn't sleep until dawn.

That evening Solovine waited impatiently for Einstein outside the Patent Office, not to complain about the trashing of his room but to catch up on the subject of the meeting he'd missed. Einstein attacked him with mock fury: "You miserable wretch! You dare miss a regular meeting and listen to violin playing? Barbarian! Boor! If you ever again indulge in such folly, you will be excluded and expelled from the Academy!" They then discussed the previous night's subject, David Hume's *An Enquiry Concerning Human Understanding*, until past midnight. Solovine concluded that "what stamped our Academy, as we jokingly referred to our meetings every evening, was the burning desire to broaden and deepen our knowledge and our affection for each other."

In the fall of 1903 Habicht left Bern to teach math and physics at a Protestant school in Schiers, a small town too far away for him to attend Academy meetings. Einstein was sorry to see him go but missed his intellectual sparring partner, Michele Besso, even more. Besso was now in Trieste, running an engineering consulting office.

Just seven months after joining Bern's Association of Scientists, Einstein was asked to address them—a sure sign that he was beginning to impress fellow scientists. He accepted the invitation and read a paper, "Theory of Electromagnetic Waves," to his fellow members on December 5, 1903.

The following spring, anticipating a visit from Habicht, Einstein promised to receive him "with pleasure and the rest of my feelings—we have put that up in preserve jars for appropriate occasions. We are expecting a baby in a few weeks." His other big news was that he had discovered "the relationship between the magnitude of the elementary quanta of matter and the wavelengths of radiation in an exceedingly simple way."

Einstein was also moved to contact his friend and benefactor, Marcel Grossmann, who was still teaching at Frauenfeld. In a letter he sent to Einstein accompanying a copy of his paper on non-Euclidian geometry, which he hoped to get published, Grossmann mentioned he had just become the father of a girl, Elsbeth. Einstein was struck by the coincidence, writing on April 6, "There is an extraordinary similarity between us. Next month we are also going to have a baby. And you will also receive a paper from me, one that I sent to Wiedemann's *Annalen* a week ago." (Max Planck was editorial director of *Annalen der Physik*. Eilhard Wiedemann was one of the several editors.)

The Einsteins' child, a son, was born on May 14, 1904; they named him Hans Albert. Four months later Albert's boss told him that he was no longer working on a trial basis, but had been made a permanent staff member for having proved "very useful." Haller also increased his salary to 3,900 francs (a modest $600 a year) but declined to promote him from third- to second-class technical expert; that would have to wait until Albert mastered mechanical engineering.

What he had mastered was the art of simultaneously coping with a baby and studying. A visitor discovered Albert's method, after gasping his way through an obstacle course of damp clothes hanging to dry, stale pipe smoke, and a leaking wood stove. There was Albert, ignoring or oblivious of the acrid pollution, lost in a book while casually rocking the baby's cradle with his foot. He even worked while taking Hans Albert for an airing through the city streets, using the baby carriage as a mobile desk. He kept a notebook under the blankets and stopped occasionally to take it out and record his latest thought.

Einstein and Besso still discussed their mutual passion, physics, in a frequent exchange of letters, but it wasn't enough for Einstein. Questions that had haunted him for almost ten years seemed close to resolution, and he needed Besso to help bring the answers into focus. Besso's business in Trieste was hardly booming, and when Einstein suggested that by working together they could resume their around-the-clock conversations, he jumped at the chance. He applied for a Patent Office job and, toward the end of 1904, was hired as a technical expert. Again following Albert's advice, Besso and his wife (a Winteler) and young son moved into an apartment in Bern within shouting distance of the Einsteins.

While walking to and from work together, and sometimes at the office, Besso willingly joined Einstein on an imaginary flight alongside a beam of light. It had been established that light traveled through empty space at some 186,000 miles per second. If they caught up with the light beam and then traveled with it at the same velocity, would the light appear to be at a standstill? Although Einstein thought it might appear to stand still, he wasn't convinced. What did Besso think?

As he got out of bed one morning, Einstein had another thought that stuck with him: Were two events occurring at the same time to one observer also simultaneous for a second observer in a different location? Everyone seemed to think so, but that didn't mean they were right.

In later years, Einstein explained to reporters why he had asked himself what could be considered naive questions for a man in his twenties. He said that children who wonder about things like light, time, and space are satisfied with the stock answers, and never give them another thought as adults. But, because he was a late developer, he first pondered such "simple" questions as an adult, and so probed them more deeply and tenaciously than any child would do.

Besso played along with him, arguing for the established views. He took Albert's speculations seriously but rebutted him with Newton's concepts of time and space, which had been accepted by the scientific community for more than two hundred years. Now, however, Newton was not beyond criticism. He had described light, for instance, as a stream of particles ("corpuscles"), but recent experiments had revealed its mysterious dual nature. It was also wavelike. Furthermore, both Faraday and Maxwell had shown that Newton's ideas did not adequately explain electromagnetism.

The Olympia Academy continued to flourish during this period, even when some members moved out of town. Others took their place, among them Habicht's younger brother, Paul. Nor did being a parent deter Einstein from his self-imposed studies. In fact, this third-class technical expert was on the verge of discoveries that would rank him among the greatest scientists of all time.

CHAPTER 11

The Special Theory
of Relativity

1905

---

26 years old

Cartoonists picture Einstein calmly contemplating the stars as he puffs on his pipe and occasionally writes the odd equation. Friends, too, spread this fiction, believing it to be true. Apparently that was what Einstein wanted, because he did his best to hide his turbulent nature. Once, however, an interviewer caught him off guard. Then he admitted that his urge to understand the universe kept him in a state of "psychic tension . . . visited by all sorts of nervous conflicts . . . I used to go away for weeks in a state of confusion, as one who at that time had yet to overcome the stage of stupefaction in his first encounter with such questions." He soon regretted this admission and unsuccessfully asked the journalist not to publish it.

Friends who never saw evidence of mental turmoil assumed he had been misquoted or joking. After all, they knew him to respond to serious occasions and pompous individuals with sardonic or tongue-in-cheek remarks or an explosive laugh. Even his confidant, Besso, was apparently unaware of the intensity of Einstein's emotions.

One balmy day in the early spring of 1905, Einstein told Besso that he needed just a few more pieces to complete the cosmic jigsaw puzzle. For this task, Besso would need to be more than a sounding board. They would have to battle as a team. Besso, as always, was willing. Putting aside their Patent Office work, they discussed the missing pieces exhaustively. It became obvious that it would take more than a day of brainstorming for answers to mysteries that had obsessed Einstein since his youth, and he returned home in despair, feeling he would never discover "the true laws . . . based on known facts."

There is no record of how late he went to bed that night, what he had for supper, or whether his sleep was disturbed by nightmares. He woke next morning in great agitation, as if, he said, "a storm broke loose in my mind." With it came the answers. He had finally tapped "God's thoughts" and tuned in to the master plan for the universe.

Banesh Hoffmann confirmed this account: "Einstein said his basic discovery came on waking up one morning, when he suddenly saw the idea. This had been going around and around at the back of his mind for years, and suddenly it wanted to thrust itself forward into his conscious mind. We know brilliant ideas come at crazy times. You may be just waking, or lost in a forest, and suddenly the idea comes—almost as if it's coming from somewhere. Einstein once said something very interesting to me when we were trying to think: 'Can we get another idea that will solve this problem?' And he said, 'Ideas come from God.' Now he didn't believe in a personal God or anything like that. This was his metaphorical way of speaking. You cannot command the idea to come, it will come when it's good and ready. He put it in those terms: 'Ideas come from God.'"

Einstein's sudden revelation as he woke that morning reached beyond electricity, magnetism, matter, and motion to the nature of light, space, and time itself. It even touched on the secret of creation. But when he saw Besso later that day, he gave his friend no idea of the scope of his breathtaking discovery, nor of its dramatic birth. Instead, he greeted Besso with a casual, "Thank you. I've completely solved the problem."

Physicist Stanley Goldberg explains Einstein's discovery this way: "The 'thing' into which Einstein had sudden insight in the spring of 1905 is that judgments of simultaneity are intimately involved in both spatial and temporal measurements." He now realized that events can only be regarded as simultaneous in one's own immediate environment. As Einstein put it: "The solution came to me suddenly with the thought that our concepts and laws of space and time can only claim validity insofar as they stand in a clear relation to our experiences; and that experience could very well lead to the alteration of these concepts and laws. By a revision of the concept of simultaneity into a more malleable form, I thus arrived at the special theory of relativity."

Einstein gave great credit for his inspiration to the ideas of Ernst Mach and even more so to those of David Hume, "whose treatise on understanding [A Treatise of Human Nature] I studied with fervor and admiration shortly before the discovery of the theory of relativity. It is very well possible that without these philosophical studies I would not have arrived at the solution."

As if possessed, he devoted every spare moment over the next few weeks to expressing his ideas on paper, filling thirty-one pages with his small, neat handwriting. The final paper, titled "Zür Electrodynamik bewegter Körper" ("On the Electrodynamics of Moving Bodies"), was strangely free of footnotes or references, as if the inspiration had indeed come, if not from God, from some otherworldly source. He did acknowledge Besso's assistance, though.

Later, Einstein said that the ideas of his idol, nineteenth-century chemist Michael Faraday, the brilliant son of a blacksmith, had also played a guiding role. Faraday had supposed that the electromagnetic force takes time. Einstein went further, assuming that so did electromagnetic interaction. Eventually, he would conclude that all interactions at a distance take time: that, in a sentence, is Einstein's special theory of relativity. Einstein also cited the work of physicists Hendrik Lorentz and George Francis Fitzgerald as helpful. Yet, wrote physicist George Gamow,

> Einstein was probably the first to realize the important fact that the basic . . . laws of nature, however well established, were valid only within the limits of observation and did not necessarily hold beyond them. For people of the ancient cultures the Earth was flat, but it was certainly not for Magellan, nor is it for modern astronauts. The basic notions of space, time, and motion were well established and subject to common sense until science advanced beyond the limits that confined scientists of the past. Then arose a drastic contradiction, mainly due to Michelson's experiments concerning the speed of light, which forced Einstein to abandon the old "common sense" ideas of the reckoning of time, the measurement of distance and mechanics, and led to the "non-commonsensical" Theory of Relativity. It turned out that for very high velocities, very large distances, and very long periods of time, things were not as they "should have been."

Gerald Holton, professor of physics and of the history of science at Harvard University, takes a different view. He has shown that Michelson's experiment had at most a marginal effect on Einstein's theory of relativity, that he probably read about it like everyone else but didn't take it seriously. Holton uncovered evidence to support his view in the Einstein Archives. According to Holton, Einstein did not depend on Michelson's experiments to produce special relativity, and said so repeatedly and emphatically in his own correspondence. Holton cited these letters in his *Thematic Origins of Scientific Thought* (Harvard University Press, 1973). He points out that:

> Faraday, however, played a large role [in Einstein's thinking]. He appears in the first paragraph of the 1905 paper. All of Einstein's contemporaries always referred to the latest news from the laboratory. Einstein starts his article with an experiment which was over 70 years old, an 1830s experiment. And he says, "Faraday has absolutely shown that all that counts in order to produce an electromagnetic induction is a relative speed between the magnet and a coil. It doesn't matter whether you keep one stationary or the other; so it's only the relative motion between them [that produces the electromagnetic induction]." There's nothing absolute in mechanics. However, in Maxwell's theory of electromagnetism of the time the ether was privileged — speeds were measured with respect to it. Therefore there was an absolute meaning of a velocity with respect to the ether. And Einstein said that Faraday's experiment shows that one has to reformulate science in such a way that, as in mechanics where there are no absolute velocities, there also will be no absolute velocities in electromagnetism. That's the connection between Faraday-Maxwell velocities — that is to say, movements in time — and Einstein.

Einstein mailed his relativity paper to *Annalen der Physik* in June 1905. It was the fourth paper he had sent to the scientific journal so far that year — an astonishing, if not unique, creative output. Then, exhausted, he took to his bed for several days, anxiously waiting to see if any of his theories would be published.

When he was back on his feet, he invited Habicht, who was still teaching at Schiers, to attend the meetings of the Olympia Academy. Getting no reply, Einstein sent him a bantering letter in which, almost as an afterthought, he mentioned his own recent prodigious output and promised to send Habicht all four works, including one — a paper on the photoelectric effect — which "is very revolutionary."

It was indeed, for it smashed the belief experimentally confirmed over the past two centuries that light was a wavelike phenomenon. Planck had put a dent in this view with his recent disturbing claim that matter emitted and absorbed light in particles or quanta. Einstein went further, now suggesting that light itself, freed from matter, traveled as separate particles. By applying Planck's quantum theory to light, Einstein helped to establish quantum mechanics.

Experiments had shown that when light hits certain metals, electrons are "torn from the metal and a shower of them speeds along at a certain velocity." This was clearly demonstrated when electricity flowed through wire attached to the metal after its exposure to light. Einstein thought that particles of light "bombarded" the metal sporadically, like bullets, and not smoothly and continuously as would be expected from waves.

In his seventeen-page paper, titled "Über einen die Erzeugung und Verwandlung des Lichtes betreffenden heuristischen Gesichtspunkt" (On a heuristic point of view concerning the generation and conversion of light), Einstein explained how the bombardment by light quanta, or photons, caused the outgoing shower of electrons, or photoelectric effect. Why did he call it "heuristic" — meaning a valuable guide but unproved or incapable of proof? Because he viewed his quantum hypothesis concerning light as "irreconcilable with established principles . . . perhaps even ultimately untenable," and at best an incomplete account of optical phenomena.

The practical application of his theory has been the "electric eye," used for opening and closing doors by remote control, for detecting intruders, and for counting and sorting goods. It also made radio and television possible. In time, it — not relativity — won Einstein the Nobel Prize.

The sight of sugar dissolving in water had prompted Einstein's second paper, "Eine neue Bestimmung der Moleküldimensionen" (A new determination of the sizes of molecules). It addressed the size of sugar molecules, which he calculated as about one twentieth of a millionth of an inch across. He had originally submitted this paper to the University of Zurich as his Ph.D. thesis, but Professor Kleiner, who had rejected his first thesis in 1901, rejected this one as too short. Einstein then added one sentence and it was accepted. (He found that very funny.) Professor Kleiner had recommended it, bolstered by a second opinion from Professor Heinrich Burckhardt, who wrote that despite "crudeness in style and slips of the pen in the formulas . . . [the paper showed] thorough mastery of mathematical methods." He reworked this paper before sending it to *Annalen der Physik* and considered it a very worthwhile project, judging by a letter in which he wrote, "A precise determination of molecules seems to me of the highest importance because Planck's radiation formula can be tested more precisely through such a determination than through measurements of radiation."

His third paper was titled "Über die von der molekularkinetischen Theorie der Wärme geforderte Bewegung von in ruhenden Flüssigkeiten suspendierten Teilchen" (On the motion — required by molecular kinetic theory of heat — of small particles suspended in a stationary liquid). It was inspired by the Scottish botanist Robert Brown, who, staring through a microscope, saw pollen dust executing haphazard, zigzag movements in water, as if alive. Intrigued, Brown substituted organic and inorganic substances and noted the same curious, non-stop dance. His observation of what had seemed impossible—perpetual motion—bewildered fellow scientists because it had no apparent cause and defied known natural laws, including the second law of thermodynamics. Einstein, too, was fascinated and began to investigate. He decided that "Brownian movements" were actual collisions between invisible molecules of the water itself and the visible particles. This was a daring conclusion, because many doubted the existence of molecules and atoms. He persisted, and produced a formula stating that the average displacement of the visible particles in any direction increased as the square root of the time. Then, by measuring the distance the particles had traveled in that time, he calculated the number of incessantly moving but invisible molecules in a certain volume of both liquids and gases. That's how he estimated that a gram of hydrogen consists of 303,000,000,000,000,000,000,000 molecules! (The number is read as 303 sextillion in the U.S. system, 303 thousand trillion in the British system.)

Not only had Einstein found evidence that atoms of a definite finite size really exist, he also had created a statistical method to chart their behavior. Experiments by French physicist Jean Perrin verified Einstein's work, confirming the perfect accuracy of his equations and demonstrating the physical reality of atoms. (Among contemporaries opposed to the atomic hypothesis were Wilhelm Ostwald, George Helm, and Ernst Mach.) Poincaré had previously underscored the value of Einstein's papers by calling Brownian motion and the photoelectric effect two of the most pressing scientific problems of that time.

His fourth paper, "On the Electrodynamics of Moving Bodies,"— the relativity paper — proved to be the most sensational, for it challenged Newton's view of the universe, one that had endured for two centuries. Newton saw space as a fixed, ether-pervaded, physical reality, through which stars and planets move and against which their movements should be measured. Time, too, he regarded as an unvarying absolute flowing from an infinite past to an infinite future. British astronomer Arthur Eddington believed such preconceived ideas about location in space came from our apelike ancestors. When Newton was pressed to justify his views, he cited a more august authority—God Himself. Einstein used intuition and math to correct, extend, and qualify Newton's "inspired" guesses. He came up with a cosmos in which stars, planets, and galaxies move in relation to each other and not to an exclusive, God-appointed space.

This meant, said Einstein, that one's relative position in the universe controls one's viewpoint—sometimes to comic effect. To a person on Mars, for instance, people at the earth's North Pole would appear to be standing upright, those in

China and Britain would seem to be standing sideways, and those at the South Pole would appear to be hanging upside down and about to drop off into space.

Einstein also pointed out that the illusory nature of relative motion is often experienced by passengers in a train standing in a railroad station. When a train on an adjoining track starts to move slowly and smoothly it is impossible for them to tell which train is moving until a third, stationary object provides a point of reference.

Having demolished Newton's idea of absolute space, Einstein now demolished the idea of absolute time, saying there is no such thing as a "now" that is independent of some system of reference. "We have to take into account," he wrote, "that all our judgments in which time plays a part are always judgments of *simultaneous events*. If, for instance, I say, 'That train arrives here at 7 o'clock,' I mean something like this: 'The pointing of the small hand of my watch to 7 and the arrival of the train are simultaneous events.'" A New Yorker, for instance, phoning a friend in California is justified in believing they are talking simultaneously, even though his clock shows 7 P.M. and his friend's shows 4 P.M., "because they are both residents of the same planet, and their clocks are geared to the same astronomical system."

But what if we tried to communicate with another system "right now" and chose Alpha Centauri, the nearest star to us aside from the sun? Because it is 26 trillion miles from Earth and radio waves travel at the same speed as light, some 6 trillion miles a year, it would take 4.3 years for a message to get through to Alpha Centauri and another 4.3 years to receive a reply. Furthermore, when we look at Alpha Centauri "now" we are really seeing the star as it was 4.3 years ago. In fact, any star we see "now" may no longer exist. No one, Einstein concluded, can assume that his or her subjective sense of "now" applies throughout the universe. Instead, every reference body (or coordinate system) has its own particular time.

For years he had wondered how things would look to him if he rode on a light beam. Recently he had glanced back at the Bern's clock tower while riding a streetcar and thought, What if the streetcar were moving at the speed of light? Applying his new theory, he decided that the clock would appear to him to have stopped, while the watch in his pocket would continue to run at its usual rate. This confirmed his idea that time is not the same for all observers when objects approach the speed of light.

The behavior of light remained the great paradox. Einstein assumed that it consisted of a steady stream of small particles called photons moving through empty space at some 186,000 miles per second. This speed of light, in fact, was the prevailing view, frequently confirmed by experiments. But Einstein's daring and unique conclusion was that it *always* moved at 186,000 miles per second and was completely unaffected by the motion of its source or of its observers. This defied logic and common sense. Wasn't Newton's view more reasonable, that light reaches an observer more rapidly if the observer is moving toward rather than away from it? How could Einstein possibly discount the movements of the sources of the light and of those observing it? To demystify this weird phenomenon, Einstein turned to mathematics and the work of Fitzgerald and Lorentz.

Trying to account for Michelson's failure to find any movement of the earth in relation to the ether, Irish physicist George Francis Fitzgerald suggested that the measuring instruments Michelson used had contracted slightly and distorted the reading. He then produced equations showing that matter contracts in the direction of its motion, the contraction increasing as the speed increases. The Fitzgerald contraction, as this phenomenon is known, is "exceedingly small in all ordinary circumstances . . . If the speed is 19 miles a second—the speed of the earth around the sun—the contraction of length is 1 part in 200,000,000, or 2½ inches in the diameter of the earth."

Dutch physicist H. A. Lorentz stated that a flying charged particle foreshortened in its direction of travel would increase in mass. Einstein, in turn, applied Lorentz's equations, known as the Lorentz transformation, to all objects, including clocks and measuring instruments. Einstein showed that objects moving at great speeds and over vast distances decreased in size and increased in mass. Strangest of all, he proved that at those speeds time slowed.

Lincoln Barnett, an interpreter of his theories whom Einstein respected, wrote:

> It explains why all observers in all systems everywhere, regardless of their state of motion, will always find that light strikes their instruments and departs from their instruments at precisely the same velocity. For as their own velocity approaches that of light, their clocks slow down, their yardsticks contract, and all their measurements are reduced to the values obtained by a relatively stationary observer . . . The greater the speed, the greater the contraction. A yardstick moving with 90 percent the velocity of light would shrink to about half its length; thereafter the rate of contraction becomes more rapid; and if the stick could attain the velocity of light it would shrink to nothing at all. Similarly a clock traveling with the velocity of light would stop completely. From this it follows that nothing can move faster than light, no matter what forces are applied.

From these calculations would arise the humorous scenario of a young astronaut who goes on a long flight at close to the speed of light. On returning to earth, still young, he finds that his twin brother is an old man. This scenario, impossible in fact, was the result of following math equations to the outer limits of absurdity (as Einstein would explain later in a conversation with journalist Alexander Moszkowski). It was not unlike entertaining the possibility of an irresistible force meeting an immovable object.

Mileva almost met an immovable object in trying to persuade Albert to take a break from his work. Eventually he agreed to visit her parents in Novi Sad, and to take one-year-old Hans Albert with them. Albert faced this first meeting with some trepidation, especially as Mileva's mother had once threatened to give him a "thrashing." When they arrived, though, past animosities seem to have been overcome. Mileva introduced him to friends and relatives, and was his tour guide in Belgrade and at Kijevio, a lakeside vacation resort.

He endured weeks of anxious waiting on his return home. Finally, much to his delight and relief, *Annalen der Physik* published his relativity paper on September 26, 1905. He then anticipated a lively, if critical, response to his work, but to his dismay there was none, either positive or negative. He looked in vain for any mention of it in subsequent issues of the journal. No one seemed even mildly curious about his extraordinary new view of the universe. He even sought some response from a former fellow student at Zurich Polytechnic, Margarete von Uexküll, the young woman whose "impossible" professor Einstein had helped to placate. She listened politely, but then told Mileva that Albert's theories were "utterly fantastic." The loyal if somewhat premature Mileva replied, "But he can prove his theory."

He was briefly elated some months later when the great Max Planck wrote for clarification of obscure points in the theory. Einstein replied, but that seemed to be the end of anyone's interest in it.

If Einstein had cited sources in his "Electrodynamics of Moving Bodies" paper, he would surely have aroused more response. As Abraham Pais points out, "Relativity includes the completion of the work of Maxwell and Lorentz . . . Einstein's oeuvre represents the crowning of the work of his precursors, adding to and revising the foundations of their theories." Einstein's assistant and biographer, Banesh Hoffmann, was surprised that the brilliant French physicist, Henri Poincaré, "failed to take the crucial step that would have given him the theory of relativity [before Einstein], so close did he come to it . . . Yet when he came to the decisive step, his nerve failed him and he clung to old habits of thought and familiar ideas of space and time. If this seems surprising it is because we underestimate the boldness of Einstein in stating the principle of relativity as an axiom and, by keeping faith with it, changing our notions of time and space." According to Einstein, it was a failure to "get it" rather than lack of nerve. Poincaré, he said, was "hostile [to the theory of relativity], and despite all his acumen, he hardly seemed to understand what we were doing."

Einstein, meanwhile, was pursuing relativity to further and even more remarkable effect. It had occurred to him, he wrote to Habicht, "that the relativity principle in connection with the Maxwell equations demands that the mass is a direct measure for the energy contained in the bodies; light transfers mass. A remarkable decrease of the mass must result in radium. This thought is amusing and infectious but I cannot possibly know whether the good Lord does not laugh at it and has led me up the garden path."

This latest breakthrough turned out to be nothing to laugh at, even though it eventually met with plenty of secular scoffing for defying contemporary scientific thought. Fellow scientists believed that energy and mass were unchangeable. However, Einstein had recently heard of experiments in which speeding electrons acquired increased energy. This encouraged him to marry his previous electromagnetic equations with the velocity of light, a union begetting the discovery that a body's mass diminishes after releasing energy in the form of light.

From there Einstein leaped to the extraordinary deduction that all energy has mass — a theory that he suggested could be tested in the case of radioactivity, for example, because of its high energy release. He put his thoughts into a three-page paper which he sent to Paul Drude, now editor of *Annalen der Physik* under Planck's editorial direction. It appeared in the November 1905 issue, as a "mathematical footnote" to his first relativity paper published two months before. As if expressing his uncertainty, the title was a question: "Ist die Trägheit eines Körpers von Seinem Energieinhalt abhängig?" (Does the inertia of a body depend upon its energy content?) Two years would pass before Einstein came to an even more astonishing conclusion, that matter and energy are intimately connected as different aspects of the same thing — matter approaching the speed of light becomes energy, and energy slowed becomes matter.

As his sister, Maja, recalled, Einstein "imagined that his publication in the renowned and much-read journal would draw immediate attention," and relished the prospect of "sharp opposition and the severest criticism." But he looked in vain for comments of any kind in several subsequent issues of *Annalen der Physik*. Had the "good Lord" indeed led him up the garden path?

Deeply disappointed by the apparent lack of interest in his ideas, Einstein sought to revitalize the Olympia Academy by trying to tempt Habicht to quit his distant teaching job and join him at the Patent Office. He also missed the lively Solovine, who, after a brief stay as a student at the University of Lyons, was now working as an editor and writer for a scientific journal in Paris. It was going to be a cold winter.

CHAPTER 12

"The Happiest Thought of My Life"

1906 to 1911

27 to 31 years old

There was good news in the new year. The great Max Planck wrote from Berlin wanting more information about relativity, and there were signs that Einstein's theories had encouraged others to investigate further. Heinrich Zangger, professor of forensic medicine at the University of Zurich, was particularly interested in Brownian motion. During their meeting to discuss it, he and Einstein became friends.

Life at home was going smoothly, too. Despite his preoccupations, Einstein willingly helped Mileva with household chores, splitting wood for the leaky stove and carrying coal up several floors to heat their small apartment. They often laughed together over the haughty, impertinent manner of their two-year-old son, Hans Albert.

On April 1, 1906, Einstein had more good news to report: his promotion to technical expert second class, and with it a raise. The increase of 1,000 francs to 4,500 greatly relieved Mileva, who shortly before had wondered if they could even afford a visit from their friend Helene Savic.

Yet Einstein was discontented. He feared that at twenty-seven his creative days were numbered. He referred to himself as a "venerable federal ink shitter" and hankered for the more carefree times and company of old friends. Having failed to entice Habicht back to Bern, he tried to persuade Solovine to join him at the Patent Office. But Solovine preferred to remain in Paris, where he was soon to become the official translator of all Einstein's works into French.

He brightened at the prospect of a visitor from Berlin. Spurred by Planck's interest in relativity, Planck's assistant, Max von Laue, arranged to meet Einstein at the Patent Office. On his arrival, in the summer of 1906, he announced himself and sat in the waiting room. Moments later a young man strolled in and glanced around. He was so casual and dressed so informally that von Laue ignored him. Then, in a Chaplinesque scene, Einstein went out—and came back in again. This time, von Laue took a chance and introduced himself.

Einstein invited him home and on the way, as a friendly gesture, handed him a cigar. Walking across the bridge over the Aare River, von Laue surreptitiously dropped it in the water. He knew a bad cigar on contact, even if some aspects of Einstein's theory puzzled him. Still, he was sufficiently impressed after their first conversation to write a positive article on the subject. Like Zangger, von Laue became a lifelong friend and supporter.

Einstein remained on affectionate terms with the Winteler family, especially the parents, even though he had badly hurt Marie Winteler. In November 1906, his sister Maja, who was engaged to one of the Wintelers' sons, Paul, had devastating news to tell: Paul's brother, Julius, back home from a trip to the United States as a ship's cook, had gone berserk, fatally shooting his mother and a brother-in-law and then killing himself. Baffled and distressed, Einstein sent a letter of condolence to the survivors. He was to be permanently tied to this tragic family through Maja, who eventually married Paul Winteler, and through his friend Michele Besso, already the husband of the formidable Anna Winteler.

Einstein found escape from personal tragedies in his work, which, fortunately, he could do at any time, in any place. His office, after all, was under his hat. Asked once where his laboratory was, he held up his fountain pen. So it was hardly a surprise that, while vacationing in the mountains during the early summer of 1907, he thought up a machine to measure minute changes of electricity. He sketched it on a postcard he sent to the Habicht brothers, and within a month they had produced a prototype. Mileva helped perfect the device, all of them sustained by the ever-ready Turkish coffee made in a coffee machine in the Einsteins' apartment. But manufacturers showed no interest in the invention.

By this time the scientific community's growing interest in his theories, especially relativity, had renewed his desire to pursue an academic career. It would provide more time and opportunities to develop his own ideas and to confront his critics on their level. The customary way to join the staff of Bern University was as a *Privatdozent* (unsalaried lecturer), a sort of apprentice, which would allow him to keep his Patent Office job as well as pocket the token fees paid by students attending his lectures.

Alfred Kleiner, who already had his eye on Einstein as an up-and-coming physicist, impressed on him Bern's value as a steppingstone to a professorship at the University of Zurich, where Kleiner headed the physics department. In June 1907, Einstein submitted his application to Bern, together with his special relativity thesis and seventeen other published works.

To his dismay, the department head, Aime Forster, turned him down flat, dismissing the relativity paper as "incomprehensible." Einstein blamed his rejection on ignorance. Nevertheless, he was over the disappointment a few days later, when he began to fine-tune an astonishing aspect of relativity.

Soon after his rejection by Forster, Einstein was invited by Johannes Stark, editor of *Jahrbuch der Radioaktivitat und Elektronik,* to write a review of the relativity theory. He took the opportunity to send what he had been mulling over for the past two years — his momentous conclusion that everything in the universe is a repository of enormous, latent energy. In the review, he revealed this secret of all creation in six strokes of the pen, the equation $E = mc^2$ (read as energy equals mass multiplied by the speed of light squared). The formula implied that mass is frozen energy and predicted that converting a small amount of mass would release an enormous amount of energy — as eventually happened with the atomic bomb. Though Einstein neither revealed how to split the atom nor suggested the possibility, his formula would be used to uncover the mystery of how the sun and stars radiate light and heat for billions of years (through nuclear reactions). It explained, too, Marie Curie's recent discovery that just one ounce of radium emitted four thousand calories an hour indefinitely.

Einstein also raised a question: "If every gram of material contains this tremendous energy, why did it go so long unnoticed?" His answer was that the energy had never been observed until radium's recent discovery by the husband-and-wife team of Pierre and Marie Curie. He compared the situation, before the Curies appeared on the scene, with the existence of a fabulously wealthy man who kept his riches secret because he never spent or gave away a cent.

"Imagine the audacity of such a step," said his later assistant and biographer, Banesh Hoffmann. "Every clod of earth, every feather, every speck of dust becoming a prodigious reservoir of entrapped energy. There was no way of verifying this at the time. Yet in presenting his equation in 1907 Einstein spoke of it as the most important consequence of his theory of relativity. His extraordinary ability to see far ahead is shown by the fact that his equation [$E = mc^2$] was not verified . . . till some twenty-five years later."

At home he dealt with a proliferating correspondence from fellow physicists but still found time to entertain his son, once delighting the three-year-old by converting a piece of string and a couple of matchboxes into a working cable car. He also reserved one evening a week for his own entertainment, joining an attorney, a math teacher, a bookbinder, and a man of private means for a few hours of Haydn, Mozart, and Beethoven.

After the Bern failure, he reconsidered his options. Perhaps he had been too ambitious and should start at a lower rung of the academic ladder. There was a vacancy for a math teacher in Zurich. If that fell through, he'd give Winterthur another try. He had enjoyed his brief spell there as a temporary teacher six years before. But he regarded it as a Teutonic stronghold where there would be two strikes against him: he didn't speak Swiss-German and he looked Jewish. He hesitated to apply in person.

Neither plan worked out, fortunately. The failures were fortunate because it was while he was still in the Patent Office that Einstein got what he called "the happiest thought" of his life, which was to lead, after years of intense, exhaustive work, to a revolutionary new picture of the universe — eventually published as his general theory of relativity. The happy thought occurred as he tried to extend special relativity, which applied only to a hypothetical universe where objects moved with constant velocity in gravity-free space. What occurs in the real universe, he wondered, where objects are subjected to gravity and acceleration? He was astonished by the sudden idea that a man falling freely—and accelerating—would not feel his own weight. Einstein would later say that he took "the first step towards the solution of this problem [of acceleration] when I endeavored to include the law of gravity in the framework of the Special Theory of Relativity."

At the back of his mind was Faraday's unified field theory, which held that everything in the universe is connected with everything else and that gravity is one of the conditions (corresponding to the model of the atom). Newton saw gravity as a force attracting objects to one another. Not so, said Einstein; objects move in a gravitational field, their paths determined by the curved structure of space. Although gravitation is still called a force, Einstein saw it as an effect of the distortion of space by matter. He concluded that the motion produced by gravitation is equivalent to the motion produced by acceleration—his principle of equivalence.

Planck's first reaction was discouraging: "Everything now is so nearly settled," he told Einstein. "Why do you bother about these other problems?" He bothered, suggests historian of science I. Bernard Cohen, because "Einstein was a genius, far ahead of his contemporaries. He knew that special relativity was incomplete, that it did not deal with accelerations and with gravity."

Now the renowned Professor Hermann Minkowski came back into his life. As Einstein's math teacher at Zurich Polytechnic a few years earlier, he had dismissed Einstein as a lazy dog. Lecturing in Göttingen in December 1907, however, he enthusiastically supported Einstein with a dazzling new presentation of special relativity's significance. "From now on," he said, "space and time separately have vanished into the merest shadows, and only a sort of combination of the two preserves any reality." This union was to be known as space-time. Having introduced the fourth dimension, Minkowski set about giving space-time sophisticated mathematical underpinnings, as well as promoting Einstein's views before a wider professional audience. Einstein welcomed this influential ally with reservations, believing that Minkowski was pushing the mathematical abstraction too far. Einstein's immediate need, however, was for a sounding board to replace Besso. His supersensitive friend was apparently still shattered by the tragedy in his wife's family and unable to concentrate on their talks. As Einstein wrote to Solovine, "Even my conversations with Besso on the way home have stopped." Luckily for Einstein, a replacement was on the way.

While taking an oral exam in Germany early in 1908, math student Jakob Laub had argued with Professor Wilhelm Wien over several aspects of relativity, with neither giving way. Go ask Einstein, Wien suggested. And Laub went.

When he entered the apartment—the Einsteins had moved to the Kirchenfeld district where they had briefly lived before—Albert was trying to warm the cold room by vigorously poking the fire in the oven. He welcomed Laub's questions, and they found so much to discuss that for several weeks, at midday and during the evening, Laub fetched his new sparring partner from the Patent Office, to make use of every free moment. Laub was still with Einstein in the spring of 1908, when Mileva took Hans Albert to visit her parents in Novi Sad. While she was away, Einstein felt moved to assure her that though he and Laub spent a lot of time together, he was lonely and waited lovingly for her return.

During their three-month collaboration, Einstein made good use of Laub's math prowess, and they produced two papers. Later in the year, Laub left to become an assistant to Philipp Lenard in Heidelberg. Einstein advised him to tolerate Lenard's whims because he was a masterful and original thinker. Mileva had briefly studied under Lenard, who would become Einstein's most bitter enemy.

Meanwhile, Einstein was getting advice from Kleiner to reapply for a lecturing position at Bern University. He did so — and this time he was accepted.

He hardly expected an overflow crowd at his lecturing debut in the summer of 1908. It was scheduled for the ungodly hour of seven in the morning, but he certainly thought there would be more than the three friends who showed up, including Besso. He was unlikely to attract a crowd at any time, though, as he was nervous and poorly prepared.

Einstein's scarecrow appearance and lackluster lectures did not discourage Professor Kleiner from trying to recruit him. Kleiner knew an exceptional mind when he met one, and he wanted to make use of it at the University of Zurich. There was only one physics vacancy, however, and Friedrich (Fritz) Adler was the favored candidate. He was already on the staff of the university as a much admired *Privatdozent*. What's more, his father, Viktor, was the founder of the Austrian Social Democratic Party, to which most hiring committee members owed allegiance. Certainly Kleiner had considerable influence as the senior physics professor, yet Adler seemed a sure thing.

Adler, however, recognized Einstein's gifts, and was himself not eager for the job, having decided to quit academia for a political career. However mixed his motives, he behaved with magnanimity. He wrote to his father:

> The other candidate is a man named Einstein, who studied at the same time I did. We even heard a few lectures together. Our development is seemingly parallel: He married a student at about the same time as I, and has children. But no one supported him, and for a time he half starved. As a student he was treated contemptuously by the professors . . . He had no understanding of how to get on with important people . . . Finally he found a position in the patent office in Bern and throughout the period he has been continuing his theoretical work in spite of all distractions.

Adler followed up this letter with one to the hiring committee:

> If it is possible to obtain a man like Einstein for the university, it would be absurd to appoint me. I must frankly say that my ability as a research physicist does not bear even the slightest comparison to Einstein's. Such an opportunity to obtain a man who can benefit us as much by raising the general level of the university should not be lost because of political sympathies.

Kleiner knew he had a strong case in recommending Einstein as "among the most important theoretical physicists," with an "uncommonly sharp conception and pursuit of ideas." But there was a major obstacle: his candidate was a Jew. So was Adler, but he was already part of the establishment. How Kleiner finessed this tricky situation is revealed in the selection committee's report:

> These expressions of our colleague Kleiner, based on several years of personal contact, were all the more valuable . . . since Herr Einstein is an Israelite and since precisely to the Israelites among scholars are ascribed (in numerous cases not entirely without cause) all kinds of unpleasant peculiarities of character, such as intrusiveness, impudence, and a shopkeeper's mentality . . . It should be said, however, that also among the Israelites there exist men who do not exhibit a trace of these disagreeable qualities and it is not proper, therefore, to disqualify a man only because he happens to be a Jew. Indeed, one occasionally finds people also among non-Jewish scholars who in regard to a commercial perception and utilization of their academic profession develop qualities which are usually considered as specifically "Jewish." Therefore neither the committee nor the faculty as a whole considered it compatible with dignity to adopt anti-Semitism as a matter of policy and the information which Herr Kleiner was able to provide about the character of Herr Dr. Einstein has completely reassured us.

Einstein could pass for a goy, was the thrust of the message. This must have been a great comfort to committee members, it seems, because in a secret vote on May 7, ten voted for him, none were against, and only one abstained. He was in.

Einstein resigned from the Patent Office on July 6, 1909. Otto Wirz, a coworker and future novelist, overheard Friedrich Haller's incredulous response: the director thought Einstein was joking. It is not clear whether Haller was amazed by Einstein's leaving after a recent salary increase, or by his elevation to an associate professorship at a leading university.

He would doubtless have been astonished to see Einstein the next day at the University of Geneva, mingling with Marie Curie, chemist Wilhelm Ostwald, and scores of other international notables. They were there to receive honorary degrees and celebrate the 350th anniversary of the university's founding by theologian John Calvin. Einstein almost didn't make it, because he had mistaken the ornate invitation addressed to "Monsieur Tinstein" for an advertisement and chucked it in a wastebasket. Not receiving a reply, the organizers had gotten Einstein's friend Lucien Chavan to persuade him to travel to Geneva for an important unspecified event. Once there he reluctantly joined a procession, wearing his suit and a strawhat, a sparrow among over two hundred distinguished peacocks and peahens in full academic regalia: gold-braided tailcoats, colorful medieval gowns, and violet silk sugar-loaf hats.

"The festivities ended in the Hotel National," Einstein recalled for his biographer, Carl Seelig, "with the most opulent banquet I have ever attended in my life. It encouraged me to say to the Genevan patrician sitting next to me: 'Do you know what Calvin would have done had he been here? . . . He would have erected an enormous stake and had us burnt for sinful extravagance.' The man never addressed another word to me."

Plenty of others were eager to speak to him two months later at Salzburg, where he addressed his first international physics conference. There he predicted that light would be found to be both wave and particle, and introduced the idea of complementarity fifteen years before Niels Bohr hit on it. As Abraham Pais put it: "Thus Einstein must be considered the godfather of complementarity." Even the outspoken, hypercritical physicist Wolfgang Pauli (a 1945 Nobel Prize winner for his exclusion principle) acknowledged that Einstein's lecture was a milestone in theoretical physics.

At Salzburg he met science greats Max Planck, Max Born, and Arnold Sommerfeld. To Born, Einstein "had already proceeded beyond Special Relativity which he left to minor prophets, while he pondered about the new riddles arising from the quantum structure of light, and of course about gravitation and General Relativity." The excitement upset Einstein's stomach, but Sommerfeld, his host—a doctor's son—took good care of him.

After moving into an apartment in Zurich, Einstein had a happy surprise: Friedrich Adler lived in the apartment below. Adler was equally pleased to have Einstein literally within earshot. Another friendship was in the making. Adler felt closer to Einstein than to any other physicist he knew, telling his parents, "The more I speak with Einstein, and this happens often, the more convinced do I become that I was right in my opinion of him. Among contemporary physicists he is not only the clearest but the one who has the most independent brains ... The majority of scientists don't even understand his approach." The two shared their ideas at every opportunity. When they were interrupted by their rowdy young sons, rather than silence the boys, Einstein and Adler climbed to the building's attic to continue their conversations.

Albert and Mileva loved being back in a city full of friends and pleasant memories. But life wasn't the same. Mileva was pregnant with their second son, and Albert was saddled with regular teaching assignments and seminars as well as administrative chores. Mileva had lost interest in his scientific speculations, being preoccupied with the demands of housekeeping and a family. She was incensed by his attempt to renew his friendship with Anna Schmid, who had written to congratulate him on his university appointment. Mileva had intercepted their letters and frustrated a possible liaison. Einstein was miffed and confided in Besso in a November 1909 letter. Besso, however, sympathized with Mileva. This is the first hint of trouble in the marriage and of Einstein's interest in other women.

Early the following year, one man recommended Einstein for the Nobel Prize for physics—Wilhelm Ostwald, who had won the chemistry prize the previous year. Ostwald, who had rejected Einstein when he had applied to be his assistant in 1901, now cited him for producing in special relativity the most far-reaching concept since the energy principle. The prize, however, went instead to Johannes Diderik van der Waals of Amsterdam University, "for his work on the equation of state for gases and liquids." This made it possible to study temperatures near absolute zero.

At Salzburg, Sommerfeld had introduced Einstein to one of his former students, Ludwig Hopf. Needing an assistant, Einstein hired him. Hopf was also a talented pianist, an added bonus.

Einstein's teaching had improved, but not his appearance. The students' first sight of the new professor was of a scruffy young man whose pants were too short, holding what looked like a visiting card. It turned out to be a scrap of paper with the main points of his lecture on it. Despite his informality, said student Hans Tanner, "after the first few sentences he captured our hearts." Tanner, later a professor of physics at Winterthur, recalled how Einstein encouraged students to interrupt the lecture if anything was not clear, and that during the breaks "he would take one or the other of the students by the arm in a most comradely manner to discuss the subject with him. Weekly from eight to ten in the evening, we students had to propound a theme . . . and at the end he would say, 'Who's coming to the Cafe Terasse?' The discussions continued there." Once "we sat and gossiped until closing time in a cafe . . . As we left Einstein said, 'Is anyone coming home with me. This morning I received some work from Planck in which there must be a mistake.'

"In his apartment he gave us Planck's pamphlet, saying 'See if you can spot the fault while I make some coffee.'" When they couldn't he pointed it out and said, "'We won't write and tell him he's made a mistake. The result is correct, but the proof is faulty. We'll simply write and tell him how the real proof should run. The main thing is the content, not the mathematics. With mathematics one can prove anything.'"

He demonstrated his cavalier attitude toward math one afternoon in a café when he and engineer Gustave Ferrière were discussing math's rigid rules. Einstein placed five matches on the table and asked, "What is the total length of these five matches if each is two and a half inches long?" "Twelve and half inches," Ferrière replied. "That's what you say," said Einstein. "But I very much doubt it. I don't believe in mathematics."

Instead of four he now had twenty-four in his class, among them Adolf Fisch, who noticed how after his lectures Einstein was often surrounded by students asking questions, which he answered in a patient and friendly manner. While many students loved him, some colleagues resented his egalitarian ways as a threat to their privileged status. They enjoyed being treated as exalted beings by their "inferiors," while he behaved as if he had none. He spoke "in the same way to everybody. The tone with which he talked to the leading officials of the university was the same as that with which he spoke to his grocer or to the scrub-woman in the laboratory." Why, he had even moved furniture to his new apartment by pulling it on a cart, with no sign of embarrassment; and when a second son, Eduard, was born on July 28, 1910, Albert got a sack of wool for the baby's cradle and carried it home on his back.

In October 1910, Einstein completed another important paper, this one on critical opalescence, answering a question children often ask: Why is the sky blue? He concluded that dust and air molecules reflect much more of the blue part of light than of the rest of the spectrum.

Einstein's assistant, Hopf, had an interest in psychoanalysis, which had led him to Carl Jung. When Einstein expressed curiosity about the man and his ideas, Hopf arranged a meeting. Jung was equally interested to learn about Einstein's discoveries. The first meeting led to several more, usually at dinner parties. Jung "had difficulty in following his arguments" but was impressed "by the simplicity and integrity of his brilliant powers of thought."

Both were out of their depth, in strange waters and with no lifeguard around. But they managed to keep afloat—and in touch. The tenor of their conversation can be judged from Jung's comment to a friend: "Although I am no mathematician, I am interested in the advances of modern physics, which is coming ever closer to the nature of the psyche . . . So long as you keep to the physical side of the world, you can say pretty well anything that is more or less provable without incurring the prejudice of being unscientific, but if you touch on the psychological problem, the little man, who also goes in for science, gets mad." If he was referring to Einstein as "the little man," he was wrong. Einstein didn't get mad, being well aware of the psychological factor in scientific affairs; in fact, it was about to affect his career. Nevertheless he remained wary of psychoanalysis as therapy.

After only a few months in Zurich, and despite what seems to have been a joyful and productive time, Einstein began secret negotiations to move to the German University in Prague. Perhaps he was tempted by a full professorship and a salary large enough to afford a servant. When the secret got out, students who idolized him petitioned for him to stay. He did, and got a pay raise.

Even so, he went ahead with his decision to apply for the job in Prague. It seems that someone with influence was determined to get rid of him. Who and why remains a mystery. Carl Seelig believed Einstein was too proud to kowtow to anyone to advance his career—or, presumably, to keep his job. The only hint that a clash with a superior made him leave Zurich is in Kleiner's letter to an unnamed colleague: "After my statements about his conduct some time ago (after which he wanted to apologize, which I once again prevented), Einstein knows he cannot expect personal sympathy from the faculty representatives. I would think you may wait until he submits his resignation before you return to this matter." Clearly he could no longer count on the support of Kleiner, his mentor. Prague then became an attractive possibility.

As in Zurich, two applicants were after the same job. Einstein's rival was Gustav Jaumann, an Austrian. Candidates were listed in order of achievements, which put Einstein on top. Prague was Austrian territory at the time, though, and the government usually hired Austrians rather than foreigners, regardless of talent. Because Jaumann was Austrian, he was chosen over Einstein. But when he learned how he had been chosen, Jaumann refused to have anything to do with a university that did not appoint the best candidate, and virtually handed the job to Einstein — an almost exact repeat of the events in Zurich.

Before they left for Prague, Einstein took Mileva with him to Leiden, in the Netherlands, where he had been invited to lecture. The place was like a magnet to him because H. A. Lorentz, the Dutch physicist he idolized, was there, a frail-looking, white-bearded father-figure he regarded with a mixture of awe, admiration, and love.

If anyone doubted Einstein's gift for friendship, they had only to see him and Lorentz together. Their mutual friend, physicist Paul Ehrenfest, attests to that. In what is believed to be a partial draft of Ehrenfest's funeral oration for Lorentz, he noted how Lorentz always provided Einstein with

> a warm and cheerful atmosphere of human sympathy. The best easy chair was carefully pushed in place next to the large work table for his esteemed guest. A cigar was given to him, and then Lorentz began quietly to formulate questions concerning Einstein's theory of the bending of light in a gravitational field. Einstein nodded happily, taking pleasure in the masterly way Lorentz had rediscovered, by studying his works, all the enormous difficulties that Einstein had to overcome before he could lead his readers to their destination. As Lorentz spoke on, Einstein began to puff less frequently on his cigar, and he sat up straighter and more intently in his armchair. And when Lorentz had finished, Einstein sat bent over the slip of paper on which Lorentz had written mathematical formulas to accompany his words as he spoke. The cigar was out, and Einstein pensively twisted his finger in a lock of hair over his right ear: a habit several friends remarked. Lorentz, however, sat smiling at an Einstein completely lost in meditation, exactly the way a father looks at a particularly beloved son—full of secure confidence that the youngster will crack the nut he has given him, but eager to see how. Suddenly Einstein's head shot up joyfully; he "had it." Still a bit of give and take, interrupting one another, a partial disagreement, very quick clarification and a complete mutual understanding, and then both men with beaming eyes skimming over the shining riches of the new theory.

While staying with Lorentz, the Einsteins got word to an old college friend, Margarete von Uexküll, who lived nearby, that they would like to see her. She couldn't remember knowing any Einstein, but she came anyway, and instantly recognized Mileva with delight and amazement. Told that Mileva had married Albert Einstein, she was still puzzled until he approached, and then she recognized him as the fellow student who had helped her with her work. She never forgot her failure to remember him, calling it one of the most embarrassing moments of her life.

CHAPTER 13

To Prague and Back

1911 to 1913

32 to 34 years old

Mileva was not dragged to Prague kicking and screaming, but she was reluctant to leave Zurich, which she loved.

They nearly didn't make it to Prague anyway, because Einstein listed his religious affiliation on his job application as "None." Told that this would automatically disqualify him for the post, he reluctantly wrote *Mosaisch*, the Austrian term for Jewish. This was just a hint of more red tape to come, or as he eventually complained, "infinitely much paperwork for the most insignificant Dreck."

As a state official—the university being a state institution—Einstein had to swear allegiance to the Austro-Hungarian emperor, Franz Joseph, while decked out like what he thought of as a Brazilian admiral, in a blue-and-gold uniform, triangular cap, and ceremonial sword, all bought at great expense for this one occasion. He eventually sold the regalia to his successor — and eventual biographer — Philipp Frank, for half-price. For his official reception, however, he dressed as he pleased, in a workman's shirt, and was amused when the porter mistook him for an electrician come to repair the lights.

His early days in Prague brought rave reviews. The city's intelligentsia turned out in force to hear his inaugural lecture in the huge auditorium of the Natural Science Institute. Mathematician Gerhard Kowalevski, who was in the audience, noticed that the casually dressed Einstein "captured all hearts," speaking "with great naturalness and at times with refreshing humor."

Surprisingly, he began to follow the convention of calling on colleagues and their families, most of whom were eager for their first face-to-face meeting with this genius with "the dreamy look in his eyes." These visits turned into the European version of the Chinese water torture: evening after evening of the same trivial chitchat over teacups with people he didn't like. Eventually he stopped going.

Einstein was ambivalent about Prague. He was delighted with his working conditions but disappointed with the many cold and unfeeling people he met in the streets and cafés. He found them both pretentious and servile, an attitude epitomized by the porter's daily greeting: a bow and "Your most obedient servant." This struck him as phony as well as fulsome, a mockery of how Prague's citizens really felt about each other. Here, unlike Zurich's free and friendly atmosphere, he sensed an air of bitterness and resentment. Such an atmosphere was hardly surprising. The city was segregated into Czech, German, and Jewish circles. Each group despised the others, and they were united only in their fear of disease-carrying beggars polluting the medieval streets.

Einstein made the best of it, however, and looked on the bright side: his increased salary helped pay for Fanni, a live-in maid, and their spacious apartment was lit with electricity instead of the kerosene lamps and gaslights of Switzerland. On the other hand, Switzerland was much cleaner, and the water clear as vodka. In their Prague apartment it flowed in an unappetizing brown stream, compelling them to buy bottled drinking water and to cook in water from a street fountain.

Even the brown water came in useful, however, when a fire broke out in their maid's bedroom. Einstein quickly doused it, then found himself covered in fleas—probably from the maid's secondhand mattress. He hurriedly took a bath to get rid of them, an urgent precaution because bubonic plague and typhoid fever were not uncommon in the city.

His assistant, Ludwig Hopf, couldn't take the unsanitary conditions. Using his modest salary as an excuse for the move, he left for Germany, where he became a professor of hydrodynamics and aerodynamics and married a doctor's daughter. Einstein then hired Emil Nohel, a farmer's son and math major. He aroused Einstein's interest in the problems of Jews who, far from larger Jewish communities, lived totally in an alien society. Nohel told him how Jewish peasants and tradesmen in Bohemia who no longer knew Hebrew tried to keep in touch with their roots on the Sabbath by switching from Czech to German. As the next best thing to Yiddish, it served as a substitute for Hebrew.

Despite his disdain for the general population in Prague, Einstein soon made a handful of friends. One was mathematician Georg Pick, a bright, provocative, and uncompromising man in his fifties. During the walks they took together, he fascinated Einstein with his memories of the great Ernst Mach, the university's first rector, now retired and partly paralyzed. Pick had been Mach's assistant and was inclined to repeat his remarks, which implied that Mach had anticipated Einstein's theories. Pick himself had come up with brilliant ideas others had taken over and developed into independent branches of mathematics. He encouraged Einstein to use an advanced type of calculus in his elusive quest.

Einstein was intrigued with the idea and would soon take his advice. He also accepted Pick's invitation to join his music group. Like Einstein, Pick played the violin at every opportunity. The only sour note was the group's pianist, a sister-in-law of Sanskrit professor Moritz Winternitz. A retired piano teacher, she treated Einstein as if he were one of her less promising pupils. He thought she'd make a good army sergeant. Still, he often lingered at Winternitz's home, intrigued by the professor's five children, to whom he became devoted. He explained his interest in them by saying he wanted to see how a number of goods produced by the same factory would behave.

On Tuesday evenings he argued politics and philosophy and made music with other young Jewish intellectuals at the home of Bertha Fanta, an ardent Zionist. There, or in Prague cafés, Einstein met Franz Kafka, Hugo Bergmann, and Max Brod. Kafka was not yet a famous writer, Bergmann was an active Zionist and a talented musician, and Brod was a novelist whose work had historical and philosophical insights. Einstein showed no interest in Zionism but otherwise delighted in the philosophical and political arguments, and, of course, in the music.

Although he couldn't deny that everyday life was more pleasant in Switzerland, and that Swiss students were more intelligent and hardworking, he had nothing but praise for the institute where he worked and its magnificent library.

For months he had agonized over quantum theory, not convinced that quanta really existed. He had also struggled with the baffling dual nature of light and, most intensely, with the mystery of gravitation. By June of 1911, Einstein concluded that one aspect of his still unfinished general theory of relativity — that the sun's gravity bent starlight passing by on its way to the earth — could be experimentally investigated during a total eclipse, when starlight would be more visible. In his paper "Über den Einfluss der Schwerkraft auf die Ausbreitung des Lichtes" ("On the Influence of Gravitation on the Propagation of Light"), published in *Annalen der Physik*, he invited astronomers to put his theory to the test.

Meanwhile, Einstein had to defend special relativity from attack from both sides of the Atlantic, despite a recent comment by Arnold Sommerfeld that the theory was "so well established that it was no longer on the frontiers of physics." Yet in Prague, philosopher Oscar Kraus was so disturbed that some scientists could believe in an idea that he considered absolutely absurd that it gave him nightmares. Princeton's Professor W. F. Magie agreed with Kraus, complaining, in a presidential address to the American Physical Society, that the special theory failed because it did not describe the universe in intelligible terms capable of being grasped by the common man as well as the trained scholar.

Torn between explaining his old ideas at home and exploring and discussing new ones abroad, Einstein seized every chance to get out of Prague. He attended a conference in Karlsruhe in September and left Mileva behind with the boys. She ruefully remarked that he was away so often she wondered if he would recognize her.

Later during the fall of 1911, he had another chance to get out of town. Ernest Solvay, a wealthy Belgian industrialist, sponsored an international conference in Brussels for world-class scientists, asking them to focus on the crisis in physics sparked by the quantum work of Planck and Einstein. It was an enormous success. Twenty-one turned up, among them the Dutch physicist Hendrik Lorentz (who presided), Einstein, Marie Curie, Ernest Rutherford, Henri Poincaré, Paul Langevin, Max Planck, and Walther Nernst. Einstein marveled at Lorentz's tact, fluency in three languages, and scientific genius.

Einstein ridiculed reports of a romance between two of the scientists at the conference, the widowed Marie Curie and the married, though separated, Langevin — a scandal headlined in popular newspapers. Although he knew both, and liked Langevin unreservedly, Einstein thought their liaison completely implausible. To him, Curie, who had recently won her second Nobel Prize for her work on radioactivity, was "sparklingly intelligent, but despite her passion she is not attractive enough to be really dangerous to anybody."

A young British physicist, Frederick Lindemann, attending the Solvay conference as Nernst's secretary, surveyed the talent and was most impressed by Einstein and Poincaré. Lindemann later recalled that

> [al]though Einstein had already published so many masterpieces, none had been actually put to the test and his theories were looked on rather as tours de force than as definitive additions to knowledge. But his pre-eminence among the twelve greatest theoretical physicists of the day was clear to any unprejudiced observer. I well remember . . . M. de Broglie [secretary of the Solvay Conference and elder brother of physicist Louis de Broglie] saying that of all those present Einstein and Poincaré were in a class by themselves. He was a young man [thirty-two], singularly simple, friendly and unpretentious. He was invariably ready to discuss physical questions with a young student, as I then was. And this never changed though the adulation showered on him might well have turned any man's head.

At the time, Lindemann had written in the same vein to his father: "I got on very well with Einstein who made the most impression on me except perhaps Lorentz . . . Einstein asked me to come and stay with him if I came to Prague and I nearly asked him to come and see us at Sidholme [Lindemann's home in Devon, England]. He says he knows very little mathematics but he seems to have had a great success with them." Lindemann shared Einstein's high regard for Lorentz, "a wonderful all round man with extraordinarily quick comprehension and also a sense of humour."

Einstein was obviously not happy in Prague. In November 1911, his friend Marcel Grossmann threw him a lifeline. Now the influential dean of physics at Zurich Polytechnic — where he and Albert had been students together — Grossmann wondered if Einstein would like to join him. In the New Year an official offer arrived — a ten-year appointment — and Einstein joyfully told the Habicht brothers to expect the homing pigeons to return the following summer. To others he spoke of returning to *Europe,* and referred to Prague as "beyond the Pale." He could hardly wait to leave the place.

Einstein's grim view of the city failed to discourage physicist Paul Ehrenfest, one of Einstein's many correspondents, who wrote to say that he would soon be job hunting in Prague. Einstein was waiting for him on the station platform, a cigar in his mouth and Mileva at his side. They took him to a nearby coffeehouse, where they discussed everything but physics — their favorite cities was one topic — until Mileva left to prepare a special meal (Ehrenfest was a vegetarian). The men then headed for the university, arguing physics nonstop. A sudden downpour made them raise their voices, but they otherwise ignored the foul weather. Einstein had a date with a string quartet that evening, so he handed the soggy Ehrenfest over to a colleague, Anton Lampa, for safekeeping. After the musical interlude, he returned to reclaim the visitor, then took him home. At half past one the next morning they were still arguing.

"Within a few hours we were true friends," Einstein later wrote, "as though our dreams and aspirations were meant for each other." Ehrenfest matched Einstein's informality and shared his passion for physics. He echoed his host's enthusiasm in a diary entry for February 12, 1912: "Yes, we will be friends. Was awfully happy."

They stopped talking, it seemed, only to eat, sleep, or make music. The small group included Ehrenfest at the piano and Einstein on the violin, with seven-year-old Hans Albert beating time. Ehrenfest quickly took a shine to the boy; he sat next to him at meals and brought him into the conversation. One Sunday they went for a walk, with Albert pushing his young son, Eduard, in a baby carriage, and Hans Albert at Ehrenfest's side, chatting away.

Einstein suggested that Ehrenfest apply for the job he was quitting. But Ehrenfest now decided he wanted to work with or close to his newfound friend, wherever that happened to be. In any case, he disqualified himself by refusing, as a staunch atheist, to pretend to a religious belief—a necessary subterfuge enforced by Emperor Franz Joseph's rule—to be considered for a university appointment in Prague. Unlike Einstein, he was not willing to compromise.

Those who were close to Einstein believed that his marriage started to crumble when they were in Prague. If so, he and Mileva hid it from their sons. Years later, Hans Albert said he was unaware of it, and although he occasionally sensed tension between his parents, his father's merry whistling while he shaved and jaunty walk when he set off for the nearby university reassured him that there was nothing to worry about.

Nor does a letter Mileva wrote to Besso during the Einsteins' early days in Prague give any hint of trouble ahead. Hans Albert, she wrote, "loves going to school, is improving well with his piano lessons and delights in asking interesting questions of his papa about physics, mathematics and nature."

However, she was surely not thrilled by Albert's trip in the spring of 1912 to Berlin, where he renewed a childhood acquaintance with Elsa Löwenthall, his recently divorced cousin. This must have worried the normally suspicious and jealous Mileva, and with good reason.

On his return to Prague, Albert wrote to Elsa in response to a letter from her, saying that he had become very fond of her. Elsa replied, sending her letter to his office address, and made him promise to destroy all her letters — which he apparently did — but she kept and treasured all of his.

In their early correspondence, Albert confided to Elsa that he had suffered terribly because he had been unable to love his mother: "When I think of the bad relationship between my wife and Maja or my mother, then I must admit . . . that I find all three of them quite unlikable, unfortunately!" The confession then became a declaration of love: "I have to have someone to love, otherwise life is miserable. And this someone is you . . . I'm not asking you for permission. I am the absolute ruler in the netherworld of my imagination." Elsa taunted him with being henpecked, which he denied — "let me categorically assure you that I consider myself a full-fledged male" — and explained that he tolerated Mileva out of pity. A week later, on May 7, he took a step back: "I cannot tell you how sorry I am for you, and how much I would like to be something for you. But if we give in to our affection for each other, only confusion and misfortune will result."

Two weeks later, on May 21, he was in full retreat. Saying "I have misgivings about our affair," he wrote, adding that it would be his last letter and urging her to stop writing to him. But he left an escape clause, saying that if she needed to confide in him she could contact him at a new address he would give her when he returned to Zurich later in the summer.

So matters stood in the spring of 1912, when his successor and future biographer, Philipp Frank, called on him in Prague. Einstein, now thirty-three, welcomed the twenty-eight-year-old Frank in his office. The office overlooked the parklike grounds of a mental asylum, where women inmates gathered in the mornings and men in the afternoons, often engaged in fierce disputes. Einstein, who never stopped agonizing over quantum theory, pointed to an agitated group and said to Frank, "Those are madmen who do not occupy themselves with quantum theory."

On his triumphant return to Zurich in June of 1912, Albert and his family moved into a sunny apartment in the Zurichberg quarter. His old college pals Marcel Grossmann and Louis Kollros greeted him warmly at the Polytechnic. At this place where, as a student, he had been humiliated and treated shamefully, he was now the reigning star — though he never behaved like one.

He was soon absorbed in his ongoing attempt to devise the field equations that would explain gravitation's role in the universe — or, as he sometimes put it, "to know God's thoughts." Finding it incredibly difficult, he pleaded with Grossmann to save him from going crazy. Grossmann came to his aid once again, and they began to tackle the problem together.

Working around the clock, Einstein couldn't answer Ehrenfest's avalanche of letters over the following months and feared that his silence had jeopardized their friendship. To make amends, in the summer of 1913 he invited Paul and his wife, Tatiana, also a talented physicist, to Zurich.

Einstein was at the door of their boardinghouse early on the morning after their arrival, and left with Ehrenfest for a walk in the hills. When the day turned hot and hazy, they headed for some shade trees to rest and cool off. As they sat there, Einstein explained how he hoped to focus his still hazy views of the universe.

One evening, after attending a colloquium on Lawrence Bragg's theory of X-ray diffraction by crystals (Bragg and his father, Henry, were to share a 1915 Nobel prize in physics), Einstein and Ehrenfest continued the discussion with others in the garden of a coffeehouse. It was something of a free-for-all, with Ehrenfest at first arguing with Einstein and Max von Laue about Bragg's ideas, then Einstein breaking off to argue about Brownian motion with someone else.

Their reunion in Zurich proved a smashing success. Ehrenfest filled his diary with exuberant accounts of the talk, and the music, of meeting the young Hans Albert again, and of getting to know Albert's stimulating friends, Marcel Grossmann and Michele Besso. Only once in two weeks did Ehrenfest's diary include a note of regret, and that was: "A day without Einstein!"

After the Ehrenfests' departure, Einstein resumed his work on gravitation at fever pitch. "I occupy myself exclusively with the problem," he wrote Sommerfeld, "and now believe that I shall master all difficulties with the help of a friendly mathematician here [Grossmann]. But one thing is certain, in all my life I have labored not nearly as hard, and I have become imbued with great respect for mathematics, the subtler part of which I had in my simplemindedness regarded pure luxury until now. Compared with this problem, the original relativity is child's play."

Earlier that year, when Nobel Prize nominations were being submitted, Wilhelm Ostwald had again recommended Einstein. This time he was joined by three fellow Nobelists, Ernst Pringsheim, Clemens Schaefer, and Wilhelm Wien—the latter suggesting that Einstein share it with Lorentz. Instead, the 1912 award went to Swedish physicist Nils Gustav Dalen "for his invention of automatic regulators for use in conjunction with gas accumulators for illuminating lighthouses and buoys."

On March 14, 1913, a greeting card he received from Elsa for his thirty-fourth birthday took his mind off gravity for a while and renewed their clandestine contact. She encouraged a reply by asking for his photograph and a layman's book on relativity. Einstein wrote back twice, saying there was no book on relativity for the layman. He warned Elsa that Mileva was a very jealous woman, yet encouraged Elsa to visit him in Zurich, when he would tell her "all about those curious things that I discovered."

During the winter Mileva had suffered from rheumatism, which, because of her congenitally dislocated hip, made walking extra painful. She confided to her friend Helene Savic that these days Albert had little time for her and their boys, as he was devoting himself completely to his work.

However, he took her with him to Paris in the spring of 1913, keeping a promise to Paul Langevin to address the French Society of Physicists. He liked and admired Langevin, calling him "the only Frenchman who completely understands me." The Einsteins stayed with Marie Curie in the French capital and got along so well with her that they agreed to spend the coming summer vacation hiking together.

Soon after he returned to Zurich, Planck and Nernst arrived in like a couple of corporate headhunters, hoping to tempt Einstein to join them at Berlin University. They offered him a bigger salary, membership in the exclusive Prussian Academy, and freedom to pursue his work with no strings attached. The only discouraging note was Planck's response when Einstein spoke of his exhaustive efforts to fit gravitation into a general theory of relativity. "As an older friend I must warn you against it," Planck said. "In the first place you will not succeed, and even if you do succeed, no one will believe you."

Einstein felt grateful to Planck, because his early enthusiasm for special relativity largely accounted for Einstein's prestige in the scientific community. He also thought Planck an aloof and naive man who knew as much about politics "as a cat did of the Lord's Prayer." He warmed more to the plump and jovial Nernst, who was childishly vain but funny. He was also quick-witted. When Einstein quoted Langevin to the effect that only twelve people in the world understood relativity, Nernst replied that eight of them lived in Berlin. Whatever their flaws, they surely knew how to entice Einstein to Germany—he was obviously ready to make the move—and they returned home confident their quarry was in the bag.

On their summer hiking vacation with the Curies, the Einsteins took Hans Albert with them. Eduard, who was ill, stayed at home, cared for by family friends. Marie Curie's biographer, Rosalynd Pflaum, tells of an incident that occurred as the hikers, including Hans Albert and Curie's daughters, Irene and Eve, walked together that summer from the slopes of the Alps and south to Italy's Lake Como:

> There was a lot of shoptalk in German . . . and Eve was amused at the way Einstein circulated absentmindedly among the boulders, so deep in conversation that he walked alongside deep crevasses and toiled up the steep rocks without noticing them. One day, the three young people howled with laughter when Einstein suddenly stopped dead, seized Marie's arm, and demanded, peering intently at her: "You understand, what I need to know is exactly what happens to the passengers in an elevator when it falls into emptiness." The imaginary fall in an elevator posed problems of transcendent relativity—there would be no gravitational pull so they would float—and he was struggling with the problem of discovering a mathematical entity with which to represent gravitation.

Einstein remarked in a letter to Elsa that Marie Curie did nothing but complain during this vacation. He also described her as having the soul of a herring — although he may have been exaggerating to allay any fears Elsa might have about having a romantic rival.

That fall a Dutch physicist, Heike Kamerlingh Onnes, was awarded the 1913 Nobel physics prize "for his investigations on the properties of matter at low temperatures which led, inter alia, to the production of liquid helium." Einstein had been recommended for the third time; Wilhelm Ostwald and Wilhelm Wien had again proposed him and were joined this time by a German professor of medicine, Bernhard Naunyn.

While the Nobel winners headed for Sweden to receive their prizes, Einstein and Grossmann were off to Vienna to summarize their work in progress at a meeting of German scientists. There, Einstein said that they had been helped by following the lead of Friedrich Kottler, a young Viennese physicist who was the first to write the Maxwell equations in generally covariant form. ("Covariant means," explains Stanley Goldberg, "that the equations change together in such a way that the form of the equations is always the same in all inertial frames of reference. If, for example, in one frame of reference F equals MA then in another frame of reference F will also equal MA, if transformed covariantly. That is, the form of the equations is always going to be the same.") Einstein asked if Kottler was in the audience, and Kottler stood for a moment to receive the applause.

George von Hevesy, a Hungarian physicist and chemist who would receive the 1943 Nobel Prize in chemistry, was also in the audience. He spoke with Einstein afterward about a recent theory of physicist Niels Bohr, a shy, twenty-eight-year-old Dane, who had united the Planck-Einstein quantum theory with Ernest Rutherford's concept of the atom. Einstein's first response was to say that he had once had similar ideas (which Bohr later acknowledged) but had feared to publish them because, if true, they would mean the end of physics. His second response was to call Bohr's theory of the helium atom "an enormous achievement."

After leaving Vienna, Einstein spent a few days in Berlin with Elsa and her daughters, Margot and Ilse. Elsa gave him a brush to control his unruly hair and instructions to take better care of his appearance. He told her that if she couldn't accept him as he was, she should find a friend more to her taste. He didn't mean it, writing to her on his return home to Zurich, "I now have someone of whom I can think with untroubled delight and whom I can live for." He described his existence with Mileva as like living in a cemetery because of the icy silence between them, and said he treated her like an employee, had his own bedroom, and avoided being alone with her. But when Elsa suggested divorce, he said that Mileva had done nothing a court of law would consider justification for a divorce.

On December 7, 1913, Einstein formally agreed to go to Berlin, telling Lorentz he couldn't resist an offer that allowed him to devote all his time to thinking. It would also free him from lecturing, which was getting on his nerves, he told Ehrenfest. As he explained to Zangger, not only would he have stimulating scientific colleagues in Berlin, but he would also be in contact with first-rate astronomers who could test his theory that gravitation causes light to bend.

However, he understood Mileva's trepidation as she contemplated the future: "My wife howls unceasingly about Berlin and her fear of my relatives. She feels persecuted and fears that her last quiet minute arrives at the end of March [the planned time of their move to Berlin]. Well, there is some truth in it. My mother is of a good disposition, on the whole, but a true devil as mother-in-law. When she is with us then everything is filled with dynamite."

The clashes between his wife and his mother hardly distracted him for long. Not even the loudest baby-crying seemed to disturb him; he could go on with his work completely impervious to noise. He was also indifferent to the weather. One winter evening during a snowstorm he stood under a street lamp with several students, discussing a problem. He handed his umbrella to one, took out a notebook and jotted down formulas for several minutes, oblivious of the whirling snow, until he got the answer. He then stayed until they all understood it.

Could he really totally isolate himself from his environment to concentrate on his work? Not always, if we take him at his word that Mileva was driving him up the wall. Yet he rarely chose to be alone. If he wasn't bringing students home with him, he arrived with musicians. Mileva grumbled to Helene Savic that there was little time for a tranquil private life.

Late in the year, Planck proposed Einstein for membership in the prestigious Prussian Academy of Sciences — an exclusive and influential group of leading scientists. Twenty-one voted for him, and only one against, a big enough majority for Emperor Wilhelm II to give his official approval.

The prospect of moving close to Elsa doubtless played some part in his decision to take the job in Berlin, but it was hardly a major factor. He may well have wanted someone to love and an escape from what had become a miserable marriage, but it was his work that obsessed him. And Berlin offered ideal conditions for it.

CHAPTER 14

The War to End All Wars

1914 to 1919

35 to 40 years old

Mileva had no sooner arrived in Berlin than she left again, taking the boys with her. Einstein wept when his sons returned to Switzerland, but the split-up was inevitable. Mileva hated Germany—just as Albert had in his youth. Furthermore, she had an ongoing feud with his formidable mother, who had never reconciled herself to the marriage. And the marriage itself was now a mockery. They had been mismatched from the start, but time and circumstance had intensified their differences. If not certain of Albert's romantic interest in his divorced cousin, Mileva surely suspected it.

She and her sons had arrived in Berlin in April 1914. By July they were back in Switzerland, living in a Zurich boardinghouse, and Einstein was giving his inaugural address at the Prussian Academy.

When the war in Europe erupted in August, the Einsteins' separation—he in warring Germany, she and the boys in neutral Switzerland—made traveling difficult and a reconciliation that much less likely. Einstein was now involved with Elsa and set impossible terms for any renewed life together: a business relationship with as little personal contact as possible. In the meantime, he said, all he expected from her was news of his beloved sons every two weeks.

Germany's invasion of neutral Belgium aroused international outrage that the kinsmen of Goethe and Beethoven should act like barbarians, and brought Britain into the war against Germany. German propagandists responded to worldwide condemnation by persuading ninety-three intellectuals, including Max Planck, to sign a manifesto approving the action as a military necessity. It denied charges of atrocities in Belgium, characterized the enemy as "Russian hordes allied with Mongols and Negroes unleashed against the white race," and warned that without the German military "German culture would have been wiped off the face of the earth." A few days later a Berlin University biologist, Georg Friedrich Nicolai, produced a pro-peace countermanifesto calling for an end to the war and the creation of a united Europe. One hundred intellectuals were asked to sign it. Only four did, and Einstein was one of them.

The next month he joined the League of the New Fatherland, a political party dedicated to peace and the prevention of future wars. However, Berliners were more likely to support the anonymous German physicist who wrote that the happiest day of his life would be the day the English fleet was sunk and London leveled to the ground—a sentiment, Einstein remarked in a letter to Ehrenfest, that showed "what a wretched species of beast we belonged to."

That fall he was naturally delighted when his good friend Max von Laue received the 1914 Nobel Prize for physics "for his discovery of the diffraction of X-rays by crystals." Einstein had also been nominated that year—again by Bernard Naunyn—for "relativity, diffusion, and gravitation." The Nobel judges had several excuses for rejecting Einstein's relativity theory: it had not been experimentally verified, some thought he was overrated, and others, if not all, found the theory confusing.

By December, Mileva and the boys had settled into a Zurich apartment furnished by Einstein, who had stripped his own Berlin bachelor apartment of all but the bare essentials and shipped the rest to Mileva. He promised to send her financial support every three months—which he did. But it was not enough to match the rising cost of almost everything; she had to give math and piano lessons to help pay the rent.

He wrote loving, encouraging letters to his sons, pleased that his five-year-old Eduard had begun reading Shakespeare and that ten-year-old Hans Albert was interested in geometry. He missed them, and regretted that their separation prevented him from helping them to study. Yet he also welcomed the tranquil atmosphere in his bachelor apartment. Elsa and her two preteen daughters, Ilse and Margot, lived nearby, above her parents' apartment, so he could always find welcome company if he wanted it, and a home-cooked meal. In the summer of 1915, he went on vacation with Elsa and her daughters to a remote Baltic island. He followed that trip with a visit to his sons in Zurich, during which he took Hans Albert hiking and boating in southern Germany.

Before returning to Berlin Einstein called on the French writer Romain Rolland, a fellow pacifist staying at Vevey near Lake Geneva. Zangger went with him. The trio sat at the far end of the garden, the only background noise a swarm of bees. Rolland had antagonized his countrymen for not supporting the war. He left a detailed account of their meeting, noting in his diary for September 6, 1915, that Einstein

> speaks French rather haltingly, interspersing it with German. He is very much alive and fond of laughter. He cannot help giving an amusing twist to the most serious thoughts. Einstein is incredibly outspoken in his opinion about Germany, where he lives and which is his second fatherland (or his first). No other German [Rolland is wrong here; Einstein was a Swiss citizen] acts and speaks with a similar degree of freedom. Another man might have suffered from a sense of isolation during that terrible last year, but not he. He laughs. He has found it possible, during the war, to write his most important scientific work. I ask him

whether he voices his ideas to German friends, and whether he discusses them with them. He says no. He limits himself to putting questions to them, in the Socratic manner, in order to challenge their complacency. People don't like that very much, he adds.

Einstein regarded most Germans as lunatics, especially the frenzied war advocates. He was disgusted with hypocritical colleagues (among them Nernst and Haber) who, while working on poison gas production, complained that the enemy wasn't fighting fair.

Rolland summed him up as "one of the very few men whose spirit had remained free among the general servility."

Back in Berlin, after eight years of trying, with time out to conduct gyro-magnetic experiments with Wander Johannes de Haas, Einstein was closing in on the true nature of the universe and gravitation's strange role in it. He called a truce in his war of words with Mileva to concentrate on this task, telling her that their sons were the most important people in his personal life, and admitted he was wrong in accusing her of trying to poison their minds against him. That settled, he put almost everything on hold, Elsa included, declined to give or attend lectures, and left letters unanswered to focus with almost frenzied intensity on his ideas about the as yet undisclosed nature of the cosmos.

Over the next five weeks in the fall of 1915 he skipped meals and worked far into the night. When he did eat, he cooked everything together in the same pot to save time and trouble. Calling on him unexpectedly, his future stepdaughter Margot found him boiling an egg in a saucepan of soup, intending to eat both—but he cheerfully admitted he hadn't bothered to clean the eggshell. As might be expected, he suffered agonizing bouts of indigestion, but kept at his work.

With growing excitement he neared his goal until, at the end of November, his heart palpitating, and feeling as if something in him was about to snap, he got the answer he was after. It was like emerging from the dark into the light. He was euphoric for days afterwards, having achieved in the general theory of relativity what some consider to be "the supreme intellectual achievement of the human species." His physicist friend, Max Born, considered it "a great work of art," and "the greatest feat of human thinking about nature, the most amazing combination of philosophical penetration, physical intuition, and mathematic skill."

Gravity, Einstein concluded, is not a physical force of attraction acting through space, as is commonly thought, but a manifestation of the universe's geometry. Space is curved or warped by the presence of matter, and objects move through space along the shortest path following the contours of space. The concept is neatly described by John Wheeler: "Space tells matter how to move and matter tells space how to curve." Wheeler adds: "Gravity is not a foreign and physical force acting through space—it's a manifestation of the geometry of space right where the mass is."

Einstein's future colleague and biographer, Banesh Hoffmann, explains why this work of Einstein's is regarded with such awe:

What were the seeds that gave rise to this wonderfully unique structure? Such things as Newton's theory and the special theory of relativity, of course, and Minkowski's idea of a four-dimensional world, and Mach's powerful criticism of Newton's theory. And the mathematical framework already prepared . . . But after that, what? The principle of equivalence, the principle of general covariance, and—why, essentially nothing else. By what magical clairvoyance did Einstein choose just these two principles to be his guide long before he knew where they would lead him? That they should have led him to unique equations of so complex yet simple a sort is in itself astounding.

There are ten enormously complicated gravitational field equations governing space-time curvature, which,

if written out in full instead of in the compact tensor notation, would fill a huge book with intricate symbols (in one form, millions of them). And yet there is something about them that is intensely beautiful and almost miraculous. Their power and their utter naturalness in both form and content give them an indescribable beauty.

Einstein's theory explains the origin and destiny of the universe, predicts that light passing a massive object will undergo a reddening and that a clock near a massive object will run slower than one at a distance from it, accounts for the inconsistent orbit of the planet Mercury, and points to the existence of gravity waves (moving waves of energy that transmit a gravitational effect), as well as what have come to be known as black holes. He made one addition that he would later regret. Einstein held the conventional view that ours is a static universe; that despite minor changes in the movements of stars and planets, the overall picture would remain essentially the same. Yet his equations pointed to an expanding universe. So he added what many scientists call a "cosmological constant"—also known as "the fudge factor"—to bring it back to a static condition.

Einstein sent "Die Grundlage der allgemeinen Relativitätstheorie" (The foundation of the general theory of relativity), his paper explaining his theory of gravitation, or general relativity, to *Annalen der Physik* in March 1916. That same year he expanded and simplified the fifty-page thesis into a book, *On the Special and the General Relativity Theory: A Popular Exposition*, using only elementary mathematics—something of a feat in itself.

In it, Einstein writes that space is warped by objects and that the more massive the object, the greater the effect. The extent of the warping or curvature of space is greatest near the object and grows progressively less as the distance from it increases. Imagine, for example, that space is a taut rubber sheet. A rock placed on it would cause it to sag or warp in that location; the heavier the rock, the greater the sag. A ball rolled across the sheet would be deflected as it passed near the rock; the closer it got, the more it would be deflected. If it passed very close to the rock, it would orbit the rock repeatedly. Similarly, in our universe, radiation and material objects are deflected by the curvature of space near massive objects. That is why the earth orbits the sun and why the moon orbits the earth; it is following the path of least resistance created by the earth.

He had concluded in 1911 that starlight passing close to the sun should be slightly deflected toward it, following the curvature in space caused by its massive presence. Now he calculated that the deflection would be twice as great as he had originally thought. Normally these light rays would be invisible on earth, obscured by the brightness of the sun, but they could be seen during a total eclipse—one of which was expected in 1916. He hoped, despite the war, that astronomers would test his theory.

Meanwhile, Einstein had sent Mileva into a physical and mental breakdown by asking for a divorce. Apparently she was still in love with him and hoped for an eventual reconciliation. Doctors were puzzled by her collapse; it could have been psychosomatic, or caused by a heart attack or a sudden, deep depression. They ordered her to remain in bed and avoid excitement. Naturally she couldn't take care of her sons, who went to live with Helene Savic in Lausanne.

Einstein's first instinct on hearing the news from Besso and Zangger was to hurry to Zurich. On second thought, he realized that his presence would only agitate Mileva. Soon he had a theory to support his decision to keep his distance: As a cunning woman who would use any excuse to get her own way, Mileva was simply shamming to avoid a divorce. His mother agreed that Mileva was probably faking it, but thought Albert should take care of his sons.

Valuing Zangger's friendship and not wanting to be thought heartless, Einstein explained to him that leaving Mileva had been a matter of survival. Besso, however, was very sympathetic to Mileva's plight and gradually persuaded Einstein that she was genuinely and seriously ill. But Besso's efforts didn't change the situation. Einstein rebuffed his well-meaning attempts to effect a reconciliation, comparing the prospect of living with Mileva to having an odious smell stuck up his nose. Einstein was not a man to pull his verbal punches.

In 1916, Einstein's name came before the Nobel Prize selection committee for the fifth time. He was recommended by Felix Ehrenhaft for Brownian motion and for special and general relativity. But the judges decided that no one deserved the 1916 physics prize.

Late in September, he visited Leiden to see Lorentz and Ehrenfest. They sat in Lorentz's cheerful study, wreathed in cigar smoke and an atmosphere of mutual admiration, discussing the theory of bending light in a gravitational field. His two weeks with them were like a beautiful dream, and he returned to Berlin refreshed.

There he received startling news. On October 21, 1916, his idealistic friend Friedrich (Fritz) Adler, while shouting, "Down with tyranny! We want peace!", had assassinated the Austrian prime minister, Count Karl von Stürgkh, by putting three bullets in his head as he dined in a crowded Viennese restaurant. Now Adler was in prison on a murder charge, facing a death sentence. Einstein made inquiries about how he could help his old friend but found that only Adler's lawyers, family, and psychiatrists were allowed to visit him.

More bad news came from Mileva: Her brother Milosh, a doctor in the Austrian army, was a war prisoner in Russia.

There was no way Einstein could help Milosh, and little he could do for

Adler. While in prison awaiting trial, Adler believed he had discovered an elementary law that other scientists had overlooked—exposing relativity as fatally flawed. His father, Viktor, hoped this would prove to be the work of a madman and justify the defense that his son was mentally deranged (Fritz's mother, sister, paternal aunt, and maternal uncle had suffered from mental illness). Viktor sent copies of the manuscript to several physicists, including Einstein, all of whom admired Fritz Adler. They knew that to declare the work nonsense would be a terrible insult to a highly sensitive man, so they all delayed their responses. His scientific theory, according to Philipp Frank, was not the work of a madman; nevertheless, "his arguments were wrong."

Einstein wrote to Fritz on April 13, 1917, that he hoped to discuss his friend's scientific ideas after the trial, which psychiatrists had declared him fit to face, and offered to be a character witness. But Einstein was not called. Adler was found guilty and sentenced to death by hanging—a sentence later commuted to life in prison. (He was freed in 1918 and elected soon after to the Austrian National Assembly.)

Although Einstein considered Adler kindhearted and unselfish, he feared that his friend's masochistic urge for martyrdom almost amounted to a death wish. He kept in touch with Adler after the trial, and in an interview with *Vossische Zeitung*, published on May 23, 1917, praised his friend's character and said that Adler occupied himself in prison by studying relativity.

On May 7, 1915, a German U-boat sank the ocean liner *Lusitania*. Unarmed, though carrying ammunition, the ship had more than 1,000 passengers aboard, including 440 women and children. Among the fatalities were 115 Americans. Spurred by public outrage, the American government pressured the Germans into a promise to restrict its marauding submarines. Many Germans blamed the subsequent stalemate in the war on the restraint put on their U-boat commanders. Kaiser Wilhelm agreed, and on January 9, 1917, sent a message to the German navy: "I order that unrestricted submarine war be launched with the greatest vigor on the 1st of February . . . Take care however that this intention shall not prematurely come to the knowledge of the enemy and the neutral powers." It was not an easily kept secret. And Americans were again among the resulting casualties. After a struggle with his conscience, the pacifist American president Woodrow Wilson asked Congress to declare war on Germany, and on April 6, 1917, the United States joined the battle.

Einstein speculated that future generations would be astonished that such a monstrous war could ever have occurred. He called Berlin a lunatic asylum and wished that he could move to Mars to observe the inmates through a telescope. Berlin did not have a monopoly on lunatics: the British had recently captured six miles of mud in the Battle of the Somme at a cost of almost half a million men.

The strain of wartime conditions, his conflict with Mileva, the separation from his sons, and overwork—he had produced ten scientific papers and a popular book on relativity in a year—took their toll. In the fall of 1917 he collapsed in agonizing pain. In two months he lost fifty-six pounds. He believed that he had cancer and reconciled himself to dying. Then his illness was diagnosed as gallstones, and finally, correctly, as a stomach ulcer.

At Elsa's insistence, he moved into an apartment next to hers, where she could keep an eye on him, making sure he kept to a strict diet by cooking all his meals. Weak and depressed, he spent much of his time in bed. Even then he would work nonstop for long periods, once remaining in his room for three days. He had asked not to be disturbed, so Elsa left his meals on a tray outside his door. After one such burst of study, he produced the quadruple formula for gravitational radiation.

Earlier in the year, three persuasive nominations for Einstein to get the 1917 Nobel physics prize had been mailed to the selection committee: Arthur Haas proposed him for his new theory of gravitation, Emil Warburg (the founder of modern photochemistry) for his work on quantum theory, and French physicist Pierre Weiss, in a more poetic vein, for his attempt to conquer the unknown. In his letter, Weiss outlined Einstein's work in statistical mechanics, his two axioms of special relativity, his light-quantum postulate and the photoelectric effect, and, finally, Einstein's work on specific heats.

Although the judges acknowledged that Einstein was a "famous theoretical physicist," they noted that Charles Edward St. John at the Mount Wilson Observatory had failed to find the redshift predicted by general relativity in which the light from a star, or galaxy, is displaced towards the lower frequency end of the spectrum. They concluded: "Einstein's relativity theory, whatever its merits in other respects may be, does not deserve a Nobel prize." No other name was announced, however, and the prize was deferred until 1918.

During this period he kept up his correspondence with Dutch astronomer Willem de Sitter, who was also ill. The principal progenitors of modern cosmology commiserated over each other's health and criticized each other's model of the universe in sometimes weekly exchanges. De Sitter showed that by eliminating the cosmological constant, Einstein's field equations represented an expanding universe. Einstein thought this interpretation both senseless and irritating because it implied a moment of creation. He also took issue, on the same grounds, with de Sitter's model, which contained a singularity (where a point in space is distorted to such a degree that its geometrical properties collapse; now called a black hole). But before the year was out he was reluctantly admitting the possibility.

Wartime conditions, of course, made it impossible for Einstein to contact enemy scientists directly. But de Sitter, living in neutral Holland, acted as a go-between, sending Einstein's paper on general relativity to Arthur Eddington, secretary of the Royal Astronomical Society in England. Eddington, a math whiz, soon grasped its importance, and added some of his own ideas. Eddington's "Report on the Relativity Theory of Gravitation," published in 1918, was called a masterpiece.

As an invalid, Einstein was in no condition to care for his sons. Before his illness he had considered taking Hans Albert out of school, bringing him to Berlin, and teaching the boy himself. When Mileva objected, Albert promised that he would do nothing of the sort without her approval. He also thought of paying for Hans Albert to live with Maja in Lucerne. That idea came to nothing. He was even more concerned over seven-year-old Eduard, a shy, sickly little boy, whom Einstein suspected was mentally disturbed. Fearing that he would never be normal, Einstein even wondered if it would have been better had the boy never been born. He believed that the illness Mileva had been suffering from (tuberculosis of the lymph nodes) when Eduard was conceived was responsible for his mental state—a somewhat daring diagnosis from a layman.

Einstein also entertained the notion that Eduard's condition was inherited from Mileva's family: her severely disturbed sister, Zorka, had been hospitalized in Zurich's Burgholzi psychiatric clinic early in 1918. Eventually, after urging by his friend, Professor Zangger, who had a medical degree, Einstein paid for Eduard to spend some time in a children's sanitorium in the Swiss mountains, hoping that specialized care would restore him to health.

By April 1918 Einstein had regained several pounds and was allowed to resume his normal activities. But after he played his violin for an hour the agonizing pains returned. The following month he was back in bed, this time with jaundice.

Being constantly alone in the apartment with Albert, Elsa worried about her reputation and that the gossip would hurt her daughters. This situation made it seem inevitable that Albert and Elsa would marry, even though it would be a union of nurse and fragile patient rather than of two ardent lovers.

And so, that summer, Albert renewed his appeal to Mileva for a divorce. She agreed. One inducement was Albert's promise to give her all the Nobel Prize money—$32,000—a small fortune that would ensure financial security for her and the boys. It was a gamble because he hadn't yet won the prize, but both she and Albert were confident that it was on the way. He had, after all, been nominated for it six times in the past eight years, the first time in 1910, and with his contacts in the scientific community, he must have known that he was in the running. Meanwhile, he would continue to support Mileva and their sons as in the past.

The sweet and gentle Besso was horrified that they intended to divorce. He again tried to mediate, but Einstein told him bluntly that there was nothing to mediate. Besso and Zangger remained personally concerned with the welfare of Einstein's sons—acting in fact as surrogate fathers—and when Albert found communication with his children difficult for various reasons, they let him know how the boys were doing. Zangger also gave Mileva moral support and financial advice.

In 1918, Charles Glover Barkla of Edinburgh University won the deferred 1917 Nobel Prize for physics for discovering "the characteristic Röntgen radiation of the elements." Those who recommended Einstein for the 1918 award again included Warburg and Ehrenhaft. Wien and von Laue also proposed him, but suggested that he share the prize with Lorentz for their work on relativity. But the 1918 prize was deferred until 1919.

Officially the war ended on November 11, 1918, when Germany surrendered, but the fighting went on in the streets as Germans battled among themselves for political control of the country. Later that month, Einstein's friend Max Born was in bed with bronchitis when Einstein phoned to say that a group of revolutionary students and soldiers had seized the University of Berlin and imprisoned the rector and other professors. He feared their lives were in danger. A meeting was being held by extreme left-wing revolutionaries to decide their hostages' fate, and Einstein said he intended to intervene because he believed he had some influence over the students. Would Born join him?

Born immediately left his sickbed and walked to Einstein's home "in the Bavarian Quarter through streets full of wild-looking and shouting youths with red badges." Einstein and their psychologist friend Max Wertheimer were waiting, and the three of them took a tram to the Reichstag, which the students had taken over. It was surrounded by excited crowds and guarded by armed revolutionaries with red armbands. They were denied entry until Einstein was recognized, then he and his companions were allowed to go inside.

There the three visitors listened to the students vote in favor of rigid rules for running the university: "only socialist doctrines should be taught, only socialist professors and students allowed." Einstein obviously disapproved, and when asked his opinion by the chairman of the students' council replied that the most valuable thing in German universities was academic freedom. The radical students disagreed. According to Born,

> Einstein was well known to be politically left-wing, if not "red." But this was too much even for him . . . I can still see before me the astonished faces of these eager youths when the great Einstein, whom they believed wholeheartedly on their side, did not follow them blindly in their fanaticism. Yet Einstein was too kind and wise to disappoint them, and quickly turned the discussion to the . . . occupation of the buildings and the arrest of the rector and other professors. They declared that they had given over buildings and prisoners into the hands of the new socialist government and we ought to negotiate with them.

The trio found the harassed newly elected Social-Democrat president, Friedrich Ebert, in the palace of the Reichschancellor. He was too busy, he said, having to respond to the just-received "terrible" armistice terms from Versailles, to deal with trivialities like a rector locked in his office. However, he scribbled a note for them to hand to someone in authority, which they did. Soon after, the imprisoned academics were released.

"We left the palace of the Reichschancellor," Born recalled, "in the highest of spirits, with the feeling of having taken part in a historic event, and hoping indeed that the time of Prussian arrogance was finished, that it was all over with the Junkers, the hegemony of the aristocrats, the cliques of officials, and the military, that now the German democracy was victorious." In that mood Einstein apparently became a German citizen (as a Swiss he was allowed dual citizenship), a decision he later regretted.

Einstein was slowly recovering from his illnesses, and would soon be free—at least for a time—from the bitter fights with Mileva. They were divorced in

February 1919. In June, he and Elsa were married. He gave up his bachelor life, which had almost killed him, to join Elsa and her two daughters—his "harem," as he called them—in their spacious apartment.

If Elsa hoped to change Albert to become part of her bourgeois world, she was disappointed. According to Philipp Frank, Einstein "lived in the midst of beautiful furniture, carpets, and pictures; his meals were prepared and eaten at regular times. But when one entered this home, one found that Einstein still remained a 'foreigner' in such surroundings—a bohemian guest in a middle-class home." By his choice his bedroom resembled a monk's cell—no carpet on the floor, no pictures on the wall. He lived up to his belief that every possession is a burden, and owned nothing of material value. He used ordinary soap to shave, and submitted to a haircut only every few months, when Elsa was allowed to take just a little off the ends. Friends hardly noticed the difference. Some resented her way of treating him like a precocious child. At times, though, he needed mothering, as his cousin Alice Steinhardt noticed. She was on a train journey with the Einsteins on a cold day when Albert began shivering. Elsa soon spotted the trouble, poking a finger through his clothes and scolding him, "You've forgotten to put on your underwear."

His absentmindedness frightened Margot, his youngest stepdaughter, one day when he failed to emerge from the bathroom after more than an hour. Finally she aroused him from his reverie by calling his name. "Sitting in the tub all that time he had been working on some problem, and confessed to her, 'I thought I was sitting at my desk.'"

Although Einstein had been lured to Berlin in part by the promise of a lecture-free existence, he began giving them occasionally as his health improved. He also hired astronomer Erwin Freundlich, with whom he had been corresponding for several years, to assist him. At the start of the war Freundlich had attempted to test relativity from Russian territory during an eclipse of the sun, but he had hardly pointed his telescope when Russian soldiers arrested him and deported him back to Germany.

Before Freundlich could make another attempt with Einstein's encouragement, he was beaten to it by English astronomer Arthur Eddington. Alerted during the war by de Sitter to Einstein's fascinating light-bending claim for gravity, Eddington could hardly wait to check it out—by taking photographs of starlight during a solar eclipse. Because the next eclipse of the sun would not be visible in Europe or North America, two British expeditions set off for distant spots where it could be seen: one to Sobral in northern Brazil and the other, headed by Eddington, to Príncipe, an island off West Africa.

To Eddington's dismay, when he woke in Príncipe on the fateful morning of May 29, 1919, it was pouring with rain. He waited anxiously hour after hour until 1:30 P.M. before getting his first glimpse of the sun. Then, almost frantically, he began taking photographs. He was so busy changing photographic plates that he only looked up twice: one glance confirmed that the eclipse was under way and the other spotted trouble—star-obscuring clouds. But he kept going.

On June 3, Eddington wrote in his diary that because of the clouds he had

had to measure the starlight deflection—if any—in a different way from what he had intended. It was then that what he later called the greatest moment of his life occurred: his measurement on one photographic plate agreed with Einstein's theory.

Three months previously, Eddington and his assistant, E. T. Cottingham, had been discussing the prospective expedition with the Astronomer Royal, Frank Dyson. Cottingham asked what would happen if they didn't confirm Einstein's theory. Knowing that Eddington enthusiastically supported general relativity, Dyson replied, "Eddington will go mad and you will have to come home alone." Now, the exultant Eddington said to Cottingham, "You won't have to go home alone." At home, however, they realized that one positive photograph out of sixteen was hardly overwhelming evidence. And so the impatiently awaited results from Sobral, on the way to Britain for processing and measuring, took on added significance.

In early October, Einstein heard from Lorentz: "As the plates are still being measured he cannot give exact values, but according to Eddington the thing is certain . . . We should indeed rejoice."

However, the first photographs from Sobral were duds. The following seven, wrote Eddington, "gave a final verdict definitely confirming Einstein's value of the deflection, in agreement with the results obtained at Príncipe." All that remained was to convince the rest of the scientific fraternity that the experiments had confirmed relativity — no easy task. He began to reexamine the evidence.

The Nobel Prize judges were also waiting for the results. They had recently announced that the deferred 1918 Nobel Prize for physics was going to Max Planck for his services "to the advancement of Physics by the discovery of energy quanta."

Planck in turn proposed Einstein for the 1919 award for taking "the first step beyond Newton" through his theory of general relativity. Warburg, von Laue, and Edgar Meyer also nominated Einstein. Svante August Arrhenius, a Swedish Nobel Prize–winning chemist, proposed him for Brownian motion. None of them knew that Einstein's chance of success depended on the solar eclipse expedition led by Arthur Eddington that would confirm or refute one aspect of Einstein's special relativity. But the results came too late for the Nobel committee's decision, and they awarded the 1919 prize to Johannes Stark of Greifswald University, Germany, "for his discovery of the Doppler effect in canal rays and the splitting of spectral lines in electric fields."

CHAPTER 15

In the Spotlight

November to December 1919
———
40 years old

Expectant scientists and curious reporters packed a London hall to hear whether Eddington's solar eclipse findings confirmed or refuted Einstein's "weird" relativity theory. It was weird, according to Eddington, because general relativity required "non-Euclidean space." Eddington's first glowing statements had perhaps been wishful thinking; he acknowledged that conditions had not been ideal and that he had taken some starlight photos through an overcast sky. Now, after careful analysis of all the evidence, he was about to give his final verdict.

Whom did *The New York Times* send to cover the event and introduce relativity to the American public? Its golf expert, Henry Crouch! Had Einstein known, he would surely have let rip with a staccato burst of delight.

With no illusions about his handicap, Crouch decided to skip the meeting, intending to summarize the report from the next day's London *Times*. Then he was informed that the British paper planned to give the story modest play, and his paper wanted detailed coverage. By that time it was too late to attend the meeting.

Crouch was in luck, however. He got Eddington on the phone shortly after the meeting and persuaded him to give a replay of his speech. But to Crouch's dismay, Eddington might just as well have been speaking Swahili. He asked Eddington to start again and to make it as simple as possible. The amiable astronomer obliged.

His account, filtered through Crouch, appeared in the New York paper on November 10, 1919. The jaunty headlines, "Light Askew in the Heavens" and "Einstein Theory Triumphs," were printed three days after the London *Times* report appeared. The London paper's story was headed "Revolution in Science" and "New Theory of the Universe—Newtonian Ideas Overthrown," even though Einstein had stressed that his ideas were evolutionary, not revolutionary, because they augmented rather than replaced those of Newton, a scientist he idolized. Eddington succinctly confirmed the evolutionary nature of Einstein's theory: "In each revolution of scientific thought new words are set to the old music, and that which has gone before is not destroyed but refocussed."

Even the tabloid press took notice; one not inaccurate tabloid headline Einstein would have enjoyed was "Space Caught Bending." As for the serious press, *The New York Times* got his age wrong, calling him "a Swiss citizen about 50 years of age who has been living in Berlin about six years." He was, in fact, only forty, but his recent illnesses and sporadic health problems had aged him prematurely.

Throughout the world, few claimed to understand relativity. At first, philosopher and mathematician Alfred Lord Whitehead was impressed with the theory, but then wavered and covered his tracks, saying: "Einstein is supposed to have made an epochal discovery. There is no more reason to suppose that Einstein's relativity is anything final, than Newton's *Principia*." Einstein would have agreed. Joseph Thomson, president of the British Royal Society, master of Trinity College, Cambridge, and winner of the 1906 Nobel Prize for physics, confessed his confusion by stating, "Perhaps Einstein has made the greatest achievement in human thought, but no one has yet succeeded in stating in clear language what the theory of Einstein's really is." London-born physicist Oliver Heaviside, another Nobel Prize winner (the Heaviside layer, a stratum of ionized gas that conducts, reflects, and refracts radio waves, is named after him), was one of the few who seemed to understand relativity—and he denounced it as drivel.

The newspaper-reading public was not well served. The English *Manchester Guardian*, for example, had chosen an even less likely reporter than Crouch to introduce Einstein—Samuel Langford, a music critic. Normally, the editor would have assigned the job to David Mitrany; as a political economist he was the closest thing to a physicist on the staff. However, Mitrany being unavailable, Langford had to fill the breach. His strong suit was a rough grasp of German, but he surely didn't understand Einstein. When the pudgy little musicologist waddled into the office after hearing Einstein lecture, a colleague asked Langford what he thought about relativity.

"Platitudes, my boy!" he replied. "Just platitudes!"

While British critics were generally polite, the Americans treated Einstein and his theory as if they were equally ridiculous. The language of attack was often far from scientific, and much came from eccentrics and crackpots. But enough eminent scientists treated Einstein with derision to put his theory in doubt, despite Eddington's confirming photographs.

Columbia University astronomer Charles Lane Poor took a confident swipe at both Einstein and Eddington, saying, "The supposed astronomical proofs of the theory, as cited and claimed by Einstein, do not exist." Poor implied that Einstein had become unhinged, a victim of the crazy times.

It may well be that the physical aspects of the unrest, the war, the strikes, the Bolshevist uprisings, are in reality the visible objects of some underlying, deep mental disturbance, worldwide in character. This new spirit of unrest has invaded science ... I have read various articles on the fourth dimension, the relativity theory of Einstein and other psychological speculation on the constitution of the universe; and after reading them I feel as Senator Brandegee felt after a celebrated dinner in Washington. "I feel," he said, "as if I had been wandering with Alice in Wonderland and had tea with the Mad Hatter."

Poor explained the bending starlight by pointing out that all rays of light passing from one medium to another—from air to glass, for example—are bent or refracted. It was therefore no surprise to him that starlight, on entering the earth's atmosphere, would appear bent as in Eddington's photographs.

The New York Times jumped on the skeptics' bandwagon. It sided with American scientists in gently razzing the British for being—for a change—the naive and gullible guys who "seem to have been seized with something like intellectual panic when they heard of photographic verification of the Einstein theory, [from which] they are slowly recovering as they realize that the sun still rises—apparently—in the east." After quoting the Royal Society president's claim that relativity was "perhaps the greatest achievement in the history of human thought," the paper hedged its bets, wondering if Englishmen had forgotten about optical illusions, or whether starlight might have been deflected by gases in space. As for Einstein's view that the universe has a fourth dimension, time, the newspaper's leader writer floated the idea that Einstein might have pinched his concept from H. G. Wells's fantasy *The Time Machine*.

When physicist Sir Oliver Lodge, principal of Britain's Birmingham University, visited Albert Michelson, head of the physics department at the University of Chicago, a few months after hearing Eddington's announcement, he called relativity "repugnant to commonsense." Einstein regarded common sense as "a deposit of prejudice laid down in the mind before the age of eighteen"—and so it proved to be in Lodge's case. Like most contemporary scientists, Lodge took for granted the existence of the ether, supposed to be a vague, invisible substance through which light and sound were transmitted. Now here was Einstein saying that his theory abolished the need to believe in the ether, because his equations were valid in an etherless universe.

Paradoxically, although Michelson had won the Nobel Prize in 1907 (the first American to get it) largely for experiments indicating that the ether was a myth, he shared Lodge's feelings. He still hoped that the existence of the ether could be proved experimentally. Michelson acknowledged that Einstein's math was sound because his equations had been experimentally confirmed, but said he could not follow Einstein's complex reasoning.

While some questioned whether Einstein was German-Swiss or Swiss-German, astronomer and mathematician Professor Thomas Jefferson Jackson See, head of the department of astronomy at the University of Chicago, focused on Einstein's professional credentials. See charged that Einstein was neither an astronomer, a mathematician, nor a physicist. "He is a confusionist," See concluded. "The Einstein theory is a fallacy. The theory that the 'ether' does not exist, and that gravity is not a force but a property of space, can only be described as a crazy vagary, a disgrace to our age." See himself had impressive credentials, having investigated double stars, the ether (which he believed existed), the cause of gravitation and magnetism, cosmic evolution, and earthquakes. He had also established the wave theory of solid bodies.

An engineer, George Francis Gillette, scoffed that "as a rational physicist, Einstein is a fair violinist"; he sputtered his contempt for relativity as "the moronic brain child of mental colic . . . cross-eyed physics . . . utterly mad . . . the nadir of pure drivel . . . and voodoo nonsense." He predicted that by 1940 "relativity will be considered a joke . . . Einstein is already dead and buried alongside Anderson, Grimm, and the Mad Hatter."

In giving a "technical analysis of the mathematical and philosophical falla- cies of Einstein," Dr. Arthur Lynch lumped him with history's fakes and scorned his supporters as credulous dupes, concluding, "As I cast my eyes over the whole course of science, I behold instances of false science, even more pretentious and popular than that of Einstein, gradually fading into ineptitude under the search- light; and I have no doubt that there will be a new generation who will look with wonder and amazement, deeper than now accompany Einstein, at our galaxy of thinkers, men of science, popular critics, authoritive professors and witty dram- atists, who have been satisfied to waive their commonsense in view of Einstein's fallacies."

Lynch produced an impressive list of European scientists who, like the Manchester music critic, turned thumbs down on Einstein's theory: Henri Poin- caré, mathematician Gaston Darboux, Paul Painlevé, Le Roux, Curbastro Gre- gorio Ricci, Tullio Levi-Civita, and Émile Picard, a French mathematician, who said, with prophetic irony: "On the subject of relativity I see red."

American professor and physicist Dayton Clarence Miller, of the Case School of Applied Science, thought he had given relativity a knockout blow when he announced to the Western Society of Engineers that his own experiments completely refuted Einstein. And if Miller was too undistinguished for his opin- ion to carry weight, the anti-relativity faction could claim as their ally an au- thentic genius, Nikola Tesla.

An outstanding pioneer in the electric-power field, Tesla was the father of myriad inventions including robots, radar, neon and fluorescent lighting, remote control by radio, radiotelegraphy, steam turbines, high-frequency generators, speedometers, and airplanes. As his spellbound biographer related: "At a time when electricity was considered almost an occult force, and was looked upon with terror-stricken awe and respect, Tesla penetrated deeply into its mysteries and performed so many marvelous feats with it that, to the world, he became a master magician with an unlimited repertoire of scientific legerdemain so spec- tacular that it made the accomplishments of most of the inventors of his day seem like the work of toy-tinkers."

Tesla rejected Einstein's view on gravity to promote his own — that space was not curved — called atomic power an illusion, and mocked the idea that energy could be obtained from matter according to the formula $E = mc^2$. For a time,

Tesla's opinion of Einstein's theory provided a reputable banner behind which lesser mortals could rally. His credibility began to erode, however, when details of his eccentricities leaked out. He was afraid of harmless round objects such as billiard balls and pearl necklaces, and reluctant to shake hands for fear of catching a disease. He began to work on weird inventions: a camera to photograph thoughts and a device to produce death rays. But what finally brought his critical faculties into question was his confession to being romantically involved with a pigeon.

For the most part Einstein stood above the battle, declining to cross swords with either critics or crackpots. Nevertheless, he was disappointed that his early mentor, Ernst Mach, did not support him. He also hoped in vain for approval from Albert Michelson. Still, he was confident he was right.

Einstein let the bewildered off the hook by admitting that "there is no logical path to these . . . laws. They can only be reached by intuition, based on something like intellectual love of the objects of experience." Intuition perhaps explains why some scientists sensed that Einstein had indeed charted a new universe, even though they couldn't follow his directions. The consensus among such intellectuals as George Bernard Shaw and H. G. Wells was that Einstein had made an extraordinary breakthrough.

Despite or possibly because of the opposition, his sudden worldwide fame was unparalleled, especially for a physicist or mathematician. He was regarded by many as an almost supernatural being, his name symbolizing then—as it does now—the highest reaches of the human mind. British scientist J. B. S. Haldane called him the greatest Jew since Jesus. A schoolgirl, hearing him discussed with the breathless wonder reserved for Santa Claus, wrote to Einstein asking if he really existed. He replied to her, as he invariably did to children, and to almost everybody else.

From 1919 on he was without question the world's most famous and celebrated scientist, the most loved and the most hated. His photo in the December 14 issue of the *Berliner Illustrite Zeitung* was captioned "A new figure in world history whose investigations signify a complete revision of nature, and are on a par with insights of Copernicus, Kepler, and Newton."

Einstein could not prevent babies and a cigar brand from being named after him. He did decline a generous offer from the London Palladium's booking agent to include him on the bill with comedians, tightrope walkers, and fire-eaters.

He was already walking a tightrope. Along with both Jewish and non-Jewish colleagues, Einstein had begun to help Jews who had fled to Germany from eastern Europe after the war, even as a powerful lobby in the German government agitated to kick them out. Einstein arranged university courses for some and wrote to the *Berliner Tageblatt* calling those who wanted to deport the refugees inhumane.

Had he secluded himself in his Berlin attic study to contemplate the nature of the universe, he would have been less loathed and less lionized. But now he spoke out. As an ardent pacifist, he proposed an investigation of German war crimes and, as an antidote to wars, advocated world government—in a country of rabid nationalists spoiling for another fight. Above all, he was a Jew, and so became the target of anti-Semitic Germans exasperated by his worldwide fame.

They sneered at Einstein's work as "Jewish science." Physicist Philipp Lenard, a 1905 Nobel Prize winner, led the pack. He had done early research on the photoelectric effect before Einstein perfected it. In fact, as a student, Mileva had attended Lenard's lectures in Heidelberg. He typified the brilliant mind lobotomized by bigotry. An incipient Nazi, he spouted emotional nonsense such as "the Jew conspicuously lacks understanding for the truth, in contrast to the Aryan research scientist with his careful and serious will to truth . . . Science, like every other human product, is racial and conditioned by blood."

Unable to tolerate Einstein's success, Lenard falsely attributed relativity to F. Hasenohrl, "a pure German" killed in the war. Lenard and other anti-Einstein colleagues, such as Paul Weyland, Ernst Gehrcke, and Johannes Stark, sounded more like Joseph Goebbels precursors than scientists. Professor Ludwig Bierberback called Einstein "an alien mountebank," and Professor Wilhelm Muller of Aachen's Technical College said the theory of relativity was a bid for "Jewish world rule."

Einstein's less-articulate opponents waited outside his home or office in the Prussian Academy of Sciences to greet him with obscenities, or crammed his mailbox with threats. At one of his Berlin lectures, a right-wing student shouted, "I'm going to cut the throat of that dirty Jew!" He was apparently encouraged by Rudolph Leibus, an anti-Semitic Berliner who offered a reward to anyone who killed Einstein the pacifist. Leibus happened to be lecturing in the United States at the time, but was arrested upon his return to Berlin, only to be released after paying a small fine.

Einstein's confidence and sense of humor helped him laugh away the attacks. One admiring student, Ilse Rosenthal-Schneider, experienced his humor firsthand. After his lectures, she sometimes rode the tram with him from Berlin University to their respective homes. Once he showed her a book blasting relativity, which he had annotated with wisecracks. Where the author wrote, "It is completely unintelligible why Einstein said . . ." Einstein had noted in the margin, "What a wonderful admission." On another page, he had written, "Enter the donkey, beautiful and valiant."

Ilse recalled that Einstein would tease her

> whenever the opportunity offered itself. He knew I loved to read Kant . . . So he teased me, comparing Kant's intuition with the Emperor's clothes. Once when we debated for a long time some of Kant's intricate questions and had mentioned the various widely differing interpretations by the Kantians in their schools of philosophy (of which there were about as many as there were universities in German-speaking countries, sometimes several different ones at the same university) Einstein illustrated his views in the following way: "Kant is a sort of highway with lots of milestones. Then all the little dogs come and each deposits his contribution at the milestone."
>
> Pretending to feel indignation I said, "But what a comparison!"
>
> Einstein, laughing loudly, only remarked, "But your Kant is a highway, after all, and that is there to stay."

To help students visualize his idea of what seemed impossible—a finite universe without boundaries—he told them to think of how things would seem to an extremely flat bug of only two dimensions living on the surface of a spherical body and unable to escape from its surface. After suggesting that our entire universe is like the surface of a sphere that has no beginning and no end, yet is finite, he asked, "Don't you agree that it would be much more pleasant than living on an island of matter in vast, empty space? That is what we would have according to Newton's theory."

In response to a question from nine-year-old Eduard, who had inherited his bright eyes and love of music, Einstein again used the insect analogy. When Eduard asked him, "Why are you so famous?" He replied, "When a blind beetle crawls over the surface of a globe, he doesn't notice that the track he has covered is curved. I was lucky enough to have spotted it."

At the end of the momentous year, Albert was back in Berlin. From there he wrote to Mileva and his sons in Zurich, raising a laugh from Hans Albert with his account of the price of fame: "I feel now something like a whore. Everybody wants to know what I am doing all the time, and everybody wants to criticize."

He ended the year in tears and told Born why. Max Planck's eldest son had been killed at the battle of Verdun, and in 1917 one of Planck's twin daughters, Grete, had died after childbirth. The surviving daughter, Emma, then took care of Grete's baby, and in January 1919 married the widower. When Einstein last met Planck he had just received the devastating news that Emma, too, had died after giving birth. Planck's "misfortune moves me very deeply," Einstein wrote to Born about their mutual friend. "I could not hold back my tears when I visited him . . . He behaves remarkably bravely . . . but one can see that he is eaten up by grief."

CHAPTER 16

Danger Signals

1920

---

40 to 41 years old

Pauline Einstein wanted to spend what was left of her life with her son. Dying of stomach cancer, she arrived in Berlin with her daughter, Maja, who stayed to see her installed in Albert's study, which had been converted into a bedroom. Her doctor, Jonas Plesch, sedated her with morphine. It dulled the pain but left her incoherent. Emotionally exhausted, haunted by the memory of his dying father, Albert witnessed his mother's last days, telling friends that at times she suffered unspeakably, that Elsa was a great help, and that "all this has diminished further my already faltering desire to achieve great things."

There was little comfort in the outside world. Although Einstein supported the newborn liberal Weimar Republic, many Germans despised it as "the Traitor Republic," unjustly accusing it and the Jews of stabbing World War I soldiers in the back, sustaining the myth of the invincible front-line soldier and sowing the seeds for World War II. Ironically, German Jews had been fervent patriots; a higher percentage of Jews than non-Jews had served in the German military, and twelve thousand had been killed in action.

Society was disintegrating before his eyes. Army officers who once roamed Berlin's streets seeking recruits had been replaced by pimps on the prowl for transvestites, fetishists, and boy prostitutes. Women who dated occupying troops were roughed up and "traitors" were executed by "patriots," while resistance groups threw bombs and stored arms for the next revolution. Four thousand armed members of the Freikorps—swastikas emblazoned on their helmets—took Berlin by force, and the government headed by President Ebert fled to Stuttgart. Thousands faced starvation, and Einstein feared that emergency supplies sent by the British and Americans could not feed everyone. While some thought that the peace treaty was not tough enough, Einstein felt that it was too stringent and that the Allies' demand for huge reparations was impossible to fulfill.

His more immediate concern was the Freikorps, the professional assassins and terrorists who had chosen Wolfgang Kapp to replace Ebert. Kapp dithered: He had the Prussian State cabinet arrested and then released, banned exams at Berlin University, and confiscated all matzoh flour for the coming Passover. When the police refused to support him, Kapp resigned, saying that he had completed all his aims.

The Freikorps's fury spread to the university, where supporters booed and jeered Einstein at the start of one of his lectures. The demonstration was not necessarily an expression of anti-Semitism, he told a questioning reporter, but it could have been. A university spokesman tried to blame the affair on latecomers fighting for the few remaining seats—a tribute to Einstein's popularity. But Einstein didn't buy it. He had apparently discovered the true nature of the noise by the time of a second interview, in which he said unequivocally that the hostility had been directed at him.

At least his mother would never know of it; she died at the end of February. Einstein's grief reassured Kathe Freundlich, his assistant's wife. Kathe had been shocked by Einstein's earlier callous remark that no one's death would disturb him, but after his mother's death, when he "wept, like other men, I knew he could really care for someone."

Pauline's death left him sad and bewildered — as if, he said, the future was hidden behind a blank wall. Actually, it was staring him in the face—in his newspaper, which reported that Corporal Hitler's National Socialist Party was energetically moving forward with his twenty-five-point program.

Churchill would eventually call Hitler "a maniac of ferocious genius, the repository and expression of the most virulent hatreds that have ever corroded the human breast." These virulent hatreds had already infected a handful of Einstein's Berlin students. Fortunately, Einstein would soon be on his way to Leiden, where the adoring Ehrenfest and a classroom of lively and congenial Dutch students awaited him for what he expected to be an annual engagement as a visiting professor.

Ehrenfest was coaching Einstein by mail for his inaugural Leiden lecture, titled "The Ether and the Theory of Relativity," tentatively set for May, the start of what would be his residence for several weeks each year. It promised to be tricky—lecturing in a straitjacket if Ehrenfest could be believed—without visual aids or a blackboard to handle equations. Even gestures were out, because his hands would be covered by academic robes. And then, of course, there was the language barrier. Ehrenfest advised giving the old men in the audience time to wake up at the end of the lecture before thanking them for listening. The jest had a kernel of truth. Still, he assured his friend, printed copies of the talk would sell like hotcakes.

Einstein was less confident. He thought he was growing senile—a softening of the brain, he called it—because he'd made no recent progress in general relativity or in the unified field theory. He tried literature as a stimulant, and made a lucky choice: Dostoyevsky's The Brothers Karamazov. He thought it was the best book he'd ever read.

But no distraction could shield him from the horrors outside, where Nazis and Communists fought bloody battles, sometimes in the streets right below the windows of his apartment, and many starved to death. "The infant mortality is appalling," he told Ehrenfest. "No one knows in what direction the political ship is drifting. The State has sunk to its lowest ebb of impotence. In the background the main forces are at war: the sword, money and extremist socialist gangs."

The government announced strong measures to prevent an influx of Jews trying to escape even greater misery in Poland and Russia. Einstein responded in the liberal daily, *Berliner Tageblatt,* in an article suggesting that anti-Semitism was being aroused to account for social conditions, when the true cause of the problem was the devastating economic depression. Yet he turned down an invitation to a conference to combat anti-Semitism in the academic world, saying, "I would gladly come if I believed in the possible success of such an undertaking."

On April 24, Einstein told Solovine, who was still working as an editor in Paris, that he was separated from Mileva and that the boys were living with her in Zurich. Always discreet if not secretive about his private life, even among close friends, he made no mention of divorcing Mileva nor of his year-old marriage to Elsa. He did keep Solovine informed of mutual friends: After roaming the world, Besso was back at the patent office (thanks to Einstein), while Paul Winteler and his wife (Einstein's sister, Maja) were happy and still living in Lucerne.

Though unable to settle down to his own research, he encouraged Solovine and Born to write books on relativity. He sent Solovine a brief, equation-free account of it, recommended other sources, and offered to check his and Born's manuscripts.

To Ehrenfest's dismay, when Einstein arrived in Leiden that May to give his inaugural lecture, university bureaucrats had not yet endorsed the appointment. Einstein made the best of it, saving his friend from further embarrassment by junking his prepared lecture and giving a few informal talks before heading home.

Back in Berlin, he missed the Ehrenfest children and being "pampered and overestimated" by their parents. "My thoughts are often with you," he told Ehrenfest. "It's so remarkably good for both of us to be together more often, because it's just as though Nature had made us for each other."

When Ehrenfest expressed pessimism about his own work, Einstein tried to console him by saying that everyone gets lazier and more stupid as they get older. Ehrenfest was not to be consoled, replying, "Don't be impatient with me. Bear in mind that I hop around among all you big beasts like a harmless and helpless frog who is afraid of being squashed."

At last, the inaugural Leiden lecture was set for October. The authorities had made up for their foot-dragging by also inviting Paul Langevin of the Collège de France and Pierre Weiss of Strasbourg to honor the occasion. They would discuss magnetism, on which subject both were experts. Einstein had himself recently experimented with de Haas on the relation between magnetism and rotation. They had confirmed the gyromagnetic effect—the torque induced in a suspended metal cylinder when it was suddenly magnetized.

Though Ehrenfest was "burning with impatience" for the discussion, he was also "very depressed—partly because of the eternal (minor!!!) worries about money, partly because I am not working at all. What I *can* do is not science but only a bit of entertaining conversation . . . about physics—the physics done by others. How extraordinarily happy I could be, if only I were not so slack and so unproductively ambitious."

In June, Einstein escaped from Germany for a spell to lecture in Norway and Denmark; his stepdaughter Ilse went with him. The German ambassadors to those countries sent word back to Berlin that he had been well received.

On his return home, he met Niels Bohr for the first time. His initial impression of Bohr was of "an extremely sensitive lad who goes about the world as if hypnotized," while Bohr, for his part, admired Einstein's detached attitude and "rather profound humor behind his piercing remarks." Together with Planck, they talked physics from morning to night, during which Bohr challenged Einstein: "If you are so concerned with the situation in physics in which the nature of light allows for a dual interpretation, then ask the German government to ban the use of photoelectric cells if you think that light is waves, or the use of diffraction grating if light is corpuscular."

Einstein's next out-of-town engagement was at Hamburg University, with the mayor, the university's rector, and faculty members in attendance. Standing in line to get in on that hot, humid evening, a student caught snatches of conversation from people nearby: "One hears that Einstein has very serious opponents" . . . "Professor Hahn has found the atom can be split" . . . "Nationalist students booed Einstein in Berlin." As Einstein reached the lectern and began his hour-long explanation of relativity, a thunderstorm broke out as if the heavens, too, had joined the opposition.

One of the most active members of the opposition out to destroy Einstein's reputation was Paul Weyland, a professional agitator and small-time crook. He headed, or fronted for, the Study Group of German Natural Philosophers, which was bankrolled, apparently, by anti-Semitic industrialists such as Henry Ford (the German Foreign Office believed that Ford was a financial supporter of Weyland). Weyland could pay top dollar to writers and speakers willing to smear or discredit Einstein. The organization attracted people who despised Einstein as a pacifist and internationalist; experimental physicists who resented the publicity given to a theoretician; and confused philosophers who had misinterpreted aspects of his theory and were in effect attacking their own ideas. Many of them were also virulent anti-Semites.

Weyland's biggest catch was Nobelist Philipp Lenard. Einstein had based his new conception of light on Lenard's experimental observations of the photoelectric effect. Lenard had originally admired Einstein, speaking of him only in superlatives. In 1909, he had characterized him as a "deep and far-reaching thinker," and confided that he treasured a letter that Einstein wrote to him in 1905. But he had grown into a rabid nationalist, embittered by the outcome of the war and seething with hatred for the "arrogant" British and the Jews. He couldn't do much about the British, but Jews were close at hand. His bitterness increased when the Weimar government gave German Jews equal rights, at least on paper. His escalating hatred focused on the most famous and respected Jew in Germany—Albert Einstein.

Lenard ridiculed relativity as "absurd," and even talked Nobel judges out of giving Einstein the prize for almost a decade, telling them that relativity "had never been proved and was valueless." His attacks were all the more formidable because of his stature as a scientist. He was also an incipient Nazi, and so determined that Germans should dominate everything that when he found equipment in his laboratory marked in amperes, units of measure of electric current named in honor of the well-known French scientist, he changed the term to "Weber," a German scientist in the same field.

When Lenard's group, known both as the Anti-Relativity League, or the Anti-Einstein League, booked Berlin's Philharmonic Hall for August 24, Einstein turned up there with his friend and colleague Walther Nernst (soon to be awarded the Nobel Prize for chemistry). To get inside they had to walk past swastika signs and anti-Semitic pamphlets on sale in the entrance. He was looking forward to debates that he laughingly called cockfights.

According to Philipp Frank, Einstein "always liked to regard events in the world around him as if he were a spectator in a theater." In that spirit, Einstein took his seat, only to hear himself denounced as a publicity hound, plagiarist, charlatan, and scientific dadaist (dadaism was a wacky artistic cult that glorified the meaningless). He laughed at the most outrageous statements and at a distorted account of relativity by physicist Ernst Gehrcke, and sometimes applauded mockingly as if at a satirical comedy.

The liberal *Berliner Tageblatt* saw through the motives of the speakers and declared that they hadn't laid a glove on Einstein's theory. The same paper printed a letter in Einstein's defense from his colleagues Nernst, Otto Rubens, and von Laue, and Minister of Culture Konrad Haenisch. A *New York Times* reporter thought the rally "had a decidedly anti-Semitic complexion, which applied equally to the lecture and to a large part of the audience."

On the home front, Einstein was also engaged in "cockfights." Mileva had agreed to his plan to take their sons on vacation in the fall, but balked when he invited Hans Albert to Berlin. Before she would consent, he had to promise to keep Elsa out of sight while Hans Albert was in Berlin, and to eat alone with his now sixteen-year-old son. Einstein resented the sensitivities of both women that made such absurdities necessary.

It may have been in this exasperated mood that he mailed an angry article to the *Berliner Tageblatt*, which the paper printed on the front page. In it, he first scorned his critics as unworthy of an answer, then answered them; he said that they wouldn't have attacked him had he been a German nationalist, with or without a swastika, instead of a Jew with international views. He cited prominent theoretical physicists throughout the world who supported relativity, and pointed out that Lenard, the only scientist of international repute to oppose relativity, had contributed nothing to theoretical physics. He wrapped the article up by dismissing Lenard as superficial and Weyland as both insolent and vulgar.

On first reading an account of the attacks on Einstein, Ehrenfest wrote to him from Holland: "I am awfully distressed that you have been dragged in the mud. Please just spit on *all* these attacks." He understood that Einstein had declined to leave Berlin when he had had the chance the previous year, especially as he did not want to disappoint Planck, who had been supportive and encouraging. He also knew that Einstein found that working with people like Planck was intellectually stimulating. Ehrenfest ended his letter by assuring Einstein that "if the situation there embitters you so you cannot work, you can rely on us to exert all our efforts to get you here."

Ehrenfest was aghast a few days later when he read Einstein's response to his critics and immediately wrote to him:

> My wife and I absolutely cannot believe that you yourself wrote down at least some of the phrases in the article . . . This answer contains particular reactions that are completely non-Einsteinian . . . If you really did write them down with your own hand, it proves that these damned pigs have finally succeeded in touching your soul which means so terribly much to us. You must understand me correctly: I might have sinned 100 times worse, but it is a question of you and not me. And you are of such a nature that this "My Answer" doesn't correspond to you, but rather sounds like an *echo* of the filthy attacks on you. I urge you as strongly as I can not to throw one more word on this subject to that voracious beast, the "public." And now please don't be angry with me. Whatever else may happen never forget how faithfully all of us here are attached to you— from Pavlik to Lorentz.

Einstein responded that he had to defend himself against insults that had been made "repeatedly and publicly. I had to do this if I wanted to remain in Berlin, where every child recognizes me from photographs. If one is a democrat, one also has to acknowledge the claims of publicity." He swore that he would never again allow anyone to provoke him into such a response.

Max Born's verdict was that Einstein had lost his temper but won the argument.

When the press printed rumors that Einstein was being run out of Germany, Arnold Sommerfeld begged him to stay and proposed countermeasures. He was willing to read a statement at the upcoming Bad Nauheim conference of the Society of German Natural Scientists and Physicians in favor of Einstein and against scientific demagogy, and then take a vote, which he was sure would show that the majority supported Einstein. The offer sounded to Einstein like a popularity contest. He appreciated Sommerfeld's concern, but preferred to keep the discussions on a scientific level, admitting that his angry outburst to the press had been a mistake.

Sommerfeld came up with another idea: To appease Lenard, Einstein should announce that his argument was with Weyland and his gang, not scholars like himself. But Lenard was in no mood for appeasement. Instead, he made a preemptive strike; he threatened to demolish his "arrogant and insulting" opponent at the imminent Bad Nauheim conference.

The conference began in September 1920, amid great tension. Fearing violence, the organizers arranged for armed police to guard the hall's entrance. Anticipating stormy debates, Einstein was determined to control his temper.

Max Planck, who chaired the meeting, had a plan of action. Well aware of the volatile, dangerous mood of anti-Semitic demonstrations, he kept things under control by hurrying the discussions that followed the reading of each scientific paper. As soon as the talk became heated, he called for the next paper.

This strategy worked until Lenard took the floor, with noisy approval from many in the audience. He objected to relativity, he said, because it eliminated the ether and could not distinguish between a train braked to a halt and the surrounding world not moving.

Einstein replied that it was simply a matter of perspective. After that his words were hard to catch because he was jeered and shouted at during the exchange. Planck's cries for silence were ignored. The following exchange, however, was clear enough to be reported:

Lenard: Relativity violates commonsense.
Einstein: What is seen as commonsense changes over time.
Lenard: At best relativity has limited validity.
Einstein: On the contrary, an essential aspect of relativity is its universality.

Lenard said that some imaginary experiments had failed to confirm relativity, and asked, "Why were they considered invalid?"

Because, Einstein replied, only those imaginary (or thought) experiments that could, in principle, be carried out—though in practice they might not be feasible—were considered valid.

Lenard struck the last blows in the "cockfight" by insisting that the ether could not be discounted, that relativity applied only to gravitation, and that there were still difficult problems with the theory.

Max Born then spoke in favor of Einstein, remarking that relativity valued observation more than equations.

Gustav Mie followed with what later became the Nazi party line, crediting others with relativity.

Von Laue, Nernst, and Rubens came to Einstein's defense with a joint statement to the press: "We must stress that apart from Einstein's relativistic researches his work has assured him a permanent place in the history of science . . . His influence on the scientific life not only of Berlin but of the whole of Germany can hardly be overestimated. Whoever is fortunate enough to be close to Einstein knows that he will never be surpassed in his respect for the cultural values of others, in personal modesty and dislike of all publicity."

Privately, in a letter to Einstein, Planck condemned Weyland for promoting "scarcely believable filth." The German Foreign Office representative in London was more diplomatic, reporting that the attacks on Einstein had aroused indignation in Britain, and that if the rumor that he was being hounded from the country were true, it would be catastrophic for German science and foreign relations. "We should not drive away such a man," he wrote, "whom we can use in effective cultural propaganda."

That fall, Einstein was once again not among the Nobel Prize winners, although the persistent Warburg had nominated him for a fourth time, joined by a German anatomist, Wilhelm von Waldeyer-Hartz, and Leonard Solomon Ornstein, a Dutch expert on statistical physics who cited Einstein for general relativity. A joint letter signed by Lorentz, Willem Julius, Pieter Zeeman, and Kamerlingh Onnes mentioned Einstein's theory of gravitation and the positive results of the 1919 eclipse expeditions and rated him "in the first ranks of physicists of all time." Yet the 1920 prize went to French physicist Charles Édouard Guillaume, "in recognition of the service he has rendered to precision measurements in Physics by his discovery of anomalies in nickel steel alloys." The judges' excuse for not selecting Einstein was that the bending-lights result of the 1919 eclipse expeditions had not confirmed his relativity theory beyond a reasonable doubt.

Meanwhile, Einstein was undercutting the efforts of friends who claimed he was averse to personal publicity by giving a series of unusually revealing interviews as a favor to a journalist acquaintance, Alexander Moszkowski.

Einstein's relativity theories had been parodied in the press, satirized by cartoonists and comedians, and derided by those who didn't get it—which meant almost everyone else. The "twins paradox," for example, which is often used to illustrate one aspect of relativity, was frequently brought up to ridicule him as a nutty professor under the spell of Lewis Carroll. The paradox involves two young men who are identical twins. One stays on earth while the other, moving at close to the speed of light, takes a journey through space. When the traveler returns to earth he is still young, but his brother has become an old man.

Einstein had a chance to set the record straight when Moszkowski asked him about it. Surprisingly, Einstein was still an interviewer's dream, not yet wary of reporters seeking provocative quotes and snappy headlines. No one before or since persuaded him to answer such freewheeling, wide-ranging, off-the-wall questions.

Moszkowski wanted to know if the twins paradox was true. Einstein said that the effect illustrated by the paradox had been wildly exaggerated. The reason for this is that the speeds attainable by humans are so much less than light speed that the resulting age difference would be insignificant. If, for example, the young space traveler had covered 19 billion miles at 600 miles a second (which is about one hundred times faster than the greatest speed yet attained in space flight but still only $\frac{1}{320}$ of light speed), when he returned to earth he would be just one second younger than his brother.

What is there beyond our world? Moszkowski then asked.

Einstein said he believed other worlds independent of ours might exist. By independent he meant that even if we spent an eternity searching for them and theorizing about them, we would still never get to know them. He explained why by asking Moszkowski to imagine humans as two-dimensional beings living on a plane of indefinite extent, with organs, mental attitudes, and instruments adapted to their two-dimensional world. They would also have a science of two dimensions, able to give them a complete picture of their cosmos. Independent of this two-dimensional world there might be another one with different phenomena and relationships. It would be impossible for these two worlds to know of each other's existence. We are in the same position as the two-dimensional creatures, said Einstein, except we have one more dimension.

He conceded the possibility of astronomers discovering worlds beyond ours, but only those with three dimensions. And so, Einstein concluded, we are limited in our knowledge to a finite universe and leave other worlds of different dimensions to science fiction.

Moszkowski mentioned that spiritualists quoted Einstein to support their belief in the fourth dimension, a dimension Einstein attributed to time and they to the spirit or "occult" world.

I can't be expected to respond to ignoramuses, he replied, or to those who misinterpret my views. He admitted that some scientists had clouded the issue by using the term "occult" to describe a different phenomenon: Huygens and Leibniz, among others, had discarded Newton's theory of gravity because it allowed for action at a distance, which they referred to as "occult."

Moszkowski had been intrigued by astronomer Camille Flammarion's science fiction story, Lumen, in which the eponymous hero moves faster than light and achieves time reversal. As a result, he sees the end of the Battle of Waterloo before it starts and watches cannonballs fly backward into cannon barrels and dead soldiers come back to life and rejoin the fight.

"Simply impossible," Einstein replied. "Of course, we can imagine events which contradict our daily experiences without taking them seriously. Relativity shows that nothing can exceed the speed of light. Assuming that Lumen is human, with a body and sense organs, at the speed of light his body's mass would become infinitely great."

Moszkowski then brought up hypnosis. Why did scientists call it nonsense without investigating the phenomenon?

Einstein said that although he did not doubt the existence of hypnosis, he scorned the showmen and self-proclaimed psychics who used it to deceive the gullible. "Serious scientists have to avoid nonsense of this sort," he said, "since the public misinterprets even a casual interest in the subject."

Einstein went on to tell Moszkowski that the day before, just for fun, he had been to a theater in which a woman gave a thought-reading demonstration. He had been asked to think of two numbers. He chose 61 and 59, which he was asked to whisper to her manager. The woman then repeated them correctly. Although both men were too far from the stage for her to overhear, Einstein concluded that a code or signals from the manager had given her the numbers.

Moszkowski brought up Immanuel Kant, the philosopher, to bolster his argument. Kant believed in the occult powers of the Swedish scientist-inventor Emanuel Swedenborg, who was said to be able to witness events in distant places by mind power alone—once reporting a fire in Stockholm, three hundred miles away. Kant investigated and confirmed the story. Swedenborg's purported clairvoyance reached as far as Mars, which he described as being populated by intelligent beings. Perhaps his most remarkable claim was to have had conversations with angels, even though they spoke only in vowels.

Einstein let that one go, making no attempt to defend the otherwise much-admired Kant, nor to question Swedenborg's sanity.

Moszkowski then speculated that had Einstein lived during the Inquisition, his relativity theory would have been damned as a manifestation of the devil "and they would have honored you with a funeral pyre."

Without totally discounting the possibility, Einstein thought it unlikely, because "mathematical-physical and astronomical works were never attacked by the Papal courts. On the contrary they have been encouraged. A whole list of priests, particularly Jesuits, made great discoveries in natural science."

They moved on from science and the supernatural to women. Einstein had liberal views for his time, and approved of women getting the same opportunities as men to pursue scientific careers. But he said he doubted that they would reach the same heights, because they were handicapped by their physical makeup.

How about Marie Curie? Moszkowski wondered.

Einstein dismissed her as a brilliant exception, smiling as he estimated the intelligence of women at large: "It is conceivable that nature may have created a sex without brains."

Moszkowski realized that this "grotesque" remark was not to be taken literally, but "was intended as an amusing exaggeration"—though, of course, Einstein's critics took it seriously.

In those days scientists regarded personal publicity and self-promotion with distaste if not contempt. When the Borns heard that the interviews with Moszkowski were to be turned into a book, they feared it would fuel more anti-Semitism. After reading the manuscript, Hedi Born urged Einstein to stop its publication, for otherwise "the gutter press will get hold of it and paint a very unpleasant picture of you. Your jokes will be smilingly thrown back at you. A completely new and far worse wave of persecution will be unleashed not only in Germany but *everywhere* until the whole thing will make you sick with disgust." Longing for "the eloquence of an angel" to influence his decision, she stuck to hyperbole, warning Einstein that if published, the book "would be the end of your peace, everywhere and for all time."

Hedi blamed Moszkowski for trivializing Einstein and obliquely responded to his denigrating and chauvinistic remark about women by promising to keep her pleading letter a secret, for "I have heard enough how much you dislike it when women meddle in your affairs. Women are there to cook and nothing else; but it sometimes happens that they *boil over.*"

Max Born had been traveling when Hedi wrote the letter, and she told him about it as soon as he returned home. He then followed it up with his own warning, that anti-Semites such as Weyland and Lenard would triumph if the interviews were published, by accusing Einstein of self-promotion. "I implore you to do as I say. If not, Farewell to Einstein . . . In these matters you are a little child. We all love you, and you must obey judicious people (not your wife)."

Born understood that Elsa favored publication, wanting to help the Einsteins' friend Moszkowski, who was broke.

Einstein had already decided to "obey" Hedi; as he wrote to Max Born, "Your wife . . . is objectively right, though not in the harsh verdict of Moszkowski. *I have informed him by registered letter that his splendid work must not appear in print.* I would like to thank your wife most sincerely."

Despite the letter, the book was published in 1921 under the title *Einstein the Searcher: His Work Explained from Dialogues with Einstein.* Now Einstein and the Borns waited for the public reaction.

CHAPTER 17

Einstein Discovers America

1921

---

41 to 42 years old

Einstein's New Year's resolution was to save his twelve-year friendship with Max Born, which was threatened by an angry correspondence between their wives. Incensed at Elsa for not discouraging Albert from risking his reputation at the hands of interviewer Alex Moszkoswski, Hedi Born let fly. Her angry letter to Elsa prompted Albert to criticize Hedi for exaggerating the situation. The feud between their wives caused the two men to stop writing, until Einstein broke the silence with a wish to bury the hatchet. With some relief, he reported that the cause of the quarrel, Moszkowski's book, had appeared without any earth tremors so far, although he hadn't yet read it. Born responded in kind, first assuring Einstein of his great affection and acknowledging the absence of any earthshaking calamity, but deploring the advertisements for the book plastered on every billboard.

Despite this publicity, Einstein felt no need to keep a low profile. He was, as Count Harry Kessler, a writer and man-about-town, wrote in his diary for February 4, one of the many Berlin notables to attend the fabulously successful first night of Richard Strauss's ballet, *The Legend of Joseph*. Walter Rathenau, the German foreign minister, was also in the audience.

Ten days later Einstein and Kessler joined a group of fellow pacifists aboard an Amsterdam-bound train to seek support for their cause at the International Trades Union Conference. His behavior amused the worldly Kessler. Everything delighted Einstein, especially the sleeping car. En route Kessler asked if relativity applied to atoms. He recorded their conversation in his diary:

> Einstein said no: size (the minuteness of the atom) comes into it here. So size, measurement, greatness and smallness, must be an *absolute*, indeed almost the sole absolute that remains, I said. Einstein confirmed that size is the ultimate factor, the absolute that cannot be got away from. He was surprised that I should have hit on this idea, for it is the deepest mystery of physics, the inexplicability

and absoluteness of size. Every atom of iron has precisely the same magnitude as any other atom of iron, no matter where in the universe it may be. Nature knows only atoms, whether of iron or of hydrogen, of equal size, though human intelligence can *imagine* atoms of varying magnitude.

Kessler quipped, "Then man is more intelligent than God. And God is stupid, lacking even human imagination and intelligence." Einstein took him seriously, replying that the deeper one penetrated into nature's secrets, the greater became one's respect for God.

No sooner was Einstein back from Amsterdam than he was off to lecture in Prague.

Philipp Frank, his successor at the University of Prague, was anxious to protect him from the curious crowds that were likely to gather if he stayed at a hotel. The recently married Frank hadn't yet found an apartment, so he and his wife were living in the physics lab. Would Einstein care to join them there? He could sleep on a sofa in an adjoining room that had once been his office—the one with the large windows overlooking the garden of a mental hospital—and only the three of them would know he was there. Einstein gladly accepted. The next morning he compared it to being in church, saying that it was a remarkable experience to wake up in such a peaceful room.

He sympathized with the new democratic government of Czechoslovakia under Tomáš Masaryk. Wanting to feel the pulse of the city, he persuaded Frank to join him on a tour of several cafés frequented by different political and social groups. On the way home they bought liver for lunch. When Mrs. Frank began cooking it in water, Einstein objected that the boiling point of water was too low and that she should cook it in fat. She took his advice and afterward, whenever Einstein's theories were mentioned, she recalled his theory of liver frying.

The day that he was to leave Prague, his "hideaway" was discovered, and many people called on him in the physics lab. Frank kept most at bay but gave in to the entreaties of a young man who said he'd waited for years to speak with Einstein. He had a bulky manuscript and claimed to have a plan to convert the energy in the atom, revealed by the equation $E = mc^2$, to make incredibly powerful explosives. About one hundred people had already sent Einstein similar ideas. He told the young man, "You haven't lost anything if I don't discuss your work with you in detail. Its foolishness is evident at first glance. You cannot learn any more from a longer discussion." Frank agreed that the machine the inventor had envisioned to convert mass into devastating energy couldn't possibly work.

Einstein's next stop was Vienna. There, nervous at the thought of facing an audience of three thousand in a huge concert hall, Einstein persuaded his host, Felix Ehrenhaft, to sit near him.

Albert must have done the packing for this trip, because he had arrived at the Ehrenhaft home with just two pairs of pants, both badly crumpled. Mrs. Ehrenhaft carefully ironed a pair for the lecture and returned them to his room. To her dismay, he was wearing the crumpled pair when he appeared onstage with her husband. It didn't cramp his style, if he even noticed; he quickly overcame his stage fright, using striking illustrations to make a point and good-natured jokes to entertain those who missed the point.

Back home with his host and hostess, he resisted Mrs. Ehrenhaft's attempts to domesticate him. Having bought him a pair of slippers and left them in his bedroom, she was surprised to see him padding around barefoot. Hadn't he seen the slippers? Yes, he replied, adding, "they're entirely unnecessary ballast."

On his return to Berlin, Einstein heard from Chaim Weizmann. Through dogged persistence, the Zionist leader had persuaded the British government to help make Palestine a homeland for Jews. Now he wanted Einstein to join him on a lightning fund-raising tour of several American cities to finance the building of Hebrew University in Jerusalem. He hoped to raise several million dollars, although Louis Brandeis, a U.S. Supreme Court justice and Weizmann's rival for leadership of the World Zionist Organization, warned him that he would be lucky to get half a million. That was why Weizmann wanted Einstein along: to draw the crowds.

Fund-raising held as much appeal for Einstein as joining a military band. Weizmann enlisted the aid of their mutual friend, Kurt Blumenfeld, a Berlin-based Zionist activist, who got Einstein to change his mind by "making me conscious of my Jewish soul." Once committed, Einstein became enthusiastic, feeling an intense need to serve the cause. He told Maurice Solovine, who was still in Paris, that he would be acting as a high priest and decoy to help mistreated Jews everywhere find a refuge in Palestine.

He and Elsa, with Weizmann and his wife Vera, arrived in New York harbor aboard the *Rotterdam* on April 1, 1921. Because it was the Sabbath, they agreed, out of respect for their religious supporters, not to disembark until after sunset.

Anxious reporters with deadlines that couldn't wait stormed the ship and asked their questions. No one inquired about fund-raising; most wanted the lowdown on Einstein's strange theory. His grasp of English roughly matched the reporters' knowledge of relativity, but some seemed to assume that shouting at him would break the language barrier.

Elsa offered to interpret. Her high-school English was adequate for ordering toast and coffee, but she was unable to unscramble her husband's esoteric scientific speculations.

Reporters then asked Weizmann to take over. He probably spoke better English and German than anyone on board, but his field was biochemistry, and relativity baffled him too. So the reporters turned back to Einstein, the man who made the headlines.

Through Elsa, he again tried to demystify fifteen years of work in a few words: "Falling bodies are subjects independent of physical causes, and light in diffusion is bent." Seeing blank expressions, he offered another version: "Before, it was believed that if all material things disappeared out of the universe, time and space would remain. According to my new theory of relativity, time and space disappear with material things." As he flicked his hand like a magician doing a disappearing trick, he smiled as if to assure them that there was no immediate danger.

One reporter noted that the hand figuratively destroying the universe still clung to a briar pipe he seemed anxious to save whatever the fate of the universe. Another mentioned that an American scientist, Charles St. John, was conducting experiments that promised to refute relativity. Einstein nodded approval, saying his theory was not infallible and that he wanted to be absolutely fair to the theories of others.

What about Columbia University's astronomy professor Charles Lane Poor? asked a third newsman. He says Newtonian laws explain all physical phenomena and that your theory can't be proved.

"In a certain sense no theory can be proved absolutely," Einstein conceded. "Each theory tries to explain certain facts and is acceptable if those facts fit into the general conception of the theory. But no one theory can totally explain all the facts. And in that sense, I agree, a theory cannot be proved."

Those hoping to goad Einstein into counterattacks were disappointed. He even disclaimed originality, attributing his theory to the pioneering work of Galileo, Newton, Maxwell, and Lorentz. Without their ideas, he said, he could not have made his own deductions.

When reporters ran out of questions or neared their deadlines photographers took over, yelling orders like army sergeants drilling new recruits. Eventually the two men broke away and hid from the press in an empty cabin until sunset, when they could go ashore.

A wildly excited crowd waited for them, their cars parked bumper to bumper, many sporting the blue and white flags of Zion as well as the Stars and Stripes, their drivers honking horns in rough unison. Mounted police enhanced the mood of a massive wedding party on the verge of a cheerful riot. The car for the Einsteins and Weizmanns was last in the line of beflagged and honking autos. Instead of being taken directly to their hotel, they were carried, as if by a flood, in a procession stretching out of sight and headed for the Lower East Side.

They crawled in first gear through the crowded Jewish quarter, to a continuous roar of delight. Flushed with excitement, Elsa clutched a bouquet someone thrust at her and gasped, "It's like the Barnum circus!" Albert agreed, comparing himself unfavorably with an elephant or a giraffe as an attraction. It was almost midnight before they got to the Hotel Commodore, where they were waylaid by another adoring crowd.

After a City Hall reception three days later, the Einsteins lunched with their friend, Samuel Untermyer, a leading Manhattan attorney. That evening he took them in his car on a sightseeing tour up Riverside Drive and then across town.

Watching the streetlights and illuminated billboards come to life, Einstein was euphoric: it was such a dazzling contrast to drab Berlin. He was lyrical about the skyscrapers, even though they hid much of the night sky, comparing them with majestic mountain ranges. As they passed Italian, German, and Chinese restaurants he exclaimed, "It's like a zoological garden of the nationalities." He raved about everything, especially the people in the streets, who looked vibrant and healthy, so unlike Berliners—himself included—who had not yet recovered from the near starvation of the war years.

They returned not to the Commodore but to the even more opulent Waldorf-Astoria. There, reporters followed them into their suite. Only one understood German, so Elsa willingly interpreted for the others, enjoying the limelight her husband shunned. She eagerly told the reporters how much they had enjoyed their first tour of Manhattan. Asked his opinion of American youth, Einstein said he expected great things from them, searched for a metaphor, and concluded: "A pipe as yet unsmoked. Young and fresh." That led to joking remarks about his own pipe, from which he was rarely parted, and the inevitable vain attempts to get a succinct definition of relativity — or at least an admission that the theory was beyond the average reader.

Isn't your theory so complicated that only twelve men understand it?

"I think any intelligent undergraduate can. My students in Berlin and, I believe, most scientists do. Its great value lies in the logical simplicity with which it explains apparently conflicting facts in the operation of natural law . . . Two of the great facts explained by the theory are the relativity of motion and the equivalence of mass of inertia and mass of weight."

How will it affect our readers? The man in the street?

"It won't," Einstein replied. "But from the philosophical aspect it alters the conception of time and space. Before, it was believed time and space were separate from matter. My theory says time and space are inseparable."

Then he asked them a question. Why was he of such intense "psychopathological" interest to their readers, who really didn't have any idea what he was talking about?

One reporter took a shot at the answer. He suggested that there are two mysteries that most intrigue people: one is the nature of God, the other the nature of the universe. In penetrating one mystery, providing a new view of the universe, Einstein had somehow come closer to revealing the nature of God. And even though they didn't understand the details of his discovery, it had aroused the public's admiration and curiosity.

Einstein smiled as if agreeing, but obviously didn't buy it. Why would their interest in the nature of the universe make them wonder what time he woke and what he had for breakfast? He didn't realize that to many people his enigmatic achievements made him a celebrity on a level with movie stars like Charlie Chaplin, Rudolph Valentino, and Mary Pickford. Almost everyone wanted to know what *they* had for breakfast, and with whom.

His own theory about his celebrity status was that "the ladies of New York want to have a new style every year. This year the fashion is relativity." It didn't hurt that he had a sense of humor and responded to banal questions with wisecracks. He had an endearing personality and all the right humanistic instincts in the eyes of the mostly liberal reporters. One of them concluded, "So far from being the usual conception of the average man of science he has made an unusual impression of geniality, kindliness and interest in the little things of life."

Not everyone greeted them with joyous abandon. Fiorello La Guardia, president of New York City's Board of Aldermen (and later the city's mayor), wanted to honor Einstein and Weizmann with the freedom of the city, but another alderman suggested that this "enemy alien" Einstein might be a fraud: "How do we know he really discovered relativity?" A third alderman, Bruce Falconer, who had never even heard of Einstein or Weizmann before, balked at the thought of America becoming "a forum for the airing of foreign political questions," and added: "America for Americans. America first." His remarks incited a Jewish alderman, Gustave Hartmann, to respond, "It's because you are against the Jews!" "You're a liar!" Falconer yelled, while La Guardia pounded his gavel. Hartmann explained that Einstein was "the greatest scientist since Copernicus." Falconer wouldn't listen: "Write me a memorandum about it," he snapped. When New York State officials voted unanimously to give the two men the freedom of the state, Falconer finally caved in, saying, "If anyone in my hearing had said it was Professor Einstein, the scientist, who was to be honored I would not have objected. After all, Weizmann and Einstein are common names in the New York telephone book."

Meeting Adolph Ochs, owner of *The New York Times*, soon after the Falconer episode, Einstein repeated his complaint that the public's interest in him was "psycho-pathological," which only a psychologist could explain. Ochs assured him that his reporters would never ask prying questions; consequently, Einstein never refused to speak with *New York Times* reporters—especially William Laurence, who had smuggled himself out of Russia in a sauerkraut barrel, arriving in the United States in 1905, and was for thirty years science editor and reporter for the *Times*. "Einstein really did not want to talk to reporters," Laurence recalled, "but he realized somehow that I knew something of what he was talking about . . . From then on, I wrote considerably about Dr. Einstein"—and his theory.

Some eight thousand people squeezed into the Sixty-ninth Regiment Armory in Manhattan on the evening of April 12 to hear Weizmann and Einstein talk, while three thousand waited outside just for a glimpse of them.

Judge Gustave Hartmann, fresh from his argument with Alderman Falconer, introduced Weizmann as the leader of the World Zionist Organization who had obtained Palestine as a homeland for the Jewish people under a British mandate, and Einstein as "the master intellect and greatest scientist of the age."

Weizmann spoke to "my brothers and comrades" on behalf of those who had been waiting through the ages for the restoration of their native land. "The pioneers cannot wait," he said. "They are already on their way." He said the hopes of the Jewish people had concentrated on them, the Jews of America: "Here you are sitting five or six thousand miles from Palestine, a country which many of you will never see, and you are waiting to hear me speak about the country. And you know very well that you will probably have to pay for it. It is extraordinary. I defy anyone, Jew or gentile, to show me a proposition like it."

Neither Weizmann nor Einstein was a spellbinding orator. In small gatherings Weizmann had a quick, devastating wit, but before large crowds his quiet, matter-of-fact delivery and lack of dramatic gestures prompted a bewildered Frenchman to ask, "How can a Russian Jew be so British?" Yet this evening he had a rapt audience.

Kurt Blumenfeld had warned Weizmann to "please be careful with Einstein. [He] often says things out of naivete which are unwelcome to us." With this in mind, Einstein had been told that he would usually only be expected to sit on the platform near Weizmann. His presence alone would satisfy the audience. But now he was called on to speak.

Einstein stood and said, "Your leader, Dr. Weizmann, has spoken, and he has spoken very well for all of us. Follow him and you will do well. That is all I have to say." His was the shortest speech of the evening and the only one reported verbatim in next morning's *New York Times*.

At Columbia University three days later, he was greeted by Michael Pupin, surely the world's only physics professor to have started his career as a shepherd in Serbia. He introduced Einstein as "the discoverer of a theory which is an evolution and not a revolution of the science of dynamics." Einstein agreed, and then raised a laugh right away by pretending to erase chalked equations on the blackboard with an invisible eraser — until Pupin brought him the real thing.

The following week he gave four lectures at the City College of New York, translated by Morris Cohen. I. I. Rabi attended one of them and recalls:

> There I was about to enter graduate work and there was the great Einstein. I think he was talking about physics of the day, particularly quantum theory. As a lecturer he was a model of absolute clarity, with a sense of humor. By contrast, Oppenheimer was somewhat mysterious; so was Bohr, of course. And Feynman was a showman, and very amusing. Einstein gave the impression he was very naive on political matters, and very obstinate about scientific ones. I don't think he was at all naive. He was a very sophisticated man, but seemed naive because he cut to the heart of the problem. He would appear naive if one didn't approach those problems in a fundamental way. He certainly was a very great person. Of the great scientists he was certainly the best in this century. His interest was always in profound questions, and his influence will last. He didn't suffer fools gladly, but he didn't reject them.

While Einstein was lauded in Manhattan, his critics were applauded in Philadelphia, at a meeting of the American Philosophical Society. Inventor Charles Francis Brush got a big hand after announcing that his experiments showed that Einstein was wrong. A. G. Webster of Clark University agreed that if Brush's experiments were correct "they absolutely disprove Einstein's theory." "If they do," interjected another scientist, "no one will be better pleased than Professor Einstein." He said that Einstein had told him he wanted someone to either prove or disprove his theory.

Einstein meantime had a ringside seat at the ongoing battle between Weizmann supporters and Brandeis partisans who wanted their man to head the World Zionist Organization. Most American Jews favored Brandeis; Europeans, Weizmann. Vera Weizmann recalled how at one meeting, when Brandeis complained "that Palestine was full of malaria and therefore it would be wrong to encourage immigration into Palestine until malaria had been eliminated, our much admired friend and magnificent orator, Schmarya Levin [pro-Weizmann], burst out ironically, 'So you mean, Justice Brandeis, that the first immigrant should be allowed into Palestine only when the last mosquito has left it.' "

Einstein turned up at the meetings at the Commodore Hotel, even one held at midnight, but took no part in the bitter arguments about the best means to help develop Palestine and who should lead the enterprise. He listened for a while instead, then retreated to another room to play a borrowed violin or discuss the Talmud, philosophy, and mysticism with Dr. Schmarya Levin, a brilliant scholar.

Zionists of both factions wanted Einstein to attract the crowds but were afraid of his independent views and caustic tongue. Weizmann claimed he could control Einstein, and did. At public meetings Einstein applauded in the right places and took his turn at the microphone simply to endorse Weizmann's words.

After a quick trip to Washington with Weizmann to meet President Warren G. Harding and to address members of the National Academy of Sciences, Einstein resumed what was to be a nonstop tour. The fund-raisers crossed America at an exhausting pace, from New York to Chicago, Chicago to Boston, Boston to New York, New York to New Jersey, New Jersey to New York, New York to Cleveland, Cleveland to Washington, D.C., Washington to New Jersey, then back to New York.

While the fund-raisers hardly underestimated the problems facing Jews in Palestine, they were not on the spot. The British colonial secretary, Winston Churchill, his assistant, T. E. Lawrence (Lawrence of Arabia), and Herbert Samuel, the first British High Commissioner in Palestine, were there and were seeing firsthand the grim shape of things to come.

The British party had arrived in Palestine on March 24 to see how best to implement the Balfour Declaration, having chosen Gaza, with fifteen thousand Arabs and less than a hundred Jews, as a good starting point. The initial shouts from the crowd of "Cheers for Great Britain!" implied that they were right. The illusion held for only a few moments; then friendly greetings were muffled by cries of "Death to the Jews!" and "Cut their throats!" Not understanding Arabic, Churchill and Samuel continued to smile. Lawrence knew Arabic and sensed the danger, but chose not to alarm the others, fearing that it might inflame the situation.

Five days later Churchill planted a tree on Mount Scopus, the site of the proposed Hebrew University, saying, "Personally, my heart is full of sympathy for Zionism." Privately he told a friend of "the splendid open-air men and beautiful women who have made the desert bloom like a rose." The settlement Churchill had visited was paradise compared to many of the forty-two others threatened by disease and marauding Arabs. One was peopled by thirty-five Polish Jews, all ex-soldiers, who worked as laborers for seventy-five cents a day. They slept in tents or the ruins of crusaders' castles, despite the threat of nightly attack by Bedouins.

Discussing the dreams of these settlers, Weizmann was eloquent before small groups or man-to-man. He spoke then "with a magnificent mixture of passion and scientific detachment" of the pioneers who had found Palestine a deserted land, neglected for generations during which the once-forested hills had been left bare of trees and the fertile earth washed into the sea. He appealed for money to help them restore the land, rebuild the country, and make a healthy environment, admitting that "when you drain the marshes, you get no returns, but you accumulate wealth for generations to come. If you reduce the percentage of malaria from forty to ten, that is national wealth."

He would continue:

> We are reproached by the whole world. We are told we are dealers in old clothes, junk. We are perhaps the sons of dealers in old clothes, but we are the grandsons of Prophets. Think of the grandsons and not of the sons . . . If you want your position to be secure elsewhere, you must have a portion of Jewry which is at home, in its own country. If you want the safety of equality in other universities, you must have a university of your own. The university in Jerusalem will affect your status here: professors from Jerusalem will be able to come to Harvard, and professors from Harvard to Jerusalem.

As Einstein traveled with the Weizmanns to various American cities, he divided his time between lectures, when he spoke freely and at length—reassuring the religious that relativity did not preclude the existence of God—and brief money-raising speeches for the Zionist cause. Making a pitch for the proposed Hebrew University before groups interested in education, he called it the greatest thing in Palestine since the destruction of the Temple of Jerusalem, and told how Jewish students and teachers knocked in vain at the doors of the universities of Eastern and Central Europe.

In early May 1921, as Einstein spoke in Chicago of the need for Jews to have a home of their own, Arabs rioted in Jaffa. Thirty Jews and ten Arabs died in the fighting. As a result, Herbert Samuel temporarily suspended all immigration, which Zionists saw as rewarding the rioters. Soon after, Churchill gave the House of Commons his impressions of Palestine: "I defy anybody, after seeing work of this kind [the vineyards and orange groves of the Yishuv] achieved by so much labor, effort and skill, to say that the British government, having taken up the position it has, could cast it all aside and leave it to be brutally overturned by the incursion of a fanatical attack by the Arab population from outside."

While in Chicago, Einstein gave three lectures at the University of Chicago and had a brief conversation with Robert Millikan, a professor of physics. In 1914 Millikan had experimentally validated the existence of the photoelectric effect, the discovery of which would earn Einstein the 1921 Nobel Prize, which was deferred to 1922.

Einstein then went to talk with Albert Michelson, head of the physics department. Michelson had a second experiment in mind to try to establish the ether as a fact. It would both test Einstein's relativity theory and bring the university much desired publicity. He was less than enthusiastic, however, anticipating a second negative result. Einstein encouraged him to go ahead anyway. He, too, expected the same outcome, but hoped this would reconcile Michelson to relativity. The American still clung to old, classical ideas, which accepted the ether as a given.

In a book published six years later, Michelson went as far as he could to support Einstein: "The theory of relativity has not only furnished an explanation of known phenomena, but has made it possible to predict and to discover new phenomena, which is one of the most convincing proofs of the value of the theory. It must therefore be accorded a generous acceptance notwithstanding the many consequences which may appear paradoxical."

While Einstein and Michelson continued to discuss the nature of the universe, their wives and a few friends were having lunch together, during which Elsa displayed her eccentric English and gave a hint of how misleading she may have been in interpreting Albert's ideas to the press.

She told the lunch party that on a recent trip to Princeton she had been mercilessly bitten by snakes. She responded to their looks of incredulity by elaborating:

"Oh yes, the snakes flew around my head and bit my face and hands. They even flew under my skirt and bit my legs and ankles." Her audience grew more puzzled and astonished as she embroidered her tale, but she told it with great conviction, and as she was the great man's wife, no one questioned it. In some ways it was no less strange than the theory of relativity, and that had to be accepted whether they liked it or not. But they looked at Edna [Michelson, who spoke German] for confirmation. "Do you really mean that in Princeton, New Jersey, flying snakes bit you?" Edna asked in German. "Ach, nein!" Mrs. Einstein explained. "Ich spreche von Schaken." (I'm talking about mosquitoes)! It was an easy mistake and everybody had a good laugh.

A week later, when Einstein was back in Princeton to give four lectures, university president John Grier Hibben awarded an honorary degree to this "new Columbus of science who sails across uncharted seas of thought." In between lectures, Einstein listened to an account of recent experiments by Dayton Miller said to refute both him and Newton.

Miller, a science professor at the Case Institute of Technology in Cleveland, Ohio, and an expert on sound—he had the largest collection of flutes in America—was creating a lot of noise as the man who saved the ether and sank relativity. His convoluted calculations were simply too tricky for Einstein to swallow and provoked his now famous line: "Subtle is the Lord, but malicious He is not." This saying, in German ("Raffiniert ist der Herr Gott, aber boshaft ist Er nicht"), is engraved above the fireplace of the faculty lounge of Princeton's math department. Miller, of course, turned out to be wrong.

When Einstein arrived in Boston on the morning of May 17, he might have expected picketing protestors aroused by a local cardinal, who had warned Americans that he was a dangerous atheist. But his recent response to the question, "Do you believe in God?"—"I believe in Spinoza's God, who reveals himself in the harmony of all being"—seemed to have reassured the cardinal and his flock. The crowd greeting Einstein at the South Station was more than friendly.

After driving through Boston's Jewish quarters, the North and West Ends, the Einstein party had breakfast in the Copley Plaza Hotel with Governor Cox, Mayor Peters, and seventy-five distinguished guests. When they'd finished eating, the men hesitated to light their cigars until Vera Weizmann signaled her approval by smoking a cigarette. Considered daring for a woman in those circumstances, her action made the headlines. As a female physician, she was also something of a rarity in the 1920s.

Einstein spent the rest of the morning at Harvard, where he helped several physics students solve problems. After attending a luncheon given by state officials, he agreed to join Mrs. Weizmann at an afternoon meeting of a local Hadassah women's group. Elsa was ill and resting.

Einstein followed Mrs. Weizmann's speech to the women at Mishkan Tefila Temple with a few words of his own. Later, they emerged from the meeting into dazzling sunlight and the inevitable eager crowd.

"Let's run away somewhere we don't have to see anyone," Einstein suggested.

Vera enjoyed his company and liked his direct, simple, and at times "flirtatious" ways. He had caught her in the right mood for adventure, exhausted with touring America as if part of a gypsy caravan, and always surrounded by crowds. They took a cab until they reached open country, where Einstein suggested they get out and walk.

As they strolled along country roads, he kept dropping something that made a metallic noise, then retrieving it quickly before she could see what it was. When she couldn't resist asking, he replied, "Oh, that's my secret." Eventually he explained that it was exactly half his luggage. Elsa, not being well enough to pack, had left him to his own devices, so he'd shoved his toothbrush and a toothpowder container into his suitcase for the Boston trip. He'd been dropping and picking up the container like a kid playing with a toy.

Einstein's childlike yet self-assured personality amused Vera Weizmann, though it sometimes hurt or offended others. His biographer and friend Philipp Frank was with Einstein when an Orthodox Jew asked for directions to a strictly kosher restaurant. Einstein named a place nearby. When the man insisted, "Are you sure it's strictly kosher?" Einstein replied, "In fact, only an ox eats strictly kosher." Obviously offended, the man hurried off. Einstein denied any ulterior motive when Frank disapproved, saying that he was being objective and had simply pointed out that an ox's diet is "the only strictly kosher food, because nothing has been done to it."

Frank concluded that Einstein was not a cynic but saw everyday life in a comic light, a common attitude in the part of Germany where he grew up. People there had a dry, sometimes crude, sense of humor.

Einstein's casual conversation was a mixture of lighthearted jokes and ridicule, leaving some wondering whether to laugh or protest. Stuffed shirts were wary of him, but the unpretentious were usually delighted. Frank observed something else that Einstein's assistant, Leopold Infeld, also noticed: Einstein sympathized with the problems of every stranger he met, but if the stranger attempted closer contact, he withdrew into his shell. This was not true of his response to several close friends, particularly from his student and early working days. And despite his expressed chauvinistic view of women, he dropped his guard and his criticisms with those he liked.

Vera Weizmann was one of those. He had flirted with her on the ship over, and Elsa had remarked to Vera that she wasn't disturbed because Albert was not drawn to intellectual women, being attracted only to those who did manual work. Time would tell that Elsa didn't know Albert very well.

When he and Vera Weizmann returned to the Hotel Touraine, one of the reporters waiting to interview Einstein handed him a pamphlet. It was a questionnaire written by Thomas Edison, the famed creator of over a thousand inventions, including the lightbulb.

Edison took a dim view of higher education in the United States, and his ideas of what should be taught were heatedly debated and reported. He scoffed at proponents of a liberal arts education, saying it failed to teach the facts needed in a practical world. He weeded out liberal arts graduates from his own company by requiring employees and job applicants to answer a list of "practical" questions, such as: Who invented logarithms? Where is Kenosha? What are leucocytes? What U.S. city makes the most laundry machines? How far is it from New York to Buffalo? What is Grape Nuts made of? What is a Chinese windlass? What is the distance between the earth and the sun?

Before Edison hired anyone he required the applicant to answer almost 150 such questions. The results had disappointed him. Of several hundred job applicants just over thirty passed his test, and few of those were college graduates. "Men who have gone to college I find amazingly ignorant," Edison complained. "They don't seem to know anything." When his employees failed the test, Edison gave them a week's pay and fired them.

"I wouldn't give a penny for the ordinary college graduate," he told interviewer Edward Marshall. "Except those from institutes of technology. They aren't filled up with Latin, philosophy and the rest of that ninny stuff."

B. Lord Buckley, head of the Buckley School in Manhattan, agreed. He said that a well-read person could easily answer eighty percent of Edison's questions, though apparently he didn't volunteer to take the test.

The test drove one young man from Holyoke to seek police protection. He claimed that several men had tried to steal a book containing his answers to Edison's questions, which he valued at a million dollars. The police concluded that he had been driven temporarily insane through studying the Edison questionnaire.

Einstein kept his head when confronted with the list, and acknowledged he knew of Edison as the inventor of the phonograph and various electrical appliances.

The group of reporters waited while an Edison question, "What is the speed of sound?" was translated into German.

Einstein's answer was translated into English. "I don't know offhand," he said. "I don't carry information in my mind that's readily available in books."

Told of Edison's view that a knowledge of facts was vitally important, Einstein disagreed: "A person doesn't need to go to college to learn facts. He can get them from books. The value of a liberal arts college education is that it trains the mind to think. And that's something you can't learn from textbooks. If a person had ability, a college education helps develop it."

The Edison affair was a diverting interlude, but interviewers soon refocused on the enigma readers wanted solved. A young *Boston Globe* reporter speculated that although Einstein might be too ignorant, by Edison's standards, to be a sound engineer, surely he could illuminate his own ideas. That led the reporter to ask: How about another attempt to unravel relativity in a simple sentence?

Though near collapse from the rigors of the tour, Einstein agreed to try when Mrs. Munsterberg, the wife of a Harvard psychologist, offered to translate. It isn't easy, he said. Newspapers throughout the world have tried, "but none has ever clearly explained the theory so that it could be fully understood by the general public." He paused for a moment, then said: "Time and space and gravitation have no separate existence from matter."

Manhattan was his next stop. He told eight hundred Jewish physicians at the Waldorf-Astoria that Jews were denied a higher education in many countries, but "the existence of the Jewish University in Jerusalem will give an opportunity to the Jewish mind to express himself . . . The Medical College will undoubtedly be the most important department of the University, as we Jews have always excelled in this particular branch of science." The doctors responded with $250,000 for the cause.

While in Manhattan he had a surprise visit from Max Talmey, his boyhood mentor, who had emigrated to America. They hadn't been in contact for nineteen years, but as Talmey entered the hotel room Einstein recognized him immediately, exclaiming, "But, doctor, you distinguish yourself, indeed, through eternal youth!" A few days later the Einsteins called at the Talmeys' home, where Albert encouraged their twelve-year-old pianist daughter, Frieda, to concentrate on the classics, and gave their ten-year-old daughter, Elsa, a ride on his back. Then he was off to Cleveland.

His appearance at Union Station on May 25 caused a near riot among a waiting crowd of three thousand. Weizmann had arrived incognito an hour earlier to avoid the crush and was waiting in a car to pick him up. Their wives were to follow. Einstein was saved "from possibly serious injury only by strenuous efforts by a squad of Jewish war veterans who fought the people off in their mad efforts to see him." Almost all the Jewish businessmen in Cleveland closed their stores or offices to march in a parade that accompanied the two men from the railroad station to City Hall — "a triumphant march, with men attempting to cling to the running board and the rear of the auto and being hurled unceremoniously therefrom by traffic patrolmen."

Einstein regretted having defined relativity in one sentence while in Boston; it left too much out. Given another chance in Cleveland, he added an important rider, saying that it could be explained to anyone simply and succinctly, *but* that "it is necessary that a person should give several weeks of studying to the underlying principles on which it is based."

At a dinner in the B'nai B'rith Club and a mass meeting in the Masonic Hall, some $200,000 was raised for Zionist funds. The total from their American tour was close to a million, short of their goal, but more than Brandeis had predicted and enough to start building the Medical Faculty of Hebrew University.

The Einsteins sailed for England at the end of May, their luggage heavy with treasure. Every chance they had to escape from reporters and formal functions, they had roamed the aisles of five-and-dime stores and emerged with kitchen gadgets, jars with tops that turned easily, colored ashtrays that closed over ashes at the flick of a wrist, and four cases of ginger ale (then unknown in Germany). Einstein also picked up a little English on the trip by listening to others speak. "I am the acoustic type," he told Paul Schilpp. "I learn by ear."

Their liner, the *Celtic*, docked at Liverpool on June 8. Einstein braced himself to explain his ideas at Manchester University, where he also spoke of Zionist aspirations to Jewish students. A dinner party awaited them in London, arranged by their host, Viscount Haldane, the former secretary of state for war (whose daughter reportedly fainted on meeting Einstein). A dazzling and diverse array of superstars turned up, representing science, the theater, the military, politics, and the church: Arthur Eddington, Alfred Whitehead, George Bernard Shaw, General Sir Ian Hamilton, Harold Laski, the dean of St. Paul's, and the archbishop of Canterbury. Einstein assured the archbishop that relativity would have no effect on people's morale, being "purely abstract—science." Elsa laughed when the archbishop's wife called Einstein's theory "mystical," implying that he and it were anything but.

On Sunday Albert and Elsa lunched at the Rothschilds' where, again, he faced the inevitable questions. His answers even confused the scientist Lord Rayleigh, who concluded, "If your theories are sound, I understand . . . that events, say, of the Norman Conquest have not yet occurred."

Afterward, Einstein placed a wreath on the grave of his idol, Isaac Newton, in Westminster Abbey, and in the afternoon gave a lecture at London's King's College. He spoke flawlessly for an hour without notes to an audience that included his friend Frederick Lindemann and astronomer James Jeans, and got a standing ovation.

The next day, Lindemann, who held the chair of experimental philosophy at Oxford, drove Einstein there and showed him around the Clarendon Laboratory. Then he returned him to the Haldanes' London home. Einstein told Lady Haldane that his first visit to England had been one of the most memorable weeks of his life: He had been stimulated by his scientific conversations with her brilliant husband and especially treasured his friendship with the couple and their children.

Before he returned to Germany Einstein wrote to friends, marveling at the ebullience of Americans and their efficient, gadget-filled lives, telling what a warm welcome they and the British gave him, and happily anticipating renewed cooperation between former enemy scientists. But at a homecoming party in Berlin, hosted by the president of the German Red Cross and attended by President Ebert and many cabinet ministers, Einstein reported that when he first arrived in the United States, the anti-German feeling was so strong that even the German language was treated like a communicable disease. During his visit, however, he sensed a change of heart. As time passed people began to speak to him in German and expressed genuine sympathy for the German scientists with whom they had "maintained so close a friendship before the war."

Even so, apart from the New York City alderman who had called him an "enemy alien," Einstein must have encountered considerable anti-German sentiment in the States. Less than three years before his arrival, more than six thousand American men had lost their lives fighting German troops at the battle of Belleau Wood. To many Americans, especially those with loved ones killed or maimed during the war (there were over three hundred thousand U.S. casualties), Germans were synonymous with poison gas and prison camps.

The New York Times, however, suggested that "if he had been in the country longer he would have discovered that the individual German gets in the United States now as always just about the treatment his manners and personal character would earn for him anywhere. Our feeling for 'Germany' is a different matter. By that word we still mean the Government that started and waged a war against the rest of the world."

His trouble with the press wasn't over. A few days after the homecoming party, a Dutch journalist asked his impressions of America. It would be some years before he realized that as a public figure he would be "called to account for everything one said, even in jest, in an excess of high spirits or in momentary anger."

According to the news account by the Dutch reporter, Einstein described Americans as so bored as a result of living in a cultural wasteland that they found his visit exciting and greeted him with extravagant enthusiasm. He reportedly rated American scientists as so inferior to Europeans that it was "nonsense" to compare them, and described American men as henpecked by wives who controlled everything in the country, treated them like "toy dogs" (he probably meant "lapdogs"), and squandered their money on the latest fashions. Considering that he had previously said how much he had enjoyed America and Americans, this remark made him sound like a hypocrite.

Asked by an American reporter for her reaction to Einstein's indictment, a Mrs. Countiss of Chicago called it "perfectly ridiculous. The professor must have met a bunch of movie actors over here. If American men make more money than other men to spend on their wives, that's America's luck." A Mrs. Baueur, when invited to comment, said "men are women's equals" and denied they were treated like toy dogs. Robert Millikan defended Einstein, who, he said, "had a hectic time here," and "saw Americans under very unfavorable conditions."

Einstein claimed he had been misquoted, which is more than likely, considering that his words were translated from German to Dutch, back into German, and then into English. What he had not yet learned was that the press feeds on feuds and that his wisecracks would be printed verbatim—and not labeled as jokes. If Americans had known that Einstein was generally amused by the everyday world, they might have laughed with him. In fact, his criticism of America and Americans was high praise compared to his sole diary entry after visiting Marseilles, France: "Bugs in the coffee."

He gave a more cautious and considered opinion of the United States for the *Berliner Tageblatt*, in which he rated Americans as far superior to Europeans in building solid, labor-saving houses, and producing better consumer goods. He praised them as friendly, kindhearted, self-confident, optimistic, and free of envy—though less engaged than Europeans in intellectual and artistic pursuits. As for American scientists, he thought it wrong to attribute their superior scientific research exclusively to the country's great wealth, citing devotion, patience, a spirit of comradeship, and cooperation as contributing factors.

That schools, the telegraph, the telephone, and railroads were mainly in private hands had surprised him. And though he had noticed extremes of wealth and poverty, he conceded that private philanthropies ensured that no one suffered great hardship.

However, he loathed "the cult of the individual." It struck him "as unfair, even bad taste, to select [individuals] for boundless admiration and attributing to them superhuman powers of mind and character. This has been my fate, and the popular estimate of my powers and achievements compared to the reality is grotesque." His admirers would see this as another example of his innate modesty; his enemies, as the plain truth.

He saw one bright aspect to glorifying individuals "whose goals lie wholly in the intellectual and moral sphere. It proves that knowledge and justice are ranked above wealth and power by a large section of the human race. This idealistic outlook is particularly prevalent in America, which is falsely criticized as a singularly materialistic country."

Prohibition was a mistake, in his opinion, proving impossible to enforce; he said that "it is an open secret that the dangerous increase in crime is closely connected with this." Finally, he urged America, as the world's most powerful nation, to take the lead in disarmament talks, warning that its present isolationist attitude would lead to disaster.

He spent the rest of July free from controversies on a happy lakeside vacation with his sons, seventeen-year-old Hans Albert and eleven-year-old Eduard, then resumed his duties at the Kaiser Wilhelm Institute and Berlin University.

Most professors there dressed formally, read their lectures from notes, rarely making eye contact with students, and treated underlings as servants. Physicist Erwin Schrödinger and Einstein were easygoing exceptions. They had been intellectual pen pals before they met and soon became close friends, sailing together on the river at Caputh or walking in the nearby woods.

On Thursday afternoons, Einstein usually appeared at seminars during which promising physics students presented their ideas and he, von Laue, Nernst, and Lise Meitner, among others, would encourage, comment, and criticize. Occasionally, members of the public slipped in to sit in the back of the room for a glimpse of Einstein's head — he usually sat in the front row; once, a heavily painted prostitute gave him a quick appraisal before leaving. It was in this somewhat theatrical, highly charged atmosphere, that "terribly nervous" physics student Esther Salaman began to give a talk. She saw Einstein in the semidarkness — the lights had been dimmed for her to present her slides — and he seemed to be looking at her "as if to say 'Don't worry.'"

Salaman began to discuss a recent problem concerning radioactivity that had arisen at the Cavendish Laboratory in Cambridge, England. A young lecturer interrupted, suggesting a solution in such a long-winded manner that Salaman was unable to follow the argument. As she recalled, "Einstein came to my rescue. 'Clever but not true,' he said, and he restated the problem, and said what we knew and did not know about it so clearly and simply that everyone was satisfied.

Another student, Max Herzburger, had "unforgettable" discussions with Einstein as they walked in a park, impressed because "he took nothing as certain truth because it was written in books, and was always asking questions which led to a deeper understanding of the problem." Fifty years later, Hungarian student Dennis Gabor could still hear Einstein's voice and repeat some of his sayings verbatim. One was "the saying of Oxenstiern — with how little wisdom the world is governed — is true also in science. What the individual contributes to it is very little. The whole is of course admirable." Gabor had "never known anybody who enjoyed science so sensuously as Einstein. Physics melted in his mouth!" Gabor later helped invent the electron microscope and won the 1971 Nobel Prize for his work with holography.

When a Hungarian student, Eugene Wigner, arrived for his first conference at the University of Berlin, he sat in the middle row of three rows of chairs, and had his first glimpse of Einstein, in the row ahead, next to Max von Laue. "I hardly understood a word," Wigner recalled. "I kept hearing the buzz of strange phrases like 'ionization energy.' Yet somehow I was fascinated." Wigner, too, became a Nobelist, winning the prize in 1963 for his contribution to nuclear physics.

Von Laue ran the conference, handing out several new physics papers and asking students to prepare oral reviews for the following week. Wigner later wrote:

> If the reviewer presented a clear picture, no comment came from the front row. But if the review of the paper was unclear, questions were sure to arise from the front row, especially from Einstein. He was always ready to comment, to argue, or to question any paper that was not impressively clear. "Oh, no. Things are not so simple." That was a favorite phrase of Einstein. He could have made a great show of his own importance. He never thought to do so. He did not want to intimidate anyone. On the contrary, he accepted the logic of a colloquium: that human intelligence is limited; that no man can find everything alone; that we all contribute. Perhaps that is why I never felt nervous at the colloquium. Einstein made me feel I was needed.
>
> He worked simply and helpfully, more like a friend than a teacher. His thoughts often turned philosophical. He told us once, "Life is finite. Time is infinite. The probability that I am alive today is zero. In spite of this, I am now alive. Now how is that?" None of his students had an answer. After a pause, Einstein said, "Well, after the fact, one should not ask for probabilities."
>
> Einstein was a solitary man who liked to meditate on the world while walking alone. Yet I was one of many students whom he encouraged to address him

personally and visit him at home. There we discussed not only statistical mechanics, but all of physics; and not only physics, but social and political problems. Einstein heard us out with great interest. His personality was almost magical.

He was almost impossible to surprise in the realm of physics. He seemed to foresee everything of major importance. Concepts seemed to occur to him, fully realized. Their flaws and implications he saw immediately.

Einstein's friend, Max Born, described this talent as "the gift of seeing a meaning behind inconspicuous, well-known facts which had escaped everyone else. The most important example is the equivalence of gravity and acceleration, known since Newton's time, but not previously recognized as a clue to the understanding of the cosmos."

Student Leopold Infeld had been briefly introduced to Einstein's special relativity theory at college in Cracow, Poland, in 1917. Two years later, when the theory was confirmed and Einstein world famous, Infeld was a teacher in a small town, giving a public lecture on the subject. Despite the freezing cold the hall was packed. Infeld moved to Berlin in 1921, where he hoped to hear Einstein teach. But when he applied to be a student at the university he was turned down. By this time he felt Berlin was a "hostile" city, and in a desperately lonely state had the nerve to telephone Einstein. To his delighted astonishment, he was invited to come over right away.

These admiring students—Salaman, Gabor, Wigner, and Infeld—either didn't know that others had disrupted his lectures the previous year, or chose to ignore it in their memoirs. But the opposition persisted, and not only in Berlin. Einstein had arranged to lecture at Munich University in November. Then he learned that student members of the Swastika Majority refused to assure the rector that they would not incite anti-Einstein disruptions, and so he canceled the lecture. This was a great disappointment to young Werner Heisenberg and Wolfgang Pauli, who had missed him at an earlier lecture in Jena, and had looked forward to meeting him in Munich.

In November, no one was awarded the Nobel Prize for physics, but Max von Laue had some inside information and advised Einstein that he might very well be next year's winner.

At the end of December 1921, Einstein was approached by a student in a quandary. Max von Laue had given Leo Szilard a tricky problem in relativity theory for Szilard's Ph.D. thesis. After working on it for six months, Szilard was convinced it couldn't be solved. Then, he told Einstein, he suddenly awoke one morning with a completely new idea, nothing to do with the problem von Laue had set. He had written it up during what he later considered "in a sense [the] most creative period of my life." Szilard revealed the new idea to Einstein, who called it "impossible . . . Something that can't be done."

"Well, yes, but I did it," Szilard insisted.

"How did you do it?" Einstein asked.

When he gave him the details, Einstein was enthusiastic, which gave Szilard the courage to submit it to von Laue. He at least agreed to read it, and the next morning phoned Szilard to say: "Your manuscript has been accepted as your thesis for the Ph.D. degree." Later Szilard realized that his thesis—that the second law of thermodynamics also covered the laws that govern thermodynamic fluctuations—was not a new idea at all, but "rather the roof of an old theory."

Szilard's biographer William Lanouette noted that the relationship between teacher and student was not always a smooth one: "Einstein really liked Szilard's lateral thinking, when he would mix together all different fields. Rather than be a vertical thinker in great depth in one field, he would pull things from all over the place and synthesize them. However, Szilard was not always the obedient student: he could be very insulting. One witness remembers cross exchanges between them at the Kaiser Wilhelm Institute. Einstein would go to the board and write something and Szilard would say, 'Stupid! It can't be right.' Einstein would think about it and then agree. Szilard usually was right but he was very blunt about it."

Einstein was hardly predictable. Although he had canceled the Munich lecture, anticipating trouble there, he accepted Paul Langevin's invitation to lecture in Paris, where he would be at more physical risk than in Munich. In France, especially because of the recent war, there was still a seething hatred of Germans. And vicious anti-Semites were certainly not in short supply.

CHAPTER 18

The Nobel Prize

1922 to 1925

---

43 to 46 years old

His intuition that nature was harmonious continued to obsess Einstein. It led him to think that electromagnetism and gravity had a common origin. Why would discovering that excite him? Because it might be the master key to many of nature's mysteries. After several years of hunting equations to unite the two fundamental forces, in January 1922 he produced a small paper on the subject, his first tentative step in a journey that would last a lifetime.

Then he turned to his upcoming travel plans. When Count Harry Kessler dined at the Einsteins' on March 20, they were still talking about their triumphant visit to America and England, even though Albert admitted he felt like a con man who had failed to give the public what they wanted. He hoped his upcoming weeklong visit to Paris would help revive cordial relations between German and French scholars soured by the war. As for German academics who opposed his visit, he didn't give a damn. He despised them and their opinions.

Aware that fame was fleeting, he told Kessler that he meant to take advantage of it while he was still in demand. He had accepted invitations to lecture in China and Japan later in the year (Palestine and Spain would eventually be added to his itinerary).

Charmed by the Einsteins as he always was, Kessler noted that the other dinner guests included "the immensely rich Koppel" and "Bernhard Dernburg (as shabbily dressed as ever)," and that "an emanation of goodness and simplicity on the part of the host and hostess saved even such a typical Berlin dinner-party from being conventional and transfigures it with an almost patriarchal and fairytale quality."

One early spring evening Chaim Weizmann and Walter Rathenau, Germany's recently appointed foreign minister, called on Einstein. Einstein had advised Rathenau to decline the job, fearing that a Jew in such a powerful position would be bad for Jews. It was certainly bad for Rathenau, who struck

Einstein as "an idealist even though he lived on earth and knew its smell better than almost anyone else." When he was appointed foreign minister in February, a right-wing paper had inflamed readers with: "Now we have it! Germany has a Jewish foreign minister! His appointment is an absolutely unheard-of provocation of the people!" Mobs in the streets had chanted, "Shoot down Walter Rathenau, the Goddamned Jewish swine!"

Rathenau received death threats from right-wing zealots and anti-Semites almost daily. Yet he felt entirely German, did not think of himself as Jewish, and was out of sympathy with Zionism, telling Einstein that Palestine was just a lot of sand.

Weizmann saw Rathenau's attitude as

> all too typical of many assimilated German Jews; they seemed to have no idea that they were sitting on a volcano; they believed quite sincerely that such difficulties as admittedly existed for German Jews were purely temporary and transitory phenomena, primarily due to the influx of East European Jews, who did not fit into the framework of German life, and thus offered targets for anti-Semitic attacks . . . By no stretch of the imagination could [the sophisticated and German-born] Rathenau be described as an East European immigrant.

Threats were made against Einstein, too, especially in Paris in anticipation of his arrival. Barricades were erected around the College de France where he was to give his first lecture. Paul Langevin and astronomer Charles Nordmann went to meet him at the Belgian border, and then all three took a train back for the four-hour journey to Paris.

Nordmann's impression of the forty-three-year-old Einstein was "one of disconcerting youth, strongly romantic, and at certain moments evoking in me the irrepressible idea of a young Beethoven, on which meditation had already left its mark, and who had once been beautiful. And then, suddenly, laughter breaks out and one sees a student. Thus appeared to us the man who has plumbed with his mind, deeper than any before him, the astonishing depths of the mysterious universe."

On the way Langevin was warned by a telegram from the French police that excited groups of students were gathering at the Gare du Nord, where their train would arrive. The students were believed to be members of a nationalist group opposed to Einstein, known as the Patriotic Youth. Langevin followed police advice to alight at a suburban station to avoid the waiting crowd. In fact, the police had been misinformed; it was a friendly crowd organized by Langevin's son.

Apparently a welcoming committee was informed of this change of plan. Ten-year-old Hilaire Cuny happened to be on the platform and spotted a group of men in frock coats, all wearing hats, surrounding a bareheaded man whose "eyes wandered, with a kind of amused irony, over his entourage and his surroundings . . . Einstein had come, almost clandestinely, to France . . . The hatred for Germany was still very strong, in spite of the victory . . . Because of this discreet arrival the next day's newspapers announced, 'Professor Einstein Has Disappeared!' In fact, he had been quietly driven to the German embassy."

Germany's ambassador to France, von Hoesch, insisted that Einstein stay at the embassy, which caused several skirmishes with the minister's valet. Einstein was wearing the only pair of shoes he'd brought with him, and the valet felt obliged to clean them several times a day. "He keeps taking them from me," Einstein complained to a friend, "even though I tell him I'm going out in the rain, and they'll get dirty right away."

It was standing room only for his first lecture at the College de France amphitheater on March 31. Paul Painlevé, the mathematician and former premier of France, had excluded potential troublemakers by personally checking tickets and admitting only invited guests, among them Marie Curie and philosopher Henri Bergson. Einstein spoke French well but slowly, with Langevin close by to prompt him whenever he was at a loss for *le mot juste*, and a blackboard at his back which he soon covered with equations. He repeated his performance at the Sorbonne, where his friend Maurice Solovine was among the overflow audience.

The French press were generally wild about him. Charles Nordmann had helped to diffuse anticipated trouble by pointing out in *Le Matin* that "in the middle of the war, Einstein rejected Prussian militarism," diplomatically neglecting to mention that he had also condemned French militarism.

Reporters may have found the French-language version of relativity somewhat esoteric, but they had no problem defining—and raving over—its exponent.

To *Le Figaro*, "This famous mathematician isn't at all austere, nor fierce, not dry; on the contrary, his demeanor is soft and his smile is sweet. Einstein has a handsome face, a high brow, hair that has been cut short since his last photographs and is naturally curly and peppered with silver strands, a mouth shadowed by a small mustache and eyes that are inspired rather than melancholic."

To *L'Humanite*, "Everyone had the impression of being in the presence of a sublime genius. As we saw Einstein's noble face and heard his slow, soft speech, it seemed as if the purest and most subtle thought was unfolding before us. A noble shudder shook us and raised us above the mediocrity and stubbornness of everyday life." And, "Oh, those eyes! Those who have seen them will never forget them. They have such depth! One might say that the habit of scrutinizing the secrets of the universe leaves indelible traces."

His mission of reconciliation largely succeeded, but the French Academy decided not to invite him to lecture when thirty members threatened to walk out if he walked in. His friend and biographer Philipp Frank explained why. People's response to Einstein, he said,

> depended greatly on their political sympathies. As a famous historian at the Sorbonne put it: "I don't understand Einstein's equations. All I know is that Dreyfus adherents claim that he is a genius, while Dreyfus opponents say he is an ass." Dreyfus was a captain in the French army who in 1894 had been (falsely) accused of treason by anti-Jewish propagandists. The affair developed into a struggle between the Republic and its enemies, and the entire country was divided into two camps, the defenders of Dreyfus and their opponents. "And the remarkable thing," added this historian, "is that although the Dreyfus affair has

long been forgotten, the same groups line up and face each other at the slightest provocation."

At his request, Einstein spent his last day in France touring the World War I battlefields with Solovine, Langevin, and Nordmann. Moved by the devastation, he said, "All the students of the world must be brought here, so they can see how ugly war really is." They had lunch at Rheims, and as they were leaving, two senior French army officers at a nearby table, recognizing Einstein, stood and bowed.

Typically, when he went back to Berlin, he forgot the few things he'd brought with him for his Paris stay. Solovine mailed them on, and in his letter of thanks, Einstein remarked, "Those days were unforgettable but devilishly tiring; my nerves still remind me of them."

In April, Rathenau signed the Treaty of Rapallo, reestablishing Germany's diplomatic relations with Russia and pledging economic cooperation. By doing so he signed his death warrant. Fanatics who already hated him as a liberal and a Jew now accused him of selling out to the Bolsheviks. For a time he accepted police protection; they guarded his home and escorted him everywhere, and he showed Harry Kessler a Browning revolver he kept close at hand. Then, without explanation, he dismissed his guards.

On June 11, 1922, Einstein was in the Reichstag preaching peace and international reconciliation to already converted members of the German Peace Federation. A few days later, he met Marie Curie in Geneva. As recent members of the League of Nations' International Committee on Intellectual Cooperation, they and a handful of others were charged with representing the conscience of the civilized world. One evening, after discussing international social and economic problems, the two took a break and sat under a street lamp on the shores of Lake Geneva. Their conversation drifted from politics to their first love, physics, as Einstein wondered aloud, "Why does the reflection in the water break down in this spot and not at another point?"—a puzzle he apparently never pursued.

Meanwhile, in Berlin, a penitent confessed to a priest details of a plot to kill Rathenau. The priest broke church rules to inform the papal nuncio, Eugenio Pacelli (later Pope Pius XII), who gave him permission to warn Joseph Wirth, the German chancellor. He, in turn, warned Rathenau. But Rathenau simply put his hands on Wirth's shoulders as if to console him, saying, "Dear friend, it is nothing. Who would want to do me any harm?"

Rathenau's attitude was incomprehensible. He behaved as if he was in no danger even after Dr. Weizmann, chief of the Prussian police, had warned him that if he persisted in driving slowly to and from the Foreign Office in an open convertible, no police in the world could guarantee his safety.

The very next day, June 24, as Rathenau was being driven at the usual leisurely pace in his open car, two young men drove alongside, riddled him with submachine-gun bullets, breaking his spine and jaw, then threw a hand grenade that almost severed his legs. He never regained consciousness.

Einstein had good reason to believe that he faced a similar end. The previous year a Berliner had been given a nominal fine and a slap on the wrist for offering a reward to anyone who killed Einstein. Now the police told Einstein that he, Theodor Wolff, the editor of *Berliner Tageblatt*, and Max Warburg, a banker, were on the death list of a group planning to assassinate influential Jews. *The New York Times* reported that on the advice of friends, Einstein had "fled from Germany temporarily because he was threatened with assassination by the group that caused the murder of Rathenau."

The report was wrong: Einstein had not left the country. He did leave Berlin for a while, however. Rathenau's murder also led him to reconsider his agreement to be the keynote speaker at Leipzig on September 18. Planck was to chair a meeting of German scientists and physicians and hoped that Einstein's talk would replace "the silly advertisements that puff relativity theory" with "a purely objective point of view." But Lenard threatened to disrupt the meeting.

Einstein assured Planck that if it had been really important, these threats would not have stopped him from giving the Leipzig lecture (doubtless true), but on this occasion, von Laue could easily replace him. "The trouble is," he told Planck, "that the newspapers have mentioned my name too often, thus mobilizing the rabble against me. I have no alternative but to be patient—and to leave the city." He asked Planck to treat the incident with a sense of humor, as he was doing.

Planck, who had always admired Einstein, called the attacks against him "spiteful . . . contemptible . . . ugly and vicious," and was furious that "a band of murderers, who go about their business in the dark, dictate the scientific program of a purely scientific body." He made the best of it, though; after getting von Laue to substitute for Einstein, he asserted that "looked at entirely objectively this change has perhaps the advantage that people who still believe the principle of relativity is basically propaganda for Einstein . . . will learn [the truth]"— presumably by having Planck and von Laue, who were not Jewish, endorse it. Chancellor Wirth sympathized with Planck, and called the Rathenau killers deluded boys in the service of a great evil. Adolf Hitler called them German heroes.

A member of the murderers' right-wing group took a bribe and revealed the assassins' identities. Police surrounded their hideout in a castle on July 17 and, during a fierce storm, shot one to death. The other committed suicide.

Over a million stood silently in the streets near the Parliament building where Rathenau's funeral service was held. Einstein appeared among the mourners, though he believed he might be the next victim. He even wrote an obituary notice, saying, "My feelings for Rathenau were and are joyous admiration and gratitude for the hope and consolation he gave me in Europe's current dark days."

Harry Kessler remarked that "the response which had been denied to Rathenau's life and thought were now accorded to his death." Not by Philipp Lenard. When classes were officially canceled in all German universities to mourn Rathenau, Lenard scheduled a lecture in Heidelberg for that day, which he urged supporters to attend. It took on the nature of a celebration; Lenard, after all, had advocated killing Rathenau on the grounds that he was destroying the country. A group of workers who tried to persuade Lenard to cancel the class were drenched with water thrown from the second story of the building, so they broke in and grabbed him. As he was about to be tossed into a nearby river, policemen appeared and took him into protective custody.

Einstein confided to Solovine that since Rathenau's assassination his life had become nerve-racking and that he was constantly on the alert.

Weizmann was having his own problems. "All the shady characters of the world are at work, against us," he told Einstein. "Rich servile Jews, dark fanatic Jewish obscurantists, in combination with the Vatican, with Arab assassins, English imperialist anti-Semitic reactionaries—in short, all the dogs are howling. Never in my life have I felt so alone—and yet so certain and confident." His instinct was right: a month later, on July 24, the League of Nations unanimously ratified the Mandate for Palestine submitted by Britain—which meant the country would be governed by the British.

That summer, student Werner Heisenberg arrived in Leipzig, thrilled at the prospect of hearing Einstein speak and perhaps meeting him afterwards. Exhausted after the long train journey from Munich, he fell asleep on a patch of grass. A girl woke him in time for the lecture, and as he entered the hall someone gave him a pamphlet. Endorsed by nineteen professors and physicians, it damned relativity as wild speculation promoted by the Jewish press and alien to the German spirit.

At first, Heisenberg thought it was the work of a lunatic: "However, when I was told that the author [Lenard] was a man renowned for his experimental work I felt as if part of my world were collapsing. And now I made the sad discovery that men of weak or pathological character can inject their twisted political passions even into scientific life. My immediate reaction was to drop any reservations I may have had with regard to Einstein's theory." Heisenberg was so upset that he "failed to pay proper attention to Einstein himself, and, at the end of the lecture, forgot to avail myself of Sommerfeld's offer" to introduce him to the speaker. Had he been introduced he would, of course, have known it was von Laue and not Einstein. The experience must have disturbed him more than he realized because, writing about it in his memoirs some fifty years later, Heisenberg still thought Einstein had given the lecture that day!

Fortunately, the Einsteins escaped the deadly atmosphere in Berlin, leaving on October 8, 1922, to fulfill his lecture commitments in Japan and China. Their ship stopped en route at Colombo, Singapore, and Hong Kong. At Colombo he rode with Elsa in the customary mode of transport, a carriage pulled by a man, although he was "bitterly ashamed to share responsibility for the abominable treatment accorded fellow human beings." At Shanghai they were greeted by a group from the Germany colony singing "Deutschland über Alles."

Japanese government officials argued about relativity as Einstein approached their capital. The minister of education said that ordinary people would understand it; the minister of justice said they wouldn't. The ministers of agriculture and commerce sat on the fence, saying that they might understand "vaguely."

This prompted the justice minister to take a pragmatic approach; opening a book on relativity, he found higher mathematics on the first page—which proved his point and won the argument.

Yet Einstein's Japanese audiences showed incredible patience; it was assumed that most were there just to see a celebrity, but they listened politely through a lecture that lasted, with translations, almost four hours. They were equally impressed by *his* endurance.

The Einsteins' suite of rooms with a balcony in the luxurious Imperial Hotel overlooked a square where thousands spent a night-long silent vigil. At sunrise they called for Einstein to appear and greeted him with a roar of approval. He bowed to acknowledge the cheers, whispering to Elsa, who had joined him on the balcony, "No living being deserves this sort of reception," and when the cheering continued, added, "I'm afraid we're swindlers. We'll end in prison yet."

Einstein met the emperor and empress (she spoke with him in French) and the German ambassador; the ambassador looked somewhat askance at his casual clothes but liked his friendly, unpretentious manner. He was enchanted by the Japanese for their beautiful manners and lively interest in everything, their intellectual integrity, artistic sensitivity, and common sense. He thought the country wonderful, too. His only serious disappointment seems to have been their music. He loved the country and its people so much that he cried when he left.

A few days away from Tokyo he heard that he had won the Nobel Prize for physics.

Why had it taken so long? Why was Einstein's nomination for the Nobel Prize rejected eight times even though he was acknowledged to be one of the greatest scientists ever? Writer Irving Wallace discovered what may be the answers. Seeking authenticity for *The Prize*, his novel about the award, Wallace wrote to Einstein to ask how he was informed of his Nobel win and if he had any criticism of the procedure and the judges' choices. Einstein replied: "I was informed that I had received the Nobel Prize by a telegram which I got while on board ship on my way back from a visit to Japan ... The prize was not personally handed to me. I was invited, instead, to attend a Swedish scientific congress at Goeteborg [sic] where I delivered an address ... I found that the procedure of selection of the prize-winners—at least in my field—is fair and conscientious."

Wanting more accurate background material, Wallace went to Sweden and to his delight found that one of the Nobel Prize judges was eager to discuss the normally hush-hush subject. Dr. Sven Hedin told Wallace that the anti-Semitic German scientist Philipp Lenard had had great influence with himself and with fellow judges. Lenard put the word in that relativity was really not a discovery (for which the award is made), that it had never been experimentally proved, and was, in any case, of no value. Hedin, for one, needed little persuading, and it was clear to Wallace where his sympathies lay: later, during World War II, Hedin supported the Nazis and was proud to call Göring, Himmler, and Hitler his intimate friends.

Wallace's wife, Sylvia, helped him with more background material for the novel by interviewing (in 1949) Dr. Robert Millikan, a 1923 Nobel laureate who was then Caltech's president. She noticed that of the two busts in his office one was of Ben Franklin and the other of Einstein. Millikan told her in confidence that there was another reason other than Lenard's pressure for withholding the prize from Einstein for so long. The committee wanted to give him the award, Millikan explained, having learned about it from a confidential source. One of the committee members then "spent all his time studying Einstein's theory of relativity. He couldn't understand it. Didn't dare to give the prize and run the risk of learning later that the theory of relativity is invalid."

A combination of anti-Semitic venom from Lenard and bewilderment on the part of the Nobel Prize judges explains why Einstein continued to be rejected for eleven years—from 1910 to 1921. By then, Lenard's influence had worn thin and the support for Einstein had become overwhelming.

In 1922, Arnold Sommerfeld wrote an enthusiastic letter of support. So did other leading scientists, including Marcel Brillouin, who obviously hoped to shame the judges by writing: "Imagine for a moment what the general opinion will be fifty years from now if the name of Einstein does not appear on the list of laureates." They were joined by Felix Ehrenhaft, French mathematician Jacques Hadamard, Edgar Meyer, Stefan Meyer, Bernhard Naunyn, Gunnar Nordström, and Emil Warburg—all giving repeat recommendations. Other Einstein boosters included Theophile de Donder of Brussels; Swiss astrophysicist Robert Emden and Ernst Wagner of Munich; his friends Max von Laue of Berlin and Paul Langevin of Paris; and Edward Poulton of Oxford, England.

More support came from Carl Wilhelm Oseen of the University of Upssala. He had been asked by the Nobel Prize selection committee to spell out the value of the photoelectric effect, for which he had vainly recommended Einstein for the prize in 1921. His thumbs-up report seems to have worked.

Einstein finally got the 1922 prize "for his services to Theoretical Physics, and especially for his discovery of the law of the photoelectric effect," rather than for relativity, which had continued to baffle the judges. Because Einstein was still on his world tour, the German envoy to Sweden accepted the award on his behalf at the Nobel ceremonies on December 10.

Homeward bound from Japan, Einstein and Elsa stopped in Palestine. From Haifa they took a train to Lydda. Einstein was tired after sitting up all night in a second-class carriage, having refused to occupy a sleeping berth in the first-class section.

The couple stayed with Herbert Samuel, Palestine's high commissioner, whose palatial residence, the prewar, German-built Augusta-Victoria hospice, was run like an upscale Buckingham Palace. A cannon fired each time Samuel left the place, and when he rode through the city he was escorted by mounted troops. This pomp certainly didn't delight Einstein, the egalitarian, but he had other things on his mind. Elsa, however, hated the formality and elaborate etiquette so much that she often made excuses to go to bed early.

When officially welcomed by the Palestine Zionist Executive on February 7, Einstein apologized for not speaking in Hebrew, saying his brain wasn't up to it.

The next morning he and Elsa were driven between lines of cheering school-children to a reception at the Lemel School, where he said "it is the greatest day of my life. This is a great age, the age of the liberation of the Jewish soul; and it has been accomplished through the Zionist movement, so that no one in the world will be able to destroy it."

They dined with Attorney General Norman Bentwich and his wife, Helen, after which Einstein played second fiddle in a quartet with Bentwich and his two sisters. He amused Helen Bentwich by advising her not to read a book they were discussing, because "the author writes just like a professor."

The highlight of Einstein's twelve-day stay was when he spoke on Jerusalem's Mount Scopus, the site of the future Hebrew University. There, on February 9, he inaugurated what he had helped make possible. In the afternoon, lecturing on relativity, he stumbled over his opening remarks in Hebrew but then continued in French, which pleased the French consul.

"The brothers of our race in Palestine charmed me as farmers, workers, and citizens," he wrote to Solovine, who was still in Paris. He said he was optimistic about their future but didn't want to join them. As he explained to Frederick Kisch, a Zionist official, it would cut him off from friends and work in Europe, where he was free. In Palestine he would always be a prisoner—an ornament.

After planting a tree on Mount Carmel, he visited Haifa's high school and technical college, and was honored with the "freedom" of Tel Aviv. Later, during his tour of a kibbutz, an embarrassed young woman jumped to the wrong conclusion when he asked about the relationship between the sexes there. As her only response was to blush, he tried again. Telling her not to be alarmed by the question, he explained, "We physicists understand by the word 'relationship' something rather simple, namely: How many men are there and how many women?"

Walking on the Mount of Olives with Bentwich, he spoke enthusiastically of the beautiful Arab peasant dress and the Arab village, which appeared to grow out of rock.

When Kisch saw the Einsteins off at Jerusalem station on February 14, 1923, he asked if Einstein had any advice to improve conditions in Palestine. Yes, he said. "Collect more money." He told Chaim Weizmann, "The difficulties are great, but the mood is confident and the work to be marveled at."

The next stop was Barcelona, Spain, and then Madrid, where he again "whistled his relativity tune," as he jokingly called it, to King Alfonso XIII and members of the Academy of Sciences. He was distressed when people fell on their knees at the sight of him, but most of his experiences seemed like a wonderful dream. As he said to Elsa, "Let's enjoy everything before we wake up."

He did not enjoy the train journey from Madrid to the French border in the luxurious royal coach provided by the king. When they got out, Albert told Elsa: "*You* can do what you like, but in future I'll always travel third class."

Soon after he arrived home, the Swedish Ambassador to Germany called on him personally to hand him his Nobel medal and diploma.

Later in the summer, he resigned from the League of Nations' Committee for Intellectual Cooperation. He was frustrated by the group's impotence, explaining: "There appeared to be no action, no matter how brutal, committed by the present power groups against which the League could take a stand."

Then he was off to Sweden to speak to members of the Scandinavian Society of Science in Göteborg—in place of his Nobel acceptance speech—before an audience of some two thousand, including King Gustav V. Although Einstein wore a dinner jacket for the ceremony, it had seen better days and better care. Shortly before he was about to make his entrance, someone offered him an immaculate replacement. He simply flicked "the sleeves of his jacket with his fingertips, and said, 'It's all right. We can put a sign on my back: "This suit has just been brushed."'" Then he walked onto the stage, where he surprised the audience by speaking about relativity and not mentioning the photoelectric effect for which he had received the award.

As promised, he sent the $32,500 in prize money to Mileva, to help support her and their sons. He could afford it. Although galloping inflation had deflated his professor's salary, and he accepted only expenses for his lectures at home and abroad, he augmented his income by acting as a patent expert for several German companies, appearing in court as a witness in patent infringement cases, and giving well-paid advice about new products. Einstein made a fortune for one company by advising the mass production of gas refrigerators, and happily took a plane ride for another employer to test flying instruments.

Einstein quite often made the ten-hour train journey from Berlin to Zurich to see his sons and advise Mileva how to handle them. Hans Albert was determined to be an engineer; Einstein, at first opposed to this plan, was now reconciled to it. Eduard's strange intensity, however, still worried him, but he was proud of the younger boy's passion for learning and terrific memory.

Hans Albert, at least, was aware of his father's affection, recalling that "while it was there, it was very strong. He needed to be loved himself. But almost the instant you felt the contact, he would push you away. He would not let himself go. He would turn off emotion like a tap."

For someone who generally found traveling tedious, Einstein seemed to be always on the move. To some extent it was evasive action, keeping his enemies off his track; or, as he once said, "Life is like riding a bicycle. To keep your balance you must keep moving." What he could not foresee were the dirty methods his opponents would use to discredit him or to goad others into trying to kill him.

A nationalist newspaper might just as well have advertised for a hit man when it printed as fact the rumor that he would visit Russia in September. This was even picked up by the liberal newspaper *Berliner Tageblatt*, which reported on October 6 that he was on his way to Moscow. On October 27, another nationalist newspaper announced that he would arrive in Petersburg the following day. A fourth paper revealed on November 2 that "Einstein is staying in Petersburg for three days."

All lies. Einstein never intended to visit Russia, nor did he ever do so. The planted story implied that he, like Rathenau, was giving aid and comfort to the dreaded Bolsheviks and deserved the same fate.

In fact, Einstein was about to leave for his yearly stay with Paul Ehrenfest as visiting professor in the tranquil city of Leiden, Holland.

It was like a month's vacation with loving friends, and a world away from the threats and insults in Berlin. He had complete freedom in the Ehrenfests' home. They set up a small table for him in their dining room with milk, bread, cheese, cakes, and fruit always on tap. He also had a place to work, a bed, his violin, and an affectionate and intellectually stimulating atmosphere. "What more can a man want?" he asked, having perhaps inadvertently left Elsa out of the idyllic picture. On this visit he and the Ehrenfests discussed, among other things, the puzzling nature of light. An American physicist, Arthur Compton, had recently completed experiments confirming Einstein's view of light quanta as particles. However, he still resisted Compton's assertion that light also came in waves.

One day he went with the Ehrenfests to Spa, a Belgian health resort. Walking around the town, deep in a spirited conversation with Tatiana, Ehrenfest's physicist wife, they were suddenly confronted by a horde of photographers. When one stopped shooting, another started, until Tatiana protested, "It's impossible to have an intelligent conversation like this!" Soon after the cameramen left, a woman approached, interrupted their resumed conversation, and asked Einstein for an interview. Not only did he answer her questions patiently, but he posed for photographs, explaining afterward that he couldn't refuse the woman, who had three children to support and had come from Germany especially to see him.

On the other hand, he never hesitated to get rid of people who annoyed him—a socialite, for example, who had invited herself to a musical evening at the Einsteins'. On leaving, she asked, "You will allow me to return, I hope, professor?" To which he simply replied, "No." When Elsa berated him for being so blunt, he said, "But why should she come back? I really don't see the necessity!" His elder stepdaughter, Ilse, confirmed that his interest in a person had nothing to do with his or her title or social class. He might speak for hours with a beggar and rudely reject an "important personality."

Einstein was still staying with his friends in Holland when he read of Adolf Hitler's coup attempt. During a political rally in a Munich beer hall, Hitler had jumped on a table, fired his revolver at the ceiling to get attention, and shouted, "The National Revolution has begun!" Holding the group at gunpoint, he threatened to kill several politicians and then himself if they didn't join him. They played along or stalled for time.

Next morning, after smashing the printing machines of a social Democrat newspaper, the *Munich Post*, and seizing the War Ministry, three thousand Nazi brownshirts marched toward the city center. Police opened fire, killing sixteen and wounding many more, including Hermann Göring. Hitler went into hiding with a dislocated shoulder. He was arrested three days later, on November 11, 1923, and charged with treason.

Planck feared that when Einstein learned of Hitler's violence and Lenard's renewed smear campaign, he might return permanently to Switzerland or stay put in Holland. There he was not only out of danger, but treated with affection. Prizing him as an "ornament for which the entire world envied Germany," Planck made Einstein a great offer. All he would be required to do was to give an annual lecture in Berlin and to list Berlin as his "official" home. That way he could stay out of the city, and away from his enemies, except for one day a year.

Einstein refused to be intimidated into accepting Planck's proposal, which would have made him a virtual refugee. He could easily have gone along with it, as he had several chances to work abroad. Lorentz had recently retired, and Einstein was offered his chair at Leiden University—which Ehrenfest eventually occupied. Spain wanted him, too. But he believed that he did his best work in Berlin, despite the danger, and he valued the small group of supportive colleagues such as Planck, Nernst, Haber, Meitner, and von Laue.

Early in 1924, James Wharton, Berlin correspondent for the North American Newspaper Alliance, asked Einstein if the rumor was true that he intended to move to Jerusalem because of German anti-Semitism. Einstein denied it, and he turned his thoughts back to science.

He told Hedi Born that he hated her husband's new idea "that an electron exposed to radiation should choose *of its own free will,* not only its moment to jump off, but also its direction." If that were the case, he said, he would rather be a cobbler, or work in a gambling casino. He admitted that his many "attempts to give tangible form to the quanta have foundered again and again, but I am far from giving up hope." He was tempted, he said, to stroke her head as a response to a "pretty remark" in her letter to him, "if that is at all permissible in the case of a married lady."

Einstein's interest in other women was not limited to Hedi Born, though with her it is unlikely to have gone beyond flirting. However, biographer Abraham Pais discovered that for some years "he had a strong attachment to a younger woman," and that his letters to her "express emotions for which, perhaps, he had no energy to spare in his marriage. This interlude ended late in 1924, when he wrote to her that he had to seek in the stars what he was denied on earth."

While Einstein ended this relationship, his stepdaughter Ilse, now twenty-seven, began one, marrying Rudolf Kayser, the editor of a literary magazine. Before long Kayser would add one more to the many hundreds of books already published on Einstein and his relativity theory. The newly married couple moved to a house near the Einsteins' apartment, which pleased Elsa, who could hardly stand to have either daughter out of sight.

In the summer of 1924, on his way to Kiel for a health cure for his chronic stomach pains, Einstein stopped for a few days in Göttingen to chat with friends and colleagues. Werner Heisenberg, who was there assisting at Max Born's seminar about the controversial new quantum mechanics, finally got to meet Einstein. His pleasure in having the great man discuss quantum theory with him during a short walk turned to dismay, however, because Einstein had scores of objections to his ideas about atoms and their stability.

That July, Einstein changed his mind and rejoined the committee for the League of Nations, believing he could use the position to improve Franco–German relations. He also knew that his resignation had pleased his enemies, who wanted the League to fail.

If anyone still thought he supported the Communists, the preface he wrote that winter for Isaac Don Levine's *Letters from Russian Prisons* would refute them. Levine's work, sponsored by an international committee headed by the American civil libertarian Roger Baldwin, provided powerful evidence of Soviet persecution of political opponents. In his preface, Einstein wrote of "the regime of frightfulness in Russia. You will contemplate with horror this tragedy of human history in which one murders from fear of being murdered. And it is just the best, the most altruistic individuals who are being tortured and slaughtered—but not in Russia alone—because they are feared as a potential political force."

In 1925, back from his annual trip to Leiden, Einstein found that his home had become a magnet for both hero worshipers and hatemongers. Those who slipped past Otto, the doorman, and Elsa's rigorous screening badgered Einstein with questions, both cosmic and comical. Others begged for handouts or job recommendations. He could never be sure that the next one through the door was not out to kill him.

Mail continued to pour in from women who had named their sons after him, proposed romantic interludes, or sought advice about intimate affairs. One tried to snare him by euphemism, offering herself as a companion in "cosmic contemplation." Elsa handled many of these missives, but Einstein was still over-whelmed by the overflow and dreaded the sound of the mailman.

Hate mail was usually anonymous, but a series of disconcerting letters from France were signed "Eugenia Dickson." What apparently set this woman off was Einstein's opinion, since hardened, judging by his preface for Levine, that the Bolsheviks were laughable, but not as bad as they had been painted. To Dickson, a dispossessed Russian refugee, they were evil incarnate, and Einstein's attitude simply outrageous. Her letters, mailed from locales progressively closer to Berlin, indicated that she was slowly but surely en route from Paris to confront him. She maintained her wild, threatening tone even as she did a complete about-face, now charging him with being a tsarist agent provocateur, Azef-Einstein, mas-querading as a physicist named Albert Einstein. She also hinted at a more per-sonal reason for her animosity. Elsa was upset, but Albert dismissed the letters as the ravings of a hysteric with an overwrought imagination.

One morning, as Margot returned from shopping, she spotted a woman behaving oddly in the doorway of the family's apartment building. Otto, who usually handled such situations, was otherwise engaged. Suspecting that the woman might mean trouble, Margot called her mother from a nearby street phone, warning her that the stranger had entered the building and could be on the way up. Margot was right. Elsa opened the door to face Eugenia Dickson brandishing a large hatpin like a dagger and demanding to see "Azef-Einstein." After a brief struggle, Elsa disarmed the woman and then called the police.

Out of sight and sound of the disturbance, probably in his study, Einstein was oblivious of the attempt on his life until Elsa and Margot told him about it. This immediately aroused his sympathy for the would-be assassin.

Meanwhile, at the Leipzig railroad station, Paul Ehrenfest waited for Albert to arrive by the morning train. As always, he eagerly looked forward to their meeting. Hours went by—and no Einstein. Perhaps he'd forgotten; it wouldn't be the first time. Ehrenfest called the Berlin apartment, but the phone was engaged or out of order. He stayed in his hotel through most of the day, hoping his friend would turn up.

Albert arrived late that evening with an extraordinary excuse: he had been to the jail to see the woman who had tried to kill him. He had to go, anyway, to press charges against her. Once there, he asked to speak with her, perhaps prompted by curiosity first aroused by the intriguing lectures he had attended years before on criminal psychopaths.

Eugenia Dickson had greeted him in her cell with an immediate apology. Mistaken identity, she said. He was not the tsarist agent provocateur Azef-Einstein, "because your nose is much shorter." But that didn't let him off the hook. Now the personal reason for her animosity came to light; though not the agent provocateur, he had been her lover, she insisted, and had deserted her after fathering a child, who had since died.

She pleaded with Einstein to save her from being locked up in a mental institution, which without his say-so seemed inevitable. He agreed to help and even went out to buy her a few things she had requested.

Because he declined to press charges, the police let her go, which agitated Elsa, who feared a second attack. She was further upset by a Berlin police investigator who said that they could not rule out the possibility that Einstein had had an affair with the woman.

A cynic might think Einstein had an uneasy conscience and was trying to buy Eugenia Dickson's silence. But as his doctor friend, Janos Plesch, noted: "Einstein is a man of great good nature . . . the sight of distress always inspires him with desire to help. He gives away what spare money he has—he never has much—to people in need of assistance."

Soon after the incident, Elsa and Albert were interviewed by A. V. Luna-charsky, a Soviet commissar for education who doubled as a journalist. When they mentioned their scary encounter with Eugenia Dickson, it turned out that he had met her in Paris some years before. She was, he explained, the Russian widow of an American and had been living as an emigré in Paris since the Bolshevik Revolution of 1917. Knowing he was a prominent journalist, she had unsuccessfully tried to pressure him into exposing Einstein as a fraud for masquerading as a physicist when, she claimed, he was really both an agent provocateur and her ex-lover. She had also accused a former tsarist minister, Pavel Milyukov, of giving her a child and then murdering it for some malevolent political reason.

If Lunacharsky suspected she was deranged, he was sure of it when, before her Berlin journey to confront Einstein, she had tried to kill the Russian ambassador to France. It had been a comic-opera attempt—her revolver was unloaded—causing a judge to assume that she was a comparatively harmless mental case. So he released her.

In Lunacharsky's subsequent article, "A Meeting with Genius," published in a Moscow magazine, he described Einstein's escape from the hatpin attack and Elsa's brave role as his bodyguard. The Russian was clearly captivated by Einstein, giving this endearing portrait of him:

> There is a dreamy expression in Einstein's near-sighted eyes, as if long ago he had turned the greater part of his vision to his inner thoughts and kept it there . . . Nevertheless, Einstein is a jolly fellow in company. He enjoys a good joke and readily breaks into peals of rollicking, childish laughter, which momentarily change his eyes into those of a child. His remarkable simplicity is so charming that one feels like hugging him or squeezing his hand or slapping him on the back—which in no way detracts from one's esteem for him. It is a strange feeling of tender affection for a man of great and defenceless simplicity mixed with boundless respect.

Lunacharsky portrayed Elsa as

> no longer young [at 51, she was five years older than Albert], with thick grey hair, but a lovely woman with a chaste beauty that is much more than physical beauty. She is all love for her great husband, always ready to shield him from the harsh intrusions of life and to ensure the peace of mind necessary for his great ideas to mature. She is filled with the realization of his great purpose as a thinker and with the tenderest feelings of companion, wife and mother towards a remarkable, exquisite, grown-up child.

In fact, Elsa once said Albert's personality had not changed since she first played with him when he was five!

He was in his study one spring morning advising student Esther Salaman about her future career when Elsa said that two men who looked like Orthodox Jews representing some charity wished to see him. He invited them in. They had long, black coats, broad-brimmed black hats, and side curls, and spoke a mixture of Yiddish and Hebrew. Neither Einstein nor Elsa understood a word they said after their initial greeting of "Shalom aleichem." Fortunately, Salaman acted as interpreter. The men were from Warsaw on business and had gladly walked a few miles to shake hands with a great Jew. When Elsa asked why they hadn't taken the subway, one replied that they hoped they would never have to travel in the bowels of the earth. Before they excused themselves for taking up his precious time, they asked Salaman to tell Einstein that it prolonged life to set eyes on a great man. She did. She was too embarrassed to translate their other comment: "It is a blessing to look at a king."

" 'That was quite nice,' Frau Einstein said after the men had gone, 'but we are never left in peace. The police rang up yesterday to say that they had arrested a tramp who claims he was drinking with Einstein all night. Could I tell them whether he was telling the truth? I told them my husband was playing the violin all evening, and never left the house. They were not satisfied, they're coming to ask me to sign a statement! I used to complain to Rathenau—he was a friend of ours—but he could never see anything wrong with the Germans. He would have found excuses for his own murderers.'"

Later that morning, Einstein walked with Salaman to the University of Berlin, where he was to attend a meeting of the Academy of Sciences. When she despaired of becoming a theoretical physicist because she wasn't creative, Einstein was hardly reassuring, saying that few women were creative and that he would never have sent a daughter of his to study science. "I'm glad my wife doesn't know any science," he added, "my first wife did." When Salaman cited Marie Curie as creative, Einstein said he had spent some vacations with the Curies, and "Madame Curie never heard the birds sing!" Another time he had implied that because Marie Curie was so tenacious and single-minded, she missed many of the joys of life.

As they were walking through the Tiergarten and onto the Unter den Linden, Einstein suggested that if Salaman intended to leave Germany, she should try to work at the Cavendish Laboratory in Cambridge. "England has always produced the best physicists," he said. "I'm not thinking only of Newton: there would be no modern physics without Maxwell's electromagnetic equations. I owe more to Maxwell than to anyone. But remember in England everything is judged by achievement."

Trying to articulate her feelings for Einstein, which both intrigued and puzzled her, Salaman later wrote: "It was not the same as respect, admiration, affection. I knew it was not pride—anyone could talk to Einstein. I had never felt anything like it before: I was in awe of his spirit. Being still young (twenty-five) I was surprised that it should make my heart beat faster, like love or fear."

In May 1925, Albert and Elsa set sail for South America. There he lectured to college audiences in Buenos Aires, Rio de Janeiro, and Montevideo. In Argentina, he was wryly amused by an effusive welcome from the German ambassador, noting in his diary that although Germans generally regarded him as "a foul smelling flower" they were constantly sticking him in "their buttonholes." He had a nervous collapse before the end of the tour; heart trouble was suspected, and he followed his doctor's advice to forgo a trip to California, to return home and take it easy.

He came back with a somewhat inappropriate peace offering for Mileva—a basket of cactuses—and a colorful collection of butterflies for their sons. His letters to her were not as prickly as his present, showing a genuine effort to make amends. He assured Mileva that he sympathized with her worries about their sons' future and would willingly discuss them, and said that he had hated the thought of breaking up with her as he had with Marie Winteler. Presumably, Einstein meant that he had been callous with Winteler and wished to be compassionate and understanding with Mileva. He also said that he hoped they could emulate divorced couples who maintained friendly relations. He even invited her and the boys to stay with him and Elsa in Berlin.

Convalescing from nervous exhaustion did not preclude his continued efforts to perfect a unified field theory. He kept his friend Ehrenfest informed, writing to him in August, "I have once again a theory of gravitation-electricity; very beautiful but dubious," followed exactly a month later by, "This summer I wrote a beguiling paper about gravitation-electricity, but now I doubt again very much whether it is true!" Two days later he gave it up as "no good."

A visitor, Bengali physicist Satyendra Nath Bose, arrived in November from Calcutta, having first contacted Einstein in a letter beginning "Respected Sir," and requesting him to glance at his manuscript, which had been rejected for publication. Einstein was intrigued but puzzled by what he read, calling Bose's derivation of Planck's law "beautiful" though "obscure."

Despite the obscurity, he set aside his own work to translate the manuscript into German and have it published in a scientific journal. Then he developed and immortalized the idea as "Bose-Einstein statistics"—in which quantum rules apply. Einstein and Bose also independently predicted that at extremely low temperatures, "atoms in a dilute, noninteracting gas would condense to the point where they fall into the same quantum state, essentially behaving like a single atom." (In 1995, experiments showed their theory to be a fact.)

Because he was also corresponding with twenty-three-year-old Werner Heisenberg, Einstein was among the first to learn of his enduring contribution to quantum physics. Called "the uncertainty principle," it stated that the wave-particle duality of matter makes it impossible to determine simultaneously a particle's precise position and velocity. Consequently, investigators of the atomic and subatomic world had entered a wonderland or crapshoot where no one could accurately predict the future from the past.

Einstein was swiftly distracted from subatomic particles by distress signals from Mileva. She wanted to take up his offer to discuss family problems. Hans Albert was the most immediate one. At twenty-two, with no job or job prospects, he was bent on marrying a woman named Frieda Knecht, nine years his senior, who lived in the same apartment building and was almost Mileva's double in looks and manner. It spelled disaster to both Mileva and Einstein. The estranged couple reunited now to oppose what Einstein warned would be a crime, because he suspected that Frieda's mother had a mental illness which his prospective grandchildren might inherit (in fact, Frieda's mother suffered from an overactive thyroid). Einstein was hardly overreacting. He didn't want Hans Albert to repeat his mistake. For several years now, Mileva's sister, Zorka, had been a patient in Zurich's Burgholzi psychiatric clinic, and Einstein feared that Eduard might have inherited her family's vulnerability to mental disease.

Rebuffed by Hans Albert when he tried to talk him out of the marriage, Einstein told Mileva that he believed their son's sexual inhibitions were the root of the problem. He suggested that she put him in the hands of an attractive and experienced forty-year-old woman he knew—presumably to get Frieda out of his system.

Eduard, now sixteen, was less of a worry, though his teachers considered him dreamy and unfocused, and Einstein sensed that something was wrong with him. Yet friends knew him as a quick-witted and intelligent young man—he had recently given a lecture on the history of astronomy to great applause—who wrote satirical poems and played the piano with a fervor matching Einstein's passion for the violin. His frequent letters to his father—filled with strong opinions about various composers and philosophers—were anything but unfocused. Einstein treasured them, though Eduard didn't know it at the time.

On December 25, Einstein was again absorbed in quantum physics, marveling, as he told Michele Besso, in the almost magic calculations that enabled Heisenberg, Max Born, and Pascual Jordan to develop their ingenious and seemingly foolproof theory. Yet it was one he would relentlessly resist, as it undermined his own strong belief in causality. He was interested to learn that Besso intended to visit Jerusalem, where, Einstein wrote, "our Jews are doing a lot and as usual quarreling all the time. Which gives me a lot of work because, as you know, they regard me as a Jewish saint."

He expressed his delight in the Locarno Treaty, which, he believed, reconciled the World War I enemies, paved the way for Germany to enter the League of Nations, and showed that politicians were more reasonable than professors. A strange assessment, with Adolf Hitler on the scene. Einstein could hardly avoid knowing of his recent exploits. When Hitler urged the destruction of Marxists and Jews, twenty-seven thousand Nazi party members had roared approval and pledged their support. Many more Germans ridiculed him as an ignorant, raucous, Charlie Chaplin look-alike. Nevertheless, the press took him seriously. His trial for treason—a grab for power by force—had made the front pages of every major newspaper in Germany for three weeks.

Dubbed "the wild beast" by an opponent, Hitler had been freed from Munich's Landsberg prison after serving nine months of a five-year sentence, getting out early by promising to stifle his cries for blood. But when his notoriety brought new recruits to his Nazi party, and four thousand of them packed a Bavarian beer garden to hear him, he couldn't resist repeating his threats of annihilation. Having broken his word, he was forbidden to speak in public for two years. He seemed to have been tamed and his Nazi party appeared to be heading for oblivion.

Now, instead of issuing a call to arms, the disgruntled ex-corporal took in the Bayreuth Festival, changing from his daytime leather shorts into dinner jacket or tails to attend the opera at night. Afterward, he chatted with the performers in the theater restaurant. He also had a book in the works, started in prison, originally titled "Four and a Half Years of Struggle Against Lies, Stupidity and Cowardice," then simplified to *Mein Kampf*—"My Struggle." One way or another he meant to spread his venom.

While Hitler thrilled to Wagner's martial music glorifying pagan heroes, Einstein joined Mahatma Gandhi, the advocate of nonviolence, in signing a manifesto against compulsory military service.

CHAPTER 19

The Uncertainty Principle

1926 to 1927

46 to 48 years old

The day after Count Kessler went to a Berlin theater to watch "a ribald farce with political highlights and lots of sturdy sex stuff," he threw a dinner party for the Einsteins. Albert entered the house looking extremely dignified in a dinner jacket, until Kessler noticed that he had spoiled the picture—but justified his "Bohemian" image—by wearing heavy boots. He had put on a little weight— perhaps after a recent trip to Paris where he lectured before the Franco-Palestine Society—but his host was pleased to see that his eyes still sparkled "with almost childlike radiance and twinkling mischief."

Elsa kept the table amused with tales of her husband's indifference to the medals he had been awarded. She had had to remind him repeatedly to go to the Foreign Ministry to pick up two gold medals awarded by the British Royal Society and Royal Astronomical Society. When they met afterward to go to the movies and she asked him what the medals looked like, he had no idea. He hadn't even bothered to open the packages. Recently Niels Bohr had won the American Barnard Medal, which was awarded every four years to an outstanding scientist. When Einstein read a news report that he, Einstein, had been the previous winner, he showed the article to Elsa and asked her if it was true. He had forgotten. Walther Nernst, amused by Einstein's reluctance to wear his Pour le Mérite medal at a meeting of the Prussian Academy, remarked, "I suppose your wife forgot to lay it out for you," adding jokingly, "Improperly dressed!" "No, she didn't forget," he replied. "I didn't want to put it on."

The dinner party was a great success, with some guests lingering until two the next morning. During the animated conversation, Countess Sierstorpff told how she had been converted to pacifism. Then "a discussion ensued . . . about the Sirius moon. Einstein explained the sensational discovery of its gravity and the significance of this for the deflection of red in the solar spectrum." A nephew of the great physicist Heinrich Hertz was also at the party. During the evening,

Einstein puzzled him and everyone else with his enigmatic remark, "Your uncle wrote a great book. Everything in it was wrong, but it was nonetheless a great book."

American physicist Dayton C. Miller of the Case Institute School of Applied Science said more or less the same thing about Einstein's work, telling an April 28, 1925, meeting of the National Academy of Sciences in Washington, D.C., that his over one hundred thousand observations proved the existence of the ether. Informed of this, Einstein casually replied in a letter to Besso: "Experiment is the supreme judge. If Dr. Miller's results should be confirmed then the special relativity theory, with the general theory in its present form, falls."

That same spring Werner Heisenberg addressed the physics colloquium at the University of Berlin on the controversial quantum mechanics. Afterward, Einstein invited Heisenberg to his home to discuss the subject in detail. As soon as they got inside the apartment, Einstein said: "What you told us sounds extremely strange. You assume the existence of electrons inside the atom, and you are probably quite right to do so, but you refuse to consider their orbits, even though we can observe electron tracks in a cloud chamber . . . You don't seriously believe that none but observable magnitudes must go into a physical theory?"

Somewhat surprised, Heisenberg responded: "Isn't that precisely what you have done with relativity? After all, you did stress the fact that it is impermissible to speak of absolute time, simply because absolute time cannot be observed; that only clock readings . . . are relevant to the determination of time."

Einstein admitted that he may have used that reasoning, but added, "It is nonsense all the same . . . In reality the very opposite happens. It is the theory which decides what we observe."

Though astonished by Einstein's attitude, Heisenberg found his argument convincing. Nevertheless, he suggested that they wait to see how the atomic theory developed. Einstein looked skeptical and asked, "How can you really have so much faith in your theory when so many crucial problems remain completely unsolved?"

This floored Heisenberg for a time, because the same was true of Einstein's unified field theory. Then he said, "I believe, just like you, that the simplicity of natural laws has an objective character . . . It ought to be possible to think up many experiments whose results can be predicted from the theory. And if the actual experiments should bear out the predictions, there is little doubt but that the theory reflects nature accurately."

Einstein agreed that experiment was essential to validate any theory but conceded that "one can't possibly test everything." He summed up his conclusion in a frequently quoted letter to Max Born: "Quantum mechanics is certainly imposing. But an inner voice tells me that it is not yet the real thing. The theory says a lot, but does not really bring us any closer to the secret of the 'old one.' I, at any rate, am convinced that He is not playing dice." In other words, Einstein was and would remain a determinist.

In July, Einstein took time off from scientific concerns to compose a seventieth-birthday tribute to George Bernard Shaw, praising him as a reformer who used his masterly wit and artistry to make foolish contemporaries see themselves for what they were.

A friend of his youth, Professor Alfred Stern, turned eighty in the fall, and Einstein sent him an affectionate letter recalling their enthusiastic discussions. Stern had been a Zurich Polytechnic historian; he believed that historians should try to be as objective as scientists. "I spent my most harmonious hours in your family circle and I often look back upon those days with pleasure," Einstein wrote. "My most enchanting memory, however, was and is the sight of a good man who, in a long, happy and fruitful life, has achieved the goal he set himself in his youth." He was referring, perhaps, to Stern's twelve-volume history of Europe.

The next task on his list was to come to Besso's rescue. Besso was in imminent danger of being fired from his Patent Office job because of low productivity. No one knew Besso's strengths and weaknesses, or the demands of the job, better than Einstein. He was careful not to deny what was apparent to all: Besso rivaled Hamlet in his inability to make up his mind. Yet he was highly intelligent and devoted to his work.

His indecisiveness, Einstein wrote to the government authorities, explained why few dossiers in the Patent Office had his name on them:

> Everyone at the Patent Office knows that one can get advice from Besso on the difficult cases; he understands with extreme rapidity both the technical and the legal aspects of each patent application, and he willingly helps his colleagues to arrive at a quick disposal of the case in question, because it is he, in a manner of speaking, who provides the illumination and the other person the necessary spirit of decision. But when it is up to him to settle the matter, his lack of decisiveness is a great handicap. This has resulted in a tragic situation: one of the most precious employees of the Patent Office, one I would qualify as irreplaceable, gives the impression of being ineffective.

The compassionate and perceptive letter saved Besso's job.

Sigmund Freud and his wife spent the New Year in Berlin with their children and grandchildren and were about to return to Vienna when Einstein and Elsa stopped by. This was the first face-to-face encounter between the great seminal thinkers of their time: the investigator of cosmic mysteries and the investigator of human mysteries.

Freud was in poor shape. He was seventy now, and a series of painful operations for cancer of the palate—the cost of his addiction to cigars—had left him unable to talk without the aid of an ill-fitting artificial palate. He was also almost deaf in his right ear. However, by sitting on Einstein's right, Freud was able to hear enough to declare the forty-eight-year-old physicist "cheerful, sure of himself and agreeable," and their two-hour talk "very pleasant," though it skirted their deepest interests because "Einstein understands as much about psychology as I do about physics."

Freud's once keen, probing eyes were somewhat dimmed, but not his "enchanting humor." His conversational style echoed Einstein's, being direct and lucid: " 'Even in the most technical discourse, [Freud's] humor and informality kept breaking through. He was fond of using the Socratic method. He would break off his formal exposition to ask questions or invite criticism. When objections were forthcoming he would deal with them wittily and forcibly.' "

Though Einstein's self-confidence and exuberance pleased Freud, it highlighted what he perceived as Einstein's smooth road to success in contrast with his own never-ending battles. Soon after this meeting, Freud wrote to his friend and colleague, Marie Bonaparte, that Einstein was lucky, "because he has had a much easier time than I have. He has had the support of a long series of predecessors from Newton onward, while I have had to hack every step of my way through a tangled jungle alone. No wonder that my path is not a very broad one, and that I have not got far on it." At other times, however, Freud would acknowledge having received help in his trek through the jungle from Jean Martin Charcot, Josef Breuer, and Wilhelm Fliess, not to mention Schopenhauer and Plato.

Though Einstein admired Freud's way with words, he was wary of psychoanalysis and not tempted to submit to his treatment, "because I should like very much to remain in the darkness of not having been analyzed."

Nor apparently did Einstein seek Freud's advice about Eduard. His often sickly younger son continued to play the piano with a strange, mechanical intensity that hinted at a serious emotional disturbance. In fact, both men had mentally troubled sons. Freud described his engineer son, Oliver, as extraordinarily gifted and with a flawless character until "the neurosis came over him and stripped off all the bloom."

Einstein also missed the chance for a Freudian explanation of why hordes of people incapable of understanding his ideas threatened the quiet contemplation he craved to pursue his work by chasing after him. Are they crazy or am I? he wondered.

The popular press reflected this insatiable interest, and reporters joined the hunt as if he were a rare animal—which, in a way, he was. With the explorer of the unconscious at hand, he might have discovered why he stood out among other prominent scientists as the focus of such feverish interest. Einstein halfseriously speculated that he himself was to blame; that elements in his makeup of the charlatan, the hypnotist, or even the clown inadvertently attracted attention. Although he suspected that he might unconsciously be inviting the hunt, he avoided any psychoanalytical probing, even from the master.

Shortly after their first meeting on January 2, they had a brief exchange of correspondence in which Freud again called Einstein lucky. How can you call me lucky, Einstein replied, without knowing the inside of my mind? To which Freud responded: Because you can work at mathematical physics and not at psychology, where everyone thinks they can have a say.

This was not the end of the Einstein-Freud correspondence; their future exchanges, however, focused on world peace and their mutual concern for fellow Jews.

After visiting the Freuds, Einstein found an effective way of avoiding crowds and unwelcome visitors—he went into hiding. Elsa, Margot, and Otto, the doorman of their apartment building, were recruited to keep reporters at bay and off his track. Elsa screened incoming phone calls, declining interview requests, while Margot and the vigilant doorman warned Einstein when he was in danger of ambush from lurking fans or newsmen. If he had to go out, he gave them the slip by using the rear entrance.

However, Dimitri Marianoff, a tenacious Russian-born reporter, was undeterred by the Einstein family's tactics. Given a tip that a dancing instructor had a studio directly under Einstein's apartment, Marianoff signed up as a prospective customer. He even took a few lessons, hoping to confront Einstein in the entrance hall or the creaky little lift. He was out of luck but not out of ideas, finally meeting his quarry by dating Einstein's stepdaughter, Margot. That was a feat in itself, as she was scared of strangers.

Marianoff was careful at this first meeting not to show his hand too soon, by discussing such subjects as the mutually admired Tolstoy and avoiding personal questions. Time and familiarity lowered Einstein's guard. Marianoff became such a frequent visitor, with Margot as his passport, that Einstein began to confide in him, admitting that the general theory of relativity came to him in a vision. After years of futile calculations, convinced that his quest was hopeless, he said he had gone to bed deeply depressed. Suddenly the answer appeared "with infinite precision, and with its underlying unity of size, structure, distance, time, space, slowly falling into place piece by piece like a monolithic jigsaw puzzle. Then, like a giant die making an indelible impress, a huge map of the universe outlined itself in one clear vision."

Marianoff wanted more than a few anecdotes; he had an Einstein biography in mind, written from the inside as only one of the family could know it. This necessitated a continued courtship of Margot and led to their eventual marriage. Through being so often on the spot he saw for himself how Einstein interacted with his stepdaughter. On one occasion he witnessed a peculiar scene. Two important visitors arrived unexpectedly, but instead of greeting them, to Marianoff's astonishment, Margot hid under a table. Einstein saw her dive for cover; rather than flush her out, he carefully arranged the tablecloth so she couldn't be seen. She stayed there out of sight until the visitors had left. Such farcical behavior reflected Einstein's empathy for Margot's almost pathological shyness. He, too, was extremely shy, and was often uneasy when meeting strangers. Among friends, mostly doctors and fellow physicists, he was lively, outgoing, and fun. He also enjoyed warm relationships with several of their wives.

One of the warmest was with Hedi Born, the playwright wife of physicist Max Born, who treated Einstein with affection and generally shared his political and social views. She believed, perhaps uniquely, that his mastery of life surpassed even his scientific achievements. This first struck her when she visited him during one of his serious illnesses and he said, "I feel so much a part of every living thing that I am not in the least concerned with where the individual begins and ends."

A telling confession, but hardly a fitting philosophy for a would-be drama critic. Nevertheless, Hedi reminded him of it when she sought his opinion of her recently completed play, *A Child of America*, a satirical look at Americans. Einstein dropped his own work to read the manuscript, then wrote back to Hedi, dismissing the characters as puppets yet comparing her favorably with Bernard Shaw. He concluded that her wit saved the play, which he promised to recommend to the Berlin State Theater manager as witty, amusing, and up-to-date. As it happened, Shaw considered Einstein to be a superb theater critic, rating his epistolary review of *Saint Joan* the best he'd ever read.

Einstein was also a marriage critic with a devastating sense of timing. On the eve of Hans Albert's marriage to Frieda Knecht in Dortmund he tried to stop the ceremony, warning them that separation was inevitable and divorce painful. But on this occasion Hans Albert proved even more stubborn than his father, and the wedding took place as planned.

After his failed mission, Einstein eagerly returned to Berlin to relax aboard the yacht of his best friend in the city, Professor Moritz Katzenstein, a dedicated surgeon and—that rarity—an Einstein confidant. They were as close as loving brothers, discussing, said Einstein, their

> experiences, ambitions and emotions. If, as was invariably the case, he had performed some dangerous operations in the morning, he would telephone immediately before we got into the boat to enquire after the condition of the patient about whom he was worried. I could see how deeply concerned he was for the lives entrusted to his care. It was marvelous that this shackled outward existence did not clip the wings of his soul; his imagination and his sense of humor were irrepressible. How happy he was when he had succeeded in making somebody fit for normal life . . . And the same when he avoided an operation . . . We both felt that this friendship was a blessing because each understood the other, was enriched by him, and found in him that responsive echo so essential to anybody who is truly alive.

Back home from the yachting trip, he made up for lost time by closeting himself in the study. At dinnertime, Elsa had to call him repeatedly before he appeared, grumbling about being interrupted, even for his favorite meal of lentil soup and sausages. "I don't care if you've got lots of work to do," she said. "Sit down and eat. People have hundreds of years to find things out, but your stomach can't wait that long."

When told she had accepted a dinner invitation, he complained, "Why do you do this? My time is so short. Why don't you ask me first?"

"If I asked you," she said, "you'd never go."

At that dinner the company discussed the plan of fellow guest Gerhart Hauptmann to add a scene to his latest play, in which a character literally died of laughter. Then someone mentioned astrology. Hauptmann thought there was something to it. Einstein rejected it utterly. He maintained that a belief in astrology was as nonsensical as a belief in demons, and pointed out that Copernicus had conclusively refuted the view of our earth as the center of the universe. "That was probably the severest shock man's interpretation of the cosmos ever received," he went on, because "it reduced the world to a mere province so to speak, instead of it being the capital and center."

Another guest, Count Harry Kessler, recorded the conversation in his diary, adding,

> Kerr, who sat listening with his vulgar little wife, constantly interrupted with facetious remarks which he thought witty but which were not even funny. The subject of God was a special butt for his derision. I tried to silence him and said that, since Einstein is very religious, he should not needlessly hurt his feelings. "What?" exclaimed Kerr. "It isn't possible! I must ask him right away. Professor! I hear that you are supposed to be deeply religious?" Calmly and with great dignity, Einstein replied, "Yes, you can call it that. Try and penetrate with our limited means the secrets of nature and you will find that, behind all the discernible concatenations, there remains something subtle, intangible and inexplicable. Veneration for this force beyond anything that we can comprehend is my religion. To that extent I am, in point of fact, religious."

This is arguably the best explanation Einstein ever gave of his religious attitude. When the talk reverted to Copernicus and Kessler suggested that Einstein's views of the cosmos were equally revolutionary, he curtly demolished that idea: "There is *nothing* so revolutionary about my observations."

That could hardly be said for the observations of Bohr and Heisenberg in Copenhagen. They were in shock from their own ideas, which shattered the once-comforting belief that in time the nature of the universe would be completely understood. Those ideas would soon confront Einstein with a challenge he would address for the rest of his life. Instead of feeling elated by their discovery, Heisenberg was in despair and Bohr alarmed, and they made repeated but futile attempts to disprove themselves.

The atmosphere at Bohr's institute was usually hilarious. In his forties, Bohr was a jovial father figure to scores of students, most in their twenties, from all over the world. He relaxed by watching cowboy films, but he always needed a couple of students to go with him to explain the complicated plots, recalled George Gamow, a student from Russia.

> His theoretical mind showed even in these movie expeditions. He developed a theory to explain why although the villain always draws first, the hero is faster and manages to kill him. The Bohr theory was based on psychology. Since the hero never shoots first, the villain has to decide when to draw, which impedes his action. The hero, on the other hand, acts according to a conditioned reflex and grabs the gun automatically as soon as he sees the villain's hand move. We disagreed with his theory, and the next day I went to a toy store and bought two guns in Western holsters. We shot it out with Bohr. He played the hero, and he "killed" all the students.

There was little opportunity or inclination now for playfulness. In the fall they were expected to present their new ideas to their most formidable critic, Einstein. For two months Bohr and Heisenberg argued, repeated experiments, invited students and colleagues to throw them tough questions, called on physicist Wolfgang Pauli for advice, and became so lost in their work that they almost had to be force-fed. The weird, even uncanny world they had uncovered seemed to be the joint creation of an illusionist and the designer of a gambling casino. It was too much for Bohr. He left in mid-argument for a skiing vacation in Norway, hoping to clear his mind and come up with something more plausible.

He came back with a new approach to the problem, bounding up the institute's steps two at a time to share the news with Heisenberg. He now believed that he could justify Heisenberg's uncertainty principle, which showed nature to be as tricky as a cardsharp, with a foolproof way of hiding its fundamental secrets: the moment you grasped one part of the puzzle, another escaped. For example, the force of the light needed to observe an electron orbiting inside an atom knocked the electron completely out of the atom. You could get an accurate reading of the electron's position or velocity, but not both simultaneously. Both the initial position and the velocity of the electron were needed to forecast its future activity. This, it seemed, was an intrinsic aspect of nature which could never be overcome.

Einstein, almost alone, resisted the pessimistic conclusion. Max Born was afraid that his friend was permanently isolating himself from the current generation of physicists. Heisenberg understood Einstein's resistance, knowing of his "conviction that the world could be completely divided into an objective and a subjective sphere, and the hypothesis that one should be able to make precise statements about the objective side of it formed a part of his basic philosophical attitude."

Bohr agreed with Heisenberg that the quest for an unambiguous definition of the energy of an atom would have to be abandoned, and he resolved the dilemma by suggesting that instead of aiming for the impossible, one should make the best of it. The knowledge of the position of a particle, he said, was to be taken as complementary to the knowledge of its velocity. A greater knowledge of these aspects when separated would give the most accurate picture possible of the subatomic world when they were combined.

Even more bewildering was the behavior of the electron, which seemed to be both wave and particle. When instruments were set up to measure it as a particle, it was obligingly a particle, but when the instruments were arranged to sight a wave, it became a wave. Instead of regarding the dual wave-particle nature of the electron as another problem, Bohr saw it as a path to enlightenment illuminated by his complementarity theory.

> "However contrasting such phenomena may at first sight appear," he said, "it must be realized that they are complementary, in the sense that together they exhaust all information about the atomic object which can be expressed in common language without ambiguity." . . . Essentially the theory which Bohr and Heisenberg came up with was that there are two truths rather than one alone . . . and furthermore that the two together offer science and man a more complete view and understanding of the atomic world than either would offer separately, and thus in the end a clearer view of the visible world built out of the invisible substratum of the atom. Instead of division, Bohr showed the parts, the

divisions, the components can be combined into a harmony greater than that of the sections. Each separate, even contradictory, aspect complements the other.

When Bohr's students saw his complementarity as a paradox rather than a solution, he quoted the Abbé of Galiana: "One cannot bow in front of somebody without showing one's back to somebody else."

The combined Bohr–Heisenberg ideas, known as "the Copenhagen interpretation of quantum mechanics," postulate not only uncertainty and duality but also the bizarre, ghostly behavior of matter that is created out of nothing, then just as mysteriously vanishes. On the positive side, it "explains why atoms and molecules keep their identity, their shapes and their patterns in spite of collisions and perturbations, why gold is gold wherever we find it, and in the last instance, why the same flowers bloom every spring."

Einstein was largely unaware of Bohr and Heisenberg's intense efforts to clinch their schizophrenic picture of atomic nature. He missed the September conference near Italy's Lake Como, where they fine-tuned their theory. Almost everyone interested in the subject was there except Einstein. He refused to attend because the conference was sponsored by Mussolini's Fascist government. However, he was expected at the upcoming Fifth Solvay Conference in Belgium, when Heisenberg and Bohr would again air their ideas. The big question was: would Einstein be won over?

Bohr arrived at the conference in October, confident that there were few if any questions he couldn't answer. He had encouraged students and colleagues to raise every possible objection to his complementarity principle, then countered them. After all, his theory, however revolutionary, was a development of Einstein's own discovery: as long ago as 1905 he had shown that a photon was both particle and wave.

Lorentz, the conference chairman, who had only a few months to live, asked Bohr to spell out the problems facing quantum theory. Bohr was not the easiest man to follow because of his soft voice and indistinct pronunciation. With those handicaps, he began "the strangest debate in the history of the understanding of the world," about what John Wheeler called "the most remarkable scientific concept of this century."

Bohr outlined the conclusions he and Heisenberg had reached after much intense thought, crowding a blackboard with diagrams and equations. The room was silent when Einstein rose to respond with a low-key but emphatic rejection of quantum theory as a poor attempt to destroy determinism. He later characterized Bohr and his supporters as walking on eggs to avoid physical reality.

Before he sat, several scientists stood and shouted in various languages for permission to speak next. As Lorentz pounded his gavel, trying to regain control, Ehrenfest walked to the blackboard and chalked, "The Lord did there confound the language of all the earth." It wasn't clear whether he meant the babble of voices, Bohr's indistinct delivery, or the uncertainty principle itself, but it brought a roar of laughter and a lighthearted conclusion to the first day's discussion.

The conference members all stayed at the same hotel, and the next morning during breakfast Einstein went on a good-humored attack, raising a series of ingenious objections to quantum theory. Bohr held his fire until that night during dinner, when he refuted all of Einstein's arguments. As Heisenberg later recalled,

> Einstein . . . look[ed] a bit worried, but by next morning he was ready with a new imaginary experiment more complicated than the last, and this time, he avowed, bound to invalidate the uncertainty principle. This attempt would fare no better by evening, and after the game had been continued for a few days, Einstein's friend Paul Ehrenfest . . . said: "Einstein, I am ashamed of you: you are arguing about the new quantum theory just as your opponents argue about relativity theory." But even this friendly admonition went unheard. Once again it was driven home to me how terribly difficult it is to give up an attitude on which one's entire scientific approach and career have been based . . . Einstein was not prepared to let us do what, to him, amounted to pulling the ground from under his feet.

Joining Einstein for a stroll one evening, Louis de Broglie, a leading contributor to quantum theory, was won over but not converted by Einstein's sweet disposition, "his general kindness, by his simplicity, and by his friendliness. Occasionally, gaiety would gain the upper hand and he would strike a personal note . . . Then again, reverting to his characteristic mood of reflection and meditation, he would launch into a profound and original discussion of a variety of scientific and other problems. I shall always remember the enchantment of all those meetings, from which I carried away an indelible impression of Einstein's great human qualities." Einstein was impressed, too, recommending the Frenchman for the Nobel Prize in physics. He won it in 1929 for finding the wave character of electrons.

After the conference, Bohr, Einstein, and a few others went to Ehrenfest's home, where Einstein resumed the verbal battle with Bohr. Ehrenfest, who loved both men, was upset to see one of his heroes unwilling to accept the rapidly developing quantum theory. With tears in his eyes, he told a colleague, Samuel Goudsmit, that "he could not but agree with Bohr."

Homeward bound from Belgium, Einstein traveled as far as Paris with de Broglie, and they had what was to be their last conversation on a Gare du Nord platform. Einstein said that except for the math involved, all physical theories should be simple enough for a child to grasp. But he feared that at forty-eight he was too old to tackle the problems of quantum physics. He seemed to be handing on the torch, or perhaps a road map, to de Broglie, leaving him with an encouraging, "Carry on! You are on the right road."

When Einstein reached his apartment, Elsa found him tired and unusually subdued. When an invitation came from Millikan to visit Caltech, he declined, and admitted he was completely exhausted.

CHAPTER 20

The Perfect Patient

1928

---

49 years old

Early in 1928 Einstein headed for Holland and the funeral of Hendrik Lorentz, who had died just two months after chairing the Solvay conference. For a self-proclaimed "lone traveler," Einstein was deeply attached to a large number of friends. He had all but worshiped Lorentz, saying at one time, "I admire this man as no other, I would say I love him." To Lorentz himself, he had written, "I feel an unbounded admiration for you." Einstein acknowledged the debt he owed the Dutch scientist for laying the groundwork for his special relativity theory, and told his future biographer, Abraham Pais, that Lorentz was the most well-rounded and harmonious personality he ever met.

Representing the Prussian Academy of Sciences at Lorentz's funeral, Einstein told fellow mourners,

> "I stand at the grave of the greatest and noblest man of our times. His genius was the torch which lighted the way from the teachings of Clerk Maxwell to the achievements of contemporary physics . . . His life was ordered like a work of art down to the smallest detail. His never-failing kindness and magnanimity and his sense of justice, coupled with an intuitive understanding of people and things, made him a leader in any sphere he entered. Everyone followed him gladly, for they felt that he never set out to dominate but always simply to be of use. His work and his example will live on as an inspiration and guide to future generations."

Just two years before his own death in 1955, Einstein said, "He meant more to me personally than anybody else I met in my lifetime."

A letter from Besso was waiting for him when he returned home from the funeral. Obviously hurt by Einstein having mentioned during one of their recent conversations Besso's failure to earn a doctorate, he spelled out what he had done instead: "I helped you work out things in 1904 to 1905 and took away part of your glory but gave you a friend in Planck. I defended Judaism and the Jewish family when your private life had taken a difficult turn, and I brought Mileva back from Berlin to Zurich." Now he said that he felt deeply and willingly involved in the welfare of Einstein's sons and implied that Einstein was neglecting them.

Zangger, who for years had also been an Einstein family friend, and, like Besso, was now a surrogate father for the boys, was also critical of Einstein's attitude toward his sons. He remarked that in trying not to appear soft Einstein went too far in the other direction. Whereas Zangger, for example, had been impressed by Hans Albert's math thesis for his doctorate, Einstein's only reaction had been to point to part of it and say, "You could have said it a little better here."

Soon after his journey to Holland for Lorentz's funeral, Einstein responded to an appeal from tuberculosis patients in a sanatorium in Davos, Switzerland, most of them young students whose education had been interrupted by the illness. Now they were eager to hear the latest scientific speculations.

He broke his journey on the way to Davos for a reunion with his high-school friend Hans Byland. Hans, his wife, and his son, Willy, a talented painter, were waiting for him at the railroad station. Willy watched his father march "gaily over to Professor Einstein who looked very bewildered. Then he let drop both his suitcases on the platform and embraced his old school comrade." They went to the nearby Byland home in Chur, where Einstein played his violin, smoked his pipe, and listened to Byland relay a colleague's complaint that Einstein's writings were incomprehensible. "If I write with too much brevity not a soul understands it," he replied, "if I'm too long-winded they can't see the wood for the trees, so I try a path between the two extremes." He regretted there was so little time that much of what he wanted to do would be unfinished.

At Davos, he spoke of Newton's ideas being the springboard for field theory and relativity. Now, he said, strict causality is "threatened . . . by the representatives of physics." Many of those same scientists, studying the structure of atoms, claimed that "all natural laws are in principle of the statistical variety, and our imperfect observation practices alone have cheated us into a belief in strict causality." This new theory, Einstein concluded, defined radiation and matter as both corpuscular and wavelike, "a new property of matter for which the strictly causal theories hitherto in vogue are unable to account." Following his talk, he helped raise funds for the sanatorium, playing his violin on stage and getting a laugh from the audience when he held up the musical score by Schubert to receive their applause.

In Switzerland he stayed with a friend, Willy Meinhardt, who once remarked as he handed out topcoats to guests departing after a dinner party, "This must be Einstein's—it's from ——— ," naming a cheap Berlin tailor. This was a statement of fact rather than a put-down, as another friend, Janos Plesch, confirmed: "[Einstein] attaches no importance to outward show and this applies in particular to clothing; any old suit and shoes will do as long as they are comfortable . . . He was not prepared to spend time on beautification, but neatness, cleanliness and a smooth face and chin were part of his duty to the rest of the world."

From Meinhardt's home Einstein traveled to Leipzig as an expert witness in a patent dispute. On his return to Switzerland in early April 1928, he arrived late at night, when none of Meinhardt's household staff was around to carry his heavy suitcase. So Einstein lugged it up a steep hill over slippery snow, until he collapsed. Meinhardt feared it was a heart attack.

Einstein was helped to the house and returned by easy stages to Berlin, where Janos Plesch took charge. Plesch had dedicated his book *The Physiology and Pathology of the Heart and Blood Vessels* to Einstein. Doctor and patient discussed the symptoms and possible causes as if both were specialists. In fact, Plesch valued Einstein's medical insight, which was informed by frequent conversations with his talented surgeon friend, Professor Katzenstein. Einstein believed he had been risking his heart for some time, and not just from overworking. Whenever the wind dropped while he was sailing, instead of waiting for it to pick up again, he invariably rowed vigorously to his destination. He assumed that carrying the case had been the last straw. Plesch agreed. He discounted a heart attack and diagnosed acute dilation of the heart through excessive physical effort, placing Einstein on a strict salt-free, no-smoking regime with indefinite bed rest.

Plesch couldn't have had a better patient: Einstein was obedient, trusting, and grateful for

> what was being done for him . . . He once explained that he quite realized "our primitive thought must necessarily be inadequate in face of such a complicated piece of mechanism as the human body and that the only proper attitude is patience and resignation, supported by good humor and a certain indifference to one's own continued existence." He willingly carried out whatever instructions I gave, at the same time watching the phenomena of his sickness and carefully observing the effect of my treatment.

But Einstein exhausted Elsa. After several days of being nurse, housekeeper, secretary, and business manager, she began looking for a relief secretary. Why not advertise for help? her friend Rosa Dukas asked. Elsa feared that an advertisement would attract celebrity hunters and reporters posing as job applicants. Rosa then suggested her younger sister, Helen.

At thirty-two, Helen Dukas was a tall, dark, attractive woman with a charming smile and a direct, lively manner. Her father was a wine merchant. Her mother had died when she was a child. Helen had recently lost her job with a publisher who had gone bankrupt. When Rosa told her of the job offer, Helen said, "You have gone mad, I can never do anything like that." Elsa urged her at least to try.

Helen was petrified at the prospect of meeting Einstein, intimidated by his fame, and afraid she'd reveal her ignorance if he grilled her about science. Her sister had to talk her into keeping her interview with Einstein. And so, on April 13, with legs atremble, Helen arrived at Einstein's Berlin apartment. Elsa welcomed her with tea and cakes, assuring her that she had nothing to worry about, because "my husband is very nice to everyone." More reassuring was the news that a knowledge of physics was not a job requirement.

Elsa led the way to the bedroom where Albert was laid up. When she introduced Dukas to the frail invalid, he took her outstretched hand, saying, "Here is an old child's corpse." His good-humored, simple manner immediately put her at ease.

"He was like that with everyone," Dukas recalled. "It was one of his great gifts. You felt at peace with him."

Dukas got the job and immediately started work on Friday the 13th — a good omen. She was with him as a kind of girl Friday par excellence for twenty-seven exciting years, until his death in 1955. She "had nothing to do," she said "with his scientific work. My function was to attend to his private correspondence. I was his chief cook and bottle washer." She was also his loving friend and confidante.

"She devoted her life to him," said their mutual friend, Thomas Bucky. "She helped with answering the flood of letters he got: he gave her the gist of the answers and she'd do the rest. She was more intelligent than his wife and was exactly what Einstein was not. She was Madam Trivia. She knew the whole world, everything that went on in the movies and on the radio which, of course, he was completely above. That knowledge wasn't of any use to him . . . She knew what film was playing at the Roxy, who was divorcing whom, what were Bertrand Russell's views on Eskimo marriage customs, in what ocean was the island of Psumbe, and she had always read the latest best-selling novel as well as the latest detective story." And one of her greatest assets was her ability to keep people away from him so he could get on with his work.

Dukas was treated as one of the family, and "understood her boss well. One night she had a dream which I think portrays Einstein's character beautifully," said Bucky. "This was the dream: Einstein was eating in a restaurant when suddenly a holdup man came in and lined everyone up against the wall, Einstein included. Then the holdup man started at one end and began to take money and watches and other valuables from people as he went down the line. When the robber reached Einstein, he said, 'Oh, no, I couldn't take anything from you, Professor Einstein!' And Einstein replied, 'That is terrible. I want to be treated like everyone else.' And in the dream Einstein furiously emptied his pockets, producing just one dime."

After Einstein had been confined to his bed for four months, Elsa allowed a few students into his room, warning them not to overexcite him by discussing physics. She monitored the meeting, standing just outside the door, to make sure he wasn't talking too much. She hadn't reckoned with her husband's ingenuity, though. When the young men left, Elsa found his bedsheet covered in ink where he had scribbled equations.

Einstein was still frail, but out of bed and tottering around in an old suit and sandals, when Dr. R. H. Furth called a few weeks later with an invitation for him to address the German Physical Society. To Furth's distress, Einstein said he did not ever expect to be fit enough to speak in public, although he still kept abreast of scientific news, especially the continued contention by Bohr and Heisenberg that the observer determines whether a photon reveals itself as a wave or a particle.

Despite his fundamental disagreement with Schrödinger and Heisenberg, Einstein recognized their genius and recommended that they share the 1929 Nobel Prize in physics. He had also recommended de Broglie, who won the award that year. Heisenberg had to wait for the prize until 1932 and Schrödinger until 1933.

Einstein liked to bounce his ideas off other physicists and to hire "young collaborators, predominately mathematicians, to help with difficult mathematical investigations." The few suitable candidates at Berlin University were concerned with launching their own careers, working on doctorates, or boning up for exams to become physics teachers. One of his assistants, Jakob Grommer, a grotesquely deformed Russian Jew, had worked with Einstein for several years and eventually hoped to become a teacher. Despite Einstein's recommendations, no one else was willing to hire him, because of his appearance. "Nevertheless," Abraham Pais later wrote, Grommer "blamed Einstein for not trying hard enough and finally quarreled with him." Grommer left Einstein in 1928; the following year, he found a job in Minsk, Russia, and was eventually elected to the Byelorussian Academy of Sciences.

In the fall of 1928, Cornelius Lanczos joined Einstein and was immediately asked to try to solve a tricky equation. This new assistant, according to L. L. Whyte, a young British physicist studying in Berlin,

> was intensely proud of being given such a task by Einstein, and went away feeling very humble and unsure. But he studied the equation and after three or four days—flash!—there came into his mind the perfect solution. It had all the three properties Einstein had asked for. Now, to Lanczos this was really something very extraordinary, because he was a very humble, religious type of personality and he felt that this inspiration had come from the heavens to him; and that he should be able to bring Einstein in his first week in Berlin such a success moved him deeply . . . He showed it to Einstein. Einstein looked at him and said, "Yes, very interesting, quite remarkable." There was a short silence, and then he exclaimed rather impatiently, "But don't you see, I gave you the wrong equation. It was quite wrong!" There was a silence. These two highly intelligent and sensitive men did not need to say anything, for they knew what a terrible thing had happened. What actually followed was that Einstein went and fetched his violin. Lanczos went to the piano, and they played Bach together for the rest of the hour. This story I heard from Lanczos within a week or two of it happening.

It was almost a year after Einstein's collapse before he was able to return to the Kaiser Wilhelm Institute, where he spent the mornings conferring with colleagues and his assistant, Lanczos. "Then he rested after lunch and had no appointments until three P.M.," said his secretary, Helen Dukas. "Otherwise he was always working, sometimes all through the night. Then all he'd eat would be a bowl of his favorite pasta, macaroni."

When the press reported that he was on the verge of a great discovery, Dukas stepped up her protective role, guarding him from unwelcome visitors aroused by the news, especially reporters. This did not apply to people in trouble. She continued steering him to those she knew he would want to help, such as Philippe Halsman, a twenty-two-year-old Jew serving a ten-year term in an Austrian prison, having been convicted of murdering his father. All of his family were certain of Philippe's innocence, and his teenage sister, Liouba, wrote to Einstein that the only reason for the sentence was the prevailing anti-Semitism.

Einstein had no doubt that an Austrian jury could send an innocent Jew to prison; the Austrians were among the most virulent anti-Semites in Europe. Nearer to home, Nazi thugs had recently looted scores of Jewish stores in Berlin and roughed up the owners. One hundred writers had ganged up on Einstein, publishing a book titled *Hundert Autoren Gegen Einstein* (One hundred authors against Einstein) that damned him a hundred different ways, but he knew that their animosity really sprang from his being a celebrated, proud, outspoken Jew. He was wryly amused at their attempt to sink relativity by the weight of numbers. As he said, had he been wrong, just one with the right ideas could have done the trick.

Some months before, responding to an anti-Zionist article by a Professor Hellpach, Einstein had written about how, over the years, he had seen "worthy Jews basely caricatured, and the sight made my heart bleed. I saw how schools, comic papers, and innumerable other forces of the Gentile majority undermined the confidence even of the best of my fellow-Jews, and felt that this could not be allowed to continue." So he could easily imagine himself in the prisoner's shoes—hadn't he been falsely accused of almost everything short of murder?

He wrote to Halsman's sister asking for details. The Austrian press had covered the story, and with these reports and further information from his sister, Einstein had the facts before him. Philippe Halsman, a university student studying electrical engineering in nearby Dresden, had been on a family mountain-climbing vacation in the Tyrolean Alps near Innsbruck. His father, Max, a dentist in Riga, had been suffering from dizzy spells, though these were not severe enough to call off an afternoon hike. Instead, he walked at a slower pace, resting frequently, while Philippe led the way, occasionally glancing back. At one point, losing sight of his father, he retraced his steps and found him lying unconscious in a brook. Apparently he had fallen from the mountain path and hit his head on the rocks. Philippe sought help from a nearby inn, where someone called the police. When they arrived at the scene, Max was dead. The police then determined that the dentist had been murdered and robbed.

Because of two recent unsolved murders in the area, the authorities no doubt felt a great deal of pressure to solve this one. The grieving family and friends were forbidden to bury the body. Investigators said they needed it as evidence for the imminent trial. Then, without warning that Halsman was even a suspect, the police charged him with murder.

The prosecutor cited greed as motive for the crime, accusing the son of killing his father to collect his life insurance. The defense destroyed that line of argument; there was no insurance policy, and Halsman had in no way gained from his father's death.

Yet the jury needed neither motive nor credible evidence to make a decision. Their attitude, an awful echo of the Dreyfus trial, was reflected by a comment overheard in a local train: "Whether or not the Jewish rogue has slain his Jewish father makes no difference! The prestige of Austrian justice is at stake. The Jews will have to take it or leave it." The jury agreed, finding Halsman guilty of second-degree murder. The judge gave him a ten-year prison sentence.

Halsman went on a hunger strike and demanded a retrial. It was refused. He had been in prison almost a year when the pleas of his family and a few sympathizers finally won him a second trial. This roused a group of Nazis to condemn "the monstrous influence and solidarity of Jewry!" and a like-minded local bishop to liken Halsman to Judas Iscariot. Although Halsman could not have profited monetarily from his father's death, the bishop called him "greedy" and "inhuman," and said that he lacked even "the moral fiber of Judas, who at least repented and did away with himself."

At the second trial, a detective testified that it was logistically impossible for Halsman to have committed the murder—if it was murder—and his defense attorney, Dr. Josef Hupka, stressed the lack of motive.

The prosecutor suggested the Oedipus complex: that Halsman's excessive love for his mother had driven him to kill his father.

The expert on the subject, Freud himself, demolished this line of attack. He told Dr. Hupka that it was dangerous to apply his Oedipus complex theory to the Halsman case, because he had uncovered the unconscious neurotic disorder during psychoanalysis. It simply could not be confirmed, especially in an adult, "without unmistakable evidence of its operation."

Evidently the jury wanted to punish Halsman for something regardless of the facts and the expert testimony; eight of the twelve settled for a reduced charge of manslaughter, which carried a four-year prison term. Led from the court, Halsman shouted out his innocence, damning the prosecutor, judge, and jury as criminals.

Halsman appealed his sentence to the Austrian Supreme Court but was turned down. A small group of Jews and Christians then demonstrated to raise funds for another appeal. Liouba's ongoing campaign to free him won support from many distinguised individuals, including German novelist Thomas Mann and the former French prime minister, now air minister, Paul Painlevé. She also contacted Einstein. He had never been known to ignore a cry for justice, though demonstrating was hardly his style. Instead, Einstein went to the top, writing to Austrian president Wilhelm Miklas, a devout Catholic with fourteen children. A former teacher, Miklas had stubbornly resisted the surging Nazi influence in his country, and he sympathized with Halsman's plight. In October 1930, he reduced the sentence to the two years already served. Halsman was then freed on condition that he leave the country immediately and never return. He did, and began a new life and career as a photographer in France.

A decade later, Philippe Halsman, now based in Paris, had become a famous photographer. In 1940, when the Nazis invaded France during World War II, Einstein once again came to his aid. Himself a refugee living in Princeton, New Jersey, Einstein obtained an emergency visa for Halsman by making sure his name was added to a list of artists and scientists in danger of being captured by the Nazis. This enabled him to come to the United States.

In time Halsman became a U.S. citizen and the first president of the American Society of Magazine Photographers. *Popular Photography* rated him "the world's most celebrated portrait photographer." He holds the record for the most *Life* covers (101), including photograph portraits of Churchill, JFK, Bertrand Russell, and Marilyn Monroe.

Halsman also photographed Einstein, the man who had helped him to freedom. He visited Einstein in Princeton several times, initially to thank him for the "miraculous rescue" from the Nazis. In 1947, Halsman took a memorable photo of Einstein; it was chosen for a U.S. postage stamp, which was issued on March 14, 1966. As the photographer arranged the lighting, Einstein "spoke of his despair that his formula $E = mc^2$ and his letter to President Roosevelt had made the atomic bomb possible." After Halsman took the photo he asked, "So you don't believe that there will ever be peace?" "No," Einstein replied, "so long as there are men, there will be wars."

Few knew of Halsman's tragic past, which he rarely discussed. It was not even mentioned in his 1972 autobiography, *Sight and Insight*, nor in his 1979 *New York Times* obituary.

"It was a suffering for him for the rest of his life," said his widow, Yvonne, "and for his mother and sister and for all of us. The whole thing was just an anti-Semitic process. It was the most horrible ordeal. He just wanted to leave it behind and move on with his life."

After World War II, journalist Hans Haider of Vienna's *Die Presse* investigated the case and discovered how drastically public opinion in Austria had changed: No one familiar with the trial now believed in Halsman's guilt. In the 1960s, the Austrian government officially acknowledged his innocence.

Except for the death of his friend Hendrik Lorentz, 1928 had been a good year for Einstein. His illness had turned out to be a blessing, enabling him to get on with his work free from the madding crowds. His secretary, Helen Dukas, was proving a treasure—and would stay with him for life. He had also helped to free an innocent young man from prison and possibly from extermination.

CHAPTER 21

The Unified
Field Theory

1929

─────

50 years old

Within a protective cocoon of friends and family, Einstein used his months of convalescence to develop his mathematical model for a unified field theory. His disappearance from the news columns fueled a rumor that he had finally tapped into God's thoughts and discovered that the Almighty had created everything in the universe out of electricity! Some papers ran the rumor to smoke Einstein out, hoping a little heat would make him talk, but he didn't fall for it, responding to press speculations with silence. To avoid even the appearance of self-promotion, he told Elsa and Dukas to refuse all interview requests. Frustrated reporters laid siege to his apartment building, hoping to catch him coming or going. He avoided them by staying put.

In fact, the rumor parodied his work. Helped by Lanczos, his "house mathematician," Einstein had crystallized his various efforts to explain the fundamental forces of nature as manifestations of a single underlying force, and he quietly submitted the result to the Prussian Academy of Sciences.

Now he was free, he told Besso, to read

with interest and pleasure a book on Socialism by G. B. Shaw.* It's a delicious, perspicacious and profound view of how men behave . . . [and] I'll try to do a little publicity for the book. As for my work, I've spent days and nights thinking about it and it is now in front of me complete and condensed in seven pages [soon reduced to six pages] and titled "A Unified Field Theory." It looks old-fashioned. You'll first put your tongue out at it because in the equation I don't once mention Planck's constant (h). I'll send you the publication. And if you

─────────

* The Intelligent Woman's Guide to Socialism and Capitalism, published in 1928.

don't put out your tongue you're a hypocrite—I know you, my boy, as a Berliner would say.

Asked if Einstein had solved the mysteries of the universe, the Prussian Academy took the easy way out and released copies of the theory without comment. British and American reporters perused the handout as if it were a secret code, then passed the buck by cabling it verbatim to their newspapers. And that's how it first appeared.

Back in Berlin, badgered by their editors to get Einstein to demystify his work, about a hundred newsmen swarmed outside his home. The most ambitious kept a nonstop vigil, their cars monopolizing all the nearby parking spots.

For a week he held them off, refusing to see anyone. When it was apparent that his silence only escalated their interest, he allowed a few privileged minutes to a reporter for *The New York Times*, a paper he had admired ever since its brilliant managing editor, Carr Van Anda, proved himself the better mathematician by correcting an Einstein math mistake. The incident was no surprise to those who knew of Van Anda's prowess in math, physics, astronomy, and Egyptology.

Wythe Williams, the *Times* reporter chosen to interview Einstein, had made his name cabling human interest stories from the frontline trenches during World War I. Now in former enemy territory, he was still on the lookout for colorful touches to enliven his copy. Einstein had already promised an explanation of his theory for the paper, but Williams hoped to reveal the private Einstein. He knew he was in luck when a harassed Elsa let him and his cameraman into the apartment, complaining that her husband had been driven crazy by the publicity. If not crazy, he was extremely tense and restless. Even the reporter's well-meant greeting, "The world awaits your explanation," startled Einstein, who put his head in his hands, and gasped, "My God!"

Einstein sought temporary solace from the anticipated questions at the living room window, staring at a recent fall of snow on the balcony, until distracted by a vase on the radiator. Snatching it up, he asked Elsa, "Why did you leave it here to get cracked?"

Elsa didn't even bother to look up from a newspaper she was reading: "It was cracked in falling three years ago."

"Oh," said Einstein.

The interview continued. Elsa eventually looked up and interjected that sending Albert's theory by cable to American readers "who won't understand it, anyway," was a terrible waste of money.

Not as wasteful as spending it on "useless clothes," Einstein suggested.

As Williams was about to leave, Einstein mentioned the lurking cameraman's request for an "action photo," and quipped, "Maybe he'd like me to stand on my head!" Then he concluded the interview with the smiling and disingenuous remark, "I can't understand why all the noise is being made over my little manuscript."

Little was right — six pages of equations, the fruit of ten years' work. His friend Max von Laue put it in perspective when he remarked, "Einstein came to mathematics rather out of necessity than predilection and yet he has developed mathematical formulae and calculations springing from colossal knowledge. It's open to question whether Einstein can prove his theory. As I know him, however, the rest of his life will be devoted to devising some method of proof which will upset the existing assumption that mathematical and electrodynamic laws are unrelated."

When a second reporter got past his guard a couple of weeks after the *Times* interview, Einstein tried to dispel the fantastic rumors by explaining,

> "My relativity theory reduced to one formula all laws which govern space, time and gravitation. The purpose of my new work is to further this simplification, and particularly to reduce to one formula the explanation of the field of gravity and of the field of electromagnetism. For this reason I call it a contribution to 'a unified field theory.' Now, but only now, we know that the force which moves electrons in their ellipses about the nuclei of atoms is the same force which moves our earth in its annual course about the sun, and is the same force which brings to us the rays of light and heat which makes life possible upon this planet."

If Einstein hadn't read God's thoughts, he seemed to have a great secondary source.

Asked what evidence sustained his hunch that the forces of the universe had a single source, he said, "The relation between electricity and gravity must be very close. Light waves, heat waves, radio waves and gravitation all go with the same velocity. In the final conception I believe these phenomena cannot be separated." He expressed this idea more vividly to a former student, Fritz Zwicky, saying that his ultimate aim was "to achieve a formula that will account in one breath for Newton's falling apple, the transmission of light and radio waves, the stars, and the composition of matter."

Not many even understood his special relativity theory, now twenty-five years old. When Edgar Ansel Mowrer of the *Chicago Daily News* said that he was confused by what seemed to be an illogical aspect of relativity, Einstein replied, "Don't bother your mind about it: mine is a mathematical, not a logical theory," and precluded follow-up questions by playing Bach on his violin. During an interview later that year, however, he did attempt to explain time as the fourth dimension in the theory's space-time continuum.

Einstein said,

> "Imagine a scene in two dimensional space, for instance the painting of a man reclining upon a bench. A tree stands behind the bench. Then imagine the man walks from the bench to a rock on the other side of the tree. He cannot reach the rock except by walking either *in front of* or *behind the tree*. This is impossible in two dimensional space. He can reach the rock only by an excursion into the third dimension. Now imagine another man sitting on the bench. How did the other man get there? Since two bodies cannot occupy the same place at the same time, he can have reached there only *before* or *after* the first man moved. In other words, he must have moved in time. Time is the fourth dimension."

Although the concept seemed clear to the reporter, Einstein warned him: "No one can visualize four dimensions, except mathematically. I think in four dimensions, but only abstractly."

That spring Einstein sent a bad-news/good-news letter to a Dr. Frosch, noting how several older mutual acquaintances were "under the sod and one or two of us 'youngsters' too. I nearly came to grief last year myself but it seems that certain weeds [probably vegetables] in my innards triumphed."

Einstein also kept in touch with an American pen pal, Upton Sinclair, a fellow crusader for social justice who was equally obsessed with mysteries. Unlike Einstein the realist, Sinclair believed that those Einstein acquaintances "under the sod" might still be available for conversations. Having studied psychic research at Columbia University, learned the tricks of phony mediums from famed magician Harry Houdini, and read up on hypnosis, psychiatry, and spiritualism, Sinclair believed that despite massive fraud there was a good case for telepathy and communication with the dead.

Using his Pasadena, California, home as a laboratory, Sinclair tested his theory on a permanent houseguest, Roman Ostoja. Sinclair was so taken by this muscular, dark-eyed psychic phenomenon that he treated him as a member of his family, and invited friends to attend séances at which the entranced Ostoja, while "tightly held by both knees and ankles, made a thirty-four pound table rise four feet in the air and move slowly eight feet to one side." Eventually, Sinclair put Ostoja on hold to investigate the psychic abilities of his own wife, Mary Craig Sinclair, who was eager to be put to the test.

As a child, Mary had seemed able to read her mother's mind and had once even had the same dream as her mother. To see if she could "perform" under test conditions, the couple stayed in adjoining rooms with the connecting door closed. Then Sinclair sketched objects or scenes that came to mind while she tried to read his thoughts and duplicate the sketches. The positive results far exceeded the laws of chance, and were the subject of Sinclair's book *Mental Radio*, in which he reproduced many of the sketches and described the test conditions.

Sinclair mailed copies to Freud and other "names," soliciting enthusiastic comments for publicity purposes. Freud did not respond, but Einstein thanked Sinclair for his "highly worthy book" and agreed to provide a preface. How Sinclair persuaded Einstein, who had never met him, to provide this favorable publicity still baffles scientists antagonistic to ESP. After all, if telepathy is true and thought is assumed to move even faster than the speed of light, it challenges Einstein's theory.

In the preface, Einstein wrote:

I . . . am convinced that the book deserves the most earnest consideration . . . The results of the telepathic experiments carefully and plainly set forth in the book stand surely far beyond those which a nature investigator holds to be thinkable . . . it is out of the question in the case of so conscientious an observer and writer as Upton Sinclair that he is carrying on a conscious deception of the reading world; his good faith and dependability are not to be doubted. So if somehow the facts here set forth rest not upon telepathy, but on some unconscious hypnotic influence from person to person, this also would be of high psychological interest. In no case should the psychologically interested circles pass over this book heedlessly.

Talking with a friend, Antonina Vallentin, Einstein conceded the possibility of "human emanations of which we are ignorant. You remember how skeptical everyone was about electrical currents and invisible waves?" When it came to the supernatural, though, according to Helen Dukas, he was a skeptic's skeptic; she remembers him saying, "Even if I saw a ghost I wouldn't believe it."

Einstein felt almost the same way about the unsubstantial, schizophrenic world of quantum physics. Even Niels Bohr himself admitted that "if you aren't confused by quantum physics, then you haven't understood it."

And so, Einstein worked with dogged persistence to complete his unified field theory—to incorporate both quantum phenomena and causality. A successful outcome to his work promised to delight travel agents, according to one futurist with a sense of humor, who saw unified field theory as promising space travel: "Since it is possible now to insulate against electricity, it follows that if electromagnetism and gravitation are the same thing it may be possible someday to insulate against gravity. If that ever comes to pass, motorless aircraft may ride the skies, people may step out of skyscraper windows without falling to the ground, and a trip to the moon becomes theoretically possible."

In February of 1929, a trip to the moon seemed an enchanting prospect to Berliners. Unemployment in the city had soared to 450,000; many of the unemployed, in despair, had joined the Communists or the growing ranks of Nazis. Schools closed for a week due to the bitter cold and because there were no funds to pay for heating. Thousands of Jewish refugees from Russia and Poland were particularly hard hit, especially intellectuals and creative artists who couldn't make a living because they didn't understand German.

Moved by the refugees' plight, Einstein persuaded his novelist friend Arnold Zweig to join him in helping them out. Zweig, in turn, wrote to Freud on February 18: "Since we cannot permit the refugees to perish among us in silence, and since the need is far too great for any single Maecenas to bring more than temporary alleviation to a few, we have decided to set up from our own resources the organization which is described in the circular enclosed . . . Our sole request to you, Professor Freud, [is] to lend the weight of your name to a cause for whose existence the need is urgent."

Freud replied: "Although I haven't much else to give, I feel almost hurt that you should ask me to give only my name. How often do I not envy Einstein the youth and energy which enable him to support so many causes with such vigor. [Einstein was forty-nine, Freud seventy-three.] I am not only old, feeble and tired, but I am also burdened with heavy financial obligations. I should like to become a subscribing member of your society." Presumably his offer was accepted.

To raise money for this and other charities, Elsa negotiated with an American agent to sell the original manuscript outlining Einstein's latest theory. Einstein also began charging a nominal fee for his autograph and for interviews, devoting the money to various charities.

As his fiftieth birthday loomed, he was anxious to avoid those hell-bent on celebrating it with him. Berlin's shoe-polish king, Franz Lemm, had the answer; he hid Einstein in a gardener's cottage on his palatial country estate. Dukas stayed behind in the Berlin apartment to receive congratulatory telegrams and gifts, telling those clamoring to see him that "Professor Einstein stole out of town quietly a few days ago to escape possible ovations. I have strict orders not to tell where." By chance, a tax collector called to discuss Einstein's tax return; he apologized and beat a red-faced retreat when told that it was his birthday.

"Gleeful in the knowledge that the whole world was looking for him and could not find him," Einstein shared with his family a birthday lunch of salad, pike stuffed with mushrooms, stewed fruit, and a tart. Because of his strained heart, he was forbidden wine and coffee, but allowed one pipeful of tobacco.

> Every pipe he lit was like a dagger thrust in Elsa's heart.
> "How many have you smoked already?" she asked timidly.
> "This is the first," he invariably replied.
> "But I saw you just now . . . "
> "Well, it might have been the second."
> "The fourth at least," continued Elsa.
> "You're not going to tell me you're better at mathematics than I am," said Einstein laughing.

Janos Plesch was often to hear that laugh, even when it seemed inappropriate. He would write that Einstein could "see the funny side of situations most people would regard as utterly tragic, and I don't mean utterly tragic for other people, but for himself. I have known him laugh even when a mishap or misfortune has really moved him . . . Life's too short to waste on disagreeable matters, is his attitude: there are so many important things to attend to. This may seem to suggest that he has no very deep feelings, but he has." But it is doubtful if headline-hunting reporters were aware of it.

One American journalist surprised him in his cottage retreat examining a microscope, a birthday gift. More impressed than angry at the man's enterprise in tracking him down, Einstein let him watch while he pricked his own thumb and excitedly called Margot to examine a drop of magnified blood. At five that afternoon the family and the journalist left Einstein alone to enjoy the rest of his birthday at his favorite occupation — cogitating.

Back in his apartment a few days later, Einstein found birthday greetings awaiting him from the German chancellor and government, the king of Spain, the emperor of Japan, and U.S. president Herbert Hoover. He also learned that Zionists had planted a forest in Palestine in his name. Several friends had sent violins, and his students and a bank had contributed toward a sailboat to replace his leaky wreck, which he had laughingly called *Drunken Lisa*. He accepted the boat with mixed feelings: guilt for owning such a luxury in troubled times, delight to have a boat with an enclosed cockpit and a toilet. Perhaps the most touching gift of all came from an unemployed laborer who had sent him a small package of tobacco with the message, "There is *relatively* little tobacco but it comes from a good *field*." It moved Einstein to tears, and he replied immediately, before acknowledging the gifts and greetings from the high and mighty.

He took advantage of greetings from Germany's finance minister, Rudolf Hilferding, to ask him to give asylum to Leon Trotsky, the Soviet leader who was on the run from Stalin's hit squads. Einstein tried a light touch, writing to Hilferding: "If, however, the Herr Minister doesn't allow the sick Trotsky to enter and give him asylum, then . . . if he weren't a Minister I would grab him by his ear." It didn't work, but the attempt showed once again Einstein's concern for people in distress.

Although the city of Berlin unveiled a bust of him in an observatory known as the "Einstein Tower," and the nearby city of Potsdam commissioned a statue, Plesch thought that Einstein deserved even more recognition, and persuaded Berlin's mayor and city council to give him a house. When Elsa went to inspect this gift, set in a city park and surrounded by fruit trees, she found it occupied. A woman living there said, "I know nothing about Einstein being given this house. It belongs to us and will for years to come." She was right. The Berlin bureaucrats had overlooked the present occupants' long-term lease. As a substitute gift, they told Einstein to choose a piece of land and they'd buy it for him. When word of this plan got out, an anti-Einstein faction protested, and city officials delayed the purchase of the land Einstein had selected. Finally, Einstein wrote to the mayor: "Human life is very short, while the authorities work very slowly. I feel therefore that my life is too short for me to adapt myself to your methods. I thank you for your friendly intentions. Now, however, my birthday is already past and I decline the gift."

Instead, Einstein used his own savings to have a summer home built at Caputh, an isolated village on the Havel River, on his own piece of land. The cottage's wooden walls were so thin that a private conversation could be heard throughout the house, but Einstein loved it, especially because the nearest railroad station was miles away and the rough roads discouraged all but the most intrepid strangers. One of the first invited visitors, Hans Albert, took a risky ride with his father in Einstein's new sailboat. Completely absorbed in discussing his theory, Einstein headed for the rocky river bank and almost crashed.

His morning paper brought news of deadly confrontations in nearby Berlin. The police had used their cars as battering rams to smash into Communists and unemployed workers who had defied a ban on May Day parades, killing several marchers and wounding scores of others. When the crowd of marchers fought back, some eight thousand police attacked with water hoses, tear gas, and guns. Three days of fighting left thirty-five dead and hundreds injured. Berlin's police chief claimed he had stopped a Communist attempt to start a civil war.

Einstein skirted the scene of the slaughter a few days later to take the train on his annual trip to visit Ehrenfest in Leiden. From there he went to visit his favorite uncle, Caesar Koch, in nearby Antwerp. He was also a guest of Belgium's Queen Elizabeth, taking tea with her under the chestnut trees on the palace grounds. When they strolled there together before dinner, he gave her a "tiny glimpse" into his "ingenious theory." It was the start of an enduring friendship and a frank, affectionate correspondence.

Hans Albert, now twenty-five and an engineer, frequently visited Caputh with his Swiss wife, Frieda, even though his father would spring on him such unwelcome advice as, "Don't have any children: it makes divorce so much more complicated." His timing was badly off on this one: Frieda was already pregnant with a son.

Einstein's grim view of marriage came firsthand: from his failure with Mileva and his now increasingly rocky relations with Elsa. He touched on the discord in his marriage once, when discussing his smoking habits. Because he was always cleaning his pipe, someone asked if he smoked just to unclog and refill it. He replied: "My aim lies in smoking, but as a result things tend to get clogged up, I'm afraid. Life, too, is like smoking, especially marriage."

The differences between Albert and Elsa were serious. Elsa cared about keeping up appearances; Albert didn't give a damn about appearances. He was spontaneous and outspoken; she was cautious and conciliatory. He shrank from fame and the attendant publicity; she embraced it. She was a faithful and loving wife, while his extramarital affairs earned him a reputation, among the few who knew about them, as a womanizer.

Souring the otherwise idyllic life at Caputh for Elsa, an attractive young Austrian woman had taken Einstein's fancy. By all accounts, including his own, Hans Albert never knew of his father's philandering or of Elsa's distress, which may explain his puzzlement at his father's less than ecstatic remarks on marriage.

A German author, Friedrich Herneck, was probably the first to reveal details of Einstein's affairs in *Einstein privat*, followed by the writing team of Roger Highfield and Paul Carter in *The Private Lives of Albert Einstein*.

Much of their inside information came from Herta Waldow, the Einsteins' live-in maid for five years starting in 1927. Questioned by Highfield's wife, Doris, Waldow said that Einstein "liked beautiful women, and they in turn adored him."

Asked to elaborate, Waldow confided that during the summer of 1931 a blonde Austrian beauty, Margarette Lebach, called at the Caputh house almost every week. Lebach would hand Elsa a box of her home-baked pastries, then head for Albert. Although Elsa was jealous of this relationship, whenever she knew Lebach was expected, she left for Berlin early in the morning and stayed away until evening. Waldow wondered why Elsa would "leave the field clear, so to speak," especially as "the Austrian woman was younger than Frau Professor, and was very attractive, lively, and liked to laugh a lot just like the Professor."

Elsa Mendel, another woman friend of Einstein's and a wealthy and elegant widow, used a similar modus operandi. She frequently called for Einstein in her chauffeur-driven car, gave Elsa a box of chocolate creams, then drove off with her prize for a night on the town, usually at the opera or a symphony concert— leaving Elsa behind with the chocolate creams. Mendel apparently sprang for the tickets, but Einstein paid for the extras. This annoyed Elsa, who otherwise "grudgingly" tolerated the close friendship, said Waldow. However, quarrels between Elsa and Albert were not ostensibly over his dates with the winsome widow but over the money Elsa had to dole out to pay for the incidental expenses.

Estella Katzenellenbogen, another rich, elegant woman, who owned a florist business, occasionally lured Einstein from his study to take him for rides around Berlin in her luxury car.

Eventually Elsa's tolerance reached its limit, and soon after Albert left the house on another outing she gave vent to her feelings. Waldow overheard her and her daughters arguing passionately about "the Austrian interloper ... The girls told their mother she must either put up with the relationship or seek a separation ... Elsa was in tears, but made her decision. The trips to Berlin continued.

"It was this kind of humiliation that drove Elsa into fits of jealous rage. Once, after a sailing trip at Caputh, Einstein forgot to bring back from the boat some clothes that needed washing," explained Konrad Wachsmann, an architect friend of the Einsteins who had designed their summerhouse. Einstein's assistant Walther Mayer (who had succeeded Lanczos in December 1929) went to get the clothes and gave them to Elsa. "Shortly afterwards, Einstein was summoned inside and the couple's guests overheard a sharp exchange of words." The bundle included a low-cut woman's bathing suit, which did not belong to any of the Einstein women. Wachsmann, who witnessed the incident, "explained that the outfit belonged to 'a good acquaintance' of Einstein, and that Elsa became 'monstrously worked up.'"

Wachsmann "believed that Einstein's extra-marital liaisons were 'without exception' platonic, but was painfully aware of the effect they had on Elsa. 'She probably suspected more than knew that she had only borrowed this husband, and that every day she could lose him ... [that] thought ... must have hung over her like a Damoclean sword.'"

Despite the affairs, his marriage endured. Few—if any—of his friends considered it ideal, and some resented Elsa for treating him like a wayward child. But any suggestion that she dominated him as some said defies the facts.

Arriving early one day to paint Einstein's portrait, American artist Samuel Johnson Woolf saw that Einstein was not a henpecked husband. Woolf chatted with Elsa and watched the maid polish the living room floor while he waited for his subject. "I'm glad you persuaded him to pose," Elsa said, "although he doesn't want publicity. Last week there were a few photographs of him in the paper and he was so disturbed he didn't work for two days. Have you seen him? Hasn't he got a wonderful head?" Elsa killed more time by showing Woolf a book of Albert's sayings and poems that friends had given him for his recent fiftieth birthday.

Woolf heard the sound of bare feet, and Einstein entered the living room in a bathrobe. Elsa patted him on the back, telling him to get dressed, then said to Woolf, with a smile, "He's terribly hard to manage." When Einstein returned in a rumpled brown suit, the men started for the upstairs study, but Elsa stopped them to fix Albert's twisted coat collar and try to press his unruly hair into place.

As Einstein led the way up a few stairs, he reminded Woolf of his promise not to have the painting reproduced in a newspaper. "That's all right for a theatrical prima donna who wants publicity," he said. He unlocked a white door and Woolf followed him a few feet to another door, which opened onto his tiny study. Light entered the room through a window in the sloping roof. The white-washed walls were almost hidden by shelves crammed with books and pamphlets, and more papers covered a table and two ladder-back chairs. Einstein sat in an upholstered chair on a dais under the window and began smoking a cigarette, without inhaling, the cigarette held vertically in a holder. He began taking notes, and as he became absorbed in his work he occasionally twisted his hair with his fingers.

Sketching Einstein's head, Woolf also took notes: "He has a perpetual quiz-zical expression . . . Often smiles in a quiet, embarrassed way . . . Has a bashful, malleable quality, almost childlike, accentuated by his wife's attitude to him. A sweet, motherly woman, she treats Einstein like a doting parent with a precocious child . . . Talking, he appears to be thinking of other things; gazing, he does not appear to be seeing the object at which he looks. These peculiarities are so marked as to appear almost abnormal."

Einstein approved the portrait and asked Woolf for a photo of it. The artist returned with one several days later, to find Elsa entertaining relatives and Albert wearing the same rumpled brown suit, absentmindedly eating a sandwich. Einstein studied the photo for a while; then, without a word, hurried into the library. A few moments later it sounded as if he was demolishing the place.

"Come with me," Elsa said to Woolf, urgently. "I know what he's doing."

When they entered the library, Einstein was manhandling an immense framed painting of himself hanging at a crazy angle on the wall. As became clear to Woolf later, his approved portrait had reminded Einstein of how much he loathed this other one. "I've often told you to take this down," he complained to Elsa. "Now I'm going to do it. And if I see it around I'll put a knife through it!" It was too much for him to manage alone, so Woolf helped remove it from the wall. After that, Einstein resumed his calm, friendly manner and saw the visitor out.

On June 28, Einstein was invited to lunch at the house of his friend Dr. Plesch. After the meal, he fell asleep on his host's couch. He awoke at four. In exactly an hour he was expected at the Institute of Physics to receive the first Planck Medal, a new award to which mathematicians and scientists throughout the world had contributed. In about twenty minutes he roughed out an accep-tance speech on the back of a bill he found on Plesch's desk. He then hurried into the formal attire he had brought for the occasion, and reached the hall on time. When Planck handed him the medal, Einstein said, "I knew that an honor of this sort would move me deeply and therefore I have put down on paper what I would like to say to you as thanks. I will read it." Then, wrote Plesch, "out of his waistcoat pocket came my bootmaker's bill with the scribble on the back, and he read out what he had written about the principle of causality. And, be-cause, as he said, no reasoning being could get on at all without causality, he established the principle of super-causality. The atmosphere was tense and most moving." After the ceremony, when Plesch claimed his bootmaker's bill back as a memento, Einstein also handed him the gold medal. Plesch handed it back.

A Dr. Chaim Tschernowitz was among Einstein's many visitors at Caputh that summer, and as they sailed on the Havel River their "conversation drifted back and forth from profundities about the nature of God, the universe, and man to questions of a lighter and more vivacious nature. Suddenly Einstein looked upward at the clear skies and said, 'We know nothing about it all. All our knowledge is but the knowledge of schoolchildren.' " When the visitor asked, "Do you think we shall ever probe the secret?" Einstein replied, "The real nature of things, that we shall never know, never."

At the end of August, while he was in Zurich for the Sixteenth Zionist Congress, Einstein looked in on Mileva and nineteen-year-old Eduard. The young man idolized his father, yet could never forgive him for deserting the family. Though treated affectionately by Albert, Elsa, and Helen Dukas whenever he stayed with them in Berlin or Caputh, he still felt rejected. Hypersensitive, with artistic ambitions but no creative talent, Eduard longed to make his father proud of him. Aware of Einstein's special rapport with doctors, he was studying hard to be one.

On this visit to Zurich, Einstein got a laugh out of Eduard by responding to his question, "Why are you at a Jewish rather than a scientific conference?" with a deadpan answer: "Because I am a Jewish saint."

During the conference, Einstein agreed with Chaim Weizmann's proposal for a modus vivendi with the Arabs and cooperation with the British. He also joined in a standing ovation when the Jewish political leader declared, "We never wanted Palestine for the Zionists; we wanted it for the Jews. The Balfour Declaration is addressed to the whole of Jewry."

Their optimism was tested soon after the conference, when Arab mobs went on a rampage in Palestine, killing or wounding almost 500 and destroying colleges, synagogues, and hospitals. A "savage attack" on Hebron "was accompanied by wanton destruction and looting," and left more than 60 Jews killed, including many women and children. In Safed, "Arab mobs" killed and wounded 45 Jews, while in the suburbs of Jerusalem more than 4,000 Jews were forced to flee their homes, leaving them at the mercy of looters. By August 29, 133 Jews had been murdered. Most of the 87 Arabs who died were shot by British troops and police trying to stop the violence.

Jews blamed Arabs for the carnage, Arabs blamed Jews, and both blamed the British. Fifteen thousand Jews marched to the British consulate in Manhattan protesting the policy in Palestine, and over a thousand volunteered to fight the Arabs. Arabs in Manhattan also protested British policy and demanded the repeal of the Balfour Declaration. In London, influential newspaper moguls Viscount Rothermere and Lord Beaverbrook saw the British role in Palestine as a no-win situation and advocated withdrawing troops and abandoning the country to its fate.

Einstein also protested with a spirited public statement: "Is it not bewildering that . . . brutal massacres by a mob of fanatics can destroy all appreciation of the

Jewish effort in Palestine and lead to a demand to repeal the solemn pledges of official support and protection?" Einstein pointed out that "Jews had paid for every acre of land they settled, that Jews and Arabs had shown they could live happily as neighbors if the gangsters were driven away, that the ideals of Zionism deserved world support."

Winston Churchill blamed the slaughter on the Arabs' envy of the Jews, and Lord Arthur Balfour, author of the Balfour Declaration, assured Weizmann that the British would not renounce the Jewish cause. Troops under General Dobbie made good on the promise by crushing the riots and driving off Arabs attacking from neighboring states.

On September 29, Palestine's high commissioner Sir John Chancellor cabled the Colonial Office in London: "The latent deep-seated hatred of the Arabs for the Jews has now come to the surface in all parts of the country. Threats of renewed attacks are being freely made and are only being prevented by visible presence of considerable force."

Einstein feared that this "considerable force" was unreliable, warning Weizmann, in a November 25, 1929, letter, "We must avoid leaning too much on the English. If we fail to reach real cooperation with the leading Arabs, we will be dropped by the English, not perhaps formally but de facto. And they will, with their traditional 'religious eye-opening,' claim themselves innocent of our debacles, and not raise a finger."

Throughout his life, Einstein did his best to avoid the press, but they were continually after him for interviews. He thought most journalists "were a pain." They printed his wisecracks as serious statements, or his outrageous remarks without mentioning that they sprang from an angry or audacious mood of the moment. Still, Einstein could live with this kind of reporting even when it disturbed his friends and gave ammunition to his enemies. What exasperated him was to be asked to answer for what others had said in his name, and that whatever his reply, the reporter would give it a negative twist.

Despite his scars, Einstein warmed to a German-born American interviewer, George Sylvester Viereck, known as a "big-name hunter" for having "captured" Freud, Clemenceau, George Bernard Shaw, Henry Ford, and the Kaiser, among others, for articles (these were later published in a book, *Glimpses of the Great*). Shaw questioned his accuracy and Upton Sinclair called him "a pompous liar and hypocrite." Freud told him that he had a "superman complex" both because of his urge to interview "the great" and because Viereck made it clear that he felt he was in the same league as other "supermen."

"There was something pathetic and inflated about him," said his grandson, John-Alexis Viereck, an Episcopal priest, "although he had a good heart and he both associated himself with these people and did very well in establishing intimate social connections with some of them. His great grandmother was the mistress of Emperor Wilhelm I of Prussia. So he was in the bastard line descending from the emperor. And, although he was a very kind man personally, he got inflated ideas and was somewhat obsessed with royalty and the emperor saying, 'Mon cousin,' to him and things like that. He was very excited by the Freudian stuff. To him, everything was glandular, and legend has it that he had the operation where they grafted monkey balls on him to reinvigorate him, which of course didn't work, But it gives you the flavor of the man."

Viereck had interviewed Hitler some years before and said prophetically, "One way or another the man will make a name for himself either as a great apostle of evil or he will accomplish something tremendous."

And this was the man Einstein trusted, and to whom he gave perhaps the most wide-ranging interview he ever granted.

Elsa welcomed Viereck to their Berlin apartment, filled two glasses with strawberry juice and provided two plates of fruit salad, then left the men to it. After finishing the refreshments, they walked the few steps up to Einstein's small attic study, this holy-of-holies into which Elsa had admitted she rarely entered, and then with trepidation. It had a spacious view of rooftops and sky, pictures of his heroes Faraday, Maxwell, and Newton on the walls, a globe of the world in one corner and a telescope in another. The study needed only a microscope to define Einstein's almost unlimited interests.

After a few moments, the interview began. The essence of the questions and answers follows.

Do you believe man is a free agent?

No, I am a determinist, compelled to act as if free will existed, because if I wish to live in a civilized society I must act responsibly. I know that philosophically a murderer is not responsible for his crime, but I prefer not to take tea with him. Undoubtedly my career has been determined by various factors over which I have no control, primarily those mysterious glands in which nature prepares the very essence of life. Henry Ford may call it his Inner Voice, Socrates referred to it as his daemon: each man explains in his own way the fact that the human will is not free . . . Everything is determined, the beginning as well as the end, by forces over which we have no control. It is determined for the insect as well as for the star. Human beings, vegetables, or cosmic dust, we all dance to a mysterious tune, intoned in the distance by an invisible player.

How do you account for your discoveries? Through intuition or inspiration?

Both. I sometimes *feel* I am right, but do not *know* it. When two expeditions of scientists went to test my theory [by photographs of starlight] I was convinced they would confirm my theory. I wasn't surprised when the results confirmed my intuition, but I would have been surprised had I been wrong. I'm enough of an artist to draw freely on my imagination, which I think is more important than knowledge. Knowledge is limited. Imagination encircles the world.

Do you consider yourself a German or a Jew?

It's possible to be both. I look on myself as a man. Nationalism is an infantile disease, the measles of mankind.

Then how do you justify your Jewish nationalism?

I support Zionism although it is a national experiment, because it gives Jews a common interest. This nationalism is no threat to other peoples. Zion is too small to develop imperialistic designs.

You don't believe in assimilation?

We Jews have been too eager to sacrifice our idiosyncracies to conform. Other groups and nations cultivate their individual traditions. Why should we sacrifice ours? To deprive every ethnic group of its special traditions is to convert the world into a huge Ford plant. I believe in standardizing automobiles, but not human beings.

Do you believe in race as a substitute for nationalism?

Race is a fraud. All modern people are a conglomeration of so many ethnic mixtures that no pure race remains.

Are you an opponent of Freud?

It may not always be helpful to delve into the subconscious. Our legs are controlled by a hundred different muscles. Do you think it would help us to walk if we analyzed our legs and knew the exact purpose of each muscle and the order in which they work? Freud's work is an immensely valuable contribution to the science of human behavior, but I don't accept all his conclusions. I consider him even greater as a writer than as a psychologist. His brilliant style is unsurpassed by anyone since Schopenhauer.

Do you believe in the God of Spinoza?

I can't answer with a simple yes or no. I'm not an atheist and I don't think I can call myself a pantheist. We are in the position of a little child entering a huge library filled with books in many different languages. The child knows someone must have written those books. It does not know how. It does not understand the languages in which they are written. The child dimly suspects a mysterious order in the arrangement of the books but doesn't know what it is. That, it seems to me, is the attitude of even the most intelligent human being toward God. We see a universe marvelously arranged and obeying certain laws, but only dimly understand these laws. Our limited minds cannot grasp the mysterious force that moves the constellations. I am fascinated by Spinoza's pantheism, but admire even more his contributions to modern thought because he is the first philosopher to deal with the soul and body as one, not two separate things.

To what extent are you influenced by Christianity?

As a child I received instruction in the Bible and in the Talmud. Though I'm a Jew I was enthralled by the luminous figure of the Nazarene. Emil Ludwig's book on Jesus* is shallow. Jesus is too colossal a figure for the pen of phrasemongers, however artful. No man can dispose of Christianity with a bon mot. No one can read the Gospels without feeling the actual presence of Jesus. His personality pulsates in every word. No myth is filled with such life.

Do you believe in immortality?

No. And one life is enough for me. I realize that every individual is the product of the conjunction of two individuals. I don't see where and at what moment the new being is endowed with a soul. I look on mankind as a tree with

---

*Emil Ludwig was a prolific biographer whose subjects included Napoleon, Lincoln, and Theodore Roosevelt, as well as Jesus Christ.

many sprouts. It doesn't seem to me that every sprout and every branch possesses an individual soul.

Einstein went on to say that if he hadn't become a physicist he would probably have been a musician, because "I get most joy in life out of my violin. I often think in music. I live my daydreams in music. I see my life in terms of music." He attributed his happiness to his few needs: "I want nothing from anyone. I don't care for money, decorations or titles. I do not crave praise. Apart from my work I get pleasure from my violin, my sailboat and the appreciation of my fellow workers." Elsa and his children were conspicuously absent from his sources of happiness.

Einstein told Viereck that the interview had been a delightful experience, that he felt completely at ease talking to him because they were fellow Jews. "But I'm not," Viereck replied. "I come from a long line of Protestants who emigrated to Germany from Scandinavia." Einstein simply wouldn't believe it, insisting that Viereck must have some Jewish blood, because he felt so comfortable with this interviewer.

Some say Einstein was a poor judge of character and trusted too easily. They point to Viereck as one of his biggest mistakes. Not only was Viereck not Jewish, he was pro-Nazi, and would spend five years in an American prison for acting as a secret German agent in the United States during World War II.

But Viereck's involvement with the Nazis, according to his grandson John-Alexis, "had nothing to do with anti-Semitism which he associated, quite rightly, with horror, but more due to his own misguided Nietzschean distortions. He was excited about Germany and the Fatherland and always supported the culture and history, and if you throw in Nietzsche and being the bastard grandson of the emperor — he was perfect for the Nazi spy ring that used him as a front."

In October, after the interview with Viereck, Einstein went to Paris to lecture on unified field theory at the Poincaré Institute. Maurice Solovine was in the audience. Einstein had made sure his old friend would get in by sending him an admission ticket. Although the Wall Street crash was reverberating through Europe with hundreds of thousands out of work, Solovine still held his job as an editor of a science journal in Paris. He could occasionally augment his income as official translator of Einstein's work into French. After the lecture, they chatted for hours about the good old days of the Olympia Academy, before fame struck.

Einstein was no sooner back in Berlin than he was talked into broadcasting a radio tribute to Edison to commemorate the fiftieth anniversary of his invention of the electric lightbulb. Einstein, too, could claim to be an inventor, having recently produced a miniature refrigerator. Leo Szilard, who had become his assistant, had helped him to create this machine, whose novel feature was that it had no moving parts. Unfortunately, this refrigerator proved much noisier than those with moving parts, and was never manufactured.

As always, Einstein was anxious to return to his own work, and had devised a neat strategy for getting rid of unwelcome visitors or those who stayed too long. Physics student Lancelot Law Whyte soon experienced the strategy firsthand.

An Englishman in Berlin on a Rockefeller Foundation grant, Whyte had been hoping for several months to meet Einstein, but was too shy to make the first move. At a party he got into a conversation about his research with Emil Ludwig, the biographer of Jesus Christ whom Einstein and Viereck had recently discussed. Whyte told Ludwig that he would like to meet the great scientist, but felt that his own ideas were too immature to interest him. Ludwig, who happened to know Einstein, disagreed. He took Whyte's address and told him to expect a response shortly.

It came in the mail a few days later. Einstein wrote, "I hear from my friend Emil Ludwig that we both ride the same hobby horse. I always like to talk with people interested in the same things. Please ring me up and come and see me. Don't be put off by Frau Einstein. She's here to protect me."

Despite the friendly invitation and warning, Whyte was extremely nervous. He became even more so when, after they'd been talking for about twenty minutes, a maid brought Einstein a large bowl of soup. Although he pushed it away and continued the conversation, the young student took it as an obvious hint for him to go.

In fact, it was a sign to stay, as Einstein explained in a conspiratorial whisper: "That's a trick. If I'm bored talking to someone, when the maid comes in I don't push the bowl of soup away. Then the girl takes whomever I'm with away. And I am free." That put Whyte at ease for the rest of the conversation and on several subsequent visits.

Whyte came to sympathize with Einstein's situation during that first meeting. "It was very uncomfortable for Einstein to remain in Germany," he remembered. "And I was deeply distressed to find that somebody of his greatness and world reputation had become a symbol for anti-Semitism. His very existence in Berlin University was dangerous."

Although it was a bitterly cold December, Albert and Elsa left the apartment for their country retreat. They weren't alone for long. Eager to question Einstein, reporter Dudley Heathcote of the *London General Press* braved bad roads, blinding rain, and misleading street signs to reach Caputh after dark. Elsa answered the door and remarked disapprovingly, "It is ten o'clock at night and you know the professor doesn't like journalists. But I will see what I can do." She returned smiling and ushered Heathcote into the study, where Einstein greeted him, saying, "I never give interviews as perhaps you know [except to those he liked and trusted]. But since you come with a letter from my friend Dr. Plesch, tell me what it is you want to know."

Heathcote's German being rudimentary and Einstein's English quaint, they compromised and spoke in stilted French. When he broached the problems of the Jews in Palestine, according to Heathcote, Einstein said: "This matter affects me more deeply than perhaps any other today. We Jewish people have acclaimed with the greatest warmth and enthusiasm the English proposal to reestablish a Jewish national home in Palestine and expect England to carry out her pledge . . . I remember how amazed I was when last I visited Palestine to see the great work already achieved by the colonists in a land hitherto largely unclaimed, work now imperilled by Arab agitators."

He suggested that workers in isolated settlements should have special protection and

"a policy be devised to bring Arab and Jew into closer contact for a better mutual understanding. The English cannot settle these differences by themselves. They cannot pretend to understand the psychology of two races very different from their own. The racial differences between the two nations are tremendous and the Arabs are easily swayed into religious frenzy by a class of politician whose interest is to keep them apart from the Jew. Still, if only the English authorities keep order and punish the rioters adequately and also keep a firm hand on all Arab nationalistic organizations, I think it would be possible to substitute peaceful understanding for the present mistrust . . . As a mandatory power the English could easily create institutions to compel Jews and Arabs to govern themselves and to realize how much it is in their interests to collaborate instead of fighting each other. And they can also act as arbitrators in the case of any dispute. My friend, Sir Herbert Samuel [first British high commissioner in Palestine, from 1920 to 1925], did a lot to improve relations between Arab and Jew.

"We are not out to found a Jewish state, nor are we Chauvinists who aspire to dispossess every non-Jew of his rights or possessions. I am convinced the great majority of Jews would not tolerate such a movement. We want to resettle the Jewish nation in the old home of their race so that Jewish spiritual values may once again develop in a Jewish atmosphere. As for the future relations with the Arabs, we anticipate the most friendly cooperation."

Sent a transcript of the interview for his approval, Einstein forbade its publication but in no way blamed Heathcote. He said that the subject was so sensitive he feared his views might be misinterpreted. Moreover, his flawed French had doubtlessly suffered even more injury in translation. It's unlikely, for example, that he said that Jews and Arabs should be "compelled" to govern themselves.

When Einstein learned that the British had given death sentences to twenty-five rioters who had killed Jews, he urged the high commissioner to commute the sentences, as did Brit Shalom, a Jewish peace organization. The sentences were reduced for all but three men, who had each committed multiple murders.

Soon after hearing of the commutations, he was incensed by a speech given in Berlin about the Palestine problem by Selig Brodetsky, a mathematics professor at Leeds University and a Zionist leader over from Britain. Einstein accused Brodetsky of talking like "Mussolini" and of showing no spirit of reconciliation.

Stung by the criticism, Brodetsky wrote to Einstein, saying that he may have misunderstood him because of his poor German:

During the greater part of my speech, I endeavoured to make clear that our work in Palestine must be based upon a friendly attitude towards our Arab neighbours . . . If you characterized my remarks as being like those of Mussolini, then I must take it you referred to the couple of sentences in which I ventured to lay down one or two fundamental principles . . . that our work in Palestine is quite impossible without the recognition and guaranteeing of the principles of the Mandate and the Balfour Declaration . . . I want to prevent the creation of a spirit of civil war between Jews and Arabs.

Einstein saw the danger of civil war, too, writing to Weizmann: "Should we be unable to find a way to honest cooperation with the Arabs, then we have learned absolutely nothing during our 2,000 years of suffering . . . But we should not fight among ourselves . . . Don't answer me now; you need your strength too much. I will keep quiet as much as I can."

CHAPTER 22

On the International Lecture Circuit

1930

51 years old

Asked to name the world's most popular figure, New York University's senior class of 1930 chose Einstein a close second to an American — flying hero Charles Lindbergh. If asked to explain Einstein's theories, however, chances are that every student would have flunked. Yet interest in Einstein's enigmatic work was still so intense that over four thousand people started a near riot trying to crash a film explaining relativity at Manhattan's Museum of Natural History. Guards had to be called to fight them off.

Einstein and Elsa began the year as unlikely participants in an investigation of the paranormal. The superskeptic who had said, "Even if I saw a ghost I wouldn't believe it" was giving serious attention to the claims of Otto Reiman, a bank clerk, whose gimmick was to read people's character by "feeling" their handwriting. When writing samples were requested from the audience at a meeting of Berlin's Medical Society for Parapsychology, Einstein wrote: "I do not believe individuals possess any unique gifts. I only believe there exists on the one hand talent, and on the other hand trained abilities."

The various writing samples were collected and, one at a time, slipped into Reiman's pocket. The writers were told to acknowledge when Reiman described them accurately. When he reached Einstein's message, he ran his fingers over it as if reading Braille, as he had with the others; calling it the work of a man with artistic pretensions but of average ability, he added, "He's probably a second-rate actor." From the silence in the audience it seemed an embarrassing failure, until Reiman opened his eyes, turned the page over, and noticed that there was different writing on that side. Einstein had written his message on the back of a letter from a Berlin theater manager who matched the description, and Reiman, it appeared, had inadvertently "read" the wrong man. Given a second chance, he scrutinized instead of feeling Einstein's writing. Still apparently unaware of the identity of its author, Reiman described him as logical, able to work intensely but in an easy, artistic manner, "starting at point A, taking a long jump to point D, then filling in points B and C later." Reiman also said the writer played the violin exquisitely, often stopping to make math notes; was impractical in daily life; often imposed upon; and extremely kindhearted.

When Einstein was revealed as the subject of the experiment, someone objected, "Nothing had been said about his being a great physicist." Einstein responded: "That's the most convincing part. It proves the reality of this man's gift. The theory of relativity, while important from a scientific viewpoint, is only of minor importance as regards my character, on which Herr Reiman concentrated." Elsa endorsed Reiman as right in every detail, although more objective witnesses might have challenged the "exquisite" violin playing. Afterwards, Einstein remarked: "I would prefer to say that everything I saw here tonight was a swindle, but I can't do that. I am bewildered."

Reiman accepted Einstein's invitation to make further tests at his home. There, a few days later, Einstein and several friends stuffed identical envelopes with different handwriting samples, including their own. The envelopes were shuffled and offered to Reiman, who inserted his fingers into one as if feeling the writing, and described a man Einstein immediately recognized as physicist Wolfgang Pauli. Right again. The bank clerk scored a high percentage of hits with the others. Then he demonstrated his telepathic powers. While he was out of sight and earshot in another room, the group agreed to think of the same person: Einstein's acquaintance, playwright Gerhart Hauptmann. When Reiman returned, he mimicked the man they had in mind, showing how he walked, ate, and spoke. It was Hauptmann, all agreed. No doubt about it. Had Reiman read their thoughts or tricked them? Einstein was baffled but not convinced. He remained a skeptic.

Other pursuits proved more enticing. He eagerly read a book about the Greek philosopher Democritus, a gift from Maurice Solovine, who had written the introduction. Democritus, a contemporary of Socrates, was thought mad, perhaps because he believed that everything in the world had been created from eternal, invisible atoms. Einstein was elated with both the subject and the introduction, telling Solovine that he especially appreciated the Greek's "firm belief in physical causality, which is not even stopped by the will of Homo sapiens. To my knowledge only Spinoza was so radical and so consistent." As for his own affairs, Einstein told his old friend: "My field theory is progressing smoothly . . . I myself am working with a mathematician [Walther Mayer of Vienna, who had just succeeded Cornelius Lanczos], a splendid fellow who would have been given a professorship long ago if he were not a Jew. I often think of the lovely Parisian days [during Einstein's last visit], but am satisfied with my relatively peaceful existence here. Do not hesitate to call on me if you think I can be of help in any way."

Having made some progress on his unified field theory, he decided to give it a public airing. His lecture audience included a somewhat arrogant twenty-two-year-old Ph.D. in physics named Edward Teller, and mild-mannered Eugene Wigner, a twenty-eight-year-old science teacher. Afterward, the pair went to a local zoo, where Wigner noticed that his friend was too depressed to discuss the lecture or enjoy the animals. When he asked what was the matter, Teller explained that not having understood a "syllable" of Einstein's theory, he felt stupid. "Yes, that is a general human property," replied the future Nobel laureate Wigner. Surprisingly, being called stupid "cheered me up," Teller recalled, explaining that being told he wasn't stupid wouldn't have helped. But if this future father of the hydrogen bomb didn't "get it," then Einstein was far from his goal of making his work transparent to a child.

Einstein's views as a social activist, however, were crystal clear, and his sympathies transcended race, religion, and national boundaries. Hearing of a Finn, for example, who faced punishment for evading military service, he appealed to the Finnish minister of war to pardon him; the minister said that he dared not risk it with the Russians just across the border armed to the teeth. Yet Einstein's reactions could hardly be taken for granted. He shocked a sponsoring Christian group when he declined to speak to their peace rally with a cutting remark: "If I had been able to address your congress, I would have said that in the course of history the priests have been responsible for much strife and war among human beings. They have a lot to atone for."

Shortly after that incident, he paid a rare visit to a synagogue. He went, according to biographer Jeremy Bernstein, because

> the more Einstein became aware of German anti-Semitism the closer a bond he felt to his fellow Jews. There is no more moving photograph of Einstein anywhere than one taken in a Berlin synagogue in 1930. Here he sits—skeptic and freethinker that he was and remained until the end of his life—his unruly hair flowing from underneath the traditional black yarmulke—holding his violin, prepared to play in a concert for the purpose of raising money to help fellow Jews. In the background one can make out the congregation and can only imagine, with grief, what was to be their fate.

Berlin, despite its massive poverty and sporadic street fighting, was still the musical capital of the world. There, on April 12, Bruno Walter conducted the Philharmonic Orchestra in an evening of Beethoven, Bach, and Brahms. When thirteen-year-old violinist Yehudi Menuhin made his debut, the audience went wild with excitement and the manager called the police to restore order. It was too late, however, to stop Einstein from bounding to Menuhin's dressing room by the shortest route—across the stage—and hugging the young violinist as he exclaimed, "Now I know there is a God in heaven!"

That summer, Einstein encountered a radically different audience while lecturing at England's Nottingham University. The puzzled frowns and politely muffled coughs were explained by a reporter who covered the event; because Einstein spoke in German, only the reporter and three of the professors had a clue as to what he was talking about.

Surprisingly, Einstein spoke scornfully of his heroes Faraday and Maxwell for lacking "the courage to admit that space was real," and, instead, inventing "a material called the ether." In reality, he said, "space is the solid, and matter only of secondary importance—an unsubstantial dream. That is to say, space has now turned around and is eating up matter. Space is now having its revenge." He also suggested that his unified field theory was gaining credibility, even though "many of my colleagues think I am crazy, and it is true my theory has not yet been fully tested, but tests so far have confirmed my views as to the unitary nature of the physical universe."

During his lecture Einstein covered a blackboard with math formulas, which the three professors persuaded him to sign, hoping to preserve it for posterity. Questions were invited, which, with the answers, were translated into their native tongue for the mostly tongue-tied natives. Asked what makes a successful theoretical physicist, Einstein virtually described himself: "If one gets hold of something that will not let go its hold on him—if one has the devotion for a great work—what more is necessary? Patience! Then a little more patience."

On a brief trip to Cambridge to pick up an honorary degree, he stayed overnight with Arthur Eddington and his sister at the university's observatory. Eddington had recently come to Einstein's support at a meeting of the Royal Astronomical Society, in response to the sensational discovery by American astronomer Edwin Hubble that the universe is expanding. This discovery seriously threatened Einstein's theory of a static universe.

To make his 1919 equations jibe with the views of most astronomers—as well as with his own feelings that the universe is static and unchanging—Einstein had divided both sides of a key equation by a figure close to zero. In fact, this "cosmological constant" represented so small a force of nature that it would only have a significant countereffect on gravity over billions of light years. Then a Russian mathematician, Alexander Friedmann, challenged Einstein by removing the arbitrary cosmological constant, making the equation indicate an expanding, or contracting, universe.

According to astronomer Robert Jastrow, "Einstein first ignored Friedmann's letter describing the new solution, and then, when Friedmann published his results in the *Zeitschrift fur Physics* in 1922, Einstein wrote a short note to the *Zeitschrift* proving that Friedmann's result was wrong. In fact, Einstein's proof was wrong. Finally, Einstein acknowledged his double error in a letter to *Zeitschrift* in 1923." Friedmann died prematurely, at thirty-six, in 1925; as a result, the subject was dropped until 1929, when Hubble, looking through a telescope in California, saw evidence that the galaxies are moving apart—that the universe is expanding.

Defending Einstein, Eddington cited the recent work of Georges Lemaître, a priest and professor of physics, who had revived Friedmann's work. Lemaître had also discovered that while Einstein's relativity equations were consistent with his conception of an initially static universe, by eliminating the arbitrary cosmological constant, as Friedmann had done, they foreshadowed the eventual expansion or contraction of the universe. Einstein was largely vindicated, especially when he admitted that introducing the cosmological constant had been the biggest blunder of his life.

No one promoted Einstein more enthusiastically than Eddington, a fact acknowledged by physicist Sir Joseph Thomson, who wrote, in a put-down masquerading as a compliment: "Eddington is the greatest authority in England on that important, evasive and difficult subject—relativity. He has . . . persuaded multitudes of people in this country and America that they understand what relativity means." The kindlier Sir Oliver Lodge, with no snide asides, also credited Eddington with Einstein's popularity in Britain, writing, "To the general public [Eddington] is best known as the apostle and interpreter of the great German genius Einstein . . . [who] without Eddington would be comparatively unknown in this country, whereas in fact the two have caught the imagination of the public to a surprising degree."

Einstein and Eddington took great delight in each other's company. Both were unpretentious, friendly, though essentially shy and undemonstrative men who shunned the limelight and were embarrassed by extravagant praise. Both men also had a keen sense of humor, as biographer Allie Vibert Douglas writes: "When Eddington was asked, 'When was the world created?' he would give the answer not in round numbers but exact to the last digit. He once said, with evident pleasure, that the expanding universe would shortly become too large for a dictator, since messages sent out with the velocity of light would never reach its more distant portions."

The two men met again in late June at the Berlin World Power Conference. Einstein merely welcomed the delegates, but Eddington made headlines by predicting the possibility of freeing the atom's vast, untapped energy and using "the limitless energy stored in matter—the same inexhaustible power which heats the sun." He reported that although engineers viewed this power source as a utopian dream, and physicists as a pleasant speculation, it was already being seriously investigated by his fellow astronomers.

Any pleasant speculation about the atom's potential was driven from Einstein's mind by a series of hysterical hate letters—not from the usual anonymous sources, but from his own son. Eduard, now twenty, blamed his father for "putting a shadow over his life" by deserting him. Bewildered and hurt, Einstein hurried to Zurich to reason with the boy. He found Mileva distraught. For both parents, it was now clear that "Eduard [had] suffered a breakdown . . . The family realized that [he] had long had intense and contradictory feeling toward his father— feelings of love, even worship, curiously mixed with a sense of rejection and personal inadequacy. It now seemed that his confused mind held nothing but hate for Einstein. Oddly enough, Eduard realized that his mind was sick. But he could not shake off the depression. His life was ruined and he blamed his father." Despite his distrust of psychiatry, Einstein sent Eduard to the best psychoanalysts in Switzerland and, when that failed, Vienna (though apparently not to Freud).

Writer Eduard Rubel gathered from several of Eduard's friends that just before his breakdown Eduard had been studying medicine at Zurich University, where he became obsessed with an older, possibly married, medical student. She tended to mother him, and he was deeply depressed when she rejected him as a lover. Einstein, in an undated letter to Eduard, recommended getting a job to ease his misery, pointing out that "even a genius like Schopenhauer was crushed through unemployment." He also suggested that there was an upside to this firsthand experience of depression for someone who hoped to become a psychiatrist: it would help him to become "a really good doctor of the soul." He did not advise Eduard to give up women, however, suggesting that instead he should get involved with a "plaything" rather than a "cunning" female. But Eduard was not consoled.

Einstein's laughter was not heard at Caputh that summer, and he aged visibly. Elsa told her friend Antonina Vallentin, later an Einstein biographer, that "Albert's sorrow over Eduard is eating him up. He has always tried to stay above personal worries, but this has hit him hard." Einstein, she said, had "always aspired to be completely invulnerable to everything that touched humanity. Actually, he was much more vulnerable than any other man I knew. But that situation was atrocious for him and he took it with great difficulty. I no longer found the crazy sense of humor in him, the taste for the unpredictable."

Shortly after Eduard's breakdown, Einstein was asked to state his philosophy of life, which turned out to be a denial of his oft-proclaimed detachment from human ties: "How strange is the lot of us mortals! Each of us is here for a brief sojourn; for what purpose we know not, though we sometimes think we sense it. But . . . we know from daily life that we exist for other people—first of all for those upon whose smiles and well-being our own happiness is wholly dependent."

"Eduard was a schizophrenic and would be institutionalized for most of his adult life," said Eduard's sister-in-law, Elizabeth Roboz Einstein, Hans Albert's second wife. "In early childhood he was a sort of genius who remembered everything he read. He played the piano beautifully [frantically, according to his father] and spoke English, which he learned in high school, perfectly. He was about to enter medical school when he had an unfortunate love affair which caused his breakdown. My husband thought his brother was ruined by electroshock treatment. Mileva devoted the rest of her life to Eduard. Once she took him back home. But he had to go to the sanatorium again. His father paid for that. I never figured out why they kept him in the sanatorium, because he never did anything to hurt anyone. He could eat, and behave normally, after all. Then one day they let him leave the sanatorium alone for a small walk, to see if he really needed to be locked up. He just crossed the street. And then he didn't know where he lived. He couldn't find his way back, although it was just across the street."

Mileva and her elder son, Hans Albert, hoped that Einstein would move back to Switzerland to help Eduard recover from his mental breakdown. They argued that he had done his best creative work there and would be free from the poisonous Nazi atmosphere. But he simply said he wanted to stay in Berlin. He felt, it seems, that he could still do good work there, encouraged by loyal colleagues such as Planck and von Laue.

Very little has been written about Einstein's younger son, Eduard. The author asked Einstein expert Robert Schulmann to elaborate.

D.B. Tributes to Eduard were published by his friends. What did you think of them?

SCHULMANN. I found them very touching and show his classmates were very fond of him. My impression from that book and from other sources is that Eduard was a very charming boy-man. And it tells you something about an Eduard who was badly treated by his father. Not only because he left him in Switzerland. It's clear that Einstein had a strange feeling about people with mental disorders and weaknesses. That's sort of his German generation. You remember how, in their early days, he was always talking about how strong Mileva was? So my impression is that he felt very awkward with Eduard and didn't do the right thing by him.

D.B. Could he have taken Eduard to America with him?

SCHULMANN. I doubt it. As soon as he showed symptoms of mental illness it would have been very difficult to get him in. I don't fault Einstein on that. I mean the efforts he made in the period when he was still in Berlin.

D.B. Not to have been more in touch with Eduard?

SCHULMANN. Right. But it's a complicated situation and I've not teased it all out to my own satisfaction. But we're talking about a situation where the boys are living with their mother. Einstein feels the boys are being poisoned against him. So that's one context we have to view him in. He also blamed Mileva's family heritage for Eduard's schizophrenia, arguing that she suffered from the same problems that her sister, Zorka, did.

D.B. That she was slightly mentally unbalanced?

SCHULMANN. She wasn't slightly, Zorka was completely wacko. There are horrible stories about her death, and living only with cats and [being] completely disheveled.

D.B. But Mileva wasn't as bad as that, was she?

SCHULMANN. Exactly. But Einstein said their behavior came from the same source. He says that explicitly. And he said very clearly he thought Eduard's insanity came from his wife's family. Whether he's right is a different question. It is a fact that he went on record in letters to his friends to say the problem with his life was in some degree Mileva's mental inheritance. And he also accused her of having alienated the boys from him after they separated.

D.B. Then what was Eduard's fate?

SCHULMANN. Carl Seelig helped to look after him. [Seelig was an Einstein biographer.] There was an emotional bond between them. Seelig was an independently wealthy journalist who had a large estate on Lake Lucerne. Seelig also championed a Swiss poet when he went into a mental institution. He played chess with Eduard and generally looked after him. Eventually, Eduard couldn't be kept at home any more. When Mileva died in 1948, Seelig frequently visited Eduard in the mental institution. Einstein felt a great debt to Seelig, who ended tragically. He jumped off a tram in Zurich and caught his foot. The tram ran over him and he died.

In Geneva, Einstein attended a meeting of the League of Nations' Committee on Intellectual Cooperation. When he was seen working on his unified field theory equations, a reporter asked, "Is this more interesting than your previous theories?" "I think so," he replied, with a smile, "but maybe when others read it they'll think I'm a fool." He was often seen in animated conversation with the frail, white-haired Marie Curie, a vice president of the committee, and her bright-eyed twenty-year-old daughter, Eve. As they talked, Madame Curie recalled that the young Einstein was "the funniest man," always thinking of the same thing, discussing the same problem of relativity, and seeking to talk with any scientist in any possibly related field.

He was hardly thought funny on this occasion, however, living up to his reputation for straight shooting by blasting the organization for practically giving "its blessing to the suppression of minorities. It has taken no stand against militarism. Scientists have become not representatives of science but of national traditions." Then he submitted his resignation, calling the committee, "despite its illustrious membership . . . the most ineffectual enterprise with which I have been associated."

At lunch that same day, a rabbi from St. Louis, Missouri, who had requested the meeting, asked for his view of marriages between Jews and non-Jews. "I believe in it in theory," Einstein replied. "But not in practice." It would have surprised Hans Albert to hear his father endorse even a hypothetical marriage, although when it came to Margot, Einstein leaped out of character to become a marriage broker.

Because Dimitri Marianoff and Margot were constantly together, Elsa feared gossip. Albert didn't, but—urged by Elsa—he somewhat bashfully suggested to them that matrimony might be a good idea. They agreed, as if only waiting for a nudge, and were married four days later. Einstein turned up "in a weather-beaten raincoat, old baggy trousers and his favorite slouch hat." His attire was no surprise to either bride or groom, who knew that he rarely changed into anything more appropriate even for VIPs.

After the honeymoon the couple moved in with the Einsteins, taking over one of the nine rooms in their Berlin apartment. This gave Marianoff a permanent vantage point from which to observe his subject. He watched him eat breakfast in an ankle-length nightshirt (though Einstein usually slept naked), shave with ordinary soap, and call out to his secretary, after spotting a manuscript in various colored inks, "Look, Dukas, it seems only crazy people send me their work."

With his doctor's approval, Einstein resumed his river jaunts, but not his rowing. If the boat became becalmed, he waited until the wind picked up. As a result, he was once marooned with a young woman aboard until two in the morning. He didn't say who she was.

Every chance he got, Einstein would stop off at Plesch's cottage beside the river to extemporize for hours on an organ. "Great crowds gathered outside on the river in boats, canoes, yachts, etc. listening . . . to his remarkable performance. It was not mere curiosity that drew them; no one knew that it was Einstein who was playing," said Plesch. "But I don't know anyone who exceeded him in fervor and sensibility."

As the depression deepened in Germany, bank closings escalated and lines outside soup kitchens stretched for blocks. The National Socialists increased their seats in the Reichstag from 12 to 107, while Hitler's sole major obstacle to power was the almost senile, eighty-three-year-old Marshal Paul von Hindenburg.

Yet Einstein, who voted for the winning Social-Democrats, felt confident that "the Hitler vote is only a symptom, not necessarily of anti-Jewish hatred but of momentary resentment caused by economic misery and unemployment within the ranks of misguided German youth. During the dangerous Dreyfus period almost the entire French nation was to be found in the anti-Semitic camp. I hope that as soon as the situation improves the German people will also find their road to clarity."

The more politically astute Freud wrote, "We are moving toward bad times. I ought to ignore it with the apathy of old age, but I can't help feeling sorry for my seven grandchildren."

Einstein turned from politics to counter the wild ideas of Bohr and Heisenberg, convinced that "if quantum physics is right, the world is crazy." The science of spectroscopy had been the most valuable tool in investigating the mysteries of stars. But to tackle this problem it was useless; Einstein had to rely on his imagination.

Playing God, at which Einstein had few peers, is a popular practice of physicists when an actual experiment isn't feasible. Scientists call this method a "thought" experiment. When Einstein arrived in Brussels for the Sixth Solvay Congress, he was ready to confront the champions of quantum theory with a thought experiment to refute the uncertainty principle.

At the meeting, he asked Bohr to visualize an enclosed box containing a radioactive element and an alarm clock. The alarm, set for a precise time, operates a cameralike shutter in the box that releases a certain amount of radiation. Imagine, said Einstein, that the box is weighed before and after the release of the radiation. (Under his equation $E = mc^2$, which states that mass is equivalent to energy and energy to mass, the loss in weight [mass] can be translated into energy loss.) So, Einstein concluded, one can determine both the energy of the released radiation and when it occurred—a direct contradiction of the uncertainty principle.

Bohr appeared to be in shock, said physicist Leon Rosenfeld. He spent the evening "going from one to the other and trying to persuade them that it couldn't be true, that it would have been the end of physics if Einstein were right; but he couldn't produce any refutation. I shall never forget the vision of the two antagonists, Einstein . . . walking quietly, with a somewhat ironical smile, and Bohr trotting near him, very excited . . . The next morning came Bohr's triumph."

After a sleepless night, he had found the answer: Einstein's own theory refuted the "thought" experiment. General relativity theory states that a clock not influenced by gravity runs faster than one in a gravitational field (This was later experimentally confirmed by testing atomic clocks in airplanes and satellites). Bohr, his worried face now radiant, told those at the meeting how Einstein had gone wrong. By exposing the alarm clock to the varying effects of the gravitational field, however minute, the exact time the shutter opened and released the energy would be uncertain. Years later, Einstein conceded that quantum mechanics "undoubtedly contains part of the ultimate truth."

Shuttling between scientific and humanitarian pursuits, Einstein spent a few days in London as guest of honor at a fund-raising event for poor European Jews. Thrilled to be invited to welcome Einstein, George Bernard Shaw got to work on a speech that the British Broadcasting Company (BBC) intended to broadcast to the United States, Germany, and the British Isles. He wrote urgently to the BBC director, John Reith:

> Your people have not quite grasped the importance of the occasion. I must, in the name of British culture and science welcome the foremost natural philosopher of the last 300 years. The job should really be done by the Prime Minister; but he has left it to me; and it should be heralded with the utmost possible réclame, or we shall be branded as Philistines . . . I cannot do it in 15 minutes, and shall be lucky if I get out of it in 25. As to limiting Einstein, it is out of the question: we must be prepared for 2 minutes or 30 at his pleasure.

He also hoped, in vain, that Einstein had "learned some English since he was here last; for my linguistic attainments are deplorable, and I cannot, like Eddington, converse with him in equations."

The banquet to honor Einstein was held at the Savoy Hotel with Lord Rothschild in the chair and H. G. Wells and Arthur Eddington among the thousand guests. Shaw didn't always stick to his script, but gave a dazzling and slightly dizzy short course in the history of physics:

> "I must [talk about] Ptolemy & Aristotle, Kepler & Copernicus, Galileo & Newton, gravitation and relativity and modern astrophysics and Heaven knows what, hailing Einstein as the successor of Newton . . .
>
> "Newton made a universe which lasted for 300 years. Einstein has made a universe, which I suppose you want me to say will never stop, but I don't know how long it will last . . . When science reached Newton, science came up against that extraordinary Englishman. If I was speaking 15 years ago I would have said that he had the greatest mind that ever man was endowed with. Combine that light of that wonderful mind with credulity, with superstition, and delusions that would disgrace a rabbit.
>
> "As an Englishman he postulated a rectilinear universe because the English always used the word 'square' to denote honesty, truthfulness: in short, rectitude. Newton knew that the universe consisted of heavenly bodies that were in motion, and that none of them moved in straight lines, nor ever could. Mere fact will never stop an Englishman. Newton invented a straight line, and that was the law of gravitation, and when he had invented this he had created a universe which is wonderful in itself, a complete British universe, and established it as a religion

which was devoutly believed for 300 years. I know I was educated in it and was brought up to believe in it firmly. Three hundred years after its establishment, a young professor came along. He said a lot of things and we called him a blasphemer. He claimed Newton's theory of the apple was wrong. He said, 'Newton did not know what happened to the apple, and I can prove this when the next eclipse comes.'

"The world is not a rectilinear world: It is a curvilinear world. The heavenly bodies go in curves because that is the natural way for them to go, and so the whole Newtonian universe crumpled up and was succeeded by the Einstein universe. Here in England, he is a wonderful man. This man is not challenging the facts of science, but the axioms of science, and science has surrendered to the challenge."

Asking Einstein to forgive them for having broken his "august solitude" in order to help "the poorest of the poor throughout the world," Shaw concluded with, "I drink to the greatest of our contemporaries, Einstein!"

Sweating in the glare of the lamps needed to film the event, and having been subjected to wild applause each time his name was mentioned, Einstein stood to respond. Speaking through an interpreter, he said:

> "The only way of really helping the Jew in Eastern countries is to give him access to new fields of activity, for which he is struggling all over the world. It is to you English fellow-Jews that we now appeal to help us in this great enterprise. Our friends are not exactly numerous, but among them are men of noble spirit and with a strong sense of justice, who have devoted their lives to uplifting human society and liberating the individual from degrading oppression. We are happy and fortunate to have such men from the Gentile world among us tonight. It gives me great pleasure to see Bernard Shaw and H. G. Wells, to whose view of life I am particularly attracted.
>
> "You, Mr. Shaw, have succeeded in winning the affection and joyous admiration of the world while pursuing a path which led others to martyrdom . . . By holding the mirror before us, Mr. Shaw has been able, as no other contemporary, to liberate us and to take from us some of the burdens of life. For this we are devoutly grateful to you.
>
> "I personally am also grateful to you for the unforgettable words which you have addressed to my mythical namesake who makes life so difficult for me, although he is really, for all his clumsy, formidable size, quite a harmless fellow . . . To you all I say that the existence and destiny of our people depend less on external factors than on us remaining faithful to the moral traditions which have enabled us to survive for thousands of years despite the fierce storms that have broken over our heads. In the service of life sacrifice becomes grace."

When Shaw apologized for messing up the science and for being perhaps too hard on Newton, Einstein replied through a translator, "What does it matter? That is not your business!"

Two weeks later, on his home ground, Einstein gave a talk to radical students who filled a large hall to suffocation. Many came directly from a Communist demonstration protesting wage cuts. He invited them to ask anything, even if it might seem foolish, but when they hit him with a flood of questions about the volatile political situation, he responded with good-natured wisecracks. Serious responses could have sparked a riot. His subject was science, he explained, the attempts to make sense of the universe. After decrying the belief system of primitive peoples who thought that everything that happened was caused by "a human, divine, or demonic entity," he admitted that modern man still had a long way to go in his search for the secrets of the universe, and that "the further we proceed the more formidable are the riddles facing us." Yet he remained an optimist.

Speaking to a large audience in his native tongue (he rarely spoke in anything else), Einstein was at his best, in the opinion of Leo Szilard: "He is a man as free from vanity as I have ever seen. And he talked to a thousand as he would talk to a few friends gathered at the fireside. That does not mean he is not shy, of course. Great achievement in the field of science frequently requires a certain kind of sensitivity, and sensitivity leads to shyness. His simplicity is perhaps the key to the understanding of his work—in science the greatest thoughts are the simplest thoughts."

In November, Einstein prepared for his second trip to the United States, lured to the California Institute of Technology (Caltech) and the Mount Wilson Observatory by the prospect of discussing cosmology with Richard Tolman (professor of physical chemistry and mathematical physics), having his latest ideas experimentally checked, and—not least—spending the winter in the land of blue skies and orange trees. He planned to skip New York City on this trip because "I hate crowds and making speeches. I hate facing cameras and having to answer a crossfire of questions. Why popular fancy should seize on me, a scientist, dealing in abstract things and happy if left alone, is a manifestation of mass psychology that is beyond me."

Einstein was shocked by the crass commercial offers from America when news of the impending trip leaked out. He laughed at one request—to send a pair of his old shoes to be displayed at some sideshow with those of Hollywood movie stars and candidates for the U.S. presidency—but he was more often disgusted by hucksters offering him tens of thousands of dollars to endorse neckties, toilet waters, musical instruments, and disinfectants—products he'd never heard of, let alone used. "Isn't it a sad commentary on commercialism?" he remarked. "And I must add that business firms make these offers with no thought of wanting to insult me. It evidently means that this form of corruption is widespread."

For Einstein, though, the news about Palestine was far more distressing. The British government, employing doublespeak, announced that Palestine would remain the Jewish homeland, but that no more Jews could settle there. The official excuse was that the country was filled to capacity. Yet it was clear that the announcement was a move to placate the Arabs at the expense of the Jews.

The Einsteins traveled by train from Berlin to Antwerp. As the German consul handed Elsa a bouquet at the Antwerp dockside, Albert protested the British decision about Palestine to a group of reporters: "We Jews are everywhere subject to attacks and humiliations that result from the exaggeration of nationalism and racial vanity, which, in most European countries, expresses itself in the form of aggressive anti-Semitism. The Jewish national home is not a luxury but an absolute necessity for the Jewish people. Therefore, the reply of the Jews to the present difficulties must be a determination to redouble their efforts in Palestine."

During their Atlantic crossing on the SS *Belgenland*, Einstein stepped up his criticism of the British government for not handling "the Palestine question with objectivity, which is the most that was expected. It is, however, gratifying that British public opinion is 'solid' and that voices are raised in favor of justice." Einstein himself was attacked by both Nazis and fellow Jews. Hitler's party organ, the *Völkischer Beobachter*, questioned his patriotism for traveling on a Belgian rather than a German ship that was also crossing the Atlantic (he chose the Belgian vessel because it continued on to California, his ultimate destination; the available German ship didn't). Meanwhile, a group of anti-Zionist Jews, the National German-Jewish Union, denounced him for using his fame to promote Zionism.

When his ship docked in New York City, a human hurricane hit dockside as scores of reporters and photographers headed for Einstein. He was so unnerved that he locked himself in his cabin to hide from "these men like wolves—each anxious to get a bite at me." Eventually he was persuaded to face the wolves in the dining room for fifteen minutes. Elsa preceded him, confiding to a reporter: "The professor is afraid of so many people. He'd rather not come out."

The fifteen-minute deadline provoked the reporters into yelling questions like frenzied stockbrokers in a falling market, their frustration fueled by the language barrier. Elsa then offered to translate, assisted by Paul Schwarz, the German consul.

One spry reporter got his question across by climbing onto a table and shouting, "Define the fourth dimension in one word!"

Einstein looked puzzled, then said with a laugh, "You will have to ask a spiritualist."

"That's not a trick question," the reporter persisted. "Would the correct answer be 'time'?" Einstein agreed.

Some reporters stood on chairs or bounced up and down to be noticed. Only those with lucky timing and loud voices got answers, among them the reporter who yelled, "Is there a relationship between science and metaphysics?"

"Science itself is metaphysics."

"Can you define relativity in one sentence?"

"It would take me three days to give a short definition."

"Will your new field theory develop into something as important as the relativity theory?"

"No one can possibly tell."

"Did you bring your violin with you?"

"I left it at home because the tropical climate we will encounter in going to California by way of the Panama Canal might harm it."

Elsa had laughed at some questions, but saw that the session was becoming an ordeal for Albert. "This reminds me of a Punch and Judy show," he said in an aside to her. Nevertheless, he had promised fifteen minutes and intended to stick it out.

"What do you think of Adolph Hitler?"

"I do not enjoy Herr Hitler's acquaintance. He is living on the empty stomach of Germany. As soon as economic conditions improve, he will no longer be important."

"Can religion bring peace to the world?"

"Up to now religion has not done it. As to the future, I am not a prophet."

When the fifteen minutes were up, he tried to escape from the crowd by going up on deck, but was trapped by a locked door. Then he remembered his promise to broadcast to the students of America over CBS radio, whose equipment was already set up in the ship's drawing room. Before he could reach it, however, he was steered to a rival NBC microphone and asked to say a few words. Einstein gave his opinion that America, "which through hard but peaceful labor, has achieved the position of undisputed pre-eminence among the nations of the world, stands forth today as the one tangible citadel of the ancient and high ideals of a political democracy."

Half an hour later, speaking on CBS radio, "the scientist disarmingly protested his innocence of the rivalries between radio networks," being unaware, he said, "that there was more than one broadcast system in the country. 'I had the same experience with the two companies that Jacob had with his wives, Leah and Rachel. He did not know which of them he should marry.'"

Although he slept on the ship all five nights it was docked, Einstein had changed his mind about avoiding the city. One taste of Manhattan was addictive. He went ashore every day; there he viewed Manhattan's skyline from the roof of the German consul's apartment, celebrated Hanukkah with fifteen thousand others at Madison Square Garden, and was a lunch guest at *The New York Times*, where editor Carr Van Anda neatly characterized his gift as "an imagination which often outstrips mathematics, leaving it to catch up when it may and render its verdict of approval."

Einstein was cheered at the Metropolitan Opera when he entered his box one night and tried to make himself invisible. He overcame his distaste for speechmaking to plead with members of the New History Society "to declare, before the World Disarmament Conference convenes next February, that you will refuse to give assistance to any war or preparations for war." American Legion members in California responded by declaring him persona non grata.

By contrast, he was all but crowned at a Manhattan City Hall reception, where Columbia University's president, Nicholas Murray Butler, introduced Einstein as "the ruling monarch of the mind." The municipal band then greeted the monarch with a mixed message, belting out with equal fervor "Deutschland über Alles" and "Hatikvah," the Zionist song that later became the Israeli national anthem. Mayor Jimmy Walker gave the main speech of welcome, muffing an attempt to be both honest and complimentary by saying, "We have a very profound appreciation of the contributions you have made to science even if we don't understand them."

There was even an attempt to sanctify Einstein, and Elsa talked Albert into examining the evidence, an unlikely image of himself—a Jewish agnostic—at the Riverside Church. Its pastor, Dr. Harry Fosdick, had emulated the churchmen of the Middle Ages who decorated their churches with statues of saints and kings, but he gave his church a modern touch by including scientists, artists, and philosophers whose lives had enriched the world. Fourteen eminent American scientists were asked to suggest scientific candidates, and Einstein made every list.

According to *The New York Times*, the pastor

> met Einstein and his wife at the door of the church. Einstein stopped longest before the window in which was embodied Immanuel Kant in his garden with his servant holding an umbrella over his head, Moses holding the tablets of the Law, and Beethoven composing. He passed John Bunyan, Milton, Hegel, Pasteur, Darwin, Thomas Aquinas, Descartes, Emerson, and hundreds of others. Then he asked Dr. Fosdick, "Am I the only living man among all these figures of the ages?" And Dr. Fosdick, with a heightened sense of this moment replied gravely, "That is true, Professor Einstein."
>
> "Then I will have to be very careful for the rest of my life as to what I do and say."

Later, Einstein joked, "I might have imagined that they could make a Jewish saint of me, but I never thought I'd become a Protestant one!"

A familiar face turned up on Einstein's last day in the city—George Sylvester Viereck, not yet an undercover Nazi agent. After the recent Berlin interview, Einstein treated him as a friend rather than as one of the wolf pack out to get a bite of him. As they strode briskly up and down Riverside Drive, Einstein chuckled over having briefly escaped from his entourage and the press waiting at the dock. "Don't you ever walk in America?" he asked Viereck. "If you don't look out your legs will atrophy in a few generations. I've had no chance to use my legs in New York. To get to know a country you must have direct contact with the earth. It's futile to gaze at the world through a car window. Here they bundle me up and shove me into a car like a parcel."

The ship then took Einstein and his party to Cuba, where he was honored by the Academy of Sciences, then headed west through the Panama Canal, where an Ecuadorean Indian gave him a Panama hat. While Elsa was in the beauty parlor or sitting under the sun, Einstein dictated scientific notes to his assistant, Walther Mayer, and letters to his secretary, Helen Dukas—including one to his pen pal, Upton Sinclair, whom he soon hoped to meet.

Some would think it an unlikely friendship but not I. F. Stone, the noted political gadfly and investigative journalist. "Don't forget, Upton Sinclair was not just a spiritualist wack," said Stone. "He was an American Zola. And Sinclair's novel, *The Jungle*, is not only the American counterpart of some of Zola's novels, but it paved the way for the Pure Food and Drugs Act."

Shortly after dawn on December 31, the *Belgenland* docked at San Diego. A pack of newsmen stormed the deck, two falling off the ladder on the way up, while five hundred schoolgirls in white uniforms waited onshore to serenade

Einstein. He was still in bed. A smiling and apologetic Elsa preceded him into the ship's drawing room to placate impatient reporters: "He was not up, but he is dressing now. He did not stop to take his bath. He will not have breakfast first. He will be here in a little while." He appeared soon after, took a seat, and gave Elsa a glance that said, "Here we go again!" But this time the shouting was restricted to a roar of protest when one reporter tried to have a private conversation with Einstein in German. Then the questions started.

In response to one of the reporters' queries, he agreed that science and religion did not conflict, then added, "Though it depends, of course, on your religious views." He said that during his visit he would conduct "thought" experiments to determine the influence of the earth's rotation on the sun's rays. Amused by the question, "Are there men living elsewhere in the universe?" he replied with a smile, "Men? Other beings, perhaps, but not men." Asked if there was any danger of another war in Europe, he replied, "If you don't do anything it may come. It is not a question of 'waiting' but of 'acting.' World peace is possible with the proper organization and the right ideals." Faced with many of the same questions that were bellowed at him in New York, Einstein remarked to Elsa, "It would take a whole library to answer these fellows."

Gene Coughlin, of the Los Angeles tabloid *Illustrated Daily News*, gave Einstein a page of equations, hoping that his response would make news. Einstein refused, saying through Elsa that it would take too long to discuss. Coughlin had been taken off a murder trial especially to cover Einstein's arrival, and he couldn't return to his editor without a few exclusive quotes. So, for the rest of the day, Coughlin and his photographer, Harry Spang, set off in hot pursuit of Einstein, waiting for the chance to get him alone to spring a provocative, headline-inducing question. They followed him down the gangplank as he was sung ashore by the white-clad schoolgirls and smothered in their flower bouquets; kept him in sight on a tour of San Diego and during a reception attended by eight thousand in Balboa Park; then parked outside the hotel, where he and Elsa took a siesta.

Coughlin found the right moment while tailing the car that was speeding the couple north on the coast road to Pasadena. It had stopped to let Einstein stroll over to a small headland known as Sunset Cliffs, where he stood gazing at the sea and sky. Seizing the moment, Coughlin leaped from his car, the question on his lips, followed by Spang, his camera at the ready.

"Doctor," Coughlin said, "is there a God?"

Einstein stared at the water's edge some twenty feet below, then turned to his questioner. Coughlin later wrote:

> There were tears in his eyes, and he was sniffing. Spang shot the picture as Einstein was hustled away before he could answer me. "Well," I said, "the way he reacted, he believes in God. Did you ever see such an emotional face?" Spang was standing on the edge of the headland, where the great scientist had stood. He looked down, then called me: "Come over here. And look down, like he did." I looked down and there, caught against the base of the little cliff, was a

shark that must have been dead in the hot sun for several days. "Make anybody cry," Spang said.

For several weeks, Einstein relaxed in "paradise," as he called Pasadena, free from stun-gun encounters with newshounds. During that time the public was not starved for news of Einstein, however. His other stepson-in-law, Rudolf Kayser, married to Elsa's eldest daughter, Ilse, had beaten Marianoff to a publisher with an Einstein biography, writing it under the pen name Anton Reiser. But it missed the mark, said its subject in a preface to the book, because it wasn't wild enough, leaving out "the irrational, the inconsistent, the droll, even the insane, which nature, inexhaustively operative, implants in an individual, seemingly for her own amusement."

CHAPTER 23

Einstein in California

1931

51 to 52 years old

Photographers mobbed the Einsteins when they arrived at Caltech on December 31, 1930. One broke away to take a distant shot of the action. A second joined him. Then a third. Finally they were all trying to photograph a crowd that wasn't there. Einstein must have loved this echo of quantum mechanics; now it's here, now it isn't. But the excitement was understandable: Caltech was to be the scene of a "meeting of the leading American men of physical science with a European who is the greatest living innovator in physical science, and probably one of the two or three greatest innovators in the whole history of science."

In the small Caltech classroom where he was to be interviewed, reporters were asked to go easy on their fragile subject, who was still convalescing from debilitating illnesses. Intending to interpret for Albert, Elsa accepted fourteen written questions. Sadly, her high-school English was as inadequate as it had been ten years before. Pleading "an unscientific vocabulary," she handed the questions to the bilingual physics professor Richard Tolman.

Through Tolman, Einstein said he hoped that scientists at Caltech and Mount Wilson would help him solve the biggest question on his mind: Are gravitation, light, electricity, and electromagnetism different forms of the same thing? Curiously, he still clung to his view that the general structure of the universe is static, though "not precisely static in detail, because we observe things changing in it." Perhaps he still held this view because he had not yet studied recent statements by Harvard astronomer Harlow Shapley that challenged the Einsteinian model of a finite universe in which matter or the galaxies were uniformly distributed. Having discovered and measured eighteen thousand new galaxies, Shapley found great deviations from uniformity in their distributions throughout space, and said that "the whole universe is expanding, galaxies are running away or scattering in space. Some day emptiness will be left. During the last two thousand million years cosmos or space-time had doubled in size."

Nevertheless, on this point Einstein was plainly wavering. He conceded that new Hubble-Humason observations about the redshift in the light from the nebulae indicated that the universe is not static, and that theoretical investigations "by Lemaître and Tolman show a view that fits well into the general theory of relativity."

Asked if our civilization was in danger, he missed the chance to blast Hitler or Stalin, replying instead that scientific investigation could not influence the international situation; only man's determination. He ducked the question "Is there any conflict between science and belief in an all-intelligent, omnipotent God?" by saying that it would take a book to answer.

Among nature's mysteries still to be solved, he included both the source and the nature of cosmic rays. (Robert Millikan, the head of Caltech, was then a leading investigator of cosmic rays. The origin of these high-energy subatomic particles remains a mystery.) When asked if he'd heard from a fourteen-year-old Los Angeles boy who was said to be an expert on relativity, Einstein speculated that the letter was among a three-foot pile of correspondence awaiting him. Facing the continual pile of mail was always a grim prospect, judging by his idea of hell, which he had once described as a menacing devil approaching every half-hour with a pitchfork loaded with a fresh bale of letters requiring answers.

Einstein soon ended the interview with a smile and a muttered remark, which Tolman translated as: "The professor hopes he passed the examination."

He spent most of January 3 in Tolman's basement lab, known as "the cell," exchanging views and trying out new equations with Tolman and Paul Epstein, both Caltech professors. That night he wrote in his diary: "Doubt about correctness of Tolman's work on cosmological problem, but Tolman turned out to be right."

The next morning, he and Elsa chose their temporary home, a sunny, white-walled bungalow equipped with a refrigerator and automatic washer, rather than a large home staffed with servants. Elsa's girlhood friend Barbara Seibert arrived with flowers, then she and Elsa went shopping, while Einstein headed straight for what was to be his study and began to write.

On January 7 he went to Mount Wilson Observatory to hear a lecture by astronomer Charles St. John, who revealed that a redshift in the sun had been predicted by the theory of general relativity. Einstein then huddled with observatory director Walter Adams, who had developed a test to confirm relativity through observing the companion of the star Sirius (also known as the Dog Star). Here was a truly unique object, Adams told Einstein; there was nothing else found in the entire universe to match it for brightness or density. Just one cubic inch of it weighed about a ton.

Einstein's diary entry that night reads: "It is very interesting here. Last night with Millikan, who plays the role of God . . . today astronomical colloquia, rotation of the sun, by St. John. Very sympathetic tone. I have found the probable cause of the variability of the sun's rotation in the circulatory movement on the surface . . . Today I lectured about a thought experiment in the theoretical physics colloquium."

He and Elsa accepted an invitation to tour First National Studio, but when a movie camera was pointed at him, he balked at being used for publicity. German-speaking film technicians convinced him that they only wanted to demonstrate Hollywood's version of relativity; all he had to do was pretend to drive a car. Einstein took the wheel as instructed, with Elsa at his side, and then the director ordered the cameraman to start shooting. Einstein had merely to follow instructions: to wiggle the steering wheel occasionally as if he was driving and then to point briefly at the studio floor. It would all be explained to him later.

That night the crew rigged up a screen in the Einsteins' bungalow and ran the film. To the Einsteins' astonishment, they were not parked in the film studio; rather, Albert was clearly driving Elsa on a sightseeing tour of Los Angeles. Suddenly, Elsa gasped and Albert roared with laughter as the car took off like an airplane while his film image pointed not to the studio floor but to the Rocky Mountains far below. Eventually, the flying machine landed in the German countryside and took them to several scenic spots.

Einstein laughed again when the crew told him that the secret of Hollywood's "relativity" was a combination of back projection and double exposure. The film was then given to the Einsteins to prevent its use for publicity, though some crafty character took a still photo of them in "flight," which appeared in the press.

To avoid playing favorites, Einstein also toured the Warner and Universal studios. At the latter, he told studio boss Carl Laemmle that he would like to meet Charlie Chaplin. Laemmle immediately phoned the actor, who hurried over to have lunch with the Einsteins in the studio restaurant. Elsa did most of the talking, because Albert's English still sounded like something he'd just invented. Helen Dukas and Walther Mayer were also in the lunch party. Chaplin, who fancied himself a shrewd judge of character, thought Einstein's jovial, friendly manner hid a highly emotional temperament. He saw Elsa as a lively and energetic woman who was obviously proud to be the wife of a world-famous genius, and found her enthusiasm endearing. When she whispered that he and Albert should meet again for a quiet chat, Chaplin agreed and invited him to dinner at his Beverly Hills home.

It was hardly a quiet chat: Elsa came, too, and a couple of Chaplin's friends joined them around the huge dining-room table. The highlight of the evening was Elsa's account of how Albert had arrived at his general theory of relativity. As she told it, he came down one morning in his dressing gown, but couldn't face breakfast because of "a wonderful idea." Instead, he sat at the piano, sipping coffee and occasionally interrupting his playing to repeat: "I've got a wonderful idea!" It just needed a little work. After a while he returned to his study, asking not to be disturbed, and stayed there for two weeks. He had meals sent up to him and emerged each evening for a short walk. After those two weeks he came down one morning looking exhausted and put two sheets of paper on the table. "That's it," he said. And that, concluded Elsa, was his theory of relativity.

Asked if he believed in ghosts, Einstein said that he had never met one, but if twelve people claimed to see the same ghost simultaneously he might be convinced—though his smile implied otherwise. He shook his head when Chaplin then inquired if he'd ever been to a séance during which a table took off and floated in the air. Psychic phenomena was a Hollywood craze—and fake mediums a cottage industry—but Einstein showed little interest, so Chaplin switched from psychics to physics. Did his theory conflict with Newton's? "On the contrary," Einstein said. "It is an extension of Newton."

Chaplin, a charmer on and off the screen, had met his match in Einstein, who, as his good friend Plesch once described, "if you happen to tell him the same joke twice will not interrupt you . . . but listen tolerantly and laugh with you again. He greatly appreciates mother-wit and is as delighted as a child with his own witticisms, even when sometimes a biting remark slips from his lips among friends. He is certainly no prude, though with most thinking men he rejects sheer filth, but if it has real wit it can be as broad as it likes."

During dinner, Elsa said she'd turned down many offers to make Albert a movie actor, but that he had posed for newsreel cameras in Berlin for a fee, which went to the poor. When she heard that Chaplin was about to start a world tour, Elsa invited him to be their guest in Berlin, making it clear that their modest apartment hardly compared with his hillside Beverly Hills mansion, which was located on a six-acre estate with a Japanese-style garden of bamboo bridges, pools, pagodas, and thousands of azaleas. They chose not to be wealthy, Elsa said; Albert had access to a million dollars from the Rockefeller Foundation, which he declined to use.

When Caltech threw a banquet to honor Einstein, two hundred of its most prominent figures turned up. Seven of them were scientists to whom he was greatly indebted: physicists Albert Michelson (now retired and living in California), Robert Millikan, and Richard Tolman; and astronomers William Campbell, Walter Adams, Charles St. John, and Edwin Hubble.

Millikan introduced Einstein as the man who had done the most to make modern science an empirical pursuit. Einstein then stressed how much his achievements had depended on the pioneering work of those present:

> "From far away I have come to you but not to strangers. I have come among men who for many years have been true comrades with me in my labors. You, my honored Dr. Michelson, began this work when I was only a youngster three feet high . . . Through your marvelous experimental work you paved the way for the development of the theory of relativity . . . Without your work this theory would today be scarcely more than an interesting speculation. Your experimental verifications first set the theory on a real basis . . . [And] the recent discoveries of Hubble concerning the dependance of the redshift in the spectral lines of the spiral nebulae on their distance, has led to a dynamic conception of the spatial structure of the universe, to which Tolman's work has given an original and especially illuminating theoretical expression. Likewise in the realm of the quan-

tum theory, I am grateful to you for your important assistance because of your fundamental experimental investigations. Here I acknowledge gratefully Millikan's researches concerning the photoelectric effect . . . full of the happy conviction that your researches will continue through the future to broaden and deepen, without let or hindrance, our knowledge of nature's mysterious forces."

Recently recovered from a stroke and soon to die, the seventy-eight-year-old Michelson was visibly moved by Einstein's tribute. He modestly admitted that in making his experiment "there was no conception of the tremendous consequences brought about by the great revolution which Dr. Einstein's theory of relativity has caused, a revolution in scientific thought unprecedented in the history of science."

Einstein later qualified his debt to Michelson in a letter to Michelson's biographer, Bernard Jaffe:

Michelson's experiment was of considerable influence upon my work insofar as it strengthened my conviction concerning the validity of the principle of the special theory of relativity. On the other side I was pretty much convinced of the validity of the principle before I did know this experiment and its result. In any case, Michelson's experiment removed practically any doubt about the validity of the principle in optics, and showed that a profound change of the basic concepts of physics was inevitable.

A few days after the Caltech banquet, Einstein went to Michelson's home, hands outstretched in friendship. Michelson tried to respond but was too weak to get up from his chair. It was an emotional meeting. Their mutual coolness because of Michelson's long-delayed acceptance of the special theory of relativity was forgotten. According to his daughter, Dorothy Michaelson Livingston, "Although Einstein's manner to the older man was one of great deference and respect, their conversation . . . had wit and humor. They had a few laughs at Millikan's expense. Outside his office, an enormous sign posted by one of the religious sects read: JESUS SAVES. Beneath this message some students had added: BUT MILLIKAN GETS THE CREDIT!"

Drinking wine at lunch, they wondered if Congress would repeal Prohibition. Einstein, all for repeal, spoke approvingly of German beer gardens where intellectuals enjoyed stimulating scholarly discussions. Mrs. Michelson said things were different in the pre-Prohibition American saloon, which was "often the scene of much conviviality and a pleasant escape from the daily grind, but it was not famous as the background for philosophical exchanges."

She warned Einstein not to discuss the ether with her husband as it overexcited him, then embarrassed herself by describing a person she had recently encountered as having "an incredibly bushy head of hair." She used both hands to show just how incredible it was. "In the middle of her sentence her eyes rested on Dr. Einstein's equally incredible hairdo, and her voice trailed off . . . Not knowing how to extricate herself, she hastily rang for the maid to pass the vegetables."

This was the last meeting between the two men. Michelson died that summer.

Though Michelson had taken so long to accept relativity, Einstein found that a classroom of thirty Caltech students bought the theory. Still, they were puzzled by the thought behind it. When he answered their questions by covering a blackboard with equations and explaining that they included ten unknown quantities, four of which had to be assumed, some gasped. "I was amazed, too," he said, "when working on a simple equation to explain the universe and its properties—meaning the unified field theory—to find it had sixteen unknown quantities with four to be assumed, as well as the laws of my own gravitational theory combined with Maxwell's law of electromagnetism." Despite so many unknowns, he believed that experimental proof of part of this theory was imminent.

The next night Edwin Hubble, the great explorer of the galaxies, gave Einstein a turn at the 100-inch Mount Wilson telescope—the world's largest. Through it, Hubble had seen farther into space than anyone, and he estimated that there were at least thirty million galaxies as big as or bigger than our own Milky Way. Each contained hundreds of billions of stars, and the average distance between galaxies was 1.5 million light years. Hubble also told Einstein that he had measured distant galaxies streaking away from the earth at or near the speed of light.

Einstein was tremendously impressed. The idea of an expanding universe had once "irritated" him and seemed "senseless," but now he could hardly deny the spectacular evidence. He was also doubtless pleased that Hubble's law—"the farther away a galaxy is, the faster it moves"—had been predicted by his own theory of relativity.

As he climbed to the platform supporting the telescope, Einstein called out nervously, "The mountain is spinning!" He was quickly reassured by Hubble's shouted response, "The dome is revolving and carrying you with it."

After an exciting but exhausting month, Einstein anticipated a lazy vacation in Coachella Valley, near Palm Springs, as the guest of his New York attorney friend, Samuel Untermyer. Packing her husband's pipe tobacco for the long weekend, Elsa told one of the ever-present reporters, "I must think of everything for his comfort, for he loves comfort."

She had warned the reporters that Einstein would be incommunicado during the weekend. And so, believing he had shaken the curious press, he took a solitary walk to the edge of the Mojave Desert. It was warm. He sat down, intending to sunbathe, and began to shed his clothes. He wasn't alone for long.

A woman reporter, Cissy Patterson, was hell-bent on getting an Einstein interview. Training herself to run her own newspaper (she would later own the *Washington Herald*, the paper she now worked for) by going into the trenches with the troops, she had already made her mark by dressing as a bum to spend a night in a flophouse for a firsthand report of life in the lower depths. Now she hoped to prove that an attractive female reporter had the edge on a male in getting a reluctant man to talk.

She traced Einstein to the Untermyer estate, where the European staff gave her the runaround. A Swedish maid, an English female secretary, and an English male secretary passed her along from one to the other. Finally, Cissy faced an Italian-looking butler. She pulled the helpless-little-woman routine on him, and he indicated a steep, winding trail as the way to the "very kind, very nice professor." She began the climb, wondering whether to hit him first with a relativity question or to break the ice by asking how he liked the California sunshine. Still undecided, she skirted a huge rock and suddenly came on Einstein, "gazing out across the wide and silent desert. A white handkerchief, knotted at each of the four corners, rested upon his famous shock of curly gray hair. Evidently, the professor was taking a sun bath, whilst contemplating, for he was relatively in the nude."

Should she emulate Anne Royall? After spotting U.S. president John Quincy Adams emerging naked from a swim in the Potomac River, Royall sat on his clothes until he promised her an interview.

Cissy simply retreated without a word, "crestfallen and wondering what a regular determined go-getter she-reporter would do under the circumstances." (Yet she had no qualms, soon after, about interviewing gangster Al Capone.) Her wacky account of how she didn't get the Einstein interview made the *Washington Herald*'s front page, titillating readers with a woman's view of genius in the raw. William Randolph Hearst, the publishing tycoon who owned the paper, rated her "non-interview a gem."

A few nights later Cissy covered the premiere of Charlie Chaplin's movie *City Lights*, at which Einstein was the guest of honor. Chaplin put his own twist on their rousing reception at the movie theater, telling Einstein, "They are applauding you because none of them understands you and applauding me because everybody understands me."

Cissy Patterson sat directly behind Albert, Elsa, and Chaplin, watching "Einstein from the corner of my eye . . . When the first show began with a zigzag of pretty girls across the stage, the professor's eyebrows shot up in semi circles of astonishment. Presently the loveliest drop curtain in the world parted and looped aside for Charlie's picture . . . Einstein's mouth fell open. He stared bewildered, utterly absorbed, like a child at a Christmas pantomime. During Charlie's picture this look of fixed astonishment never wholly left his face, even when the whole house roared with laughter . . . When the little blind girl came on the screen, Frau Einstein began gently to shake her head. 'Ach weh,' she sighed. Suddenly, just at the end of 'City Lights' . . . I began to cry, and pretty hard, too, but I choked, blinked and managed to take one last look at Einstein. An enormous handkerchief, maybe the same which covered his head while he was taking the sun bath on the Mojave Desert, almost concealed his face, but you could see two big sorrowful eyes still wide with wonderment, brimming with tears."

An incident that night highlighted Elsa's innate kindness and sense of humor. While she was in the powder room before the screening, a woman, mistaking her for the attendant, asked her to fetch something. The woman's shock and embarrassment were obvious when she was introduced to the Einsteins after the show. Elsa eased what could have been an awkward moment by embracing her and laughing about it.

Einstein enjoyed his Hollywood diversions, but Millikan resented anyone who took him away from scientific pursuits, especially the "dangerous radical" Upton Sinclair. Though physically slight, Sinclair had a big spirit and an ego to match. Chaplin, among others, likened him to Christ for his passionate work as a social reformer. Sinclair and Einstein had been corresponding for a few years, and after reading several of Sinclair's crusading and muckraking books, Einstein had written him: "To the most beautiful joys of my life belongs your wicked tongue." Sinclair replied, "I have just read you are coming to Pasadena . . . if you want a quiet garden to come to and be let alone in, we will provide it. You had better bring your fiddle along, and we will play duets [Sinclair at the piano] if you don't mind my being out of tune occasionally." (He wasn't kidding; according to reports, their duets were excruciating.)

Einstein had promised that "as soon as I have a bit of time, I shall visit you." But Sinclair forgot to warn his wife of the expected visitor, so she was surprised when her sister, Dolly, said, "There's an odd-looking old man walking up and down the street and staring. He keeps looking at the house as if he wants to come in." Glancing out, Mrs. Sinclair "saw a rather frail-looking man with bushy gray hair, a black hat and a suit which had never been pressed." She said, "Go out and speak to him. Ask him what he wants." Dolly came back and reported, "He says he's Professor Einstein, but I don't know whether to believe him." Mrs. Sinclair said to bring him in and called her husband from his study.

"Such was the beginning of as lovely a friendship as anyone could have in this world," recalled Upton Sinclair. "I report him as the kindest, gentlest, sweetest of men. He had a keen wit and a delightful sense of humor and his tongue could be sharp—but only for the evils of this world."

Mrs. Sinclair also cherished their friendship, writing that:

> from first to last the discoverer of relativity never refused a single request that we made of him. If it was a labor struggle, he would write a telegram of sympathy for the strikers, and Upton would give it to the press. If it was a meeting on behalf of free speech, he would sit on the platform, and make a few remarks when requested. If it was a demonstration of Jan's [Roman Ostoja's] psychic powers, he would attend and manifest deep interest.

Helen Dukas, Einstein's secretary, gives a different account of Ostoja's psychic skills. "I was at the séance [with Ostoja, a self-proclaimed but dubious Polish count]. And I was frightened to death. Because, before the sitting, Sinclair addressed us and said we shouldn't be afraid if suddenly the piano starts to play and flowers come down from above and so on. And the medium went into catalepsy, and that frightened me. Then, you know what happened? They all sat around the table, those scientists, Professor Einstein, Professor Tolman, and a doctor friend of ours. Sometimes Professor Einstein was in control. It was a real scary atmosphere. And, oh my gosh, suddenly the doorbell rang and I nearly jumped out of my skin."

Dukas was not in the circle, but sat just outside the room. She went to the door and it was a harmless telegram. Then she returned and listened to the rest of the séance.

"When the medium went into the trance he didn't say anything, just 'Wa-wawa!' and there was no response," Dukas recalled. "They all had to keep their hands on the table but nothing happened. Sinclair said it was because there was a counterinfluence in the room. Because they were nonbelievers.

"Einstein did it because he liked Sinclair personally very much. He was a sweet man. That, as far as I know, was the only séance Dr. Einstein ever attended. Well, listen, he really wasn't very interested at all. He had other interests. He said that he wrote the preface to Sinclair's book on telepathy as a matter of friendship. As you know, Einstein was always the victim of swindlers, the so-called clairvoy-ants. They claimed to have made accurate predictions about him—and he had to deny they were true."

The Sinclairs arranged for Einstein to meet some of their distinguished writer friends for dinner at the exclusive Town House in Los Angeles. When Einstein arrived, he somehow missed the cloakroom and appeared in the dining room wearing a "humble" black overcoat and a much-worn hat. In what might have been a scene from a Chaplin film, he removed his overcoat, "folded it neatly, and laid it on the floor in a vacant corner and set the hat on top of it. Then he was ready to meet the literary elite of Southern California." There was even something Chaplinesque in the way Einstein flirted with the attractive women, while Elsa—"my old lady" he called her—was at his elbow.

Elsa confirmed Mrs. Sinclair's view of her as a dutiful and utterly devoted German hausfrau during a discussion about God. Einstein had stated his belief in God, but not a personal God—a distinction which Mrs. Sinclair didn't get. She replied, "Surely the personality of God must include all other personalities." Afterwards, Elsa gently admonished Mrs. Sinclair for arguing with Albert, adding, "You know, my husband has the greatest mind in the world." "Yes, I know," said Mrs. Sinclair, "but surely he doesn't know everything!"

Sinclair monopolized Einstein for days, taking him to a movie depicting life in Mexico, *Thunder Over Mexico*, directed by a Russian, Sergey Eisenstein, as well as to a private screening of *All Quiet on the Western Front*, a heartbreaking picture of trench warfare in World War I from the Erich Maria Remarque novel of the same title. Both book and movie were banned in Germany as pacifist propaganda.

Back in his gift-strewn cottage Einstein found tangible evidence that "Amer-ica was prepared to go mad over him." A millionairess gave Caltech $10,000 for the privilege of meeting him. He declined the loan of a Guarnerius violin valued at $33,000 "with the apologetic explanation that he was used to his old one." He did accept a salmon flown from Alaska, a prize ham from Armour and Co., crates of oranges and grapefruit, polished stones from Arizona's Petrified Forest, rare cacti, and—from the manager of the Ambassador Hotel in Los Angeles—a saddle of mutton all the way from England. Elsa complained in mock despair that there wasn't an inch to spare in their little bungalow, "and if more gifts arrived they would have to move out." Ironically, while they were being showered with pres-ents in California, they were being robbed in Germany by thieves who broke into their Caputh summer home and stole silk embroidery from Elsa's bedroom.

Einstein hardly cared. His mind was soon elsewhere, as he explained to an enthralled audience that he had changed his mind and no longer believed in a "closed" universe. Conversations with Tolman and Hubble had convinced him his original cosmology was wrong, that whatever field equations were used, space could never be anything similar to the old spherical space theory. This change of opinion demolished his own theory that the universe was static and uniform as well as the theory of his Dutch astronomer friend Willem de Sitter that it was static but nonuniform. As for his unified field theory, he admitted he was stymied: "I offer it to you like a closed box, and as one who doesn't know what is in it." To which the more optimistic Walter Adams responded: "While it appears like a closed box, we hope to have the cover off soon and see what is in it."

Millikan, a staunch conservative, was uneasy when Einstein turned from science to politics. So were Zionists, especially when he endorsed Peru rather than Palestine as the Promised Land—as long as it proved fertile, was rich in natural resources, and had few dangerous animals and poisonous snakes, and only if "the native population is so thinly scattered that one may legitimately speak of empty land." Soon after his statement endorsing Peru, he appeased some Zionists by agreeing to appear at fund-raisers in Los Angeles and New York.

To Millikan's dismay, Sinclair had converted Chaplin to socialism, and Einstein shared many of their views, favoring a minimum wage, old-age pensions, and a cap on individual fortunes—or, as Sinclair put it: "I don't believe we should have two classes of persons—one with more dinner than appetite and the other with more appetite than dinner."

Einstein could have been reading a synopsis of Chaplin's movie-with-a-message, *Modern Times*, when he asked Caltech students: "Why does this magnificent applied science which saves work and makes life easier bring us so little happiness? The simple answer is: Because we have not yet learned to make sensible use of it. In war it serves that we may poison and mutilate each other. In peace it makes our lives hurried and uncertain. Instead of freeing us in great measure from spiritually exhausting labor, it has made men into slaves of machinery, who for the most part complete their monotonous long day's work with disgust and must continually tremble for their poor rations."

Upton Sinclair's son, David, confirmed that Einstein and his father "shared a similar socialist outlook. Dr. Millikan, who had invited Einstein to Caltech, didn't approve of my father one bit politically, calling him 'the most dangerous man in California.' And he did not approve of my father and Einstein getting together." Millikan made sure that didn't happen on Einstein's last work-free day in California by taking him for a cruise off Long Beach.

After two months of "loafing in Paradise," as he wrote to the Borns, the Einsteins headed by train for New York City, stopping en route at the Grand Canyon, where Albert was initiated into a Hopi Indian tribe as a chief named "The Great Relative"—perhaps a play on relativity?—and then posed in full Indian headdress for the inevitable photographers. In Chicago he addressed a peace group from the train's rear platform, encouraging them to resist military service.

In Manhattan, the Einsteins' lavish quarters at the Waldorf-Astoria—two suites consisting of two bedrooms, two bathrooms, two drawing rooms, and a dining room—disoriented Albert, so that when a friend called to see him, Elsa couldn't find him. "He always gets lost somewhere in these rooms," she explained, leading the search party. They found him wandering around, looking for Elsa, but in the wrong direction. "Lock the second suite entirely off," the friend suggested, "and you'll feel better." They did.

Responding to an urgent appeal from Weizmann, he spent his only evening in Manhattan attending a fund-raising dinner at the Hotel Astor to help Zionist causes. The next morning the Einsteins boarded a ship heading for home.

By mid-March they had returned to Berlin. In early May, Albert was in England, to lecture on relativity at Oxford. Professor Frederick Lindemann (who would become Churchill's scientific adviser during World War II) hoped to lure him to Christ Church permanently as a resident scholar. Elsa, who had stayed in Berlin, wrote to Lindemann that her overburdened husband "out of his endless modesty" would not ask for a secretary to help answer his mail. He saw to it that one was always available.

In England, Lindemann put his Rolls and one of his three manservants at Einstein's disposal. The son of a wealthy Alsatian father and an American mother, Lindemann had been educated in Germany and spoke the language fluently. He "was a heavyweight of personality. He proved himself beyond any possibility of argument, dangerous and vindictive in action, outside the ordinary human run." He never married, didn't touch alcohol or meat, and was a malicious gossip, especially about women. But no one could deny he had guts. During World War I he had worked out why so many British fighter pilots had crashed in tailspins, risking his life by putting the planes through death-defying maneuvers to prove his point. He and Einstein had known each other since 1911, when they met at the Solvay Congress.

Dons at Christ Church, Oxford, where Einstein stayed, asked him to rate Lindemann's talent as a physicist. "Essentially an amateur," Einstein is reputed to have answered, "with teeming ideas which he does not bother to work out himself, but he has a thorough knowledge of physics." He gave others a less critical assessment of Lindemann as "the last of the great Florentines, a man who embraced all science as his province, a great man in the Renaissance tradition."

Einstein's effect on Oxonian Roy Harrod progressed from awe to affection—a typical experience of those who got to know him:

> It was a great thrill when I first saw him. He was standing with his halo of white hair, in front of the blazing log fire in Christ Church Hall, where we used to assemble before dining at high table. I soon found that my natural trepidation was quite unnecessary, as he proved to be a most easy, friendly, unassuming, and indeed lovable person. In our Governing Body, I sat next to Einstein. We had a green baize tablecloth; under cover of this he held a wad of paper on his knee, and I observed that all through our meetings his pencil was in incessant progress, covering sheet after sheet of equations.

Einstein was not impressed with his digs or the talk at the high table, where the professors dined formally, noting in his diary: "Calm existence in cell while freezing badly. Evening: solemn dinner of the holy brotherhood in tails." He still rarely used English, and then only haltingly, but said he understood when others spoke it. He turned "for comfort and peace to his violin, and those crossing the Quad could hear the plaintive music floating through the windows."

Some five hundred students were part of the standing-room-only crowd for his first lecture, when he said that his special theory was based empirically on the experiments of physicists Armand Fizeau and Albert Michelson. A slightly smaller audience heard his second lecture, on cosmology, the curiosity seekers having checked out. As usual, Einstein spoke in German and without notes, but he often referred to two blackboards he had covered beforehand with equations.

To support his theory of "the mean density" of the universe, Einstein cited the fact that cosmic rays appeared to come from all directions, as well as the apparently random scattering of interstellar matter. He credited Hubble with providing the evidence for an expanding universe, and Freidmann and Lemaître with finding solutions to general relativity equations corresponding with either a contracting or an expanding universe. One puzzling feature of the theory was the question, From *what* did the universe begin expanding? This was particularly mystifying because, said Einstein, it could hardly have been "altogether small."

He warned his third audience that unified field theory was a highly speculative exercise in pure mathematics, to which he had been drawn "not by the pressure from behind of experimental facts, but by the attraction in front of mathematical simplicity and logical form. It could only be hoped that experiments would follow the mathematical flag."

Before returning home, Einstein made a brief trip south to Winchester College, a public school whose enduring fame depended largely on past glories and ancient traditions. By mistake, he walked "into a schoolboys' changing room where sports clothes were hanging on pegs underneath tablets recording the names of past generations of boys. 'Ach,' said Einstein, 'I understand. The spirits of the departed passes [sic] into the trousers of the living.' "

Back in Germany, he spent much of the summer on missions of mercy. He tried to save the lives of eight black youngsters in Scottsboro, Alabama, who were facing execution for allegedly raping two white girls, although one "victim" recanted and admitted that no crime had occurred. He appealed to the governor of California to free the socialist Tom Mooney from San Quentin, where he was serving a life sentence for murder despite confessions of perjured testimony at his trial. A labor activist, Mooney had been found guilty of helping to plant a bomb that had killed several people during a parade. Writing to Governor James Rolph, Einstein pointed out that many reputable people believed that Mooney and Warren Billings, his so-called accomplice, were completely innocent: "I myself do not claim my opinion is correct but I think a miscarriage of justice has taken place . . . and you would do much for real justice if you commuted their sentences."

The governor leaked the letter to the press, which alarmed Professor Millikan, who depended on conservative plutocrats to enrich Caltech. Many of his patrons regarded Mooney and Billings as murderous radicals. Still hoping to recruit Einstein permanently for Caltech, Millikan sought his assurance that he would not continue to make waves. Einstein smoothed the waters somewhat in his August 1 reply, saying that his had been a private letter to the governor not meant for publication and "solely prompted by pure love of justice." He believed that Mooney and Billings were probably innocent, but understood that "it can not be my affair to insist in a matter which only concerns the citizens of your country."

Turning his attention to matters in his own country, Einstein was incensed by German students who were agitating to dismiss a professor because he was a pacifist and opposed political assassinations. "It is terrible to see how inexperienced youth are misled," he wrote to the *Berliner Tageblatt*. "If this keeps on we are going to arrive through Fascist tyranny at a reign of Red terror." He believed that such a reign had already been unleashed in Yugoslavia and blamed that country's government for the murder of a Croatian leader, Professor Milan Sufflay, who had taught history at Zagreb University.

Einstein then focused on what Churchill called "the dark recesses of the Serbian and Croat underworld," accusing Yugoslavia's rulers, especially King Alexander I, of treating the Croatian people with "horrible brutality" and of inciting terrorism. Together with novelist Heinrich Mann, Einstein wrote to *The New York Times*, urging the League for the Rights of Man to protest. According to the two men, after the king had encouraged a terrorist organization,Young Yugoslavia, to get rid of the Croatian members of parliament, a Croatian leader, Stefan Raditch, and two others had been fatally shot "on the floor of the House." More recently, Mann and Einstein said, the murder of political and intellectual Croatian leaders had been openly advocated in the government press after the king visited Zagreb, the Croatian capital, in January 1931. On February 18, an official Yugoslav publication had warned: "Skulls will be split." That same evening, while walking home, Professor Sufflay had been beaten to death with an iron bar.

The New York Times reported frequent protests against King Alexander's dictatorship since Sufflay's murder, along with many suspicious "suicides" of Croats and Macedonians in Belgrade and Zagreb prisons. The newspaper also pointed out that Croatians were being terrorized because of their refusal to be dominated by a government recruited largely from Serbian extremists.

Einstein and Mann had not yet revealed their sources, nor did Einstein mention that his first wife, Mileva, was Serbian. In a letter to *The New York Times* responding to theirs, they were ridiculed for implicating "a government in this deplorable murder [of Professor Sufflay] on the basis of information evidently received from a few disgruntled politicians out of a job." The letter also addressed their charge that the king was behind the killing of the Croatian leader: "It is regrettable such a baseless and ridiculous accusation should receive the support of a man whose mind was trained in the exact methods of science. It is well known that it was King Alexander who ordered a speedy prosecution of the murderer, which ended in a conviction to [sic] life imprisonment. It is well known that the organization 'Young Yugoslavia' consists chiefly of Croatians and Slovenes. Why then point to the murder of Professor Sufflay as an instance of Serbian terrorism over the Croatians?"

"Baseless and ridiculous?" Not so, said another correspondent: "For fifteen years many friends of the Yugoslav minorities—Montenegrins and Macedonians—have been telling the world of the atrocious conduct of the Yugoslav government towards minorities. Now when the world at large is beginning to realize the truth of our assertions, and distinguished men are taking the part of the oppressed minorities, the oppressors and their satellites protest to the newspapers. If Yugoslavia wishes a proper place in the council of civilized nations, and the moral and material support of civilized peoples, it should conduct its internal affairs on an enlightened and humane basis."

A representative of the Yugoslavian Legation in Washington denied all of Einstein's charges, saying that the king was loved throughout the country and was extremely popular in Zagreb.

That summer, the documents on which Einstein had based his account of Professor Sufflay's murder and the murders in parliament were published in a Swiss newspaper, *Croatia*, along with a sworn statement by the fatally wounded Croatian deputy, Stefan Raditch, a few days before he died. All of those documents supported the contention of Einstein and Mann that the king was behind the killings.

King Alexander's dictatorship came to a bloody end three years later, on October 9, 1934, when he was shot to death by an assassin in Marseilles, France, while on a state visit. The French foreign minister, Jean-Louis Barthou, who was sitting beside the king in the car, was also killed before the assailant was slashed to death by a saber-wielding French army officer.

According to one source, the assassin, Dimitrov Veltichko, was a Bulgarian. Nicknamed "Vlado the Chauffeur," he probably murdered for the standard fee of twenty dollars. Velitchko was known as a strict vegetarian because he thought it cruel to kill animals. Another source, British historian A. J. P. Taylor, wrote that King Alexander's murderer was not a Bulgarian but "a Croat terrorist who had been trained in Italy." Whatever the ethnic or national origin of the man who pulled the trigger, Einstein was generally well-informed about the terrorism and mistreatment of minorities in Yugoslavia. History once more proved that Einstein was largely on the mark and refuted the worldly-wise who dismissed him as being naive or a political innocent.

Late in September, Einstein was induced to provide an audience of prominent Berlin socialites and business tycoons with a fun evening: a talk on the amusing phenomena of natural science. "The times are so grave today that everyone's task is to cheer up his neighbor, so I'll try to do my bit," he said. He did not know that he was following hard on the heels of Hitler, who had spent the summer chatting up many in the audience—in his case not to raise their spirits but to seduce them to the Nazi cause. Having declared his aim to overthrow the system, Hitler found many willing listeners. Germany was in desperate financial straits; unemployment had reached almost five million, and there had been a huge flight of capital from German banks. The British ambassador in Berlin reported to his government: "I was much struck by the emptiness of the streets and the unnatural silence hanging over the city, and particularly by an atmosphere of extreme tension similar in many respects to that which I observed in Berlin in the critical days immediately preceding the war."

In this atmosphere, Einstein chose to divert the well-heeled group with his answers to minor mysteries: why tea leaves crowd to the middle of a cup when stirred, why a heavier-than-air airplane stays up in the sky, why wet sand is solid but breaks up in water, and other scientific oddities. It wasn't exactly fiddling while Rome burned, but pretty close. In explaining why the wind dies down at sunset, "leaving the sailor helpless out in the middle of the water," he recalled his own plight, becalmed in his boat with a young woman until two A.M. He was obviously enjoying himself, and at the end of the talk regretted that there was no time left to present all the "delicacies" he'd prepared.

A week later he was at it again, as if he had suddenly developed an irresistible taste for public speaking. This time it was to a packed audience in the Planetarium. The hall was acoustically second-rate, and Einstein had "an amiable weakness of talking exclusively toward his left." Even so, the audience paid rapt attention, lulled by his easy conversational manner into the illusion that they followed him even when they were lost. He said that the aim of most physicists was not to increase human comfort but to understand the universe. He himself had switched from engineering to physics because, he recalled, "I said to myself, 'So much has already been invented, why should I devote myself to that sort of thing?' " The scientific method, he said, was similar to what everyone did in interpreting the behavior of other people in the light of self-knowledge, which in itself was only a theory—not that physicists relied solely on what they observed, though without observation "there would not be a single law in physics." On the other hand, he noted, "experience alone yields no physical science." It comes about, he concluded, through speculation "which is constantly tested through comparison with phenomena observed in nature."

Next, Einstein was off to lecture in Vienna with the blessing of the Weimar government, which was still using him to boost German prestige. No state official greeted him on arrival or later invited him to any state function. Additionally, because both the education minister and the rector of the university where he lectured pointedly missed the lecture, the German ambassador made frantic efforts to persuade some officials to at least join him and Einstein for breakfast. Reporting these incidents to his Foreign Office in Berlin, the embarrassed ambassador concluded: "It is typical of the manner in which in Vienna all things are dealt with from a party-political point of view that the official Austrian authorities observed special reserve with respect to Professor Einstein, because he is a Jew, and considered oriented to the political left." Would Einstein have felt humiliated? Not likely. Knowing how the Austrian authorities had recently mistreated Philippe Halsman, he would hardly have been eager to consort with their compatriots. Nevertheless, the snubbing in Vienna was an incredible contrast to his almost delirious reception in California, which he intended to revisit soon.

No one was more anxious for him to return than Tom Mooney, who hoped Einstein would come to see him in San Quentin. Unaware of Millikan's cautionary letter to Einstein not to interfere in America's problems, Mooney, his voice "choked with gratitude," asked a friend, Rebekah Raney, to give Einstein his "inexpressible appreciation for the most generous and kindly interest" he had taken in the case. He wanted Einstein to know that for eight years he had worked closely with Eugene V. Debs to promote the Socialist party. Debs, a self-educated railroad fireman and perennial Socialist candidate for the U.S. presidency, had been imprisoned for advocating pacifism during World War I. He had died in 1926 and was revered by his followers as a martyr.

Millikan was one of many to call Einstein naive about politics and the ways of the world. But in almost every case he championed, Einstein was proved right: Halsman, Mooney, Billings, and the Scottsboro boys were all found innocent, as he had suspected. Mooney was unconditionally pardoned in 1939 and Billings was freed nine months later.

Einstein was looking forward to another meeting with fellow socialist George Bernard Shaw, who was on his way back from a trip to Russia. The provocative playwright had interviewed Stalin, visited Lenin's widow, and inspected a Red Army barracks, and now—"redder than ever"—was returning to Britain via Berlin, where he had arranged to join Einstein for a chat. Shaw was a no-show, however, and Einstein was left waiting for him on the railroad platform. Despite Shaw's enthusiasm for a system Einstein deplored, he apparently did have one negative impression to report to the world: Soviet trains didn't run on time.

By early December, Einstein was heading back to Caltech through heavy rain and rough seas. Elsa had packed an eclectic bag of books, and he chose to read *Grunberg's Fairy Tales* before turning to his friend Max Born's hardly less fantastic *Quantum Mechanics*.

As the ship headed west, he was contemplating big changes in his life, and on December 6 confided to his diary: "Today I decided to give up my position in Berlin. Well, then, a bird of passage for the rest of my life. Seagulls accompany our ship, constantly in flight. They will come with us to the Azores. They are my new colleagues, although God knows they are happier than I. How dependent man is on external matters, compared with the mere animal! I am learning English, but it doesn't want to stay in my old brain."

Four nights later a gale changed Einstein's melancholy mood, as his diary entry reveals: "Never before have I lived through a storm like this . . . The sea has a look of indescribable grandeur, especially when the sun falls on it. One feels as if one is dissolved and merged into Nature. Even more than usual, one feels the insignificance of the individual, and it makes one happy." Einstein's stepson-in-law, Dimitri Marianoff, who was traveling with the Einstein family (the party also included Margot; Ilse and her husband, Rudolf Kayser; and Walther Mayer), was impressed with his fearlessness as he strode the sloping deck like an old seadog. Their ship docked at Los Angeles on December 29; it was too late for the quarantine inspectors, so the passengers disembarked the next day. As the Einsteins came down the gangplank, Albert realized that Millikan had gone along with his wishes. The only ones to greet them were a few-score longshoreman who burst into applause, a handful of Caltech officials, and a small group of students.

The day he landed on American soil, Caltech scientists announced new experimental evidence supporting his special theory of relativity. After nine years of intensive research with sophisticated instruments, Drs. Roy Kennedy and Edward Thorndike had confirmed that "the velocity of light is constant and independent of the velocity of the source of light . . . no matter how fast a star is moving, the light it sends down to earth reaches it at a speed no higher than 186,000 miles a second."

Einstein had little interest in the confirmation of a theory he knew was right. His current interest was another matter. At the outset, the wisecracking Wolfgang Pauli had greeted Einstein's attempt at unifying electromagnetism and gravitation with the one-liner, "What God hath put asunder no man shall ever join." Pauli continued to ridicule what he considered to be a futile preoccupation:

> Einstein's never-failing inventiveness as well as his tenacious energy in the pursuit of unification guarantees us in recent years, on the average, one theory per annum . . . It is psychologically interesting that for some time the current theory is usually considered by its author to be the "definitive solution."

CHAPTER 24

Weighing Options

1932

---

52 to 53 years old

Early in the new year, American journalist Dorothy Thompson interviewed Adolf Hitler in Berlin and summed him up as an insignificant caricature. Meanwhile, Winston Churchill lay in a Manhattan hospital looking at the ceiling; he had glanced the wrong way while crossing Fifth Avenue and been hit by a car. And Albert Einstein was taking a second look at the stars through the giant Mount Wilson telescope in Pasadena, where he had come to spend several months at Caltech.

Though wary of Einstein's outspoken radical views, Millikan still welcomed him at Caltech because of the prestige he brought to the school; Millikan even raised special funds to finance his visit. But the money came from political conservatives, like Millikan himself, who wanted Einstein to stick to science and stifle his views as a pacifist, humanist, and Social-Democrat. This was a vain hope, particularly because the Japanese had just launched a brutal invasion of China, and the Western world was still in the grip of the depression.

After his obligatory lectures on the cosmos, Einstein tackled peace and poverty on February 27. He told scholars gathered from eleven universities that an international court of arbitration would not ensure world peace unless the countries acted in concert to enforce the court's decisions. As for the problem of poverty and unemployment among great sections of the world's population, in his opinion, it would not be solved by the unbridled efforts of individuals to get power or property. The answer, he said, was a planned economy, meaning socialism.

Einstein suggested that the Japanese would call off their war in China if the governments of America, Britain, Germany, and France demanded the immediate cessation of warlike acts, on pain of complete economic boycott. "Why is that not brought about?" he asked. "Why must each person and each nation tremble for their existence? Because each seeks his miserable momentary advan-

tage and will not subordinate himself for the good and prosperity of their community."

That night, Einstein complained to his diary: "I . . . gave a speech, but alas!
this was hardly a responsive audience. The propertied classes here seize upon
anything that might provide ammunition in the struggle against boredom . . . It
is a sad world in which such people are allowed to play first fiddle."

His spirits were lifted by a surprise visit from Abraham Flexner, who was just
about to leave Caltech after seeking Millikan's advice on a new and unique
enterprise he had undertaken. He had barely an hour left before catching his
train to the East Coast when someone said he should call on Einstein.

Flexner, an educator with a mission, believed that higher education in America was rapidly going to the dogs. He had been spurred to action on hearing that
a University of Chicago student had submitted a Ph.D. thesis titled "The Comparison of Time and Movement of Four Methods of Washing Dishes." With five
million dollars from philanthropists Louis Bamberger and his sister, Mrs. Felix
Fuld, Flexner was all set to create a scholars' paradise in the United States to
attract the world's scientific elite. These scientists were to be handsomely paid
and left completely free to pursue their work and to hire research assistants
focused, he hoped, on something more challenging than dishwashing.

Flexner met Einstein at the Athenaeum Faculty Club and "was fascinated
by his noble bearing, his simply charming manner and his genuine humility."
They walked up and down the corridor of the building for about an hour while
Flexner sought his opinion of the plan. Einstein, obviously intrigued, asked questions until Elsa came to remind him of a lunch engagement. "Very well," he
said, "we have time for that. Let us talk a little longer."

Although the location for his enterprise had not yet been chosen, Flexner
was certain that it would be near a major university. Eventually, he hoped, many
such centers of higher learning would spring up throughout the country and by
osmosis raise educational standards and goals generally. Einstein was eager to
hear more; they agreed to continue their discussion in the summer, when they
would both be at Oxford.

Among Einstein's shipboard reading on the trip home to Germany was a
biography of Stalin by Isaac Don Levine. A longtime friend of Einstein's, Levine
had escaped from Russia as a boy and returned, an experienced reporter, to cover
the Russian Revolution. He believed that the Bolsheviks had destroyed the provisional government and ruined the Russian people's future by using violence,
discipline, and deception. In the 1920s Levine had produced a book documenting Soviet persecution of political opponents for which Einstein wrote a
preface, praising it for exposing Soviet repression. He seemed equally impressed
with the recent biography, writing to Levine that his

> book on Stalin . . . is undoubtedly the best and most profound work on that great
> drama that has fallen into my hands . . . I have the impression that you are an
> unimpeachable authority. I am truly grateful to you for the fine piece of knowl-
> edge which you have provided for me . . . The whole book is to me a symphony
> on a theme: Violence breeds violence. Liberty is the necessary foundation for

the development of all true values. It becomes clear that without morality and confidence, no society can flourish.

Arriving back in Germany on April 6, Einstein told reporters that he had had a great time in California collaborating with American astronomers and physicists. He also said that he had been moved by the concern Americans had shown for Germany's economic and social problems, even though they, too, were still suffering from the depression.

Einstein was pleased by the recent presidential election in Germany, which had returned von Hindenburg to office with 18.5 million votes. Hitler had come in a distant second with some 11.3 million votes, though he had gained over 5 million supporters in two years. The Communists trailed with less than 5 million votes. However, it was soon alarmingly obvious that the Communists were giving their support to the Nazis—on Stalin's order. Stalin had decided that fascism was the final stage of capitalism and that the sooner Germany became fascist, the sooner Communists would take over. Harold Nicolson of the British Embassy in Berlin tried to persuade the head of the German Communist Party that by assisting Hitler, "they were handing him the rope with which he would hang them!" Stalin knew better, though—and *no one* countermanded *his* orders. Communists even joined Nazi thugs in street brawls and riots against supporters of the Weimar Republic.

Einstein had hardly put his foot in the door before he was off to England to give a math lecture at Cambridge University and then to visit Oxford University, where he was again housed at Christ Church and welcomed at high table. Abraham Flexner arrived as planned and they resumed their discussion about the proposed Institute for Advanced Study, as Flexner's center was to be called. Instead of a corridor at the Athenaeum, they paced the manicured college lawn under a clear sky with birds providing a musical accompaniment. It now dawned on Flexner what a coup it would be to recruit Einstein—the world's greatest scientist—as a founding member of the institute. With great daring, he suggested that Einstein join the staff "on your own terms." Einstein didn't say no and proposed a third meeting later that summer, when Flexner intended to be in Germany on a recruiting drive.

Einstein then went to a peace conference in Geneva, where he urged total world disarmament. Nevertheless, it was clear he was not after peace at any price. Asked to endorse an antiwar declaration, he replied:

> Because of the glorification of Soviet Russia which it includes, I cannot bring myself to sign it. I have of late tried very hard to form a judgment of what is happening there, and I have reached some rather somber conclusions [presumably informed by Isaac Don Levine's biography of Stalin]. At the top there appears to be a personal struggle in which the foulest means are used by power-hungry individuals acting from purely selfish motives. At the bottom there seems to be complete suppression of the individual and of freedom of speech. One wonders what life is worth under such conditions.

Soon after Einstein returned to Germany, he and Elsa settled into their

Caputh home for the rest of the summer. Their friend Philipp Frank spent a day with them and another visitor and noted the contrast between the idyllic view and their conversation about the threatening political situation. Frank later wrote:

> I can still recall [that] when a professor who was present expressed the hope that a military regime might curb the Nazis, Einstein remarked, "I am convinced that a military regime will not prevent the imminent National Socialist revolution. The military regime will suppress the popular will and the people will seek protection against the rule of the junkers and the officers in a Right-radical revolution."

Among the visitors to Caputh that summer was thirteen-year-old Thomas Bucky, the son of Gustav Bucky, an American physician and inventor working in Berlin. Gustav, who was Ilse's and Margot's doctor, had invented "The Bucky diaphragm," which revolutionized X-ray photography and is still used in all X-ray apparatus. Thomas had been invited with his parents and brother, Peter, to spend the day with the Einstein family. He remembered being "very nervous and excited at first, and then, when I met him, disappointed. Because, although he was very polite, I sensed a feeling of reserve on his part—until he broke the ice. Yo-yos, those little tops that spin up and down on a string, were all the rage at the time. And Einstein had one, a puny model which he played with very inexpertly. And that's how he broke the ice: by playing with it. I told him it was off-balance and showed him a few tricks, explaining that it needed a loop at the end of the string for him to do tricks like freewheeling. He really seemed to welcome my advice, and I was surprised when he started a conversation with me and showed interest in my opinions."

By this time Thomas felt at ease and realized that what he took to be reserve was simply Einstein's extreme shyness. But, even so, he was a man who said what he thought. "After dinner, for example, when my mother [Frida] congratulated Mrs. Einstein on the meal, she replied, 'Oh, it was nothing. I didn't go to any extra trouble. We eat like this every night.' And Einstein said, 'What! We eat like THIS—EVERY night?'—obviously showing that they didn't. I think Mrs. Einstein blushed when he said that. You see, he was honest. She wasn't—quite. She was social.

"I remember her then as a sweet, mild, frowsy-haired lady in a flowing dress, and wearing long metal chains. She was kind and eager to please. Sometimes it took her a little while to catch on in a quick conversation.

"As well as being a physician and a pioneer in radiology, my father was an amateur physicist and inventor. And he and Einstein were soon in a corner of the room talking away, apparently oblivious of the rest of us. They often met after that, Einstein talking theory and my father explaining how the theories could be put to practical use. Eventually they produced several inventions together, including an automatic camera for medical purposes.

"During conversations with my father, Einstein often burst out with sudden, loud, explosive laughs—HA! HA! HA!—and wherever you were in the house you could hear his laughter. Many things made him laugh, especially people's silly behavior and their conventions."

The Bucky family became very close friends with the Einsteins and saw them frequently. Their car was put at Einstein's disposal whenever he wanted it, and that included the chauffeur. (Einstein always needed a chauffeur to get around because he never learned to drive a car and showed no interest in learning.) Once, Mrs. Einstein said to the chauffeur, "Please drive very carefully; you are responsible for the life of this great man." The chauffeur, Fritz Schlaefke, a staunch Social-Democrat, replied, "What about my own life, Mrs. Einstein?" Elsa's remark expressed her high regard for her husband as well as her customary concern for his well-being, but it was Schlaefke's response that would have brought a twinkle to Einstein's eye.

"Einstein became absolutely my second father," said Thomas Bucky, "and I didn't hesitate to discuss politics with him and express my opinions, and was never afraid to challenge him in debate, for he welcomed even the perhaps half-baked opinions of a teenager. He was very tolerant, but he despised yes-men, sniffed them out and avoided them. Einstein was a humanist, socialist, and a democrat. He saw no big bugaboo in socialism if it was not totalitarian. He was completely antitotalitarian, no matter whether it was Russian, German, or South American. He approved of a combination of capitalism and socialism. And he hated all dictatorships of the right or left.

"When I came up with an outlandish idea, he would tell me how crazy I was, but he still respected me. I mean, he would *explain* by saying, 'Ah you're nutty!' and then patiently point out why. But if you expressed your opinion arrogantly or emotionally he would cut you down with logic and be annoyed by your attitude. He wasn't faultless in that respect. He could have something logically explained to him, agree it was true, and then revert to his former views. Although he may have rationalized by saying he was merely behaving that way to keep his life simple."

Thomas recalls that Einstein used to shave with a safety razor and just water. Thomas persuaded him to try shaving cream. He thought it was wonderful; for the first time he could shave himself painlessly. "You know, it really works," he said. "It doesn't pull the beard." From that point on, he used it every morning until he finished the tube Thomas had given him. Then he went back to his old, simpler way of using just water.

"He loved to sail," Thomas said, "not only because he enjoyed the sun and sea but because it was the simplest method of transportation. He only used an outboard motor for emergencies and many evenings he would lie becalmed a few hundred feet from shore—with dinner and Mrs. Einstein burning in the kitchen.

"I never saw him lose his temper, never saw him angry, or bitter, or vain, or jealous, worried, impatient, or personally ambitious. He seemed immune to such feelings. But he had a shy attitude toward everybody. Yet he was always laughing, and he often laughed at himself. There was nothing stuffy about him. Still, he was aloof, always shy, hesitant. Einstein had a shell around him that it was not easy to penetrate."

As Bucky's mother, Frida, put it,

a kind of thin wall of air separated Einstein from his closest friends and even from his family—a wall behind which, in his flights of fancy, he had created a little world of his own . . . People sometimes glanced at Einstein as if something mystical had touched them—and they smiled at him. And when Einstein caught this smile, he happily returned it. He loved the simple, uncomplicated person who, even not recognizing him, felt his humanity.

Einstein liked coffee, which he knew he was not allowed to drink. In spite of this Mrs. Bucky often tempted him. She later wrote: "Once, when he refused a cup of coffee, I said, 'It didn't do you any harm yesterday.' He smilingly answered, 'Oho, Mrs. Bucky, but the devil keeps books!' "

Frida believed that "Einstein was absolutely conscious of what he gave to the world in his revolutionary ideas, but all his life he suffered for the injustice to those who deserved a better fate, either in science or in everyday life." He also felt almost ashamed that life had been so good to him. On one occasion he said to her, "You all overestimate me—I am neither especially clever nor especially gifted. I am only very, very curious. Sometimes I think I am a swindler."

He had a cynical view of physicians, which included Dr. Bucky. "In a discussion with my husband on this theme, Einstein suddenly interrupted with: 'After all, you can die even without a physician.' "

"You had to know how far you could go with him, and I could go just so far," Thomas Bucky remarked. "I once told him he should get rid of his frayed sweater, that he'd been wearing it too long, and he responded with a cool silence. And kept the sweater. Sometimes my parents would tug at me and say 'Shush!' but I usually knew when to stop.

"If Einstein had been sailing during the day, he'd come home to supper or dinner and sit around discussing any subject that came up. About nine on summer evenings he retired to his room to read, or would sit in his chair for hours quietly writing on small pads of paper. He always carried these pads with him, and whenever he had a free moment wrote these little, precise formulas, puffing away at his pipe until the air was tinted blue. Eventually my father advised him to stop smoking for health reasons. He didn't want to hurt my father's feelings— but smoking was one of his very few diversions. So he stopped smoking when my father was with him.

"He was also sensitive to his sister Maja's feelings," said Bucky. "She was almost a carbon copy of Einstein. They looked alike, with the same frizzy mane of hair, and laughed and thought alike. She was extremely intelligent and more accessible than her genius of a brother, who towered remotely from Mount Olympus. She was gentle and considerate and had a warmer personality, according to those who knew them both. She was a vegetarian because she hated the thought of hurting any living creature.

Late in the summer of 1932, Abraham Flexner traveled to Caputh to continue his efforts to recruit Einstein for his institute. When he arrived, it was drizzling steadily, and the weather was so cold that he wore his winter clothes and a heavy topcoat. He was surprised to find Einstein sitting on his porch in summer flannels, and asked, "Aren't you chilly?" Einstein denied that he was cold, explaining with a wacky logic that he dressed "according to the season, not according to the weather. It's summer."

Nazi students were protesting the presence of Jewish students and teachers in German universities, and Flexner was sure that the outspoken Einstein would soon be forced to flee the country. But in his sales pitch, Flexner never once mentioned Hitler or his rabid followers as a reason for Einstein to start a new life in America. Eager as he was to recruit this great prize for the institute, he wanted the decision to be made on its own merits. Otherwise he pulled out all the stops, painting an idyllic picture of life in Princeton, New Jersey, where the institute was to be located, and again telling Einstein to name his own price and conditions.

Their marathon conversation lasted from three in the afternoon until eleven at night. They kept talking even when they took a brief break for supper. Finally, Einstein said he was full of enthusiasm for the idea and would give Flexner his decision soon. When he did the next day, it was to decline the offer; he couldn't bring himself to leave his home, friends, and colleagues.

Bitterly disappointed, Flexner left for Berlin to cable his partners back in the States about the failure of his mission. Later, Flexner ruefully admitted that he had tried to read Einstein's works but that he hadn't understood a word Einstein had written. Nevertheless, he had no doubt that he had been speaking with one of the world's great men, "who was simple and unpretentious as any child, and as easy to get along with as any good child."

A few days later, Antonina Vallentin, a family friend and foreign correspondent for England's liberal *Manchester Guardian*, came to see the Einsteins in Caputh. She was greatly agitated, and when Elsa told her that Albert had turned down the offer from America, Vallentin burst out in alarm, "You can't allow him to refuse! To let him stay here in Germany is to commit murder!" She explained that the day before she arrived, General Hans von Seeckt (commander in chief of the German army) had advised her to warn all her Jewish friends to leave the country, especially Einstein, because " 'his life is not safe here anymore.' "

Hearing raised voices, Einstein came to investigate. Vallentin repeated the warning. The Einsteins' maid then reinforced her view; trembling with rage, she said that she'd just been told by the local baker that he couldn't understand how she could live in a "Jewish house." When Elsa asked if she thought Albert was in danger, however, the maid was reassuring. "No," she said. "Everyone loves the professor except the baker." That seemed to settle the matter for everyone but Vallentin.

On the train back to Berlin, she shared a carriage with several young Nazis. Early the next morning she phoned Elsa to tell her that the men had called Berlin's police superintendent "a swine of a Jew" and spoke of murdering him and raping his wife. She followed up this report with a letter imploring Einstein to accept Flexner's offer and escape to America before it was too late.

Elsa was prepared to do so, but then realized how "divinely happy" Einstein was in Caputh and had second thoughts. "Everything is more violent, frenzied, and ferocious in the United States," she replied. The Lindbergh baby had just been kidnapped and Al Capone was beginning a brief prison term. "Here in Caputh the villagers are devoted to him. Even the local Nazis treat him with respect."

Still, Elsa begged Albert to be cautious, to stop signing peace manifestos, and to keep quiet about his anti-Nazi views. He simply refused, saying, "If I was what you want me to be I would not be Albert Einstein." Although afraid for him, she admired his spirit, writing to Vallentin: "He does not know the meaning of fear."

In the end Einstein chose the best of both worlds — to accept Flexner's offer but to keep open the possibility of returning to Germany to make his permanent home if the Nazis, as he hoped and suspected, were just a temporary craze.

Flexner was so relieved that he raised Einstein's salary to $10,000, up from the modest $3,000 he had requested. Other financial details were left to Elsa. She arranged for Albert to be eligible for retirement at age sixty-five, which could be deferred by mutual agreement. He would get an annual pension of $7,500, and if he died first, she'd get $5,000 a year.

Einstein refused the pension offers as too generous, so they were reduced to a minimum of $6,000 for him and a minimum of $3,500 for her. He also got another unexpected salary boost. When Louis Bamberger, a backer of the institute, heard that math professor Oswald Veblen, a nephew of Thorstein Veblen, had been recruited for the faculty at a salary of $15,000, he insisted that "Einstein's salary and retirement annuity with its contingent commitment to his wife as survivor be made equal to Veblen's."* Einstein had a change of heart, and accepted the enhanced offer, saying "it was clear he would need the additional money to help friends and relatives in Germany."

Then he changed his mind again, saying he would not go to Princeton without Walther Mayer, his assistant. He wrote to Flexner: "My only wish is that Herr Dr. W. Mayer, my excellent coworker, will receive an appointment formally independent of my own. Until now, he has suffered very much from the fact that (as a Jew in Germany) his abilities and achievements have not found their deserved recognition. He must be made to feel that he is being appointed because of his own achievements and not for my sake."

Einstein wrote that same day to Sigmund Freud, with whom he had been exchanging ideas for a League of Nations publication dedicated to world peace. Einstein suggested that men went to war and sacrificed their lives "because man has within himself a lust for hatred and destruction . . . It is a comparatively easy task to call [this passion] into play and raise it to the power of collective psychosis. Here lies, perhaps, the crux of all the complex of factors we are considering, an enigma that only an expert in the lore of human instincts can resolve . . . Is it possible to control man's mental evolution so as to make him proof against the psychoses of hate and destructiveness?"

Freud thought their exchange tedious and sterile and quipped that he didn't expect it to win him the Nobel Peace Prize. He was suffering from a cancer in

---

* Veblen's pension was to be $8,000 annually, and his wife's $5,000 should he predecease her.

his mouth at the time, which might account for his cynicism. However, he persevered, replying:

> We assume that human instincts are of two kinds: those that conserve and unify, which we call "erotic" (in the meaning Plato gives to Eros in his Symposium), or else "sexual" (explicitly extending the popular connotation of "sex"); and, secondly, the instincts to destroy and kill, which we assimilate as the aggressive or destructive instincts. These are, as you perceive, the well-known opposites, Love and Hate. Transformed into theoretical entities, they are, perhaps, another aspect of those eternal polarities, attraction and repulsion, which fall within your province.

His response didn't answer Einstein's question: Can man's mental evolution be controlled?

Unlike some people who are intensely occupied with world problems but are indifferent to those of individuals, Einstein was so concerned for his assistant's fate that he jeopardized his own future by insisting on Walther Mayer being part of the Princeton package—a condition that Flexner resisted.

Robert Millikan had hoped to entice Einstein to move to Caltech permanently, but lost out because he made it plain that he would not take on Mayer as well—a decision he later regretted. Einstein still intended to visit Caltech, as promised, for the coming winter semester, but it would be for the last time. He also meant to keep his options open, retaining the Christ Church College appointment as a research assistant at Oxford and his membership in the Prussian Academy of Sciences.

Elsa wrote to Paul Ehrenfest, saying that "Albert's will is unfathomable." That was not true of his deepest commitments, though. His support for a United States of Europe and for world peace through disarmament was never in doubt. For several years now he had been a director of the German League of Human Rights, which, since 1914, had advocated a united Europe. It was to the League's members that Einstein had given one of his most self-revealing talks. In a matter of minutes, he explained himself and his credo:

> "Our situation on this earth seems strange. Every one of us appears here involuntarily and uninvited for a short stay, without knowing the whys and the wherefore. In our daily lives we only feel that man is here for the sake of others, for those whom we love and for many other beings whose fate is connected with our own. I often worried at the thought that my life is based to such a large extent on the work of my fellow human beings, and I am aware of my great indebtedness to them. I do not believe in freedom of the will. Schopenhauer's words: 'Man can do what he wants, but he cannot will what he wills,' accompany me in all situations throughout my life and reconcile me with the actions of others, even if they are rather painful to me. This awareness of the lack of freedom of will preserves me from taking too seriously myself and my fellow men as acting and deciding individuals and from losing my temper. I never coveted affluence and luxury and even despise them a good deal. My passion for social justice has often brought me into conflict with people, as did my aversion to any obligation and dependence I do not regard as absolutely necessary. I always have

a high regard for the individual and have an insuperable distaste for violence and clubmanship. All those motives made me into a passionate pacifist and antimilitarist. I am against nationalism, even in the guise of patriotism. Privileges based on position and property have always seemed to me unjust and pernicious, as did any exaggerated personality cult. I am an adherent of the ideal of democracy, although I well know the weakness of the democratic form of government. Social equality and economic protection of the individual appeared to me always as the important communal aims of the state. Although I am a typical loner in daily life, my consciousness of belonging to the invisible community of those who strive for truth, beauty, and justice has preserved me from feeling isolated. The most beautiful and deepest experience a man can have is the sense of the mysterious. It is the underlying principle of religion as well as all serious endeavor in art and science. He who never had this experience seems to me, if not dead, then at least blind. To sense that behind anything that can be experienced there is something that our mind cannot grasp and whose beauty and sublimity reaches us only indirectly and as a feeble reflection, this is religiousness. In this sense I am religious. To me it suffices to wonder at these secrets and to attempt humbly to grasp with my mind a mere image of the lofty structure of all that is there."

Later, whenever asked his religious views, Einstein's answers were variations of this speech given in Berlin.

Elsa had reason to agree with his being "a typical loner." He even had his own bedroom. Her daughter Margot and son-in-law Dimitri Marianoff occupied the bedroom adjoining Elsa's, while Albert's was way down the hall. When he wasn't in his bedroom, he was often secluded in his study. "He goes to his study," Elsa explained to a reporter, "comes back, strikes a few chords on the piano, jots something down, returns to his study. On such days Margot and I make ourselves scarce. Unseen, we put out something for him to eat and lay out his coat. Sometimes he goes out without his coat and hat, even in bad weather."

But the loner was never alone for long. Invited and uninvited visitors routed him out of his study or, when at Caputh, out of his reveries. Hoping for fewer distractions, Einstein spent the rest of the summer working on a short article on cosmology to be translated into French by Maurice Solovine. He had decided to skip a peace conference at Geneva, convinced that he was more effective by writing in its support than by making speeches.

But distractions were inevitable. When William Fondiller and his wife turned up at Caputh in a car, with a letter of introduction from American physicist and inventor Michael Pupin, Einstein was seeing several other visitors off on a bus. He wheeled around and saw the new arrivals from America. Fondiller recalled that he "put out his arm and said, 'Einstein.' He took us into the house where we paired off. I talked with the professor and my wife with Frau Elsa. He asked me quite simply, 'What are you doing in Europe?' I said, 'I'm attending an international congress on intercontinental and international communication.' 'What are some of the problems?' he asked."

One problem recently solved was how to make telephone conversations clear between people thousands of miles apart. Formerly, the speakers' words had been echoed back to them. Einstein wanted to know how it was solved. By "an echo suppressor," replied Fondiller. Intrigued, Einstein asked for a sketch of the circuits and Fondiller drew them. Einstein sighed with satisfaction. "I thought of this problem twenty years ago when I was a patent examiner," he said, "but I didn't know how to solve it. Now you've solved it."

Elsa reminded Einstein that government delegates sent by President von Hindenburg were on their way and that he should change from his open shirt and well-worn corduroy trousers to receive them. He nodded absently and continued talking. Elsa returned later to say it was time to get changed. "If they want to see me," he answered, "here I am. If they want to see my clothes, open my closet and show them my suits." He then resumed his conversation with Fondiller, explaining how he had once been hired by a large electrical company to find "the maximum possible permeability in a mass of very finely divided magnetic particles . . . They have their own mathematicians and physicists working on it and could not get a satisfactory answer. They did not come to me until they had to, because they don't like Jews too much." Fondiller was struck by the accuracy of Einstein's figure for permeability, which had been "calculated on purely theoretical grounds. It was remarkably close to the one we had achieved after some years of experimentation." The American couple didn't stay to see if Einstein changed his clothes to meet the official delegation. Chances are he didn't.

Hans Albert and his wife came for a visit, but Mileva stayed home. She wouldn't consider joining them in a country where atrocities were reported to be occurring every day. Besides, Eduard needed her constant attention.

Einstein was reminded of Eduard's problems in a rambling letter from Michele Besso to "my dear and old friend." Besso confided that he himself had been heading for a nervous breakdown over long-standing conflicts with his mother, but thanks to the emotional support of his son, Vero, he could once again "rejoice in the sky, leaves on the trees, sun and rain, the sweet look in Anna's eyes [his granddaughter], and the happy greeting of my little dog." Now that he was no longer depressed, Besso said, he could concern himself with the suffering of others—which brought him to the sensitive subject of Eduard, for whom he felt great affection.

Besso became nostalgic over their youthful days, when Einstein's genius "brought you discoveries by the armful, and from which you extracted with effort and tenacity the valuable ones. And the pure joy I felt, and my numerous objections of all kinds." He went on to recall how tormented they had both been to see Eduard's mental breakdown "almost under our eyes." He agonized over the thought that despite his sincere efforts to reconcile Mileva and Albert, in the end he was the agent of their separation. (It is not clear why he felt responsible.)

Then he came to the main purpose of the letter — to bring Albert and Eduard closer together:

> Someone asked me why he only sees photos of Einstein and his daughter. "I thought he had sons," this person said. "You never see anything of his sons, do

you?" Eduard has an extraordinary father, a brave mother. He is talented and attractive, although reserved as some young people are. He has a good friend devoted to him and even an old friend who understands him. He is well educated, and could get the job he wants if he goes about it the right way. [Eduard wanted to be a psychiatrist.]

What are you doing, oh child with white hair, about the pain of one who is searching for answers? And all that is put on my shoulders? Help if you can. Take Eduard with you on one of your long journeys. Once you've spent six months with him you'll end up understanding him and putting up with a lot of things you wouldn't accept from others. Then you'll know what unites you and, unless I'm terribly mistaken, it will lead the way to a flowering of your son's personality.

When after a month Einstein hadn't replied, Besso wrote: "Dear Albert Senior, Major, Maximus, You haven't answered me but I can't believe you're angry."

Einstein responded immediately:

How could I be angry with you? I've never seen you do or say anything with any other intention than to do good. It's the same thing in the case of Tetel [Eduard's nickname]. I've invited him to America, at Princeton, next year. This year it wouldn't have been a good idea because I had too many commitments. For Tetel it would have been more of an ordeal than a vacation. Unfortunately, everything indicates that strong heredity manifests itself very definitely. I have seen that dementia praecox [now called schizophrenia] coming slowly but irresistibly since Tetel's youth. Outside circumstances and influences only play a small part in such cases, compared to internal secretions, about which nobody can do anything.

Einstein hardly had time to focus on personal problems, though. A flurry of experts always seemed to be in wait to challenge him. Now they pounced on his October estimate that the earth was 10 billion years old. Although Caltech's finest minds were reluctant to join the pack before carefully studying his recent paper on cosmology, the one that Solovine had been translating, Professor Fritz Zwicky said that geological studies indicated that the earth was much older. Edwin Hubble, speaking for astronomers, believed that Einstein had simply multiplied by five the expansion time of observed nebulae, but he didn't want to comment further until he had seen Einstein's calculations. Sir James Jeans, a visiting British astrophysicist, daringly suggested that Einstein had been too moderate in his estimate. Jeans concluded from the radioactivity of rocks that the earth was 30 billion years old. (The planet enjoyed a spectacular rejuvenation in his 1942 book *The Mysterious Universe*, in which he put the earth's age at 2 billion years!)

Scientists in Washington, D.C., complained that Einstein had been too vague: Did he mean the earth had aged 10 billion years since its birth—probably as a mass thrown off from the sun—or since it solidified? Yale physicist Alois Kovarik and colleagues on the National Research Council Committee stuck to their figure of 1.852 billion years, though Yale astronomer Frank Schlesinger was more comfortable with 3 billion. Schlesinger suggested that a reporter had misquoted Einstein by adding 7 billion years to the true age of the earth. No one agreed with Einstein. Princeton geologist Richard Field said, "Many geologists will be surprised to learn that the earth has existed 10 billion years," because geological evidence pointed to a comparative youngster of only 4 billion. (In 1995, the generally accepted figure is 4.5 billion.)

Einstein turned from thoughts about the age of the earth to whether he should leave Germany permanently. Reluctant to cut all ties with his homeland, he announced in October: "I have received a leave of absence from the Prussian Academy for five months of the year for five years. Those five months I expect to spend at Princeton [starting in the fall of 1933]. I am not abandoning Germany. My permanent home will still be in Berlin." Flexner put a different spin on it, adding, "Professor Einstein will devote his time to the institute, and his trips abroad will be vacation periods for rest and meditation at his summer home outside Berlin."

The New York Times greeted the announcement with giddy anticipation of colorful copy: "Great news. While Einstein was in California last winter with Willem de Sitter and other compeers, the universe expanded, collapsed or oscillated on the front pages of the Los Angeles papers for weeks." Among the other well-wishers was U.S. Supreme Court Justice Benjamin Cardozo, who wrote to Flexner, "I have read with delight that Einstein has been captured for the faculty of the Institute." It was indeed a great coup. Einstein's French friend Paul Langevin remarked, "The Pope of Physics has moved and the United States will now become the center of the natural sciences." Science writer Ed Regis hardly stretched the metaphor in later concluding: "It was as if God himself were going to take up residence at Flexner's new place in Princeton."

Still, the excitement was somewhat premature, because Flexner had not yet agreed to take Mayer on Einstein's conditions. Einstein remained adamant; he wouldn't go to Princeton without Mayer, though he still meant to go to California. Just before he left, a woman wrote from Vienna asking him to comment on a news report that a medium had converted him to a belief in spiritualism. "It's utterly untrue," he replied. "I do not believe in the spirits or spiritualism, and would be happy if all my fellowmen were of the same opinion." According to Helen Dukas, fortune-tellers, mediums, and others of their ilk constantly lied about Einstein sharing their views and confirming their "supernatural" gifts.

Spiritualists were the least of Einstein's problems. A small but active group of American women were agitating to keep him out of the country. Led by Mrs. Randolph Frothingham, president of the Woman's Patriot Corporation, and co-president Mary Kilbreth, their "patriotic" organization had failed in a campaign to deny women the vote, and now focused on banning "undesirable" aliens from entering the United States. Einstein topped their danger list; they labeled him "a Communist and menace to American Institutions" who was "affiliated with more Communist groups than Josef Stalin himself." They also charged, in a sixteen-page legal brief prepared by Frothingham, that he was, as a revolutionary radical, leading an anarchist-communist band which aimed to destroy the military might of various countries as a prelude to world revolution.

As for Einstein's scientific achievement, Frothingham scoffed, "His frequently revised theory of relativity is of no more practical importance than the answer to the old academic riddle, 'how many angels can stand on the point of a needle if angels do not occupy space.'" He was also a threat to the Church, she went on, by bringing confusion to the laws of nature and the principles of science, which were revised with every new report of an Einstein theory. Among their other complaints were his apparent inability to speak English (he was improving), and that he "advises, advocates, or teaches individual 'resistance' to all accepted authorities *except Einstein*, whether it be a question of peace or war, government or religion, mathematics or anthrology" (whatever that was; perhaps she meant anthropology).

Einstein's first reaction was to laugh it off; then he responded in a statement filled with high humor, bound to infuriate his earnest opponents:

> Never yet have I experienced from the fair sex such energetic rejection of all advances; or if I have, never from so many at once. But are they not quite right, these watchful citizens? Why should one open one's doors to a person who devours hard-boiled capitalists with as much appetite and gusto as the Cretan Minotaur in days gone by devoured luscious Greek maidens, and on top of that is low-down enough to reject every sort of war, except the unavoidable war with one's own wife? Therefore give heed to your clever and patriotic womenfolk and remember that the Capitol of mighty Rome was once saved by the cackling of its faithful geese.

Dana Bogdan, chief of the State Department's Visa Division, did take heed, and informed the U.S. Consulate in Berlin of Frothingham's petition to ban Einstein as an undesirable alien. It is not clear if Einstein, when applying there for a visa on December 5, was asked more searching questions about his political beliefs than was customary. George Messersmith, the American consul, was away, and the task of interviewing Einstein fell to his assistant, Raymond Geist. On previous occasions, Einstein had been given a visa without being interrogated. This time he was questioned for almost three-quarters of an hour.

When Geist asked about his political beliefs, Einstein simply remained silent, but he conceded that he was a member of the War Resisters International, which did not require party membership. He also said that he intended to go to America for scientific work. When asked, "Are you a Communist or anarchist," however, he angrily replied, "What's this? An inquisition?" He then issued an ultimatum: If his visa wasn't granted by noon the next day he'd remain in Berlin. He wasn't angry for long, telling a reporter, "Wouldn't it be funny if they wouldn't let me in? Why, the whole world would laugh at America."

Far from laughing, many Americans were outraged, especially a group of women at the home of Mrs. Gerard Swope, wife of the president of General Electric. Protesting "in the name of the intelligent American people against the action of the State Department in forwarding to the consulate in Berlin the absurd document of a so-called patriotic society," they demanded the recall of Consul General Messersmith for humiliating America and making it "a laughing stock second only to the Scopes [monkey] trial in Tennessee. Every intelligent person in the world knows that Communists and pacifists differ fundamentally. Communists believe that violence has its place as a political method. Pacifists believe that use of force is always a mistake."

The *New York World-Telegram* summed up Einstein's treatment as "a new, all-time low for official stupidity."

Einstein did not expect the consulate to have time to issue a visa by the noon deadline, so he and Elsa stopped packing in their Berlin apartment and returned to Caputh. He awoke the next morning unhappy, convinced that his trip to the States was off. His phone rang at eleven; the call was from the American consulate, apologizing. His visa had been issued. By noon the Einsteins were back in Berlin, preparing to leave in four days for California.

Elsa was delighted, but she admitted that "the professor and I were quite determined last night to turn our backs on America forever. Now that the visas have been straightened out, we can leave on the *Oakland* as scheduled. From the deluge of cables reaching us last night and this morning, we know Americans of all classes were deeply disturbed over the case."

Attacks on Einstein's political and social views came from the usual suspects; assaults on his science came from the most unlikely individuals, such as Irish physician Arthur Lynch. Lynch had lobbed verbal hand grenades at Einstein for years. Commander of the British army in Ireland in 1918, and a former member of Parliament, layman Lynch had expertise in politics, the human body, and military maneuvers. Still, he had the gall to write *The Case Against Einstein*, published in 1932, and mail the book to fellow countryman George Bernard Shaw, hoping for a plug. Shaw was one of Einstein's most unlikely defenders, having had no science training. Using playful dagger thrusts and fancy footwork, he told Lynch that his effort was doomed to fail.

In a note to his secretary, Blanche Patch, on December 6, 1932, Shaw told her how to answer another Einstein critic, Wallace Salter:

> Say I am not an adherent to the general theory of Relativity. I am not a mathematician and do not understand it. I rank Einstein as a man of genius from personal observation and from certain propositions of his which are within my comprehension; but you and Colonel Lynch and the other anti-Einsteinians must fight it out with him yourselves: I must not make myself ridiculous by interfering in a discussion that is outside my province. As the whole progress of mathematics from its ancient simplicities to what we call its "higher" modern developments has been effected by assuming impossibilities and inconceivabilities, your line of argument does not seem to me conclusive . . . And it is not possible to disassociate mathematical nonsense from physical nonsense when the problem is one of astro-physics. Newton, Leibnitz, Young, Fitzgerald, Lorentz and Einstein are all Nonsensists; but you cannot dispose of Einstein by this statement any more than you can dispose of Michael Angelo [*sic*] by calling him an Illusionist.

As the Einsteins left their Caputh home, Albert told Elsa to take a last look, because she'd never see it again. The shape of things to come was hardly a mystery. The eighty-four-year-old von Hindenburg was on his last legs, and in the German parliament on December 7, "fists swung, chair legs were brandished, inkwells and books flew all over the chamber of the Reichstag. The Hitlerites and the Communists, finding words too feeble to express their feelings, took off their coats and just let themselves go." Storm troopers roamed the streets roughing up Jews, leftists, and anyone who was slow or reluctant to return the Nazi salute.

As Einstein headed for California, he was being discussed in a Berlin law court, where a banker was on trial for falsely using his name in a stock swindle. He claimed that Einstein had endorsed his mathematical system as a flawless forecaster of stock exchange fluctuations.

Einstein was well on the way to America when he heard momentous scientific news that had not yet reached the newspapers: working at Cambridge University's Cavendish Laboratory, "the nursery of genius," James Chadwick had discovered neutrons at the heart of the atom. With astonishing prescience, the lab's director, Ernest Rutherford, had anticipated the existence and nature of such particles "with zero nuclear charge" twelve years before. His gift of prophecy failed him now, however, when he was asked if the neutron could be be used to split the atom and release its enormous latent energy. His answer was an emphatic No. It was assumed that atom splitting (still just theoretically possible) would require incredibly powerful projectiles; the neutron, with no electrical charge, was a poor candidate for the task. Rutherford even ridiculed those who suggested that it was possible, telling them that they were "talking moonshine." Einstein agreed with him, calling the idea a fantasy.

Above: The only smiler in the group of fifty-two schoolboys in Munich, Germany, in 1889, is ten-year-old Albert Einstein (front row, third from the right). Albert wasn't happy in school, but he put up a brave front. (Courtesy of Stadt Ulm Stadarchiv)

Left: High-school graduate Einstein in Aarau, Switzerland, in 1896, when he was seventeen and in love with Marie Winteler. (Courtesy of ETH, Zurich)

Einstein's first wife, Mileva, with their sons in Zurich in 1914 when Eduard (left) was four and Hans Albert (right) was ten. She had returned with the boys from Berlin, where Einstein was director of the newly created Kaiser Wilhelm Institute for Physics. It was the beginning of the end of their marriage. (Courtesy of ETH, Zurich)

They all got the joke. During his first visit to America in 1921, Einstein met Rabbi Stephen Wise (center) and Fiorello La Guardia, then president of New York City's Board of Aldermen, and later mayor of New York City. The photo gives some idea of Einstein's exuberant laughter. (Courtesy of American Jewish Archives)

World-famous professor: Einstein in Berlin in 1922 keeps to his habit of fastening only one button on his topcoat. (Courtesy of Library of Congress)

First visit to California. The new arrivals from Germany smile for the camera in early 1931. From left, Einstein's secretary, Helen Dukas; Elsa; Albert; and Einstein's assistant, Walther Mayer. (Courtesy of Library of Congress)

Pasadena. Albert and Elsa stand on the grounds of the cottage they chose to live in during their California stay in 1931, when he was at the California Institute of Technology. (Courtesy of Library of Congress)

Einstein and Winston Churchill in the garden of Churchill's home, Chartwell, in Kent, in 1933. During this visit Einstein warned Churchill that Germany was secretly preparing for war. (Courtesy of Lady Soames)

Cycling in Pasadena in 1933. Einstein said that life is like cycling—you have to keep moving to keep your balance. (Courtesy of Caltech Archives)

The last meeting of twenty-three-year-old Eduard Einstein and his father in Zurich, Switzerland, in May 1933. (Courtesy of Margot Einstein)

Pittsburgh press conference, December 28, 1934. Asked if he thought it might be possible to release enormous amounts of energy as shown by his equation $E = mc^2$ by bombarding the atom, Einstein replied that it wouldn't be practical: "Splitting the atom by bombardment is like shooting at birds in the dark in a region where there are few birds." (Courtesy of American Jewish Archives)

The triangle: When playwright Clifford Odets and his wife, Oscar-winning actress Luise Rainer, visited Einstein in Peconic, Long Island, one summer afternoon in 1937, Odets became so jealous of what he thought was Einstein's romantic interest in his wife that he later decapitated Einstein's image in one of the photos taken during that visit. The one above, of the couple with Einstein, and the one on the right, showing Rainer and Einstein together in a chummy pose, escaped the scissors. (Courtesy of Luise Rainer)

Trenton, New Jersey, 1940. Einstein and his stepdaughter, Margot, walk to the Federal Building to apply for their citizenship papers, accompanied by a newspaper reporter. (Courtesy of National Archives)

Above: Manhattan, 1942. Thomas Bucky is at the wheel of his first Model-A Ford, with his mother sitting between him and Einstein's sister, Maja Winteler-Einstein. Einstein is in the rumble seat with Gustav Bucky. Einstein never learned how to drive. "Why should he?" said Thomas Bucky. "My brother and I drove him whenever he had to go someplace." (Courtesy of Thomas Bucky)

Left: Outside 112 Mercer Street, 1942: from left, Dukas, Thomas Bucky, Margot, Einstein, Gustav Bucky, and Frida Bucky. (Courtesy of Thomas Bucky)

Willing model: Einstein, holding a forbidden cigar, poses in his Princeton garden for Gina Plunguian in 1948. Several bronzes were made of the head that she modeled in clay. (Courtesy of Mark Plunguian)

The godfather: Einstein plays with Lincoln Logs, the Lego bricks of its day, one of his birthday gifts to his godson Mark Abrams, in April 1953, while four-year-old Mark searches for the other, more tempting gift—candy. (Courtesy of Mark R. Abrams)

Mind changer: Anthropologist Ashley Montagu chatting with Einstein in his study in early June 1953. Montagu persuaded Einstein that he was wrong in believing that human beings have instincts. (Courtesy of Ashley Montagu)

Artist's view: This striking portrait by Winifred Rieber shows Einstein in 1934 when he was fifty-five, and vividly conveys her impression of him as a force of nature. "Everything about him is electric," she said. "Even his silences are charged." (Courtesy of Library of Congress)

CHAPTER 25

Einstein the Refugee

1933

54 years old

"He's a radical and an alien Red!" charged Mrs. James Gray to an equally irate group of women trying to stop Einstein from landing on American soil. They hadn't much time; his ship was nearing the California coast. Gray represented the Woman's Patriot Corporation, the group that had opposed him before. She hoped that her protests would alert the authorities to this dangerous man and bar his entry. Why, he's not even a real scientist, she scoffed, quoting scientist Thomas See's definition of relativity as "a crazy vagary, a disgrace to our age." Robert Watt, a lonely voice at the meeting, representing the Massachusetts Federation of Labor, called the "Red Menace" issue a lot of hooey and suggested that Gray's group would be better occupied working for social reforms.

The women failed to keep the threat at bay. The next day, January 9, when the yellow flag of quarantine was hauled down from the ship, reporters and photographers scurried aboard, a rowdy welcoming committee. Einstein refused to discuss Prohibition, promised to give written answers to complicated questions later, and said, "I am sure the universe is expanding, but we have not yet been able to gather a complete history of the expansion." He told the reporters that he hoped to advance the unified field theory during his two months at Pasadena, having already successfully incorporated Lorentz's formula—that any rapidly moving object becomes shorter in exact proportion to its speed.

The Einsteins landed with thirty pieces of luggage and a violin and were soon installed in their rooms in the Athenaeum on the Caltech campus, where writer Fulton Oursler—introduced by their mutual friend Upton Sinclair—questioned him for more than an hour. Oursler explained that he was from New York City, the home of three million Jews wary of Hitler's aims and concerned about the fate of Jews in Germany. Einstein was optimistic, saying that minorities were always treated badly in difficult times, just as the Filipinos and Mexicans in California and minorities in Africa were currently being treated. In such times, he said, when people are fearful, it is easy for agitators to inflame them against minorities. When pressed, Einstein was less optimistic, and admitted that a pogrom against the Jews was possible in Germany, adding, "They have police, but if the police are in the hands of national soldiers, as they are now, it could be very dangerous without any change of law."

Einstein called Hitler a very ordinary man who aroused people's emotions by capitalizing on their prejudices; he rated Mussolini as the more intelligent of the two dictators, as he had better plans to reorganize his country.

Oursler asked if Einstein knew the story of the Jewish man in Russia who had to go to Moscow: "He was afraid to travel," said Oursler, "so he sat crouched in the corner of a third-class carriage, in an effort not to attract any attention. A big Cossack got on the train. 'Damn the Jews!' he roared. 'The Jews made all the trouble. The Jews made the famine. The Jews made the war.' The Cossack turned to the little man and demanded, 'Didn't they?' 'Yes,' said the Jew. 'The Jews and the bicycles.' 'Why the bicycles?' inquired the Cossack. 'Why the Jews?' replied the Jewish man."

"Yes, of course," Einstein said. "It is well known in Germany, but it is impossible to tell it there."

Some weeks later Einstein telegraphed Oursler, withdrawing permission to run the interview. No explanation was given. Oursler assumed that Einstein had changed his mind about the plight of Jews in Germany and feared that his comments might endanger them.

Despite the protests of the Woman's Patriot Corporation, Einstein proved to be the biggest attraction Caltech had ever had: a record number applied for tickets to a dinner held in his honor. Leon Watters, a wealthy biochemist, was among the lucky ones. He got a great seat at Einstein's table, from which he leaned across the socially prominent woman separating them to offer Einstein a cigarette. To Watters's surprise, Einstein demolished it in three strong puffs. When coffee was served, Watters offered him a cigar. "No thanks," Einstein said, "I prefer my pipe." He then read a card Watters handed over and burst out laughing. Watters had written on it:

> When men are calling one another names and making faces
> And all the world's a jangle and a jar,
> I meditate on interstellar spaces
> And smoke a mild segar [sic]

That broke the ice, and they chatted until Einstein left to give a nationwide radio talk titled "America and the World Situation."

The world situation took a turn for the worse a week later. Despite the prescient warnings of General Erich Ludendorff that "Hitler will plunge our Reich into the abyss and inflict immeasurable woe on our nation," on January 30, 1933, the doddering President von Hindenburg was persuaded by a coalition of Nazi and right-wing politicians to appoint Hitler as chancellor. Einstein soon discovered that the new chancellor sympathized with the American Woman's Patriot Corporation, accusing him on March 2, through his mouthpiece, the

Völkischer Beobachter, of "cultural intellectualism, intellectual treason, and pacifist excesses."

That same day, reporter Evelyn Seeley, of the *New York World-Telegram*, arrived in Pasadena to question Einstein and found him "wandering about in his garden, not sure whether it was yesterday or to-day." She got a good interview, though, and drove off happily towards Los Angeles, where she stopped at a Sunset Boulevard traffic light. Suddenly her car wobbled mysteriously, fashion dummies in a nearby shop window did the rumba and then hit the floor, and oranges rolled "madly from the fruit stands." Los Angeles was having a major earthquake.

Professor Herbert Dingle, over from England on a Rockefeller scholarship to learn about relativity, felt the shock in his office at Caltech and hurried to the Athenaeum Faculty Club to make sure his wife was safe. On the way, he passed Einstein and Beno Gutenberg, standing shoulder-to-shoulder, peering intently at a large sheet of paper. Gutenburg, a seismologist visiting from central Europe, was hoping to experience an earthquake, an unlikely event back home. But he and Einstein were so absorbed in studying the plan of a sensitive new seismograph that "they had failed to notice the earthquake."

Einstein only realized something was wrong when he saw people running from the campus buildings. Fifteen more shocks followed in the next seven hours. Long Beach was hardest hit, and all told, 116 people were killed and some 5,000 injured.

Meanwhile, the German press was focusing on man-made disasters, such as the burning of the Reichstag on February 27, 1933. Hitler blamed the Communists and declared martial law, freeing his followers to embark "unhindered on wholesale blackmail, robbery, violence, and murder." Despite the violence, he got a healthy forty-four percent of the votes cast in a March 5 election — almost twenty million Germans wanted him as their leader. Co-opting conservatives in a subsequent election and preventing all Communist and some Socialist deputies from running, he got a whopping majority. He then "brushed aside the Constitution and made himself dictator."

The short-lived liberal Weimar Republic was dead, and a fascist police state, the Third Reich, was born. Hitler, the former choirboy who had toyed with the idea of being a priest, showed his heavy hand from the start. In a matter of months, he abolished all civil liberties and rival political parties; built concentration camps for those opponents he didn't execute; and sent his secret police—the Gestapo—to hunt down those who had escaped.

Einstein was out of the Gestapo's reach—but only temporarily. Having completed his two months at Caltech, he was heading back toward Germany—and danger—with brief stops en route at Chicago, Albany, New York, and Manhattan.

He woke on the morning of March 14 as his train approached Chicago, where photographers, as usual, were waiting to snap him. He stood patiently for a while, but resisted a request to break his rule and smile for the camera, even on his birthday. Then he and Elsa were whisked to an official lunch. There he congratulated a fellow guest, attorney Clarence Darrow, for his masterly defense of John Scopes, who had defied Tennessee state law in 1925 to teach the theory of evolution.

At the next train stop—Albany, New York—he heard that storm troopers had ransacked his Berlin apartment, terrifying his stepdaughter Ilse. He hardly needed the friendly warning he got from the German consul, Paul Schwartz: "If you go to Germany, Albert, they'll drag you through the streets by the hair." He knew his probable fate would be more drastic than that and swore not to set foot on German soil while the Nazis were in power.

Reaching Manhattan, he left Grand Central Station via a freight elevator to dodge the waiting crowd, then headed for a fund-raising dinner for Hebrew University. Harlow Shapley, the Harvard astronomer, sat next to him during the dinner. Einstein was explaining a cosmological theory to Shapley and making a sketch on a place card to clarify a point when someone leaned over his shoulder and snatched the card as a souvenir. Einstein looked up in absolute astonishment, not yet inured to the ways of the New World.

His speech to fellow pacifists soon after incited a Berlin newspaper publisher to complain:

> Einstein has been hardly a day in New York before he has twice thrown his "powerful personality" against Germany. At a pacifist meeting he was reported to have called for the moral intervention of the entire world against Germany and Hitlerism. Now at a time when America is mendaciously "informed" about Germany and egged on against her by unclean Marxist and Democratic prop-agandas . . . this puffed up bit of vanity dares to sit in judgment on Germany without knowing what is going on here—matters that forever must remain in-comprehensible to a man who was never a German in our eyes and who declares himself to be a Jew and nothing but a Jew.

The Nazis took quick action. Storm troopers raided Einstein's Berlin apart-ment five times in two days and came up empty-handed. They had been out-witted by Margot, who had smuggled his most important papers to the French Embassy in Berlin.

When her husband, Dimitri Marianoff, phoned Margot from out of town, she warned him not to come home but to leave the country at once for Paris, where she would meet him. Ilse and her husband, Rudolf Kayser, had already escaped to Holland.

Storm troopers surrounded Einstein's country home in Caputh to search for arms and ammunition. They had been told by an informer that he had allowed militant Communists to stockpile military stores on his property. The storm troop-ers found nothing at first. At last they made an inch-by-inch search of the house, emerging with the evidence—a bread knife!

Berlin newspapers reported that Einstein was spreading atrocity stories abroad and lying about the mistreatment of Jews. His friends, the Buckys, were still in Germany; at school Thomas was sharply questioned by teachers and children "who knew of my family's friendship with Einstein . . . this 'fiendish liar,' as they called him. A friend of my father's even asked him to sign a petition to be sent to the foreign press, stating that Einstein's account of life in Germany was made up out of the air. He refused, of course. My father was also very high on the list of the Nazis' enemies, and they arrested him. But he was an American citizen so he was freed."

Planck was no Nazi, but as a staunchly patriotic conservative he reluctantly concluded that Einstein's criticism of the new regime disqualified him from holding a state position. He asked Albert to resign from the Prussian Academy, "afraid that a formal proceeding for expulsion would place Einstein's friends in a most severe conflict of conscience." Despite their disparate political views, Planck had a clear, unbiased view of Einstein's worth, which Heinrich von Ficker, the academy's presiding secretary, conveyed in a letter to Einstein, quoting a statement by Planck: "I am completely certain that in the history of future centuries the name of Einstein will be celebrated as one of the most brilliant stars that have shown in our Academy."

Einstein responded quickly. On March 29, the day after his boat docked in Antwerp, he resigned from the Prussian Academy. Bernhard Rust, the Nazi minister representing the academy, demanded that other members denounce Einstein as a traitor. Max von Laue immediately called a meeting to kill Rust's proposal. Fourteen turned up, but only two supported von Laue. Later, a few colleagues did urge Planck to protest the dismissal of Jewish and Socialist professors. Though pessimistic, Planck decided to confront Hitler.

In the meantime, he and von Laue charged Einstein with breaking the unwritten rule that scientists must keep out of politics. Albert replied to their accusation from Belgium: "Where would we be had men like Giordano Bruno, Spinoza, Voltaire, and Humboldt thought and behaved in such a fashion? I do not regret one word of what I have said and believe my actions have served mankind."

Nazi outrages continued. On April 1, mobs led by brown-shirted storm troopers stopped customers from entering Jewish shops and roughed up all who resisted. Einstein's property was seized and a reward offered for his capture as an enemy of the state. Marianoff believed that the Gestapo had hoped to kidnap Ilse and Margot as hostages to silence Einstein's protests, but the two young women had frustrated those plans by leaving the country.

Hearing that the Nazis had seized Einstein's property, Dutch astronomer Willem de Sitter and his colleagues offered financial help. Einstein replied, "In times like these one has an opportunity to learn who are one's true friends." He declined the offer with warm thanks, because things were going so well for him that he could even "help others keep their heads above water. From Germany, however, I will probably not be able to rescue anything because an action is being taken against me for high treason." What most surprised him, he added, was the failure of Germany's so-called intellectual aristocracy to oppose the Nazis.

Einstein told Planck that he still valued their friendship, and recalled the years of persecution he (Einstein) had suffered almost in silence, "but now the war of annihilation against my defenseless Jewish brothers has forced me to place whatever influence I have in the world on their side of the balance."

Philipp Lenard, the Nobel Prize–winning physicist and Einstein's bête noir, thought the time was ripe to strike again. He wrote to the leading Nazi newspaper, criticizing Einstein's science and patriotism: "The most important example of the dangerous influence of Jewish circles on the study of nature has been provided by Einstein with his mathematically botched-up theories . . . Even scientists who have otherwise done good work cannot escape the reproach that they allowed relativity to get a foothold in Germany, because they did not see . . . how wrong it is, outside the field of science also, to regard this Jew as a good German."

On May 10, tens of thousands of "good Germans" poured into Franz Joseph Platz, a large public square between Berlin University and the State Opera House. There, they gave a frenzied welcome to a torchlight parade of students and beer-hall bruisers escorting a caravan of vehicles loaded with books. The parade stopped at a huge bonfire in the center of the square. Gleeful mobs joined them at the fire, grabbing books looted from private homes and public libraries. As the crowd roared, the authors' names were called out and their books tossed into the flames: "Einstein . . . Thomas Mann . . . Heinrich Mann . . . Freud . . . Jack London . . . Rathenau . . . Arnold Zweig . . . Stefan Zweig . . . Proust . . . Hemingway . . . H. G. Wells . . . Helen Keller . . . Gide . . . Zola . . . Dos Passos . . . Upton Sinclair . . ."

Propaganda chief Joseph Goebbels gave the book burning an official stamp of approval by climbing to a rostrum, his face flushed by the flames, to yell, "Intellectualism is dead. The German national soul can again express itself."

Hemingway, Wells, Sinclair, Dos Passos, the Mann brothers, and Keller must have been surprised to learn that they were linked with Freud, Einstein, and Rathenau as Jewish intellectuals. They did have one thing in common, though — all were antifascists.

More than sixty bonfires blazed throughout Germany — an orgy of burning books.

Planck courageously responded the next day at a crowded meeting of the Prussian Academy, when he compared Einstein favorably with Kepler and Newton.

Not yet aware of the book burning, the Einsteins had settled into what Albert called a very pleasant exile in Le Coq sur Mer on the Belgian coast, where they were joined by Ilse and Margot; Albert's secretary, Helen Dukas; and his assistant, Walther Mayer. Frederick Lindemann was among their visitors during their stay. He was touring Europe, trying to place refugee scientists in British universities. He was too late, however, to get Einstein, who was already committed to America.

After his recruiting drive, Lindemann wrote to Einstein on May 4, 1933: "I was in Berlin 4 or 5 days at Easter and saw a great deal of many of your colleagues. The general feeling was much against the action taken by the Academy . . . Everybody sent you their kind regards . . . especially Schrödinger . . . It seems the Nazis have got their hands on the machine and they will probably be there a long time."

Einstein immediately replied: "I think the Nazis have got the whip hand in Berlin. I am reliably informed that they are collecting war material and in particular aeroplanes in a great hurry. If they are given another year or two the world will have another fine experience at the hands of the Germans."

Later in May, Planck spoke with Hitler, hoping to convince him "that the forced emigration of Jews would kill German science and that Jews could be good Germans . . . [Hitler replied:] 'But we don't have anything against the Jews, on the contrary we protect them.'"

Another version of this meeting gives evidence of Hitler's increasingly volatile personality:

> When [Planck] extolled [Fritz] Haber's and other Jewish Germans' contributions to science, Hitler replied that he had nothing against Jews as such, but that they were all Communists. When Planck tried to argue, Hitler shouted, "People say that I get attacks of nervous weakness, but I have nerves of steel"; he slapped his knee and whipped himself into a rage that continued until Planck took his leave. Planck told [Max] Born . . . that this interview extinguished any hopes he might have had of openly exerting influence in favor of his Jewish colleagues.

Far from protecting Jews, as a London *Times* correspondent had revealed, the Führer had taken his first steps to destroy them:

> What the Nazi *Völkischer Beobachter* calls "the dress rehearsal for the permanent boycott of the Jews" was ruthlessly enforced yesterday between 10 A.M. and midnight. It was completely effective . . . At Kiel it led to the first case of mob lynching . . . The boycott completely paralysed Jewish business life. On the stroke of 10 uniformed Nazis took position outside every Jewish shop, store, café, or other undertaking. The order of the Nazi boycott leader, Herr Streicher, that there should be neither violence nor force was not respected. In the small shops the Nazi picket—often wearing a revolver—usually stood with his legs planted astride in the doorway. Your correspondent saw several people forcibly prevented from entering or thrust violently away.

Although Einstein was concerned by these events, he had his own personal problems to deal with. He delayed a trip to England in order to see his schizophrenic son, Eduard, in Switzerland. Hans Albert blamed the mental illness on the shock treatments Eduard had been given in the Zurich psychiatric hospital to pull him out of the depression caused by his unrequited love affair three years before. The true cause has not been established.

For what would be the last meeting between father and son, Einstein took his violin to comfort both of them. What they discussed has never been revealed. It was also the last time Albert would see Mileva, who showed that she had overcome her animosity by offering him and Elsa the use of her apartment until they found a permanent home. Her offer surprised and pleased him, though he declined it. He responded by promising that he would always take care of her and Eduard.

He then left for England to lunch with Winston Churchill, an influential Conservative member of Parliament, at his country home, Chartwell Manor, in Westerham, Kent. He warned him, as he had warned Lindemann, that Hitler was secretly preparing for war. A few days later he dined with David Lloyd

George, a former British prime minister, writing as his address in the visitors' book what was also true for so many others—"without one."

Churchill confirmed these contacts, writing: "I saw a great deal of Frederick Lindemann . . . I had met him first at the close of the previous war . . . We had many talks in the small hours of the morning about the dangers which seemed to be gathering upon us [and] several visitors of consequence came to me from Germany and poured their hearts out in their bitter distress."

From Oxford, Einstein wrote to Max Born. He said that he had become an "evil monster" in Germany, and that he "never had a particularly high opinion of the Germans (morally and politically speaking). But I must confess that the degree of their brutality and cowardice came as something of a surprise to me." This cowardice was later remarked upon by the Earl of Birkenhead:

> There is no known instance in which a professor of physics or chemistry without any Jewish family ever made an open protest against Nazi activities. Even in the early years of Nazi power when opposition was not yet suicidal the scientific establishment, led by Planck and Nernst, washed its hands of the growing terror and concentrated on defending its own special privileges. The only notable scientist who was conspicuous in his disapproval of the Nazis was Max von Laue, and even his actions were taken within the physics establishment and not in open criticism of the regime.

However, one non-Jewish scientist, Otto Hahn, who was soon to discover nuclear fission, did suggest that thirty distinguished professors should jointly protest the treatment of their Jewish colleagues. Planck squashed the idea, noting that "if you bring together thirty such gentlemen today, then tomorrow one hundred and fifty will come out against them because they want their positions."

Several of Einstein's former assistants were among the many Jews thrown out of their jobs. Rudolph Goldschmidt found work in England; Ludwig Hopf, in Ireland. Leo Szilard left Germany with his life savings hidden in his shoes. In Vienna, en route to England and finally the United States, Szilard discussed Hitler with Gottfried Kuhnwald, the adviser on Jewish affairs to the Christian Socialist party. Kuhnwald predicted that Hitler would expel great numbers of Jews from Germany and "the French would pray for the victims, the British would organize their rescue, and the Americans would pay for it."

Max Born was incensed by the "parting anti-Semitic accusations and epithets" hurled at him in Göttingen, when he left for his summer home in northern Italy. He informed Einstein that to prevent his children from becoming second-class citizens in the country of their birth, he wanted them "to become citizens of a Western country, preferably England, for the English seem to be accepting the refugees most nobly and generously."

Einstein probably would have seconded that opinion. As always, he got a rousing reception in England. Had Planck been at Einstein's Lady Margaret Hall lecture he would have joined in the applause, for there was not a word in the speech about the Nazis. Instead, Einstein gave the Oxford audience a succinct history of atomic theory, concluding that there may never be a final answer: "The deeper we search the more we find there is to know, and as long as human life exists, I believe it will always be so."

Back at Le Coq sur Mer, Einstein renewed his friendship with Queen Elizabeth of Belgium, joining her almost every week to play in a string quartet and to discuss the gloomy political situation. To Elsa's alarm, he took death threats with the same indifference Walther Rathenau had shown before his assassination. A paper for German émigrés, the *Freie Presse*, in reporting the murder of Professor Theodore Lessing by Nazis, named Einstein as next on the Nazis' death list for approving a pamphlet exposing Hitler's terror tactics. Hearing of it, and of a rumor that there was a $5,000 bounty on his head, Einstein touched his head with a cheerful, "I didn't know I was worth that much!"

The Einsteins' well-meaning friend Antonina Vallentin added to the tension by showing them a German magazine that featured enemies of the Nazi regime. Einstein's photo was on the first page with a list of his "crimes," starting with relativity, and a note at the end reading, "Not yet hanged." After seeing this, Elsa couldn't sleep at night; she lay on her bed fully dressed, fearing that every floor creak was an assassin. She begged Einstein to be careful and not to take part in public demonstrations, but he refused. "We have violent arguments," she confided to Vallentin. "He has reproached me for being a contemptible coward."

The Belgian royal couple supported Elsa and arranged for detectives to guard Einstein. As he described the situation, "Safety is impossible here. Detectives sleep on the stairs, there are Secret Service men in the grounds, and my wife is terribly worried." He added that his two hobbies were peace and the welfare of the Jews, and that the Nazis were threatening both. He told Belgian pacifists: "You will be astonished at what I am going to tell you . . . If I were Belgian I'd not refuse military service under the present circumstances. Today Germany is preparing for war by every means." He went on to say that every citizen should accept military service "to contribute towards saving European civilization." This was a blow to pacifists, and a rebuttal to those who dismissed Einstein as politically naive.

Yet even though Elsa grumbled that their home had become "an asylum for the unfortunate, invaded from morning to night by people who need help," life went on. Einstein took a daily walk, amused by the young girl who waited in the same spot to greet him each time with "You dope." He responded to a comic question from a Long Island reporter, who asked: "With the earth revolving, part of the time a person is standing upside down [with Australia in mind, no doubt], upheld by gravitation . . . Would it be reasonable to assume that it is while a person is standing on his head—or rather upside down—he falls in love and does other foolish things?" Einstein replied that falling in love was hardly the most foolish human activity—"but gravitation cannot be held responsible."

He declined an offer of asylum from Turkey, wavered over an invitation to Caltech, and accepted spring lecture engagements in Paris, Madrid, and possibly Oxford. He might have gone to Hebrew University full time had they fired the man in charge, whom he considered incompetent. Einstein finally decided to join Princeton's new Institute for Advanced Study, because Flexner had finally given in and agreed to take his assistant, Walther Mayer, as well. He planned to keep his options open for the other six months of the year.

Meanwhile, Elsa encouraged an acquaintance, Commander Oliver Locker-Lampson, a spirited and daring member of Parliament, to invite Albert to England. There at least he would have the North Sea between him and the Nazis. Albert accepted, not knowing Elsa had engineered the offer. She would join him later.

Locker-Lampson took the Nazi threats as seriously as Elsa did, hiding Einstein in his vacation cottage on a windswept heath near the Norfolk coast. There he was guarded by two women secretaries, a farm laborer armed with rifles, and two detectives who questioned all visitors. During this period, Einstein was driven to London to watch from the visitors' gallery of the House of Commons as Locker-Lampson introduced a bill to help Jews become British citizens.

"Germany," said the Englishman, "has turned out her most glorious citizen, Albert Einstein. The most eminent men in the world have agreed he is the most eminent. He is beyond any achievement in the realm of science, and stands out as the supreme example of the selfless intellectual. Today Albert Einstein is without a home . . . The Huns have stolen his savings, plundered his place of residence, and even taken his violin [also his sailing boat and motorboat]. How proud this country must be to have offered him shelter." The bill got provisional approval.

Despite the demands on his time, Einstein was a sucker for anyone who asked to paint his portrait or make a bust of him. While he was hiding out in Locker-Lampson's cottage, sculptor Jacob Epstein prevailed on him to sit for a week of two-hour sessions; he later described Einstein's glance as "a mixture of the humane, the humorous, and the profound." Einstein's pipe created such a smoke screen that Epstein could hardly see his subject; finally, Einstein agreed to limit his smoking to breaks outside. The impressive result is now in London's Tate Gallery.

On this visit to England, Einstein agreed to speak in London's Albert Hall to raise money for refugee scholars from Nazi Germany. Scotland Yard got wind of a plot to silence him for good and gave him an escort of two carloads of detectives. The hall itself was peppered with them. He sat on the stage between astronomer James Jeans and Winston Churchill. Ernest Rutherford had come from Cambridge to chair the meeting. Loud, sustained cheers from a crowd of some ten thousand greeted Einstein when he stood to say: "How can we save mankind and its spiritual acquisitions of which we are the heirs? How can one save Europe from a new disaster?"

His friends saw him as both the world's greatest scientist and as a brave symbol of the resistance of Jews and democrats to Nazi oppression. His enemies berated him as a fraud, a Communist, a pacifist, and a traitor. None of the abuse distressed him nearly as much as the news of Paul Ehrenfest's suicide. During a bout of depression this devoted friend had shot and wounded his mentally retarded young son, then fatally shot himself.

Einstein might have had Ehrenfest in mind when he explained what drove him to oppose the Nazis:

"Striving after knowledge for its own end, love for justice bordering on fanaticism, and striving after personal independence—these are the incentives of the Jewish people's tradition, which make me regard my belonging to them as a gift from fate. Those who today work against the ideals of reason and of individual freedom, and want to enforce soulless slavery to the state by means of brutal despotism, justly see us as their implacable adversaries. History has imposed a difficult struggle on us, but as long as we remain devoted servants of truth, justice and liberty, we shall not only continue as the oldest living peoples, but also as contributors to the ennoblement of humanity."

He defused—though never entirely stopped—attempts to brand him a Communist by admitting that he had been fooled by organizations posing as pacifist or humanitarian groups "which are in truth nothing less than camouflaged propaganda in the service of Russian despotism . . . [I] would now like to state that I have never favoured Communism and do not favour it now . . . any Power must be the enemy of mankind which enslaves the individual by terror and force, whether it arises under a Fascist or a Communist flag."

On October 7, he and Elsa were on the Atlantic Ocean, bound for America, with his secretary, Helen Dukas, and assistant, Walther Mayer.

New York City mayor John O'Brien waited in the rain at the Twenty-third Street pier to greet Einstein as he disembarked on October 17. The mayor was involved in a reelection campaign against Fiorella La Guardia, and here was his chance to get the Jewish vote *and* press exposure. Cheerleaders stood by with a parade of supporters ready to escort Einstein to his hotel, and reporters and photographers were there to record the occasion. But Einstein never appeared. Abraham Flexner had arranged for Einstein and his party to dodge the reception; they were whisked away in a small launch to the New Jersey shore. Curious crowds still milled about the pier, hoping to catch a glimpse of him, while Einstein was in Princeton, shopping.

It was a great occasion for Princeton. As science writer Ed Regis noted, "Virtually overnight, Princeton was transformed from a gentleman's college town into a world center for physics." Einstein's assessment of his talents, however, was characteristically more modest. When he was first shown his office at the institute and asked what equipment he needed, he answered, "A desk or table, a chair, paper and pencils. Oh yes, and a large wastebasket . . . so I can throw away all my mistakes."

The local press saw him as the Invisible Man or the Scarlet Pimpernel, headlining one report with "Einstein Wears an Invisible Cloak" and promoting him as the subject of the "greatest man-hunt Princeton has ever seen . . . Persons who were not seeking him out saw him here, there and everywhere. Persons who looked high and low for him were unable to locate him."

Reporter Carl Peterson happened to be walking along Princeton's Nassau Street when he noticed coming towards me a bulky gentleman with wildly flowing hair. Then it dawned on me that it was Einstein. I turned to be sure but he had disappeared. Rushing to the Press Club, I called my editor at *The New York American*. He screamed, "So that's where he is! He eluded us at the dock! Find him, dammit, and get an interview!" I searched in vain. Realizing that I had lost him in front of Woolworth's, I went into the store and asked a clerk if she had seen a man with great shaggy hair. A customer fitting my description had been in the store. A clerk at the notions counter said he had purchased a comb. Einstein's first act in Princeton was to buy a comb!

Peterson didn't get an interview, but *"The American* was first with my comb story, and it was soon picked up by newspapers all over the world."

And they all had it wrong. Before Einstein bought the comb, a divinity student, John Lampe, saw him buy a vanilla ice cream cone dipped in a bowl of chocolate sprinkles from a local ice cream parlor and polish it off with obvious pleasure.

The Einsteins stayed at the Peacock Inn for a couple of weeks. Otto Nathan was among the first to greet them. He had arrived as a refugee from Germany only a short while ahead of the Einsteins, but he offered to help them acclimatize to life in Princeton. He was now teaching at the university—one of the three Jewish faculty members. Nathan had had a promising career in Germany as an economic advisor to the Weimar Republic; in 1947, he would become a delegate to the World Economic Conference in Geneva. He and Einstein were both Social-Democrats and anti-Nazis and quickly became close friends, although Nathan was almost everything Einstein was not: a dry, almost humorless man who rarely laughed—except when among close friends. To them, he was "an extraordinarily sweet gentle man . . . mild and full of humor." He devoted much of the rest of his life to Einstein, helping to answer his mail, dealing with the press, protecting him from unwelcome strangers, and offering financial advice. Nathan soon rivaled Dukas in his devotion to Einstein.

When the Einsteins moved into a rented house at No. 2 Library Place—where Library Place ends at Mercer Street—near the university campus, they celebrated their new home with a musical evening of Haydn, Mozart, and Beethoven. Albert played second fiddle in a string quartet; Elsa and two friends formed the enthusiastic audience.

Soon after the move, Elsa accepted the invitation of Mrs. J. Ross Stevenson, wife of the president of the Presbyterian Seminary, to meet her neighbors. Five or six women sat around the Stevensons' fireplace with Elsa, chatting and nibbling on toasted cheese and crackers. Afterward, Carolyn Blackwood, the wife of a Presbyterian minister, Andrew Blackwood, helped Elsa, who was shortsighted, to find her umbrella in the hall rack. Because it was raining heavily, she also gave her a lift home. After discussing where to buy groceries and such things, Elsa put her hand on Mrs. Blackwood's arm and said, "You and Dr. Blackwood must come and visit us." The Blackwoods thought the couple needed their privacy more than a visit, and didn't go. Later, they got to know the Einsteins well and were of great help to them.

Elsa loved her new home: "The whole of Princeton is one great park with wonderful trees," she told a friend. Einstein agreed, but he had harsh words for its denizens, calling the intellectual community "puny demigods on stilts" in a letter to his friend Queen Elizabeth. He quickly warmed to the neighborhood children, though, responding to one trick-or-treat demand on Halloween by treating them to a tune on his violin, which had been given to him by an admirer.

As for Helen Dukas, she soon became acquainted with the local police. While shopping for herself and the Einsteins in Bamman's grocery store, according to Freeman Dyson, she gave her order in a pronounced German accent and then handed over a wad of twenty-dollar bills which Elsa Einstein had drawn from a European bank. In minutes the police arrived with screaming sirens and the shop was surrounded. "Helen, meanwhile, stood calmly wondering whether she had arrived by mistake in Hollywood instead of Princeton. 'Anyway,' she said as she told the story, 'living with Professor Einstein, I was accustomed to things turning into a circus wherever we went.'" After the chief of police had questioned Helen Dukas and examined all her twenty-dollar bills, he said she was free to go. As she walked home with groceries she wondered if this was going to happen every time she went shopping in Princeton. A few days later she discovered the cause of the police interrogation. The ransom money for the recently kidnapped baby of aviation hero Charles Lindbergh had been paid in twenty-dollar bills, and New Jersey shopkeepers had been asked by the police to report any suspicious customers who might try to pass the ransom money.

Despite efforts to protect him from the outside world, Einstein was not allowed to forget Germany. The charismatic Rabbi Stephen Wise, incensed because President Franklin D. Roosevelt "has not lifted a finger on behalf of the Jews," thought Einstein was the man to get some action. So Wise contacted a Roosevelt adviser and soon an invitation to the White House was in the mail. Flexner, however, had been handling all of Einstein's mail, and he declined the invitation, explaining to Roosevelt that

> Professor Einstein has come to Princeton for the purpose of carrying on his scientific work in seclusion, and ... it is absolutely impossible to make any exception which would inevitably bring him into public notice. You are aware that there exists in New York an irresponsible group of Nazis. In addition, if the newspapers had access to him or if he accepted a single engagement or invitation that could possibly become public, it would be practically impossible for him to remain in the post which he has accepted in this Institute or in America.

Furious when he learned of Flexner's interference, Einstein said he wanted to meet the president; soon a second invitation, for a visit in the new year, was forthcoming. He also insisted that all future mail should be delivered directly to him, which included the almost nonstop anonymous hate mail, such as: "You are a Jew faker. You are a publicity seeker like all other fakers. You are a Communist and should be barred from the United States. We must start some pogroms here." Another read: "The Jews are just as dangerous today as they were during the time of Jesus and during the time of Titus the Roman Emperor. They deserve to be starved to death by any meanse [sic]. We must get rid of the insulters of the human race. They are the helpers of Satan and like Satan they must be done away with. To the river with the decievers [sic], the swindlers, the snackes [sic] in the grass. No wonder the whole world detests theme [sic] like scowks [sic]." Surprisingly, Einstein—or more likely his secretary, Dukas—kept these out-pourings from the illiterate and deranged.

Aware that Einstein was the focus of so much irrational hate, Flexner advised him to lie low and keep his politics to himself. He did neither, and found Prince-ton—despite his comments to Queen Elizabeth—a friendly place, where he could walk the tree-lined streets without any need for protection.

Most days he walked to and from his office in Fine Hall, the math depart-ment of Princeton University, said to resemble a church from the outside and a dungeon from inside. His four colleagues at the new Institute for Advanced Study were Oswald Veblen, a former Princeton math professor; the independently wealthy James Alexander, a topologist who had failed in his bid to be elected mayor of Princeton on the Communist ticket; John von Neumann, a political conservative whom some colleagues considered the world's greatest mathemati-cian; and the scrawny, high-strung, twenty-seven-year-old Kurt Gödel. Already famous for his shattering proposition that mathematics—the bedrock of scientific experiments—is inevitably imperfect and incomplete, Gödel was in no mood to create more mathematical mayhem. He had fallen for a Viennese cabaret dancer, of whom his parents disapproved. Torn between her and them, he was on the brink of a mental breakdown.

Einstein had problems, too, but not of the heart. As he was about to resume work on his unified field theory, Walther Mayer, whose hiring he had made a condition of accepting the job, abruptly left him. Why he quit is not clear, but apparently, soon after crossing the Atlantic, Mayer chose to go his own way. Several of Einstein's former colleagues, including Banesh Hoffmann and John Kemeny, suggest that he had lost faith or interest in the theory. Veblen, who advised Flexner, was no help, taking a narrow view of the contract—that Einstein had no right to a replacement.

Which didn't, of course, stop him from lecturing — much to physicist John Wheeler's delight. "My first chance to see and hear Albert Einstein came one afternoon in the fall of 1933," Wheeler recalled. "I was in my first year of post-doctoral work with Gregory Breit [at New York University's uptown campus in the Bronx] and he told me there would be a quiet, small, unannounced seminar by Einstein that afternoon. We took the train to Princeton and walked to Fine Hall. Unified field theory was to be the topic, it became clear, when Einstein entered the room and began to speak. His English, though a little accented, was beautifully clear and slow. His delivery was spontaneous and serious, with every now and then a touch of humor.

"I was not familiar with the subject at that time, but I could sense that he had his doubts about the particular version of unified field theory he was dis-cussing. I had been accustomed before this to seminars in physics where equa-tions were taken up one at a time, or dealt with in 'retail trade.' Here for the first time I saw equations dealt with wholesale. One counted the number of unknowns and the number of supplementary conditions and compared them with the number of coordinate degrees of freedom. The idea was not to solve the equations but rather to decide whether they possessed a solution and whether it was unique. It was clear on this first encounter that Einstein was following very much his own line, independent of the interest in nuclear physics then at high tide in the United States."

At year's end Einstein was moved to tears during a recital by the Westminster Choir School of Princeton, and in his confusion complimented the wrong young woman for her solo performance. He also attended a banquet at Manhattan's Hotel Roosevelt to commemorate the hundredth anniversary of the birth of Alfred Nobel, the inventor of dynamite and the originator of the Nobel Prize. There, perhaps taking a swipe at Hitler, he said he agreed with Schopenhauer that will power and intelligence were antagonistic attributes. Sinclair Lewis, the prickly Nobel Prize–winning novelist, sat on the dais next to Einstein. During dinner, irritated by photographers who were snapping him while he was eating, Lewis walked out in a huff. He showed *his* will power by staying away for twenty minutes. The overeager cameramen didn't bother Einstein, who sometimes jokingly referred to himself as a photographer's model.

Einstein expected to stay at Princeton for six months, then return to England to take up residence at Christ Church, Oxford, as a British citizen, moves actively encouraged by his friends and sponsors Lindemann and Locker-Lampson. Instead, Princeton was to be his permanent home for the next twenty-two years. He would become an American, and never once return to Europe — not even for a visit.

CHAPTER 26

A New Life
in Princeton

1934

---

55 to 56 years old

Thrill-killer Nathan Leopold's most cherished possession was a letter from Einstein. Leopold was in prison for life plus ninety-nine years. At nineteen, he and an eighteen-year-old pal, Richard Loeb, had murdered a fourteen-year-old boy in Chicago to prove that they could commit the "perfect crime," and they almost got away with it.

Leopold already spoke fifteen languages. In prison he learned twelve more, reorganized the library, wrote articles for sociology and criminology journals, and volunteered as a guinea pig for malaria research. He had been a Joliet, Illinois, prison inmate for ten years by 1934, when his studies in sociology widened to an interest in mathematical physics and then to relativity. Wanting to master the subject, he wrote to Albert Einstein. "It took some nerve to address myself to the man I consider the greatest thinker alive in the world at that time, but . . . I knew that Professor Einstein was a kindly man. I wrote to him. And he answered my question! I have only to raise my eyes to see his letter [dated January 4, 1934], for I framed it and it has followed me wherever I have gone in prison."

Einstein's advice was to start with short scientific works that used calculus, such as Planck's published lectures, before reading a collection of the original statements by Lorentz, himself, and Minkowski, as well as Eddington's book on relativity.

While Leopold was marveling over the letter, Albert and Elsa were finally meeting Franklin and Eleanor Roosevelt in the White House, where they spent the night of January 24.

Leopold Infeld, who later worked with Einstein at Princeton, said that Einstein never disclosed the contents of his secret conversations with the president. Another account has them discussing their "mutual love of sailing, Einstein's friendship with the queen of Belgium, and the desperate plight of Hitler's victims." Later, Elsa remarked to a friend that Roosevelt had asked Albert to accept what two U.S. congressmen were proposing—honorary United States citizenship—but he declined special treatment.

On March 30, the Einsteins visited their now mutual friend, Leon Watters, in his Fifth Avenue apartment overlooking Central Park. He and Albert had first met in Pasadena. Before lunch, when they were in the living room, Watters put a music roll in his automatic piano. As it began to play at the start of lunch, Watters recalled how "Einstein stopped eating and listened intently and asked, 'Was that Schubert?' I told him it was Chopin's 'Romance.'" Then Watters's niece joined them, and Einstein congratulated her for playing so well. He was astonished to learn it was done mechanically. Elsa teased him, reminding him that he had rejected the offer of a similar piano because he thought it would sound mechanical.

A few weeks earlier he and Elsa had attended a lecture in Princeton by the revolutionary composer Arnold Schoenberg, who announced he had been directed by "the Supreme Commander" to convey through his music "a prophetic message revealing a higher form of life toward which mankind evolves." Music critic Harold Schonberg called him a bitter egomaniac and compared him with Einstein because he had destroyed "the ages old concept of tonality as effectively as Einstein had destroyed Newton's macrocosmos." Schoenberg had been a Christian convert, but with the rising anti-Semitism in Germany he felt compelled to return to his Jewish religion, and like Einstein become deeply concerned with the fate of Jews in Europe. For several years he had wanted to meet Einstein to discuss their mutual interests. They met again on April 1, when Einstein was the guest of honor at a Carnegie Hall "Tribute of Music to Science," held to raise money for the Settlement of German-Jewish Children in Palestine and the New York Zionist Region. Einstein thought Schoenberg and his music "crazy"—but that was hardly disparaging. Friends noticed that he seemed drawn to the so-called crazies, and to all artists.

Winifred Rieber left a record of Einstein as an artist's model in letters to her husband. Although she had painted several famous men, including John Dewey and William James, she was nervous as a schoolgirl at the prospect of several days alone with a genius, and as her train approached Princeton she agonized over how she could keep up with his conversation.

Einstein greeted her in bare feet and sandals, an oversize sweater, and baggy pants. Rieber wrote to her husband in California that Einstein "seemed more bewildered than pleased to see me. But in a moment he was himself again, all kindness. The next day my model arrived dressed stiffly in a dark suit, white shirt with winged collar, and polished shoes . . . Everything about him was electric. Even his silences are charged. 'They say I am great,' he comments, 'and so they pursue me. Everywhere I go people are there to stare at me.' He says he is the victim of a strange mob curiosity."

During a break in the painting, Elsa brought Einstein a letter from a woman whose son thought he was Jesus Christ. He had been living on a mountain for several weeks and refused to come home. His mother begged Einstein to help, as he was the only person for whom her son would leave his mountain retreat.

Although he might be taking on a dangerous lunatic, Einstein replied to the brokenhearted mother that he was willing to see the young man. Several days later, Rieber was having lunch with the Einsteins and Helen Dukas when a young man with an intense but empty expression arrived. "I think I would have left him on the mountain," Rieber wrote to her husband. "But the kind Herr Einstein agreed to see him and talk with him while they walked in the woods."

When Einstein returned that evening, the women were anxious to know what had happened. "I didn't argue with his delusions," Einstein said, "but I reminded him that Jesus came down from the mountain to be a fisher of men." He thought the young man was quite mad, "yet when I was with him I felt the peace of high places and wondered if he was sane and the rest of us mad."

He might well have contemplated joining the young man on the mountain because of the ceaseless demand for his attention. A shoe salesman, for one, boasted that his philosophy was "as irrefutable as any mathematical equation— in which every question of the day that man can possibly propound is answered. I honestly believe I say more in the 200 pages than all humanity has said since the beginning of civilization, (and make no apologies for saying it)." He wanted only fifteen minutes of Einstein's time to sell his ideas, long enough "to know whether I'm a nuisance, a nut, or the guy I think I am. You'll know whether to give me your hand, or your foot."

Dukas kept such correspondents at bay by guarding the door. Elsa helped, too, by declining requests for him to attend public functions and by locating mosquito-free vacation spots where he could sail away from the madding crowd. She in particular had suffered from mosquito bites on an earlier visit to the States.

In the spring, Leon Watters took the Einsteins for a drive through the New Jersey countryside, where "the professor said with some glee that the houses and the towns reminded him of the region in which he was born in Germany. We all got out and walked about in the fields. The Einsteins posed for some Leica pictures, my chauffeur took pictures of the three of us and then Einstein took my camera and took pictures, one of me with Mrs. Einstein and one of me alone. He said that was the first time he had ever taken a picture."

Back in Princeton, Watters said, Mrs. Einstein reminded him of his promise "to make a day of it, [so] I remained for supper. Just as we were all seated at the table Einstein rose from his chair, went outside, came back with my chauffeur [Martin Flattery] and seated him at the table next to myself. Flattery, a decidedly modest person, felt ill at ease, and just as soon as the meal was finished, he made the excuse that he had to do some work on the car and thus escaped. This was a second time he had been an unwilling guest at their table. I can readily understand Einstein's liking for my chauffeur for he was an ideal of an English servant, a well-trained, well-built and handsome man." Einstein, of course, would have asked him in had he been ugly.

The Einsteins looked forward to a summer vacation with the Buckys in a rented cottage at Watch Hill, Rhode Island. (Being American citizens, the Buckys had no trouble leaving Germany, which they did soon after the Einsteins). But, late in the spring, Elsa learned that her eldest daughter, Ilse, was dangerously ill in Paris and being nursed by Margot. Elsa decided she had to go there alone; it would be too risky for Einstein to return to Europe with a price on his head.

Ilse was her chief concern, but she also wondered how to bring back Einstein's books and papers that Margot—helped by the underground—had smuggled out of Berlin. This was on her mind as she walked past the home of her neighbors, the Blackwoods. Sixteen-year-old James Blackwood, who was in the front yard with his mother, remembered the conversation: "I was digging up the jonquils and dividing them up under duress when Mrs. Einstein came by. Despite her shortsightedness, she recognized mother and they began to chat. My father had been thinking of asking Dr. Einstein for an introduction to people at Hebrew University and to members of the Zionist movement, but never got around to it. So mother asked Mrs. Einstein, and she said, 'Oh yes, yes. Albert would be glad to do that.' Then she said, 'How will you get there?' And mother said they were taking a tour of Palestine and Europe."

At that, Mrs. Einstein became very interested and asked, " 'On what ship will you return?' And mother told her, 'The Westernland.' And Mrs. Einstein kept muttering, 'Westernland, Westernland,' and 'the date, the date.' And I thought, 'This gal may be married to a genius but not much has rubbed off on her.' From that conversation, she chose the date of her visit to her daughter, Ilse, and she arranged to return on the same ship as my parents, the Westernland.

"When my mother had mentioned that she and my father wanted to meet Zionists in Palestine, Mrs. Einstein said, 'My darling, I did not know you were Jews.' And my mother said, 'We are not. We are Christians and Presbyterians on top of that.' Mrs. Einstein was dumbfounded why anyone (other than Jews) would associate with, even seek out the Jewish community in Palestine."

Mrs. Blackwood then spoke about the Jewish heritage and the Christian faith and the very close bonds between them. She concluded by saying, " 'And besides, Jesus was a Jew.' And Mrs. Einstein was amazed, replying, 'No Christian has ever said that to me in my life.' Then she hugged my mother affectionately. That was the beginning of an intimacy between those two. Mrs. Einstein said early on that they believed in the creative force, but not in a personal God who took any interest in the people on earth.

"Elsa explained that Albert regularly read the Bible, [both] Old and New Testaments, for the literary value and stories, not for the specifically religious message. But they had lost their Bible in their move from Berlin, and Elsa asked my mother, 'Would you get me one?' So my mother got her a copy of Luther's translation, the one they knew," said James Blackwood, "and she hugged it to her heart. And the feeling was not only acted out, but verbalized: 'I wish I had more faith.' " Especially, of course, with her daughter dangerously ill.

Elsa threw her arms around Leon Watters and burst into tears when he arrived at the dock to see her off to Europe. He gave her a corsage of white orchids, tried to comfort her, and promised to make Albert rest before lunch. After the ship sailed, the two men went to Watters's Fifth Avenue apartment, where, saying that he wasn't tired but that he would "not be insubordinate," Einstein lay on the sofa. While he rested, Watters played Liszt's "Lorelei" on the mechanical piano.

"Was it restful?" he asked as Einstein got up. "The sofa, yes," he replied. "The music, not too much. Too sugary." After lunch and a puzzling conversation with two visitors, a deposed German official and a German journalist—Einstein wasn't sure what they wanted—he told Watters to rest on the sofa, and then played the piano, reversing their roles.

Einstein later asked Watters, "Would you do a favor for me?" He felt guilty about having left Samuel Untermyer standing on the New York dock waiting for him, while he was being whisked away to New Jersey. "Untermyer is an old man, I would like to meet him and explain how it happened."

Watters arranged by phone for them to visit the lawyer's summer home, Greystones, near Yonkers. Driving through the gates they discovered that Untermyer had turned over his grounds to a group of women for a charity garden party. They were thrilled to meet Einstein.

The men finally broke away from the women and found Untermyer in his study. He and Einstein began to discuss Hitler's persecution of the Jews and agreed that, as a protest, the recently initiated boycott of German goods by Jewish organizations in America should be expanded. Einstein was pessimistic, however; he was afraid that since Hitler was eliminating the opposition he would be unbeatable. On June 30, the "Night of the Long Knives," he had ordered the execution of several hundred "enemies of the state." Had Einstein stayed, he might well have been one of them.

He moved for the summer to Watch Hill, Rhode Island, as originally planned. Dukas and the Buckys went with him. Shortly after they arrived, Watters and Kurt Rosenfeld, the former Prussian minister of justice, called on them. Rosenfeld was in the States to take part in a "Lawyers Trial" of the Nazis. The two Bucky boys had rigged up a shortwave radio and tuned in to the chief Nazi's hysterical ranting.

While the boys had fun, Helen Dukas worked and sometimes grumbled. She did most of the cooking—usually macaroni and noodles—and complained to Watters that Einstein "thinks no more of me than if I were a table or chair." James Blackwood agrees, with reservations: "She was part of the scenery, yes. But she also lived with the family, joined in conversations, had meals with them and went on vacations with them. She didn't go on the trips when my parents drove the Einsteins to the Farmers Market in Pennsylvania and the artists colony in New Hope. Maybe that's why she thought she was part of the furniture—you could leave that at home."

Freeman Dyson, who got to know Dukas after Einstein's death, believes "the point is that Helen was treated extremely well by him as far as the big questions are concerned. She always said it was an enormous pleasure to work for him, that he was extremely considerate and friendly in their working life. It's also true that she was a general factotum, and I'm sure there were times when he treated her like the furniture. I don't think it's inconsistent, considering she was with him twenty-four hours a day. So there were certainly times when this was true. I don't think it was true as a general rule."

Like a would-be biographer, Watters not only recorded Dukas's complaint in his diary but also noted Einstein's elastic muscles and quick movements, and "kindly, expressive eyes which seem to be gazing into the distance. He walks with his feet pointed straight ahead, Indian fashion. If he likes you he grips your hand firmly and gives it a twist. It was the custom to meet on the porch of his cottage every morning at nine. If I failed to show up, he had them telephone to my hotel to see if I was all right."

While Einstein and his party settled into the cottage in Rhode Island, Elsa had arrived in Paris just in time to see her daughter Ilse die of cancer. Heartbroken, she prepared to return to America with her surviving daughter, Margot. Even in her grief, though, she undertook a mission for Albert: among her luggage as she boarded the *Westernland* were several trunks and cases of his smuggled papers.

The Einsteins' Princeton neighbor, Carolyn Blackwood, who was returning from the tour of Europe and Palestine with her husband, Andrew, was surprised to see Elsa emerge from a cabin, her eyes tearful. "Frau Blackwood," she said, "I thought you would be on this ship. Come in." Mrs. Blackwood asked, "What happened?" "It is my Ilse," she said. "She is dead. I cannot bear it."

In recalling that day, Mrs. Blackwood told her son, James, "I put my arms around her and let her sob out her grief. Mrs. Einstein and Margot had left Ilse's ashes in Europe but were bringing her death mask back to America. Both were devastated by the loss of one so young, so beautiful, so sweet."

For the rest of the voyage the women met every day. One morning Elsa drew Mrs. Blackwood into the deserted cocktail lounge and whispered, "If you see anyone at the door or windows, give me a signal and I will quit speaking." She said that after the Nazis had confiscated their Berlin home with all their possessions, members of the underground had broken in and brought out some of Dr. Einstein's books and papers. These (in addition to what Margot had gotten out through the French embassy) had been smuggled from Germany into Belgium, and were now in the hold of the *Westernland*. Not being an American citizen, Mrs. Einstein feared that the materials might be seized by customs officers and cause an international incident. After all, the Nazi government, which regarded Einstein as Public Enemy Number One, still had full diplomatic relations with the United States.

Blackwood agreed to say that they were his. The Presbyterian minister told a white lie in a good cause, but wrote the simple truth on the customs declaration form: "Material acquired in Europe for scholarly purposes." (Much of it is now in the Einstein Archives.) Albert met the ship in New York, and when Elsa explained how they had succeeded in getting the material through customs he was deeply moved that his neighbors had been willing to help.

At intervals several boxes and trunks were delivered to the Blackwoods' home and crammed in a shed awaiting Einstein's return from his resumed summer vacation at Watch Hill, where Elsa and Margot had joined him. Inconsolable over the loss of her daughter, Elsa made one room into a shrine, with Ilse's death mask on display. Seeing Elsa so disturbed, Einstein insisted that she remove it because it only prolonged her distress.

From Watch Hill, Elsa wrote to Carolyn Blackwood:

> Please do not think that I am so terribly unmodest to send you once more some trunks. I am sorry that you must think of me that I abused your great friendliness. And this makes me sorry. Please excuse thousand times for all the troubles you had. When you come to eat "Spatzle" in my house you forget all this bad things and you will have a better impression of me. When I came back I felt such a homelike-feeling for this country. And not longing for Europe at all, where I have to go through the most cruel events. The only thing I left there and which is dear to my heart is the ashes of my beloved poor child. Thanks once more from the bottom of my heart. *Auf wiedersehen,* do not be angry with me on account of the accumulation of trunks.

Thomas Bucky, then fourteen, recalls the lighter side of life that summer: "Einstein had an unlimited pass to a movie theater, but rarely went. So he let me have his pass, which delighted me. He preferred to go sailing or to walk on the beach.

"Once he accepted his friend Leon Watters's invitation to dinner at the elegant Ocean House. When we entered the plush lobby, Watters suggested that we sit there for a while before dinner. 'Oh no, I can't,' Einstein said. 'I don't have any socks on.' You see, he didn't give a damn if strangers saw him sockless but he didn't want to embarrass those with him."

Einstein loved to sail, but often became becalmed when the wind died down because he refused to use outboard motors. Once he got stuck on the rocks with a storm on the way. Fifteen-year-old Harry Darlington and friends were sailing nearby. "I happened to see this boat in trouble," said Darlington. "I knew Einstein was in the area. He used to go sailing all the time. We knew it was him because we always figured he was going to get in trouble someday anyway, because we didn't think he knew how to sail. Just to watch him told us that. He was in a rented catboat that has a single stick and no shrouds on it. He'd hit a rock pile offshore and they were aground. A storm was coming up and the tide was coming in. He had some Frau with him. They were in rough shape. I didn't think they'd make it if we hadn't come over. They could have drowned. So we went over with our sailboat and jumped over and pulled them on the boat. They never would have made it without us. I think we saved their lives. He asked us who we were and later he wrote to thank me and sent me his picture."

Watters conceded that Einstein had never studied navigation, didn't know nautical signals or even the names of the parts of the boat, and, though he couldn't swim, didn't carry charts or life preservers. "But he had a good sense of direction and never looked at a compass. He could forecast a storm with uncanny accuracy, and had no fear of rough weather even though he has had to be towed several times when his masts were blown down. Once when out sailing with him, I suddenly cried out, 'Achtung!' for we were almost upon another boat. He veered away with excellent control and when I remarked what a close call . . . he started to laugh and sailed directly toward one boat after another, much to my horror, but he always veered off in time and then laughed like a naughty boy. One Saturday morning a light drizzle made sailing out of the question, so Einstein asked me if I would stroll with him in the rain. We started out along the beach toward the lighthouse. There, sitting on the seawall, he commented on the fact that I had never asked for an autograph and would I like him to write part of the equation of relativity. I took a piece of paper out of my notebook and he wrote the equation. After a time, he complained of a sore back which he attributed to a faulty mattress and said that he and Bucky had some ideas as to how a mattress should be made, a diagram of which he made on a second sheet of paper."

Too engrossed in their conversation to notice an approaching rainstorm, they were soaked before they reached the shelter of a ramshackle café on the pier. Watters asked the waitress for a towel to dry Einstein's dripping hair, but he resisted. He also refused to take off his waterlogged shoes.

The next day Watters and Einstein went sailing and got stuck at low tide on a sandbar near Stonington, Connecticut. "We discovered that the centerboard had gotten caught and we could not lift it to decrease our draft. A yacht passed and [the people on board], recognizing Einstein, offered to tow us in, but he refused. Next a boy, a clam digger, appeared in a rowboat and tried to raise the board without effect. We lay there till high tide when the people at Watch Hill, getting worried, sent out a boat which we met on the way in. It was nearly eight when we landed. Einstein had an engagement to play the violin at the home of friends that evening, and he begged me to accompany him. I protested that I couldn't go as I was, nor could he; to which he replied, 'I play the violin with my hands, not with my breeches.'" Still, to humor Watters, Einstein changed into a white linen suit.

It was raining heavily the night when Einstein, back in Princeton from his vacation, called on the Blackwoods to look at the documents Elsa had said were "very dear to him." James Blackwood, who answered the door, recalled that Einstein "wore a raincoat but no hat. His hair looked like a wet mop. Father and mother greeted him and offered to take his raincoat, but he preferred to wear it. Dripping wet, he followed me through the hallway, dining room, butler's pantry, and kitchen, and took one step down into the shed. Tense moment! Deliberately, Dr. Einstein raised the lid of an old camelback trunk. He picked out a black-bound volume of what looked like articles from his early years. He opened it, glanced down the page. His expression changed; his eyes twinkled. Dr. Einstein looked sideways at father, held the volume at arm's length, and asked in aston-ishment, 'Did I write this drivel?' We all laughed, and that broke the tension. He was obviously deeply moved to have his books and papers in his hands once more, and lingered over them for half an hour, then, refusing an umbrella, walked home hatless in the rain.

"That fall my parents had dinner at the Einsteins' home. Helen Dukas and Margot were also there. Afterward, Dr. Einstein brought a hat from the closet to prove he *had* one. He cocked the Stetson on his head at a comic angle. Going through a scrapbook filled with cartoons of himself, he chuckled in telling father and mother, who sat on either side of him, about the ones he liked best. During the conversation, Mrs. Einstein leaned over to mother and in a low, husky voice, whispered '1606,' the Einsteins' unlisted phone number."

Einstein began to reminisce about Marie Curie. When she had died in July he had said that of all the celebrated people he knew, she was the only one not corrupted by fame. Now he said "that when he was with her in Geneva they had some free time and he invited her to go sailing on the lake with him. When they got far out, she said, 'I didn't know you were a good sailor.' And he replied, 'Neither did I.' Then she said, 'But what if the boat should overturn? I can't swim.' And he said, 'Neither can I.'"

Einstein took a pipe from his pocket and put it in his mouth unlit, then bit the stem and sucked on it. He played with a match, but didn't strike it. Finally, James recalled, Mrs. Blackwood said, "'If you wish to smoke, Dr. Einstein, please do.'" "'Thank you,'" he replied, "'but I will not smoke.'

"Elsa, Margot, and Miss Dukas laughed. My parents did not catch the joke. Laughing still, Mrs. Einstein gave her account of a domestic tiff. 'I told Albert he smokes too much. He said he could quit any time he wanted to. I told him he couldn't quit. He said he wouldn't smoke until New Year's Day. I said he couldn't keep his word. Then he said he'd show me. And he hasn't smoked since Thanksgiving.'"

Then, said Blackwood, "Dr. Einstein's eyes sparkled. 'You see, I am no longer a slave to my pipe, I am a slave to *dat voman.*'" Everyone laughed, including Einstein.

(After New Year's Day, 1935, Mrs. Blackwood asked Elsa, "Did your husband keep his word about not smoking?" "Ja, Albert kept his word," she replied. "But he got up at daylight on New Year's morning, and he hasn't had the pipe out of his mouth since, except to eat and sleep.")

James Blackwood's picture of the Einsteins was that Elsa adored Albert and he had great respect and affection for her. They also seemed to have lively quarrels, judging by the fact that Elsa asked Mrs. Blackwood somewhat incredulously, "How do you keep your four boys from quarreling?"

Blackwood believes that Einstein's image as a remote loner, one who kept others at a distance, was protective coloring. Without it he would have been overwhelmed by those eager to know him or to use him. Blackwood's evidence for this is that "Einstein had a public face and a private face. Walking along the street past our home the face was impassive. He didn't even look left or right to see if anything was coming when he crossed Alexander Street. It was as if he was on rollers. He did not swing his arms. He did not walk fast. It was slow, sedate, contemplative. Einstein looked as though he was thinking about things in the next galaxy. It was quite remarkable.

"When he walked along Mercer Street with the public face he was impassive. I've seen him with a group of three men walking with him, one on either side, both leaning in front of him, gesticulating in high-voltage argument in German, while the third pranced back and forth trying to get into it. And he just glided along, unflustered and serenely above the fray. I don't know if he was listening or not. If that was your picture of who he was, he would not only have been remote. He would have been extraterrestrial."

Blackwood saw Einstein's private face express the gamut of emotions from mirth to terror. He had driven Einstein to a musical event, Haydn's *Passion* at Westminster Choir School, where he was to play his violin. The full orchestra was supposed to join in after a few chords, but Einstein came in too soon and played an arpeggio all by himself. "Wincing, he looked at the conductor apologetically while Helen Dukas in the front row nearly passed out trying to suppress her laughter."

Afterward, he met a couple of friends and talked with them. "They spoke rather excitedly in German," said Blackwood. "And he asked if I could drive them home, because it was a very rainy night. So they all got in the car and I drove to the street where they lived when suddenly one shouted, 'Turn in here! Turn in here.' Well, there was a high crown on the road and a steep climb up the driveway that I turned into. And the rear bumper caught on the crown and made the most god-awful noise. It scared all of us. But I knew what it was and Einstein didn't. He must have thought it was an attempt on his life. Looking in my rearview mirror I saw him in the backseat under the streetlight and his face was aghast. He burst out with, 'What was that?' I never drove him again."

Einstein's son-in-law, Dimitri Marianoff, saw him remain completely calm while he walked on deck during a stormy North Sea crossing. Antonina Vallentin reported that when Elsa wanted him to stop criticizing Hitler and to accept a bodyguard, he refused and called her a coward. Their picture is of a morally and physically courageous man. Despite his revealing story, Blackwood agreed: "I think that's true. But all of us can be startled."

He gave another example of Einstein's public face. "Mother was in the Einsteins' living room," recalled Blackwood, "talking with Elsa. Einstein was in the music room improvising on the piano. The music stopped and Einstein came past them, hair straying in all directions, no shirt or undershirt on, trousers sadly drooping and, I think, barefoot. He walked past them as if in a trance. It was the public face," Blackwood explained. "Impassive. No sense of embarrassment, no recognition of mother's presence. He just drifted past and walked upstairs, while Mrs. Einstein clasped her hands and said, 'Oh, Albertle!'" Blackwood assumed that Einstein didn't want to interrupt their conversation and just wanted to get on through. "But who am I," he concluded, "to judge what passes through the mind of a mole or a genius?"

Blackwood saw the private face after *Time* magazine published an account of someone who had challenged the accuracy of Einstein's predictions, and he asked him, "What about this feller?"

"Ach!" said Einstein, anything but impassive. "He is already officially crazy."

He was impassive while listening with the Blackwoods to a festival of contemporary American music, jotting his comments on the program: "Brutally cruel . . . That was half cultivated . . . If Spring were so I would prefer always

Winter . . . That may be 'Zephyrs' but it sounds like a thunderstorm to me." Each time he handed his remarks to the Blackwoods, the pew shook with Einstein's laughter, which nevertheless did not show on his face.

Einstein was also surprisingly accessible. Blackwood recalls how his brother, Andrew Jr., was training to become a Presbyterian minister "and had entered the Princeton Seminary, which has four eating clubs, each with a house of its own, where students ate. Andy belonged to Benham Club, diagonally across from Einstein's home, 112 Mercer Street. After lunch one day, Andy said, 'I think I'll drop in on Dr. Einstein.' His friends hooted with laughter. So Andy marched across Mercer Street. Miss Dukas answered the door. Andy gave her his calling card, and she promptly admitted him, to the utter amazement of the slack-jawed fellow Benhamites."

When Andrew then discussed Hitler with Einstein, "he spoke without rancor and in a soft voice, but there was this compensating fact that when Einstein spoke everyone shut up and listened. He blamed the dictator's actions on his glands and my brother asked, 'What about the devil?' And Einstein chuckled and said, 'Same thing. Same thing.' He had a mild form of echolalia in which he would repeat a word or phrase: 'Same thing. Same thing,' 'Bach is deep, deep.'"

Most say that Einstein was a good sport, responding to comments on his wild hair with a smile. Take the Princeton Triangle Club skit, for example. The scene was the interior of a barbershop with the barbers lined up and customers in the chairs. There was a big glass window at the back, and Einstein just walked by the window. That was it. It brought down the house.

Visiting an art gallery with him, Mrs. Blackwood saw how he handled an intrusive stranger. As they walked from the parking lot, a "rather masculine" woman strode towards him, extended her hand, and said, "I know who you are, and I want to shake hands with the greatest living scientist." He courteously removed his pipe, bowed his head slightly, and solemnly shook hands. On the way home Mrs. Blackwood asked him, "Does it ever get monotonous being 'the greatest living scientist'?" "I'm not great," he replied. "Anyone could have done what I did. Besides, what I have is a gift." "A gift from God?" asked the minister's wife. "I express it differently," he said. "I believe down here"—he put his hand on his heart—"what I cannot explain up here"—he put his hand on his head. "But I believe it all. I believe it all."

"What I think Einstein meant by that," says James Blackwood, who heard of the conversation from his mother, "is that he had a religious dimension to his thinking. He read both of the testaments regularly, and his early training was in both, so if something was said he could identify that it was from John or from Paul. And he has been quoted as saying late in life that he was expecting to meet some of his friends on the other side. That could have been a joke but he's reported to have said it seriously." (It was almost certainly a joke. As Einstein wrote to reporter George Seldes, "Many things which go under my name are badly translated from the German or are invented by other people.")

He was already missing one of his friends on this side. Kurt Gödel, with whom he had often walked to work, was in an Austrian sanatorium being treated for depression, the consequence of submitting to parental demands to drop the Viennese dancer he was mad about. The good news was that Einstein now had a collaborator-assistant. Brooklyn-born Nathan Rosen had walked into Einstein's office to get an opinion of his master's thesis—and stayed.

That fall the still-grieving Elsa was bedridden with arthritis. Marianoff had left the scene, claiming he could no longer support a wife. According to Thomas Bucky, the Einsteins, Margot included, breathed a sigh of relief, because Marianoff "was considered a bit of a con man." His departure meant that Margot was free to help nurse her mother.

Learning of Elsa's illness, Watters sent her a Frigidaire icebox and Dr. Bucky gave her a small electric machine to ease her pain. She thanked them for the welcome gifts and regretted that she was too sad and sick to consider any social engagements.

Elsa made an exception in October when she and Albert called on the Buckys in Manhattan. Dr. Bucky then phoned Watters to say that the Einsteins wanted to see him, too. He arrived after dinner, greeted Albert, who was extemporizing on the piano, and continued upstairs, where he found Elsa "greatly overwrought and depressed over the loss of her daughter and herself afflicted with arthritis. She overwhelmed me with thanks 'for your many favors.'"

Bucky told Watters that Einstein wished to go to a Harlem night spot, the Cotton Club. Einstein corrected him: "I don't ask, but if I am asked I go." Watters, however, "had no desire to go and dissuaded him from insisting on going. I recalled that a group of colored actors were producing a play, 'Roll Sweet Chariot,' by Paul Green, so I suggested we go there. Finally, the Professor, his stepdaughter Margot, Dr. Bucky, his son Peter, and myself crowded into a taxicab and rode to the Cort Theatre. It was a weird play . . . intended to typify the life and thoughts of the present day southern Negro, corrupted by his own freedom from guidance and authority, into a slovenly, abject object blindly seeking an explanation of himself in superstition, disguised as religion. The novelty of it intrigued Einstein, particularly the music. The players and the audience came to know Einstein was there and as we left the theater they crowded about him, but only a few approached him for an autograph."

Einstein wanted to walk home, even though it was a cold night and he had forgotten his topcoat at the dentist that afternoon. So they strolled up Fifth Avenue, past Rockefeller Center, with passersby stopping to stare at him. He also stopped to comment on almost everything, calling a statue of World War I soldiers near Sixty-sixth Street "too lifelike to be artistic." It was almost midnight before they reached the Buckys' apartment.

Watters celebrated his birthday on November 13 by going with the Einsteins, Margot, and Dr. and Mrs. Bucky to see *Men of Aran* at the Westminster Theatre. The movie was a beautiful depiction of the lives of simple people eking out a bare existence on an island off the Irish coast. Einstein loved it. The group then split up, Elsa and Margot going with the Buckys to their apartment, Einstein going with Watters to his.

There the two men smoked their pipes, ate apples, and talked. As he peeled a second apple, Einstein asked, "You consider New York your homeland?" When Watters said "Yes," Einstein "lamented, 'I have never known a place that to me was a homeland. No country, no city has such a hold on me.' He said this with a sense of regret and longing."

Discussing Judaism with Watters, Einstein said, "I never felt myself part of the Jewish race till late in life, when I saw and felt the sting of anti-Semitism, particularly in Germany. Anti-Semitism is growing in all countries; it will always be increasingly hard for Jews. Their survival up to now indicates the value of preserving their culture." He added, "There is anti-Semitism at Princeton."

Their conversation continued far into the night. After looking intently at a portrait of Watters's deceased wife, Einstein said, "The individual counts for little. Man's individual troubles are insignificant. We place too much importance on the trivialities of living."

He implied that Elsa was not easy to live with, saying, "The more delicate the scientific instrument, the more difficult to handle. Woman is more delicate than man and more sensitive. So she must be handled with care. An excitable woman is like an electric instrument which is suddenly short-circuited."

Einstein explained that he took a rest most afternoons because he'd had two critical illnesses (a strained heart and a nervous collapse) and had to be careful of his heart, even though it was not giving him any trouble at the moment. But his "boss," Abraham Flexner, was: "When I first came to Princeton I thought I understood Flexner; since then I find him an enigma. I feel that I am not kept advised as to what is going on."

The two men went to bed at about one. Einstein declined the offer of both a dressing gown and pajamas, explaining that "when I retire, I retire as nature made me." He did ask for pencil and paper, however. Watters assumed that he worked late, because the light was on in his room for a long time.

Einstein was up early and whistling cheerfully the next morning. After breakfast, the men were about to join Elsa and Margot when Watters pointed out that Einstein had left his small case in the bedroom. He retrieved it with a laugh, and said, "When I was a very young man I visited overnight at the home of friends. In the morning I left, forgetting my valise. My host said to my parents, 'That young man will never amount to anything because he can't remember anything.'"

In November, Dukas handed Einstein a letter from a New Yorker who had

just invented a remarkable hair restorer . . . It is guaranteed to cure baldness, dandruff, and itchy scalp. As you are known the world over to possess a truly wonderful head of hair, I am going to name my product 'ALBERT EINSTEIN HAIR RESTORER,' and also, I plan to print your picture on the label of the bottle. I am quite sure you will not refuse this honor, so would you please write me an endorsement? If you desire, I will mail you a complimentary bottle.

Einstein politely but firmly squelched the proposal. It was also thanks but no thanks to the man who asked Einstein to mail a pair of his shoes for an insert that would make walking comfortable, because as Einstein explained, "I have no problem in walking."

Meanwhile, in California, Einstein's crusading friend Upton Sinclair was running for governor on an EPIC (End Poverty in California) program at a time when one in seven people in Los Angeles County was on relief. Einstein supported him, along with Charlie Chaplin, James Cagney, Jean Harlow, and Dorothy Parker. Nevertheless, Sinclair faced the onslaught of the rich and powerful, especially movie mogul Louis B. Mayer. All of Mayer's MGM Studio employees were pressured into "donating" a day's pay to fight the threat of Sinclair's bid to end poverty. Some of the money financed dirty tricks, such as a fake newsreel concocted by Irving Thalberg in which extras dressed to look like hoboes were filmed pouring into California to share in the wealth Sinclair was expected to gouge from hard-pressed taxpayers if he became governor. The Hearst press damned Sinclair as a "Communist," and the later-discredited evangelist Aimee Semple McPherson staged a pageant in which Sinclair, as anti-Communist as his opponents, was demonized as the "red devil." When he lost, Einstein sent a consoling message: "In economic affairs the logic of facts will work itself out somewhat slowly. You have contributed more than any other person. The direct action you can with good conscience turn over to other men with tougher hands and nerves."

Isaac Don Levine now brought to Einstein's attention politics of a more deadly sort. Stalin had recently responded to the assassination of his heir apparent, Sergey Kirov, by executing sixty-six political prisoners. The slaughter spurred Levine to write to Einstein, asking him to follow his example and protest: "Where are the hundreds of liberal and radical voices which so properly raised a storm of protest last June upon the bloody Hitler 'purge?' Why are these professed champions of human rights so inexplicably silent in the face of the medieval bloodbath improvised by Stalin?"

Einstein replied by return of mail on December 10, 1934:

> You can imagine that I, too, regret immensely that the Russian political leaders let themselves be carried away to deal such a blow to the elementary demands of justice by resorting to political murder. In spite of this, I cannot associate myself with your action. It will have no effect in Russia, but in the countries which directly or indirectly favor Japan's shameless aggressive policy against Russia. [The Japanese army had invaded and occupied Manchuria in 1931 and many Russians feared they would be the next victims.] Under these circumstances I regret your action and suggest that you abandon it altogether. Take into consideration that in Germany many thousands of Jewish workers are driven to death systematically by depriving them of the right to work, without causing a stir in the non-Jewish world. Consider further that the Russians have proved that their only aim is really the improvement of the lot of the Russian people, and that they can in this regard already show important achievements. Why then direct the attention of public opinion in other countries solely to the blunders of this regime? Is such a choice not misleading?

Levine answered in his next letter a few days later,

I was grieved to read your statement that the only aim of the Soviet rulers is the improvement of the people's condition. How can one reconcile that belief with the fact that in 1933 from three to five million peasants were deliberately starved to death by the Stalin regime? . . . Nor can I agree with you that the horrible policies of Hitler against the Jews have not aroused a storm of protest in the Western non-Jewish world. That this storm has not been great enough is perhaps due to the fact that the Western intelligensia has dulled our sense of indignation by condoning the Red Terror and by falling for the Leninist dogmas instead of adhering to the old cries of justice, human rights and freedom. Jewish emancipation owes its birth to these issues. The modern Jew owes his present status of a freeman to the English concept of the state, a concept which the American Revolution helped make a reality in half the world. Even the comparative rights won by the Jews in Germany under the Kaisers and the pitiable liberties allowed the Jews in the last years of Czarism were all due to the triumph of the English libertarian idea of the state. How then can the Jew fail to fight for that idea to the last drop of blood in him? I fear that the fact that so many advanced Jews swear by liberty and condone dictatorship is a grave omen for our future. I fear that the American Jews may make the same mistakes as some of the German Jews did—the mistake of not foreseeing events when the handwriting is already on the wall.

Einstein did not reply, and their friendly ten-year association ended.

"During the 1920s," said I. F. Stone, "Einstein had taken strong stands against Soviet persecution and had participated in a book with Alexander Burkman, Roger Baldwin, civil libertarians, and anarchists, deploring and documenting the persecutions in the Soviet Union's labor camps, among other horrors. He did not contest most of Levine's statements, but felt the Nazis were the main threat and that all forces—including the Soviet Union—should unite against them. He had not become, as Levine believed, a virtual apologist for the Stalinist reign, though Einstein's silence gave him reason to think so."

Stone's explanation for Einstein's attitude was this: "Isaac Don Levine had gone over to the far right, and was writing for the Hearst papers in that period. The anti-Semitic angle didn't crystallize until some time later in the Soviet Union. It's not easy to read history backward. The parallel with Germany was not very good. A lot of Jews, a lot of liberals, a lot of radicals, still had hope in the Soviet Union. You know, it was not unlike the French Revolution. Nobody mistook the direction of Hitler, except a few foolish German Jews. I remember writing anti-Hitler editorials in 1931 for a paper in Camden, New Jersey, and a German Jewish reader came to me and said, 'Hitler is only against the Ostjude.' That means the Jews that came into Germany after the First World War, mostly from Poland."

On Christmas Eve, after visiting the Buckys in their Upper East Side Manhattan apartment, Einstein set out to see *Dealers in Death*, a movie on Broadway about international arms dealers. He must have booked his seat in advance, because reporters were out in force. One even followed him down the aisle, asking if he believed in Santa Claus. Einstein laughed, but didn't reply. When he left, the same reporter waylaid him with a persistent, "Do you, professor?"

"My boy, Santa Claus is a personal thing," said Einstein. "I'll think it over and issue a statement some time." If he did, none has been reported, perhaps because he was preoccupied with preparing for an imminent lecture in Pittsburgh on December 28, 1934.

Elsa felt much better several days before the lecture — Dr. Bucky's electric machine seemed to have eased the arthritis pain — and, although still heartsick at the loss of her daughter, she took charge of the details of the trip. Several people had invited him to stay with them overnight in Pittsburgh; he accepted the first offer, from J. Edgar Kaufmann. Elsa arranged for Leon Watters to act as Albert's companion and bodyguard, telling him, "I should like my husband to have rest the day before the meeting, for in Pittsburgh there are many people who will try to run in on him, a terrible trial which would tire him greatly. I would like, therefore, that on the journey he could be kept quiet, if possible." But Watters was uneasy on the train from New York to Trenton, where he expected to pick Einstein up, because "Mrs. Einstein had suddenly been put to bed with a severe case of the grippe with a temperature of 104. And I knew she was in a highly nervous state." Would Einstein have canceled the trip to stay with Elsa?

To Watters's surprise, Einstein was waiting on the Trenton platform with two colleagues, Professors Howard Robertson and S. Lefshetz, who he said were going with them. Soon after they settled in their seats, Einstein felt the need to explain why he left a sick wife, saying that he "couldn't disappoint an audience which had counted so much on his presence." Another surprise for Watters was to see Einstein neatly dressed.

After lunch on the train, the three persuaded Einstein to lie down for half an hour. Afterward, he sat next to the window and watched a train on an adjacent track headed in the same direction, which kept up with them for awhile. He remarked that this was his first view of a locomotive's connecting rods in action. Watters recalled how Einstein "spoke of the physics taught us in our school days when the smallest particles of matter were molecules and atoms. 'Now,' he said, 'we have the electron, the proton, and the quanta. Our children (i.e., the particles) get smaller and smaller.'"

In Pittsburgh, Watters stayed at his friend Nathaniel Spear's home. There he arranged for some forty reporters to interview Einstein the next morning at the Kaufmanns' home, where Einstein spent the night. When the reporters arrived, Watters led them into the living room and told them to write out their questions. Einstein joined them soon after. He "was in a jolly mood and handled himself with great charm," noted Watters. "The question posed by a reporter which now has the greatest interest, since the explosion of the atomic bomb, was this: 'Do you think that it will be possible to release the enormous energy shown by your equation [$E = mc^2$], by bombardment of the atom?' Einstein's answer was classic: 'I feel it will not be possible for practical purposes. Splitting the atom by bombardment is like shooting at birds in the dark in a region where there are few birds.'"

He also said that there was a good chance the universe was infinite, and that although most physicists believed in Heisenberg's principle of indeterminacy, he was not one of them.

As agreed, Watters ended the interview at ten, and found that most "of the reporters were faithful to their promise not to attempt to gain any advantage over their fellows. However, one girl reporter ran in front of us and asked Einstein if he ever conversed about other subjects than physics. Einstein brushed her aside and answered, 'Yes, but not with you.'"

Following Elsa's instructions, Watters "put the Professor to bed in a spare room," to make sure he was rested for the big event that evening—his lecture, which was titled "An Elementary Proof of the Theorem Concerning the Equivalence of Mass and Energy."

Watters woke him at four and they drove to the little theater of the Carnegie Institute of Technology, entering by the stage door to avoid the crowds. The theater held about 450 people, and there had been requests for tickets from several thousand scientists alone; while laymen, according to newspaper reports, were offering as much as fifty dollars for a seat just to say that they had been there, in spite of the fact that they would not understand a single word.

Einstein stood onstage behind the closed curtain, waiting to give his talk while two students chalked complicated equations from his manuscript on two blackboards. Professor Robertson and Watters sat in the front row to translate if Einstein got stuck for an English word.

While Einstein was still behind the curtain, Watters recalled, "a group of newspaper photographers broke through the stage door and rushed to the platform like stampeded cattle. Just then the curtain was raised. The audience laughed at the confusion and applauded Einstein enthusiastically for being so good-natured about it. As he began his presentation the silence of the audience was intense and the leading mathematicians and physicists cupped their hands behind their ears so as not to miss a word, for they were about to hear demonstrated by indisputable mathematical formulae a truth which converted their lifelong beliefs. At one point, he said he would not bother to explain one of the steps, 'because it was so obvious,' but the scientists called out, 'No, no,' for it was not so obvious to them. He spoke throughout in English and never once hesitated for a proper word."

Einstein acknowledged the enthusiastic applause, then ran a gauntlet of autograph hunters and well-wishers before leaving with Watters for an exhibition by amateur astronomers at the Mellon Institute.

As they were driving back to the Kaufmanns', Einstein said to Watters, "I don't like your cough. It's too deep. Come home with me and I'll put you to bed." Einstein had a suite at the Kaufmann mansion; arriving there, he sat Watters in front of the fireplace and, as Watters recalled, "rang for the butler, ordered orange juice and a little whiskey and asked the butler to tell Mr. Kaufmann that we were not to be disturbed until dinner time. He had promised Mrs. Einstein that he would not smoke while away from home and he fondled his pipe longingly but did not light it. He admitted that he felt better when he didn't smoke."

Einstein took a copy of *Hamlet* from a bookcase and read aloud to Watters, commenting as he did on the beauty of the language. "Einstein knew from previous conversations that I had read Goethe's *Faust*, and he spoke of this as unusual for an American," said Watters. "It was delightfully peaceful with only the light from the burning logs in the fireplace and we were sorry when dinner was called. Downstairs we found a gathering of Pittsburgh dignitaries, the only woman being Kaufmann's wife. They all were curious to meet Einstein and he encouraged them to explain their businesses to him. He refused cocktails and any other liquor from which he abstained regularly. After the guests had departed he asked his host to put a car at my disposal and he shook hands with this admonition, 'You go back to where you are staying and go to bed, and don't get up in the morning till I come over. I will sit with you.' True to his word he came over to Spear's in the morning and sat with me in my room. 'No, you are not coming with me,' he said. 'I will call for you this evening to attend dinner [held in Einstein's honor by the American Mathematical Society].'"

At midday Watters made a second visit to the Mellon Institute, where he met Caltech's Robert Millikan, who was looking forlorn, as Watters remembered it, "for in the excitement over Einstein, he, a [1923] Nobel Prize winner and a scientist of note, was neglected. He asked if I would bring Einstein over to his hotel to see him that night." Watters did. The next morning, Robertson and Lefshetz escorted Einstein back to Princeton, and Watters went to Cleveland, where he had an appointment, having followed Elsa's instructions to the letter.

CHAPTER 27

Settling In

1935

---

56 to 57 years old

The Nazis had confiscated Einstein's boat and summer home in Caputh, but allowed him to keep furniture from the Berlin apartment, which pleased Elsa. But Mrs. Bucky thought it would be a waste of money to ship it over. She told their mutual friend, Leon Watters, that it was unsuitable for the Colonial-style house the Einsteins had decided to buy on Mercer Street, and urged him to talk Albert out of the idea. "You have more influence with him than anyone else," she said. "He doesn't just like you, he loves you." However, Watters was reluctant to try to influence Einstein against Elsa's wishes.

Watters steered clear of the subject on a visit to them in late January. Instead, he recalled amusing incidents in Pittsburgh. Once Elsa interrupted him, exclaiming, "Albert, you never told *me* anything about that." When Watters revealed how Albert had kept his promise and resisted the temptation to smoke while he was away, she said, "But as soon as his time was up, he began again!"

Elsa admired Watters's new jacket; he told her it cost seventy-five dollars. Albert called such a price "an outrage," adding, "Look at me, I wear the same clothes all year around," implying that Watters had been taken by his tailor and could have spent the money for a more worthwhile purpose.

"For his first wife, he dressed up," Elsa complained, "but for me he will not."

Clothes were obviously not of compelling interest to Einstein, although he surprisingly mentioned having worn his comfortable American-made shoes for six months, noting that they still "looked as good as new." Einstein's usual carefree attitude about what he wore was demonstrated in April, when Elsa and Margot spent several days with the Buckys and Einstein stayed with Watters at his Manhattan apartment. He had arrived there with a large suitcase containing, to the maid's surprise, just one collar and very little else.

That evening, Einstein was in a confiding mood, and Watters recorded the gist of what he said: "I find my physical powers decreasing as I grow older; I find that I require more sleep now. I doubt if my mental capacity has diminished. I grasp things as quickly as I did when I was younger. My particular ability lies in visualizing the effects, consequences, and possibilities, and the bearing on present thought of the discoveries of others. I grasp things in a broad way easily. I cannot do mathematical calculations easily. I do them not willingly and not readily. Others perform these details better. Though I lived in Germany for many years, I never became a German citizen. I made that a condition of my going there. I never met the Kaiser, probably on that account. Since I was not a German subject, I was never called upon to sign the pronouncement of the so-called German intellectuals which sought to justify Germany's position during World War I." (This last statement is at odds with most other accounts, in which he refused to sign the document.)

Watters asked why he was going to sit for a portrait by Rabbi Stephen Wise's wife, who both agreed was no artist. "Because," Einstein replied, "she is a nice woman." The next day he persuaded Watters to walk with him across Central Park for the first sitting. A flight of birds passed overhead and he renewed a previous conversation about the mystery of migration and homing instincts. "It's possible," said Einstein, "that they follow beams so far unknown to us." Watters left him at the Wise home, then picked him up later that afternoon to take him to the Greenwich Village studio of a Russian sculptor, S. Konenkov, "who was finishing a really excellent bust of Einstein."

That evening, Einstein joined Elsa and Margot for dinner in the Bronx as guests of a clothing manufacturer. When he returned to Watters's apartment late that night, according to Watters, he "could not restrain his disgust. 'Oh, such people and such dinner! They had the whole family there. I ate too much. I couldn't get my wife away earlier.' " He never explained exactly what the trouble was, "but he was surely angry over it." Again he was in a talkative mood. What Watters most remembered of that night was that Einstein spoke in defense of free speech and expressed his resentment over anyone trying to censor him.

He blamed the Bronx dinner for his feeling unwell the following morning; nevertheless, he went to lunch with Henry Morgenthau, the U.S. secretary of the treasury. The next day, after several appointments, he dined with banker Henry Goldman, probably trying to obtain financial guarantees for refugees.

On Sunday morning Einstein again sat for his portrait by Mrs. Wise. Watters "warned him against overdoing it, especially by going to places which gave him no pleasure."

Einstein rested all afternoon, then got up to go to another affair, explaining that "he had been beguiled" by a woman into attending a Plaza Hotel dinner for charity. Einstein asked Watters if he would like to join him. Watters replied "that if my going depended on the word 'liking,' my answer was No. He then said he would withdraw the word 'like' and asked instead if I would go to please him. I consented, and then with a quizzical look such as one sees on the face of a naughty boy, he said, 'We will arrange to escape from that woman early.' "

As they were about to leave for the dinner, Watters mentioned that Einstein had forgotten to put on his socks. "Aha!" he said. "The long arm of my wife extends even to you!"

At the Plaza, the hostess embraced Einstein and then led him, like a trophy, to the reception line. Watters, who remained in a far corner of the room, could hear her loudly introducing him. Einstein caught Watters's eye and winked. The hostess got him to pose with her for the photographers, and several of her guests shook Einstein's hand furiously, shouting out his or her name while doing so. Former New York mayor John O'Brien, the then mayor Fiorello La Guardia, and an Austrian prince were among the celebrity guests on the dais.

During dinner, Einstein grinned at Watters "as though he looked upon the proceedings as a joke. Promptly at ten he took out his watch and motioned to me to come up to his chair, as he arose to excuse himself. He had previously told the hostess that he had to leave at that time. All the way home he chuckled at the thought that he had gotten some fun out of being forced against his will to attend that dinner."

The following morning Einstein again sat for the sculptor before picking up Elsa and Margot from the Buckys. The family then returned to Princeton.

Walking to and from Fine Hall, Einstein was liable to be badgered by eager beavers bent on face-to-face confrontations. Dukas guarded his privacy at home, but he was on his own outside—as fifteen-year-old Henry Rosso soon learned. A C-average student, Rosso took heart when his journalism teacher at Princeton High School offered an A grade to anyone who got an Einstein interview.

A local tradesman tipped Rosso off to Einstein's route to the office, and the next morning the hopeful student waited for him with a friendly greeting. Einstein seemed disconcerted and kept moving. Rosso followed, hurriedly explaining his purpose. Einstein stopped to explain that if he gave Rosso an interview it would set others after him. Rosso said reporters didn't read high-school papers, and offered to copyright the piece. Einstein then capitulated.

Because Rosso had put all his energy into getting the interview and none into what to ask, Einstein had to lead "the way, offering topics for discussion, suggesting emphases for the interview." Avoiding all personal revelation, he said with a smile, " 'My life is a simple thing that would interest no one. It is a known fact that I was born, and that is all that is necessary.' He concluded with, 'I discovered that nature was constructed in a wonderful way and our task is to find out that mathematical structure of nature itself. Even nature is simple if we happen to look at it in the appropriate manner . . . but it is not a belief of other investigators. It is a kind of faith that helped me through my whole life not to become hopeless in the great difficulties of investigation.' "

Highlights from Rosso's interview were later lifted from his high-school paper, *The Tower*, and printed in a Trenton newspaper and *The New York Times*. Einstein didn't blame Rosso—who called to apologize—saying that he now understood the ethics of journalism.

Although Einstein hadn't covered politics with Rosso, he did so soon after with student members of Princeton's International Relations Club. During this discussion, John Oakes recalled, Einstein was wearing "his famous sweater with a hole in the sleeve. For almost three hours in a basement room he gave an informal account of what was happening in Hitler's Germany. I remember his bright eyes, and how unpretentious and informal he was. It was a very exciting and moving experience."

Events in Germany were indeed very much on his mind. As he wrote to Frederick Lindemann, "The German situation interests me," because "of the danger which it represents to the rest of the world . . . People are gradually recognizing the full import of this danger. Two years ago it could so easily have been stopped, but at that time nobody wanted to hear about it."

Warned that his criticism of Hitler and the Nazis would harm the Jews remaining in Germany, Einstein was cautious about what he said or wrote for publication. Yet he left no doubt about his views in an unpublished manuscript found among his papers. In it, he characterized the Germans as the result of centuries of indoctrination by schoolmasters and drill sergeants. He realized that the democratic Weimar Republic after World War I had fitted them "about as well as the giant's clothes fitted Tom Thumb," and that Hitler had taken advantage of the financial depression and fears of the populace to seize power. He saw Hitler as a bitter, envious man with a mediocre mind and "a desperate ambition for power" who had shrewdly exploited the German people's taste for "drill, command, blind obedience, and cruelty . . . and played up to them with the kind of romantic, pseudo-patriotic phrasemongering to which they had become accustomed . . . and with the fraud about the alleged superiority."

On April 30, the Einsteins dined at the home of their neighbors, the Blackwoods. The other guests included physicist William Houston and his wife. Houston, president of Houston's Rice Institute, had first met the Einsteins several years before at Caltech.

It happened to be the seventeenth birthday of the Blackwoods' son, James, who opened the door to Einstein and shook his hand. Einstein's hand was "warm and soft, but he had a muscular grip. His hair looked as if it might have been combed. Instead of his usual baggy pants and leather jacket, he wore a neatly pressed business suit. Entering the parlor he gazed at the other men who had decided to wear tuxedos. 'Ach!' he said. 'I did not know the occasion was so . . . so, ah . . . *serious!*' That evening got off to a laughing start."

In a more serious mood, Mr. Blackwood handed Einstein a clipping from a magazine in which he had answered three questions as follows:

To what extent are you influenced by Christianity?

"As a child I received instruction both in the Bible and in the Talmud. I am a Jew, but I am enthralled by the luminous figure of the Nazarene."

Have you read Emil Ludwig's book on Jesus [a popular book everyone was reading and discussing at the time]?

"Yes. But Jesus is too colossal for the pen of phrase-makers. No man can dispose of Jesus with a bon mot."

You accept the historical existence of Jesus?

"Unquestionably! No one can read the Gospels without feeling the actual presence of Jesus. His personality pulsates in every word. No myth is filled with such life."

When Mr. Blackwood asked Einstein if the article was accurate, he read it carefully, then answered, "That is what I believe."

Mrs. Blackwood had baked an angel food cake for dessert. James recalled that "she held a cake breaker in her hand, an instrument for dividing a light, fluffy cake without crushing it. The handle attached to a long stem [that] held a series of thin metal prongs about a quarter of an inch apart and five inches long. In wide-eyed fascination, Einstein stared at this long-pronged cake breaker and said, 'That is what I've been looking for. It would do nicely to comb my hair.' We all laughed, Einstein loudest of all."

James Blackwood reports that "in the front room before and after dinner, my Grandma Philips had not said a word. She occupied herself examining Dr. Einstein's feet. After the Einsteins left, Grandma broke her silence: 'I don't care if he *is* a genius, *he ought to wear socks to dinner.*'"

Early that summer Mrs. Blackwood drove the Einsteins to the Jersey shore; they were hunting for a permanent summerhouse. The youngest Blackwood son, five-year-old Bill, went along with them. James Blackwood heard what happened from his mother and young brother: "Since business details did not attract Dr. Einstein, he let his wife inspect the cottage, which, as it turned out, they did not rent. He relaxed in a chair on the porch smoking his pipe. Bill stood facing him. They were silent. But they began to play 'footsie.' Slowly, still facing each other, they moved their feet—one an inch or two forward, the other an inch or two back—taking turns. Apparently the object of the game was to see how close one could come to the other's toes without stepping on them. Soon the shuffling became so lively that Dr. Einstein hoisted himself from his chair and, holding Bill's hand, began to jig around the porch. Every so often he let go to grab his pipe or, since he wore no belt, to hitch up his pants. Then he danced more vigorously than ever. The white-haired man and the dark-haired boy spun around until they were dizzy and plopped down in chairs breathing hard. Mother asked, 'Wouldn't Hollywood give a million dollars for a movie of that?' Taking his pipe from his mouth, Dr. Einstein nodded in mock-solemn agreement. 'Ja, I expect they would.'"

Now that the Einsteins, Margot, and Dukas had decided Princeton would be their permanent home, they wanted to become American citizens. To do so, they had to apply while living abroad. They chose Bermuda, sailing there on the *Queen Mary*, and were met at Hamilton by the governor and mayor of the island. The governor gave Elsa the name of two luxury hotels, both of which Einstein vetoed on sight. So they walked to the other side of the town, where they saw a modest cottage advertising rooms to let. "We will go in there," Einstein said. "I like the place."

During their week's stay in Bermuda, Einstein declined invitations to parties, banquets, and receptions; he spent much of his time exploring the island. Ambushed by an excited group of schoolgirls, he willingly posed to be photographed.

One day, while the Einsteins were eating in a small restaurant, the German chef invited Albert to his home for dinner and then to go sailing. Einstein accepted enthusiastically. When the chef returned to the kitchen, "Elsa, very much provoked, said, 'How can you do such a thing, Albert? You refuse all the invitations from the Governor and the Mayor and now you go to have dinner at the home of a chef.'" Einstein replied, "They will understand I am here to enjoy myself. Besides," he added blithely, "he has a boat."

He was gone for seven hours and it was growing dark. Elsa feared that the German chef was a Nazi sympathizer who had kidnapped Albert. Alarmed, she hurried to the chef's home, where she found the two of them eating on the verandah. "He cooked for me all his masterpieces," Einstein said. "We had a fine day."

The Einsteins chose Old Lyme, Connecticut, for their summer vacation, at a spot where the Connecticut River flows into Long Island Sound. There they rented an old Colonial frame house with twenty acres of land, a tennis court, and a swimming pool. The White House, as it was called, cost nine hundred dollars for the season. Elsa called it paradise, and their neighbor, Mrs. Joseph Copp, a professional pianist whose husband played the violin, recalls that the Einsteins "were so overwhelmed by the luxury that for the first ten days they ate in the pantry." Einstein enjoyed the comparative solitude, though there were always visitors, including Italian playwright Luigi Pirandello. He told Max Born that his life would have been indescribably enjoyable except that Elsa was ill. Even so, she often traveled to Princeton to supervise changes to the 120-year-old house on Mercer Street, especially the installation of a picture window in what would be Einstein's second-floor study at the rear of the house.

That summer, while Einstein relaxed, Harvard University doubtless stood alone as the only American institute of higher learning not clamoring for his presence; an ad hoc committee agonized over how to exclude him from the university's tercentenary celebrations in 1936 without insulting him. Committee members feared that if he came he would steal the show from the other scholars. Harvard astronomer Harlow Shapley suggested giving Einstein an honorary degree in 1935; he surely wouldn't expect to be invited to Cambridge two years running. That way, as Shapley put it, they could get "him out of the way for the tercentenary, leaving our celebration undisturbed by the madness which at that time hung around the name of Einstein."

The school's administrators agreed, and Shapley invited the Einsteins to stay at his home, proposing as the inducement an evening of chamber music. It worked. Elsa, commuting between Connecticut and Princeton where she was supervising the Mercer Street renovations, was too busy to join him. She advised Shapley to "take good care of Albert. He is a sensitive plant. He should smoke no cigar. He can have coffee for breakfast, but in the evening he must have Sanka; otherwise he will not sleep well."

However, Shapley did tempt Einstein into smoking an after-dinner cigar, which he took with a murmured "Ach, mein Weib!" (Oh, my wife!). His host was more conscientious about seeing that Einstein slept, telling him, after three hours of music, that it was time for bed. He returned to Elsa well rested.

Whenever Einstein spotted a stranger's car coming up the drive to his summer home, he escaped in his sailboat. He had already taken this escape route a week after the Harvard visit when a welcome visitor, Leon Watters, arrived. When Einstein identified him, he turned back and came ashore. Elsa was eager to show Watters around their "paradise," but Albert claimed him first because, he said, they had lots to talk about.

Einstein was in a buoyant mood, chewing a wisp of straw as they walked, and saying that his recent work "had been very fruitful and satisfying; that he had developed some new equations . . . He discussed science, economics, politics, and personalities, and seemed glad to be with someone with whom he could unburden himself freely."

"It was a memorable summer," said the Einsteins' neighbor, Mrs. Copp. "I think Einstein liked to be alone, but he adored music and was eager to meet us when he heard there was a pianist and a violinist here. My husband had to get him because Einstein wouldn't drive a car, and was terrified because he had this *precious* cargo and felt he had to be *so* careful. We saw him practically every day that summer."

Mrs. Copp recalled how "he'd come in the music room and put his violin case on the sofa and open the case in a great rush to get to his violin. He was so naive and charming and utterly guileless, and he didn't put on airs." She was surprised to find that he spoke English very well. "In fact, when he learned I had a grandson, he said to me, 'You look very young. They must have kept you in a refrigerator!'

"His wife was charming, but I couldn't make out their relationship. There was an intimacy but I don't think it was a love match. She was like a guardian, very careful to see he was properly clothed and didn't catch cold, especially when he went sailing. He was crazy about his little boat."

He told the Copps that he had had a magnificent boat that Hitler took from him, but said, " 'I have much more fun in my little boat.' Even though he almost drowned one day when he fell out of it into several feet of water."

The near miss occurred one day that summer. Andy Bloomberg, sitting in a nearby rowboat, saw Einstein step onto his eighteen-foot Cape Cod Knockabout, slip, and slide off the bow into the water. "I had to pull him up and fend off the sailboat at the same time," said Bloomberg. "And he was just sagging, limp-like! I caught ahold of him by his collar and hauled upwards . . . I had to get young Mort Tiley to come help me. We heaved him into the rowboat and got him ashore. He just said 'Thank you, boys,' real quiet-like and went up to the house to change and told us to come too." There Elsa rewarded them with fresh raspberry juice.

That was not the only boating accident he suffered that summer. On another day Einstein was rescued after he got the mast of his sailboat caught in a bridge, and on a third day, when it was blowing hard, the mast cracked and he had to be towed ashore.

He mentioned none of these escapades to his sister, Maja, who was now living in Florence, Italy, with her husband. He told her instead that, after a promising start, he was working slowly and painfully on a new paper. What he had come up with, helped by his assistant Nathan Rosen and colleague Boris Podolsky, was a paper describing an apparent flaw in quantum mechanics, which he called "spooky action at a distance." Now known as the EPR paradox—using the initials of the three authors—the theory pointed out that in an imaginary experiment pairs of particles originating from the same source but widely separated in space—even light years apart—seemed to influence each other, so that the observation of a particle in one region of space would instantly affect the state of a second particle. By measuring the position of one you could know the position of the other, and by measuring the momentum of one you could tell the momentum of the other. This appeared to give objective reality to both the position and the momentum of the distant particle, which had not been measured independently. Yet quantum mechanics, according to Jeremy Bernstein, "tells us that no quantum-mechanical description is possible in which both position and momentum are precisely specified." That these descriptions cannot be complete is what Einstein, Podolsky, and Rosen concluded in their paper "Can Quantum-Mechanical Description of Physical Reality Be Considered Complete?"

Austrian physicist Erwin Schrödinger applauded Einstein, writing on June 7: "I was very happy that in the paper . . . you have evidently caught dogmatic quantum mechanics by the coat-tails."

Rabbi Harry Cohen called on August 30 to discuss almost everything but quantum mechanics. He feared that public protests by American Jews might serve as a pretext for Germans to intensify their persecutions of German Jews. Einstein disagreed, saying that although people like Rabbi Stephen Wise were sometimes too emotional and that non-Jews would be more effective, the Germans were afraid "of what the world will say about them." He went on to say, "No people in the world take such delight in cruelty. The Germans are, by nature, cruel." He blamed this aspect of their character on "their system of education, the authoritarianism, and the militarism that govern the system. The automatic obedience to orders from above."

After fifteen minutes Elsa came into the living room with a tray of chocolates and whispered to Einstein about another appointment; this was obviously a subterfuge to enable Einstein to end the conversation if he wanted to. Instead, he motioned for his visitor to stay.

Cohen showed Einstein a newspaper feature titled "Believe It Or Not!" by Robert Ripley, which read, "Greatest living mathematician failed in mathematics," and asked if it was true. "The story has been completely distorted," he replied, with a laugh. "When I applied for admission to the Polytechnic School in Zurich at sixteen I was not completely prepared in French and botany. I never failed in mathematics. Before I was fifteen I had mastered differential and integral calculus."

When Cohen brought up the plight of Negroes, "there came into Einstein's eyes that look of moral indignation I had seen when he had spoken of German cruelty. His reply was brief and emphatic: 'There should be no discrimination. It is wrong.'"

Discussing discrimination against Jews in colleges, Cohen implied that the quota system was justified when he said: "If of every one hundred applicants for admission to a college, forty are Jews, this is surely too many. Colleges do not want forty percent Italians or forty percent Lutherans."

Einstein disagreed. "Discrimination is not right," he said emphatically. "Admission should be based on ability."

The sun was setting as they concluded their two-hour conversation. Before leaving, Cohen asked Einstein to write something in his diary to commemorate the visit. Einstein glanced through it and said, as if he'd been tricked, "It's completely empty. Nothing is written in it."

"I want you to be the first," Cohen explained.

Einstein then wrote, "To be free means to be independent, not to be influenced by what others think and say."

Like James Blackwood, Cohen noticed that Einstein repeated his phrases, spoke in a low, pleasing tone, and "did not once raise his voice or show anger or bitterness, even when he spoke about German cruelty and racial discrimination. However, his face became set, almost stern, and his eyes would light up in a way that filled me with awe. When we spoke he always looked straight at me, his large eyes directed at mine. But whenever he paused before making a reply, he no longer appeared to be seeing me. His eyes looked straight forward, as though into a great distance."

Cohen was lucky to get to Einstein; some twenty newsmen eager for interviews had recently failed. However, Einstein took pity on photographer Jack Layer, who waited on his doorstep for eight hours one day, "and agreed to pose. He ended up rowing Layer about the river in his boat for special candid shots."

During the vacation Elsa had enjoyed reasonably good health, but shortly after the Einsteins returned to Princeton to start life in their new home, she felt weak and short of breath and one eye became swollen. Einstein did not tell her that the doctor's diagnosis was inflammation of the heart, "but Elsa noticed he was unusually concerned about her." Gradually Helen Dukas took over more of the duties of protecting Albert from unwelcome visitors.

Although most Princetonians didn't bother Einstein, James Blackwood witnessed an exception. "Dr. Einstein was walking toward his home and we were about to pass each other. On the street a Studebaker that wasn't yet obsolete, but was getting there, stopped with a squeal of brakes. The middle-aged woman driver had the car angled crosswise on Mercer Street. She flung its doors open on both sides, and called out to Dr. Einstein, 'Come, I giff you a ride home.' Not knowing what to do, he stood still in great perplexity. 'It's all right,' she insisted. 'I know who you are!' Dr. Einstein turned to me, his voice a rasping whisper, 'Who is dat voman?' 'Mrs. Georges Barrois,' I said. 'A professor's wife.' 'Come!' she ordered him. Hesitantly, perhaps to avoid an accident, he came. She drove the short way to his home. He got out of the car backward, and dashed up the steps to his front door, the fastest I ever saw him move."

For Hiram Haydn, a street encounter with Einstein was an epiphany that left him speechless. As Haydn, editor of *The American Scholar*, was being driven by Christian Gauss, the dean of Princeton University, to a tea in Einstein's honor,

> Christian pointed ahead and said, "Why, there are Einstein and Eisenstaedt! Let's pick them up." Two figures walking slowly on the road, one with a great shock of hair, a turtle-neck sweater, and baggy Chaplinesque pants. We stopped beside the two men, and I got out of the car to fold up the seat on which I'd been sitting. I looked directly over the top of the car into the face of Albert Einstein. It is no secret that he sprouted whiskers, as do other men. Yet it struck me forcibly that there was light coming out of his face—that light grew there, as hairs do on the faces of all men. It seemed to me that this was not a man in the ordinary sense, that the face belonged to another, different species. And then he smiled at me. This act constituted the most religious experience of my life. At the tea that followed I did not participate at all; I sat alone in a corner, shaken by the meeting. Others came over and urged me to come and meet Einstein. I said that I had met him, offered various vague excuses. When Christian had driven me back to the inn, he detained me briefly with his hand on my arm. "Such moments," he said, "tear a rent in ordinary perceptions, cut a hole in the fabric of things, through which we see new visions of reality."

Future meetings with Einstein, Haydn wrote, did not "tarnish the first experience."

Einstein would have been aghast at the thought that meeting him was a mystical experience, astonished that anyone would think so—especially the editor of a scholarly journal. He was even embarrassed when described as "a great man." Yet Princeton student Joseph Ceruti rated him even higher. He was "leaving college to attend church and met Einstein where our respective paths crossed. When I said 'Good morning, Dr. Einstein,' he replied in kind and immediately asked what I was studying. Learning that I was a graduate student in architecture, he told me how impressed he had been by the mission-style architecture in southern California. He found the style simple, honest and organic. He thought some American architecture lacking in character, with not enough emphasis on marrying buildings and nature. When we parted, I felt I had talked to God."

Some of his correspondents certainly believed he could work miracles. He received many appeals for help, including one that read: "Dear Professor Einstein, I can't have a baby. My wish is to adopted [sic] a baby, and I don't have money. Please help me! I want to be a mother! Help me!"

Other correspondents obviously thought him more diabolic than divine. These writers, whose missives Helen Dukas continued to preserve along with serious letters demanding his attention, ranged from inane to insane. Typical of their attacks, which probably amused him, was this letter: "Dr. Einstein: You are the prince of idiocy. The count of imbecility. The Duke of cretinism. The Baron of moronity. The King of stupidity. You are not a scientist, you are an anthropological error of birth, a colossal liar, and a hypocrite. You should wear a mask when you speak to a scientist!!!" Unsolicited advice came from a retired dentist who wished to supervise Einstein's lifestyle to ensure he had "the proper diet, elimination, exercise and sleep."

At year's end, in a speech celebrating the opening of the Philipp Lenard Institute—named for Einstein's archenemy—German physicist Johannes Stark, a 1919 Nobel Prize winner, wailed,

> "Einstein has disappeared from Germany but, unfortunately, his German friends and supporters still have the opportunity to continue their work in his spirit. His principal promoter, Planck, still heads the Kaiser Wilhelm Society, his interpreter and friend, Mr. von Laue, is still permitted to play the role of adviser for physics in the Academy of Sciences in Berlin, and the theoretical formalist, Heisenberg, the spirit of Einstein's spirit, is even to be distinguished with a university appointment."

As Germany's Nazis continued their attacks on him, they were joined by Nazi sympathizers in America — adding to his growing piles of hate mail.

CHAPTER 28

Family Matters

1936

---

57 years old

Answering the door of his Berlin home, Dr. Janos Plesch faced a young woman who claimed to be Einstein's illegitimate daughter. At first he thought it unlikely, though not impossible. She was very persuasive, however, and the "intelligent, wide-awake and attractive" young boy with her did look remarkably like Einstein.

In time she convinced Plesch, and

> with the assistance of friends, who were also convinced, we set to work to help her, found her a position and sent the boy to school. Then I wrote a tactful letter to Einstein explaining the situation and giving him news of his daughter and grandchild. To my mystification Einstein showed no proper interest, and so in order to move his parental and grandfatherly heart I sent him the one or two really clever and delightful little colored sketches the boy had made and a photo. There! I thought, the features of the boy will move him.

But what moved him was another and more urgent message, this time from England.

A woman there had obviously made similar claims, telling her story to various academics, including Einstein's friend, Professor Frederick Lindemann of Christ Church College, Oxford. More skeptical than Plesch, and scenting blackmail, he wondered how to warn Einstein without destroying Elsa. She was suffering from heart and liver disease, and even the hint of a scandal might have killed her. So he addressed a telegram to Einstein's Princeton colleague, Professor Hermann Weyl:

MRS. HERRSCHDOERFFER PRETENDING TO BE EINSTEINS DAUGHTER TRIES TO FIND SUPPORT IN HIGH CIRCLES NOT BEING ABLE TO ASK FOR HELP ON ACCOUNT OF STEPMOTHER. [Had the claimant been genuine, Elsa would have

been her stepmother.] STOP FEEL SUSPICIOUS PLEASE ASK EINSTEIN PERSON-
ALLY AND CABLE IMMEDIATELY. LINDEMANN CHRISTCHURCH.

Weyl passed the warning on to Einstein, who confided in Helen Dukas. She
immediately hired a well-informed and enterprising friend to investigate the
woman.

In August, Dukas got a report from her amateur-detective friend. His research had paid off. According to his findings, Herrschdoerffer, the woman in
question, was actually Grete Markstein. She had worked in a Berlin opera house
before leaving for Paris en route to England. She left behind a "not very good
reputation." Her mother, Helene Markstein, born on July 5, 1863, had given
birth to Grete on August 31, 1894. If Einstein was her father, he must have had
an affair with Helene Markstein when she was thirty and he fourteen—a highly
unlikely prospect.

Satisfied that Grete was not his daughter, Einstein at long last responded to
Plesch. He said the joke was on the doctor for being taken in by a con woman.
In the same jocular vein, Einstein ended his letter with this bawdy doggerel:

> All my friends make fun of me,
> To help stop my family!
> I have enough of the truth
> Which for long I have carried.
> But that I put eggs sideways
> It would be cute to hear
> If it didn't hurt other people.

Another truth—"which for long [he] had carried"—was the existence of his
daughter, Lieserl. He had concealed it from the world, Plesch included, for some
thirty-odd years.

The author questioned Dr. Robert Schulmann of Boston University about
the obvious discrepancies in these accounts. Schulmann is director of the
Einstein Papers Project and an editor of *The Collected Papers of Albert Einstein*.

D.B. There's still a puzzle. The "detective" investigated a Grete Markstein.
Lindemann warned Einstein of a Mrs. Herrschdoerffer. Are these definitely
the same woman?

ROBERT SCHULMANN. No. They could be two women.

D.B. Of course Herrschdoerffer might be Markstein's married name. But that
doesn't account for the woman who appeared at Plesch's door with a little
boy she claimed Einstein had fathered. So there are possibly three women
involved at this time.

SCHULMANN. Three Einstein groupies.

D.B. From what we know of his life that isn't impossible.

SCHULMANN. No, it isn't.

D.B. In fact, one of them might have been Lieserl.

SCHULMANN. Now you're leaping tall buildings at a single bound. It is possible, though. There are two possibilities: that Herrschdoerffer was an alias Markstein used. Phonies tend to assume other names. Or Markstein and Herrschdoerffer were two different women. In which case, if Herrschdoerffer was younger than Markstein, it's more plausible that Einstein had an affair with *her* mother. Presumably he was older then. [For example, if Mrs. H. was twenty-one she would have been conceived around 1914. Einstein was then thirty-five and living a bachelor life in Berlin while his estranged wife, Mileva, and their sons remained in Switzerland.] The only thing that interests me personally is the psychological element. Whatever the identity of the woman or women, Einstein reacts in a very strange way.

D.B. By hiring the amateur detective, you mean? It implies that he had not been following the fate of his real daughter, Lieserl, but believes she may still be alive. He's saying in effect [to Dukas], "This might be her. Look into it."

SCHULMANN. Yes. I don't think he kept track of Lieserl, though his first wife [her mother] may have done so. But then again he wasn't discussing these matters with Mileva any more.

D.B. But we agree that his wanting the woman investigated implies he thought Lieserl might still be alive in 1936.

SCHULMANN. Definitely.

D.B. Because, if he knew she was dead he'd be certain this woman, or women [*sic*], was an impostor and say, "Why bother to investigate her?"

SCHULMANN. I agree. From what we know it makes a lot of sense, doesn't it?

D.B. Do you find it odd that knowing he has an illegitimate daughter Einstein writes in that doggerel, if "I put eggs sideways it would be cute to hear"?

SCHULMANN. I'm not sure that's so strange. He knows in this particular case it's not his daughter. He's not going to reveal his hand. He knows, of course, that there is one; but they haven't got the right one.

D.B. So his secret is safe. How do you explain the bawdiness of the doggerel?

SCHULMANN. Barnyard language is used throughout southern Germany. And "eggs" throughout all Germany means the balls of the male. It's likely he picked up the talk from schoolmates in Munich rather than from his parents, because they were quite proper and assimilated Jews who wouldn't talk like that.

D.B. What's the chance of your hiring a private detective to search for Einstein's daughter?

SCHULMANN. I've thought about it. But the outbreak of fighting between Serbians and Croatians put any such plan on hold.

While Einstein had been trying to avert a scandal over his illegitimate daughter, American and German newspapers were giving prominent play to the efforts of fellow scientists to destroy his work as fatally flawed.

Throughout February, *The New York Times* carried news of a triple threat to torpedo the theory of relativity, mounted by Dr. Leigh Page of Yale, Dr. Ludwig Silberstein of the University of Toronto, and Dr. William Cartmel of the

University of Montreal. The paper's front page that month looked progressively more devastating: New Evidence Held to Upset Einstein . . . Existence of Ether Seen . . . Difference in Speed of Light Observed from Earth Urged as Relativity Challenge . . . New Evidence Which Might Result in a Great Upheaval in Scientific Thought.

Einstein countered with cool self-assurance: "I have already informed Mr. Silberstein that his result is based on an error, which, unfortunately, he has so far failed to realize." He dismissed Page as a superficial thinker, and Cartmel as not worth a reply.

The torpedo was a dud. Silberstein tried to revive the attack, saying he would eventually disprove Einstein. The world is still waiting.

When the North Americans retreated, the Nazis brought forth six Nobel Prize–winning physicists, with Lenard and Stark as leading lights. All followed the Nazi party line and denounced theoretical physics as worthless "Jewish science."

But the voice of reason had not yet been completely stifled in Germany. Werner Heisenberg, an Aryan German, defended theoretical physics—and obliquely Einstein—by pointing out that Max Planck (also "unsullied" by Jewish blood), head of the Kaiser Wilhelm Institute, had done work in theoretical science which stimulated new experiments and positive results. Moreover, when German physicist Wilhelm Lenz suggested to Max von Laue at the Prussian Academy of Sciences that they attribute the theory of relativity to Frenchman Henri Poincaré, thus "cleansing" it of any Jewish connection, von Laue dismissed the idea as both shameful and stupid.

Three years earlier, at the annual Physicists Conference, von Laue had compared the Nazi government's attacks on Einstein to the Inquisition's treatment of Galileo, ending his address with Galileo's stubborn, "And still it moves!" He was applauded by the audience and reprimanded by the Prussian Ministry of Education for his courageous stance. Einstein had heard about von Laue's speech and when asked by a friend who was returning to Germany if he had a message for anyone there, Albert said, "Greet Laue for me." Anyone else? asked the friend. "Greet Laue for me," Einstein repeated.

Otto D. Tolischus, *The New York Times*'s Berlin correspondent, concluded that the "weight of numbers seem to be on the side of the theoretical physicists, but the 'German' physicists are winning out because they have greater party orthodoxy on their side." With chilling foresight, he also noted: "Nothing interests the German public more than a fight."

As for Einstein, he had only to glance at the headlines to see that the Nazi creed was continuing to turn brilliant scientists into lunatics. In one letter to his friend, Michele Besso, he referred to the Nazis as "the madmen in Germany."

If, as he sometimes said, he was pursued by a mathematical goblin that never let him rest, the goblin also kept his mind off personal problems and to some extent the horrors of the world. Writing to Besso, for example, he invariably discussed his latest scientific speculations, but never mentioned the woman, or women, posing as his daughter. Nor did he ever discuss Elsa's critical illness. He did say that he found Americans less prejudiced about class and less bound by rigid traditions than Europeans, and chuckled over the fact that the word "socialist" was taboo among Americans, yet everyone in the United States was benefiting from Franklin Roosevelt's socialist programs.

Though uncomfortable with most of the American press, especially journalists who posed dumb questions or distorted his answers for spicy quotes, he could hardly refuse Robert Smith's interview request. It was the young man's first assignment as a cub reporter on the local Princeton campus newspaper. He was sent to ask Einstein if he indulged in the latest craze among eggheads, three-dimensional chess.

"No," Einstein replied. "When I relax I want something that does not tax my mind." He recalled playing chess once or twice as a boy, but said that now he had no time for such games. He had never played bridge, and hadn't even heard of Monopoly. How was it played? When explained to him, he chuckled and remarked, "A very American game."

About this time he wrote to his sister, Maja, in Italy: "As in my youth, I sit here endlessly and think and calculate, hoping to unearth deep secrets. The so-called Great World, i.e. men's bustle, has less attraction than ever, so that each day I find myself becoming more of a hermit."

For a self-proclaimed hermit, Einstein maintained a crowded hermitage, sharing it with his wife; a live-in secretary-housekeeper; his stepdaughter; and a trickle of relatives and a nonstop flow of visitors. Most weekends and vacations he spent with friends, especially the Buckys. He was rarely physically alone, except when in his bedroom or study. It was not unknown for him to leave his office door open. Had he wished, he could have worked in solitary splendor, but when two young men, Banesh Hoffmann and Leopold Infeld, offered to join him in his quest for the unified field theory, he accepted with alacrity. So much for the hermit, though it is true that he avoided crowds, enjoyed solitude, and beat a quick retreat when uninvited visitors approached.

Banesh Hoffmann had reached Princeton via England's Oxford University. While he was studying there, accounts of Einstein's theory were blazing in the headlines. Intrigued, Hoffmann had neglected math to study something called projective relativity. This brought him to the attention of a Professor Beldow, briefly over from Princeton. Surprised to find Hoffmann the only Oxford student involved in the subject, he invited him to be a scholar at Princeton's Institute for Advanced Studies. Just the thought of being in the same building as Einstein was intoxicating. Hoffmann gleefully accepted. He met Leopold Infeld at Princeton; they collaborated on a joint paper, then wondered what to tackle next.

Infeld had known Einstein in Berlin and said to Hoffmann, "Let's see if he'd like us to work with him." But Hoffmann was bashful. He had almost to be shoved through the door to meet Einstein, who immediately said, "Please write your questions on the blackboard. And please write slowly. I don't understand things quickly." Hoffmann later recalled, "I don't know what the magic was but suddenly we were partners and mathematics was the common enemy. He suggested two things we could work on. We selected one, luckily, because the other one has still not been solved."

Hoffmann and Infeld often found the world so difficult that they felt like quitting, but Einstein wouldn't give up. He thought of all sorts of possibilities and new ideas so they could keep going.

"We did all the dirty work of calculating the equations and so on," Hoffmann recalled. "We reported the results to Einstein and then it was like having a headquarters conference. Sometimes his ideas seemed to come from left field, to be quite extraordinary. We often had heated arguments, not heated in anger but we were like equal partners. I remember just once Einstein being really angry. The reason I remember is that it made such an impression on me. He was bringing a chair up to the table and the leg was a little loose. And he said, 'Ah, kaput!' I'd never seen him so annoyed before or since.

"He made me conscious of various things that help in thinking, namely that you must be bold and you mustn't give up.

"Einstein treated me as an equal, absolutely. That's the most remarkable thing. He made me feel at home, that my views would be listened to respectfully and that no one would laugh at me. Infeld and I would sometimes go to his office and then walk home with Einstein and continue working at home, sometimes vice versa."

Infeld, his other assistant, said he and Einstein discussed "hundreds" of subjects, including the Spanish civil war (they both supported the embattled Republicans) and the Jewish problem, yet he contends that Einstein was unaware of "the real life around him."

To feel the pulse of common humanity, Einstein had only to chat with Helen Dukas, who knew all the latest news about everything. She and the letters that poured in were his links to "common humanity." In time Dukas would collaborate with Banesh Hoffmann on two delightful, adoring, though well-informed books about Einstein.

Einstein had another reliable connection to the everyday world—his sense of humor. "With Einstein there was always humor," said Thomas Bucky, who knew him as well as anyone. "Almost everything was turned to be funny. The face was always smiling, and when he read for fun Emily Post's book on etiquette I'd hear him laughing in his room. Then he'd come downstairs having read some of it, roaring with laughter, and he'd repeat those 'Thou shalt nots' of hers. He read Freud's book on dream analysis, too, and discussed it as he read it, chapter by chapter. He was interested and impressed."

On April 21, 1936, Einstein sent eightieth-birthday greetings to Sigmund Freud, praising him as one of the generation's "greatest teachers" and admitting that

until recently I could only apprehend the speculative power of your train of thought, together with its enormous influence on the *Weltanschauung* [world-view] of the present era, without being in a position to form a definite opinion about the amount of truth it contains. Not long ago, however, I had the opportunity of hearing about a few instances, not very important in themselves, which in my judgment exclude any other interpretation than that provided by the theory of repression. I was delighted to come across them; since it is always delightful when a great and beautiful conception proves to be consonant with reality. With most cordial wishes and deep respect.

Freud replied on May 3:

I really must tell you how glad I was to hear of the change in your judgment—or at least the beginning of one. I always knew that you "admired" me only out of politeness and believed very little of any of my doctrines, although I have often asked myself what indeed there is to be admired in them if they are not true, i.e. if they do not contain a measure of truth. By the way, don't you think that I should have been better treated if my doctrines had contained a greater percentage of error and craziness? You are so much younger than I am but I hope to count you among my "followers" by the time you reach my age. Since I shall not know of it then I am anticipating now the gratification of it.

All that summer Elsa was critically ill with heart and liver disease, and confined to bed in their isolated, rented Saranac Lake vacation cottage, in the Adirondack Mountains, where they were less bothered by strangers. There she was lovingly nursed by her daughter, Margot.

The Einsteins' widower friend Leon Watters was about to remarry, and Elsa wrote to him on June 12, saying that she knew he would treat his wife as something precious. Einstein also wrote to his friend; his letter, which followed twelve days later, wished Watters "all the best of luck for your coming marriage. With your wealth of family-love and your urge to share your feelings and thoughts, this change in your life will assuredly be a good step." Perhaps those succinct comments reveal what Elsa missed in their marriage and Einstein apparently never wanted.

Assured that Elsa was in good hands with Margot and Dukas caring for her, Einstein got away for short visits to friends, among them Judge Irving Lehman in Port Chester, New York. Upon his return, he found a pile of mail, among which he noticed a black-edged envelope. He opened it and learned that his old friend Marcel Grossmann had died after suffering for years from multiple sclerosis. He immediately wrote to the widow:

You are one of the few women of my generation for whom I have a really true admiration and respect. To be resolute and devote one's life without hope—that is the hardest thing a man can ever decide. I remember our student days. He, the irreproachable student, I myself, unorderly and a dreamer. He, on good terms with the teachers and understanding everything, I a pariah, discontent and little loved. But we were good friends and our conversations over iced coffee in the Metropole every few weeks are among my happiest memories. Then the end of our studies—and I was suddenly abandoned by everyone, standing at a loss on the threshold of life. But he stood by me and thanks to him and his father I obtained a post later with Haller in the Patent Office. It was a kind of salvation and without it, although I probably should not have died, I should have been intellectually damaged. And then, ten years later, our feverish work together on the formalism of general relativity. It remained uncompleted because I went to Berlin, where I continued working on my own. And then came his sickness.

During my son [Hans] Albert's studies at Zurich [in the 1920s] the signs were to be seen. Often and with great pain I thought of him, but we only met occasionally when I came to Zurich on a visit. That his suffering would be so long drawn out I could not possibly have imagined, although I knew of this suffering from a friend in Berlin. And yet he did not die until I too have become an old man—inwardly solitary, having lived through the whole gamut of destiny and with still perhaps a few years of peaceful existence.

But one thing is still beautiful: we were and remained friends throughout our lives. You, however, I respect for what you have done and because you did it for him. From the bottom of my heart I send you my sympathy, wishing you peace and consolation.

Your affectionate Albert Einstein.

While Einstein could, as this letter proves, show affection and concern for others, Elsa felt that he neglected her. Being confined to her sickbed all summer, often in pain, exacerbated her sense of deprivation. And so she confided, almost desperately, to the newly wed Leon Watters: "I think you are the most considerate, loving husband. How gladly would I send Albert to you to be taught. But, oh, God, it can no longer help."

To get an insider's view of the relationship between Albert and Elsa, this writer asked Thomas Bucky to elaborate.

BUCKY. It's another myth that Einstein was cold toward his wife.

D.B. That's what she implied in her letter to Watters. Did he treat her more as a servant than as a wife?

BUCKY. If I say yes, you'll misinterpret. Einstein had a shell around him that was not easy to penetrate. He was a god and he knew it. He was not pompous about it. But he and his wife were very different persons. I'm trying not to be misinterpreted.

D.B. Let me try to help. Was it that she realized he was a genius, a great man, and was responding to that?

BUCKY. Yes, of course.

D.B. Someone challenged him: "She waits on you hand and foot. What do you do for her?" And Einstein is supposed to have replied, "I give her under-standing."

BUCKY. It may be true. But I wouldn't describe his attitude to his wife or to his son, Hans Albert, as cool. He had a shy attitude toward everybody. He was aloof and reserved, always shy, hesitant. He didn't have a "slap on the back" relationship with anybody, including his wife and son. He was gentle, con-siderate of others, and the opposite of pompous. But I never heard even a close friend call him by his first name. When someone did treat him with undue familiarity he would shrink back. And although he and my father were close friends they always addressed each other as "Professor Einstein" and "Doctor Bucky."

Back in Princeton after his summer vacation, Einstein discussed his unified field theory with a class of some twenty graduate students. One of them, Stanislaw

Ulam, was amused by Einstein's eccentric English phrasings, such as, "Oh, he is a very good formula."

Ulam also liked Einstein's cheerful, wisecracking assistant, Leopold Infeld, who was a marked contrast to the previous assistant, Walther Mayer—"a strange personality." Infeld was a Polish Jew who had spent time in England before landing in America. Ulam—himself of Polish origin—relished Infeld's remark that "in Poland people talk foolishly about important things, and in England intelligently about foolish things."

That year, as well as continuing his research and giving occasional lectures at Princeton, Einstein, the eternal student, attended classes on topology, along with twenty-year-old Henry Bach. The young man was taking every math and physics class offered. He recalled walking into the course on topology given by Professor Tucker in Fine Hall and asking, " 'Am I the only undergraduate taking the course?' And Tucker said I was. One graduate [student] also took it for a time then dropped out. That left just Einstein and myself.

"Einstein didn't come in until the second or third day. He was talking to Tucker, and I moved to the back of the room so they wouldn't think I was trying to overhear them. Einstein was monitoring the class. There were no airs about him. He seemed just like a regular fellow, someone who would come up to you and say, 'Slow down a minute and I'll walk with you,' you know? Wouldn't you feel very much at ease with a person like that?"

Bach obviously felt at ease with Einstein, once asking him, "Are you going to cut your hair when you can't see?" Einstein was silent for a moment, then got it and laughed. He must have heard similar comments many times. After a little boy called him "the man with soap in his hair," Einstein signed a letter to the youngster with that name. Another boy asked in passing, "Is that Mrs. Einstein?" And a third, thoroughly confused, asked, "Why doesn't *he* cut *her* hair?" One visitor prepared her son in advance by describing Einstein's leonine appearance. The boy said a quick "hello" to Einstein then ran back to his mother, calling out, "He *does* look like a lion!" Einstein was apparently never annoyed by these comments and sometimes responded by wiggling his ears.

One unexpected visitor, test pilot Luis de Florez, literally dropped in on Einstein. Knowing that his friend Starling Burgess, the leading aircraft designer, was working on a project with Einstein, de Florez flew Burgess to Princeton on the off chance of meeting the great man. His plan worked. De Florez met Einstein in his office, where he was "sitting with his bushy hair and his little leather jacket. And nearby was a blackboard filled with symbols."

Burgess broke the ice by saying, "Tell the Professor about your flying device" (a sonic device to allow pilots to fly blindfolded).

De Florez then gave Einstein some background information. He started by saying, "Men can't fly without a reference in space. And as you know, birds can't fly if they're blinded or hooded."

Puzzled, Einstein said, "They can't?"

De Florez assured him they couldn't.

After a moment's thought, Einstein said, "Of course."

De Florez was amused that he had made a remark Einstein needed to think about before he could admit that it was a fact. "But," de Florez concluded, "very few people realize this fact and he was no exception."

The test pilot then explained how he had outdone birds by flying blindfolded. He had been spurred by the statement of two army pilots who had pioneered instrument flying. They claimed that no one could fly a plane without eyesight; that is, without looking at landmarks outside or instruments inside the plane. For his daring experiment in the "impossible," de Florez attached a small generator with a propeller to the wing of his plane. When he pushed the nose of the plane down in flight, the propeller picked up speed, increasing the pitch and changing the sound. Using this difference in sound between diving and level flight, he learned to fly straight and level "by ear." Once, he even flew for over an hour with his head in a book.

Einstein enjoyed visitors who, like de Florez, had esoteric information or experiences, especially when they overturned established views. He depended on Helen Dukas, though, to handle the lovesick, the mercenary, and the mentally deranged who tried to engage his interest, waste his time, or pick his pockets. Dukas rescued a good sampling of such correspondence for posterity—from Einstein's wastebasket.

In one letter, an East Coast publisher offered to produce his biography and so ensure him "lasting prestige and standing," as well as "a great financial future." The author was apparently waiting in the wings. Einstein was promised "tremendous publicity" that would be "a great asset" to his future career. All he had to do was send $2,500.

Another letter writer, "Elizabeth," got straight to the point. "Dear Mr. Einstein," she wrote. "I love you." This was followed by a sketch of a wreath of kisses. "You look sweet and full of love. It is just the picture I wanted. [Dukas may have been persuaded into sending the woman Einstein's photo.] I just kissed you. I knew you knew sweetheart of mine. When are you coming over here dear?" There was more in this vein, followed by a love poem and twenty-eight kisses.

Dukas usually selected letters that required a response, took brief notes from Einstein of what he wanted to say, then fleshed them out as she typed them up. Einstein probably did not reply to the East Coast publisher, or to Elizabeth, or to any of the following:

> I would appreciate very much if you could tell me what Time is, what the soul is, and what the heavens are.

> How do you enjoy your work? How did you come to choose it? How did you become famous?

> I shall be very happy to come out to your place any time you designate and demonstrate my way of bodily care.

> I would be sincerely honored indeed to wash your feet.

When asked to right some wrong, though, he almost always came through. Take Buckminster Fuller (later famous as the inventor of the geodesic dome).

Expelled from Harvard for "general irresponsibility" and later dismissed by the establishment as a crackpot, he was alive with bright ideas but bedeviled by bad luck.

In the mid-1930s, Fuller invested all his capital in the car of the future, hiring Starling Burgess as chief engineer and living on a diet of coffee and doughnuts. By 1935 he had produced three prototypes of the three-wheel "Dymaxion," a streamlined eleven-seater capable of 120 mph, 40 miles to the gallon, and able to turn in its own length. H. G. Wells took a test drive in the car on Fifth Avenue, and later described a similar vehicle in *The Shape of Things to Come*. It looked like a low-slung modern plane without wings. Leopold Stokowski bought one, but a second car was in a fatal accident and the publicity killed the project.

Fuller switched from cars to write *Nine Chains to the Moon*. The book's title came from his calculation that if all the people then alive stood on each other's shoulders, they would make nine complete chains between the earth and the moon.

Fuller's own philosophy, expressed throughout the book, had been influenced by Einstein's nonanthropomorphic concept of God, and he shared his belief that the two primary motivating forces of human activity are fear and longing. He also adopted Einstein's view that "scientists were rated as great heretics by the church, but they were truly religious men because of their faith in the orderliness of the universe."

In his book, Fuller pinpointed the Patent Office as the catalyst for Einstein's relativity theory. Because Einstein was often faced with inventors' competing claims to have produced the most accurate clock ever, he became acutely time conscious, Fuller said. Einstein finally concluded that the perfect timekeeper was a pipe dream, because there is no "absolute" time. He therefore challenged Newton's theory, replacing "absolute" with "relative."

Fuller made one remarkable prediction in the book, which he completed in 1936: because relativity derived from the measurement of the speed of light, it presaged "a chain reaction ultimately altering altogether the patterning of man's everyday world." In other words, he anticipated the atom bomb and atomic energy.

The publishing firm of J. B. Lippincott agreed to publish the book, and Fuller thought his luck was about to turn. Then a staff editor took seriously somebody's wisecrack that only ten men in the world grasped the relativity theory. He obviously didn't know that Einstein himself had ridiculed the "only ten men" remark. But the editor had "documentation"—a popular magazine that had repeated the "only ten men" statement and printed their names. And Fuller was not on this exclusive list.

The publisher promptly withdrew his offer, afraid of being accused of promoting a charlatan, and implied that Fuller was a fraud or self-deluded for pretending to discuss relativity with authority. Fuller was devastated. Though tempted to call the editor and publisher idiots, he suggested almost facetiously that they mail the manuscript to Einstein at Princeton for his appraisal. Then he tried to forget his disappointment.

Some months later Dr. Morris Fishbein, an editor-author-physician acquaintance of Einstein's, phoned Fuller from Manhattan with astonishing news: "Dr. Einstein is coming in from Princeton to spend this weekend with me in New York. He has your typescript and would like to talk to you about it if you are free to meet him here on Sunday night." The editor at Lippincott had taken Fuller's flip remark seriously and had mailed the manuscript to Einstein.

That Sunday evening, Fuller went to Fishbein's Riverside Drive apartment. It was standing room only, with an overflow filling the hall. Einstein sat hidden from sight at the far end of the immense living room, with an adoring group huddled around him. When Fuller got through and was introduced, he "felt as I had never felt of anyone before: there seemed to be an almost mystical aura about him."

They moved to the study, where Einstein said he had read Fuller's manuscript—now on the desk between them—and would recommend that Lippincott publish it. "Young man, you amaze me," Einstein added. "I cannot myself conceive of anything I have done ever having the slightest practical application. I evolved all this in the hope that it might be of use to cosmologists and astrophysicists, people who were thinking of the universe in a big way."

Einstein's imprimatur restored the book to the publisher's list. After that, with his confidence renewed, Fuller went on to even greater glory. He was to invent many marvels, including "the first basic improvement in mobile military shelter in 2,600 years." This accolade came from the U.S. Marine Corps.

After giving Fuller a helping hand, Einstein traveled to Albany, New York, to fulfill a speaking engagement. There, on October 15, preaching what he practiced, he decried those who promoted Darwin's "survival of the fittest" to justify selfish competition between people rather than cooperation.

Soon after he got home, he and Gustav Bucky were granted U.S. Patent No. 2050562 for their automatic camera. Bucky's son, Thomas, calls that camera "the grandaddy of the present-day automatic exposure camera." (One of the myths built up about Einstein was that he was not interested in the practical application of his ideas. Judging by his comment to Buckminster Fuller, he helped to perpetuate the myth. The fact that he was granted this patent should help to dispel that myth.)

That winter, Elsa's illness worsened, and when it was feared that she would not recover, Einstein finally expressed his devotion by spending hours at her bedside, reading and talking to her. She died on the evening of December 20, 1936, during a heavy snowstorm.

Janos Plesch acknowledged that some Berliners disliked Elsa, but he remembered her as

> a loyal and understanding wife who did her utmost to smooth Einstein's path and attend to his physical needs . . . It is no easy task to be the wife of a great man . . . She did him good service as a Cerberus [a role Helen Dukas took over] to save him from the constant molestation to which a great man is subject . . . Famous men are besieged, slandered, insulted, led into traps—and worshipped. There is no trick admirers won't get up to. The Cerberus needs a great deal of

tact, stoicism and even heroism to resist it all. Elsa Einstein performed the task superlatively well.

Banesh Hoffmann agreed with those who said that Einstein did care about his second wife. He and Infeld saw Einstein shortly after Elsa died and found him "utterly ashen and shaken." They suggested quitting work for a week until he felt a little better, but he replied, " 'No, now more than ever I need to work. I have to go on.' That doesn't sound like a man who didn't care about his wife," said Hoffmann. "I never saw him in tears, but he was certainly emotional in that he felt things deeply.

"On the other hand Einstein's friend, Max Born, was shocked because he seemed to take her death lightly, writing to Born something like, 'I think I've lost a leg. It was a little crippled. I limped a bit on that leg, but I still lost a leg.' Yet their friend Antonina Vallentin said that the death of Elsa severed the strongest tie he had with a human being. I believe it was one of the strongest."

Margot, who was very close to her mother and had helped to nurse her through her illness, was desolate. Albert tried to lose himself in his work, but he was often distracted by the situation in the Soviet Union. Through daily doses of *The New York Times* and Howard K. Smith's radio broadcasts, Einstein learned of more trouble brewing for Jews in the Soviet Union, signaled by the Moscow trial of the Jewish political leader Leon Trotsky and other "conspirators." They were accused of attempting to assassinate Stalin and of being in cahoots with Hitler and British and Japanese secret agents intent on destroying the Soviet regime. Trotsky had recently escaped to Norway, where he sought asylum, and Einstein tried to help him with a note to Stalin—but it had no effect.

Einstein was luckier in saving others in deadly peril in Germany. There, after three years of Nazi rule, the malicious campaign against Jews had intensified. Signs over the doors of some stores read, "Jews Not Admitted." Many German towns had posted signs saying, "Jews Strictly Forbidden in This Town," or "Jews Enter This Place at Their Own Risk." At a sharp bend in one road, a sign read, "Drive Carefully! Sharp Curve! Jews 75 Miles an Hour." In some communities Jews couldn't get milk for their infants, and pharmacists refused to sell them medicine.

The Nazi Commissioner of Justice, Hans Frank (who was also Hitler's personal attorney), advocated "doing away with [the Jews] in one way or another . . . We must annihilate the Jews." He left no doubt who would drive the murder machine, declaring on May 20, 1936, "There is in Germany today only one authority and that is the authority of the Fuehrer." Frank was hanged at Nuremberg as a war criminal on October 16, 1945.

Einstein stopped publicly voicing his contempt for Hitler, afraid of harming Jews stranded in Germany and of arousing an anti-Semitic backlash in America. The Führer was idolized not only in the Fatherland; sizable crowds of Americans were enthusiastic and vociferous supporters. Meanwhile, Einstein quietly continued his efforts to help Jews escape from the looming Nazi death trap.

Among those he helped was a friend, Boris Schwarz. They first met in pre-

Hitler days when Boris, a young violin virtuoso, gave a performance in the Einsteins' Berlin apartment. Albert had cried out, "Ah! One can see he loves the violin!" and showed equal ardor by eagerly joining in. They played for hours, switching from Vivaldi to Bach with hardly a pause. Elsa grew concerned that Boris was becoming exhausted and that Albert was too absorbed in the music to notice. She called time out by firmly announcing tea. That evening was the start of a close friendship, many more duets, and occasional trios when Boris's concert-pianist father, Joseph, was free from his engagements. Now, some ten years later, Boris's only hope of surviving was to leave Germany. The Nazis had made that difficult—as they had for all German Jews—by revoking his citizenship and with it his passport.

Einstein and a friend signed a sponsoring affidavit, promising if necessary to support Boris should he be allowed to enter the United States. Unfortunately, it only worked for relatives or, at best, longtime friends. Because Boris had no proof of his friendship with Einstein, American Embassy officials in Berlin took him for just another stranger in distress the softhearted scientist was trying to help.

Boris returned home, afraid he had lost his last chance to escape from Germany. Then he came across a photo, autographed by Einstein, of himself and his father with Einstein during a musical evening—solid evidence that he knew the man. That photo turned out to be his passport to the United States.

A job awaited him in Eugene Ormandy's Philadelphia Symphony Orchestra, again thanks to Einstein. Soon after Boris arrived in the States, his parents were also admitted. The families celebrated their joyful reunion by making music in Einstein's Princeton home.

It had been a rough year. Einstein had lost his wife and his close friend, Marcel Grossmann. The year had also seen the escalating barbaric treatment of Jews in Russia and Germany; the start of the Spanish civil war, with butchery on both sides; and the invasion of Abyssinia by Mussolini's legions, during which a heroic Italian pilot boasted of flinging prisoners from his plane to their deaths.

During the same year, the University of Pennsylvania faculty named Einstein one of the world's ten best teachers. He headed novelist Ludwig Lewisohn's list of the greatest living Jews; Sigmund Freud, Henri Bergson, and Martin Buber followed. Readers of a British newspaper had an even more exalted view of him. Asked to name the world's greatest man, most chose Einstein.

But he was in no mood to bask in his fame or popularity. When "the world's greatest man" was invited to contribute to a time capsule, he sent this message, written on paper guaranteed to last a thousand years:

Dear Posterity!
   If you have not become more just, more peaceful, and generally more rational than we are (or were)—why then, the Devil take you.
   Having, with respect, given utterance to this pious wish,
   I am (or was),
   Your,
   Albert Einstein.

CHAPTER 29

Politics at Home
and Abroad

1937 to 1938

58 to 59 years old

To Sidney Hook, chairman of New York University's philosophy department, the Moscow Treason Trials (also known as the Moscow Show Trials) had been a fiasco. He later wrote, "The charges against the defendants were mind boggling . . . that," under Trotsky's orders, "they had planned the assassination of Stalin," and "had conspired with Fascist powers, notably Hitler's Germany and Imperial Japan, to dismember the Soviet Union . . . Despite the enormity of these offenses, all the defendants in the dock confessed to them with eagerness and at times went beyond the excoriations of the prosecutor in defaming themselves." They were all condemned to death. Feeling sure that the men had been tortured to make such abject confessions, Hook, a former Communist moving steadily to the right, asked Einstein to support an international investigation of the trials.

Einstein feared that a public hearing would be used by Trotsky, Stalin's rival, for propaganda purposes, and recommended a private investigation by intelligent jurists; if they came to a convincing conclusion, he would welcome its public disclosure. Hook disagreed, and early in March, Einstein invited him to Princeton to talk it over.

"I took Benjamin Stolberg with me," said Hook in an interview many years later. "He was a labor journalist who spoke German. As it happened Einstein spoke good, though heavily accented, English. He heard me out without interruption. His objection to our proposed commission was that it would appear one-sided even though I said we would invite the Soviet government to send witnesses and even have someone to question or cross-examine Trotsky. Einstein was unconvinced, although he hated the Communists for making common cause with the Nazis in 1932, and collaborating with them to defeat his party, the Social-Democrats.

"Finally, Einstein said, 'From my point of view both Stalin and Trotsky are political gangsters.'"

Hook replied, "That may be, but in a civilized community it is important to see that even gangsters receive justice."

"He smiled and said, 'You are perfectly right. But I am no policeman.'"

"Both Stolberg and I were very disappointed in Einstein's attitude. But he was very friendly and insisted on walking us to the railroad station. As we walked along he was humorous about some things, but became deadly serious when he mentioned Hitler's persecution of the Jews. He was grim when he said to us: 'If war comes, Hitler will realize the harm he has done Germany by driving out the Jewish scientists.'

"I was very deferential to Einstein; I felt it was a great privilege to be talking to him. But I held my ground. I think he liked me and knew I was a Social-Democrat. I think he would have called himself a Social-Democrat, too; or a Socialist. I certainly wouldn't call him a Communist. He discovered I had once been a fellow traveler and in one of his letters he made an ironic reference to that. But he was inconsistent. In one letter to me he said that if he were in the Soviet Union, in order to make the social gains he would reluctantly accept the dictatorship. And another time he said one should never do that, because if you accept dictatorship you are accepting an evil means to achieve a good end, and that's impossible."

Einstein later explained to Hook that in saying he would reluctantly accept dictatorship in the Soviet Union at that time, he was referring to a painful *temporary* renunciation of personal independence to achieve positive goals.

Einstein's then assistant and later biographer, Banesh Hoffmann, remembered that "in those days dedicated Communists were saying, 'We've got to have these trials,' and, 'They're not a put-up job,' and, 'There's no torture.' Einstein would have nothing to do with that. He condemned the Communists and said that they must have tortured these people on trial. And, of course, he was right. So I don't think you can say, as some still do, that he was a Communist dupe."

However, Hook pointed out that "in a 1938 letter to Max Born, Einstein said he had changed his mind about the Moscow trials and had been persuaded by those 'who know Russia best' that they were authentic and not staged."

Another contemporary Einstein assistant, Valentin Bargmann, recalled: "In my presence there was nobody who persuaded Einstein one way or another. I came from Europe. And it doesn't make sense in the following way: one heard that purges were started and what it implied. However, Communist purges were not specifically against Jews. With the Nazis it was quite different, and anti-Semitism was one of the major planks; therefore the response was quite different. It took quite some time before a more or less clear picture of Russia emerged. And then, I feel, Einstein would not hesitate to support Jews persecuted by the Communists."

Hook thought that Einstein's assistant, Leopold Infeld, was a Communist and that he may have influenced Einstein into believing that the Moscow trials were genuine. But John Stachel, editor of the early volumes of the Einstein papers, doubts it: "I don't think anyone had much influence on Einstein. He was pretty much an inner-directed person. There was no evidence in the letters of any evil genius influencing him."

When Stachel's comment was reported to Hook, he replied, "Einstein was naive about the Soviet Union and in his pacifism. He believed that if you followed Gandhi's nonresistance to evil you could have peace in the world. He thought that what would work against the British would also work against the Germans and the Japanese and the Russians. And that was absolutely naive."

"I wouldn't call him naive," answered Stachel. "I'd call him direct, in that he very often cut through what to others seemed inextricable complications to what he saw as the heart of the matter. That might have struck some people as naive."

Because Hook also thought that Einstein's radical friend, Otto Nathan, had helped to form his political views of Soviet Russia and the Moscow trials, he wrote to Nathan and asked if that was so, but got no reply.

Even today, nearly sixty years later, experts argue over what exactly happened in those Moscow trials. From Einstein's correspondence with Born it is clear that Einstein knew that people were being persecuted in the Soviet Union for working on the theory of relativity, but there is no reference to them being persecuted because they were Jews. One might expect Einstein to be almost as incensed over the persecution of fellow scientists, and he did write about it in indignant terms. On the other hand, he seems to have regarded it as peripheral—a stupid act concerning a few people, fueled by Stalin's suspicion of relativity. He did not regard it as a mass cultural phenomenon.

However, Hook remained skeptical of Einstein's political acumen: "Why should one expect a great scientist to be sophisticated in politics? Have you read Newton's book on the Catholic Church? *The Book of Daniel?* You don't judge Newton by that. And we don't judge Einstein by his remarks on politics, although a lot of people like to cite Einstein as an authority."

Historian Theodore von Laue partly agrees with Hook about Einstein's naivete. Theodore is the son of Einstein's loyal anti-Nazi friend, Max von Laue, and author of *Why Lenin? Why Stalin? Why Gorbachev?* (HarperCollins, 1993), which covered the period Hook and Einstein were arguing about. "Yes, Einstein was naive," von Laue said, "but he had a basic gut reaction in favor of peace. I think that commitment was positive. Tolstoyan and Gandhian pacifists are naive. It's still a conviction that has its effects as a philosophy of life, if not an immediate application to power politics. There's a lot of celebration of Gandhi fifty years after his death.

"Stalin determined to mobilize and industrialize his country at whatever the price to human beings. It was an incredible experiment with unprepared human beings to mobilize that vast country in the shortest possible time. The human beings he worked with were pretty rough ones. He himself was a pretty rough guy. The price was extraordinary, but it would have been higher if the Germans had conquered Russia. So he had to build up the loyalty and cohesion. The danger was widely perceived in 1931 when Stalin said, 'If we don't mobilize, in ten years we'll be wiped out.' Ten years later Germany attacked. Stalin also considered international relations very dangerous generally and worked with Hitler up to a point, trying to gain some security. But Stalin's interpretation of the international situation was one of incredible danger to the persistence and continuity of Russian independence. Some consider it madness that he should have killed off so many of his military officers in the purges. But Stalin was a suspicious person and where would the internal opposition come from? Possibly the military."

Viewed in that light, Einstein's reluctance to join in a public protest seems realistic rather than naive. Max Born confirmed that in 1937 Einstein believed that the threat from Hitler left the Russians no choice but to destroy as many homegrown enemies as possible (so did many in Russia). But Born found it difficult to reconcile this fatalistic attitude with the humanitarian man he knew Einstein to be. Few outside Russia knew then that many of the victims executed or banished to the gulags were entirely innocent of any crime.

This might explain why Einstein had recently written to Vyacheslav Molotov, chairman of the Council of People's Commissars, to get his former assistant, Brooklyn-born Nathan Rosen, a job in the Soviet Union, where Rosen was living at the time. Fortunately, Rosen survived the purges; he eventually settled in Israel.

Yet Einstein's commitment to democracy was clear, as witnessed by his reaction to the Spanish civil war, which had erupted the previous summer. General Franco had led an armed revolt against the democratically elected Republican government, which had tried to curb the stranglehold held on the country by the rigidly conservative troika of the army, the Roman Catholic Church, and the wealthy aristocracy. In what turned out to be a rehearsal for World War II, Hitler and Mussolini supported Franco's forces with arms, warplanes, and troops, while Stalin had contributed weapons and technicians to the Loyalist left-leaning government. With the exception of many Catholics, the British, French, and Americans generally sympathized with the embattled Loyalists, but ensured their defeat by remaining neutral and sustaining an arms embargo against Spain. President Roosevelt told cabinet member Harold Ickes in confidence that he opposed raising the embargo and helping the Loyalists because it would lose him every Catholic vote.

On February 4, the Spanish Embassy in Washington, D.C., released a letter of support from Einstein: "I can assure you how intimately united I feel with the Loyal forces, and with their heroic struggle in this great crisis of your country. But I feel ashamed that the democratic countries have not found the necessary energy to comply with their fraternal duties."

The Spanish civil war was weakening Einstein's staunch pacifist views. He warned the American League against War and Fascism that British pacifists were endangering democracy by delaying their country's rearmament. This had left the British defenseless against the rapidly rearming and bellicose fascist countries: exactly the views of Winston Churchill. Einstein also thought it shortsighted to shrink from involvement in international conflicts. He called isolationists in the democracies selfish and shortsighted, and urged pacifists not to oppose rearmament but instead to create an effective international organization for peace.

Meanwhile, in Nazi Germany, scientists were forbidden to mention Einstein's name in lectures or scholarly papers, although not all of them were following orders. Arnold Sommerfeld told Einstein that he had twice mentioned his name during lectures and each time his students had responded positively. When he announced further lectures on relativity, he said his students had voiced their approval. "You can see from this," he concluded, "that you have not been expatriated from the German lecture halls."

Einstein's most immediate concern was office politics, however. He and his colleagues feared that their director, Abraham Flexner, was risking the financial solvency of the institute and with it their professional futures. Flexner had also given Einstein's assistant, Leopold Infeld, a hard time, allowing him only half the usual grant for 1936—a mere six hundred dollars. When Einstein requested the same amount for Infeld in 1937, it was refused. He told his assistant: "I tried my best. I told them [presumably the board of trustees] how good you are, and that we are doing important scientific work together. But they argued they don't have enough money. [In fact, ample funds were available.] I don't know how far their arguments are true. I used very strong words which I have never used before. I told them that in my opinion they were doing an unjust thing. Not one of them helped me."

Infeld was desperate. He was reluctant to return to Poland—there was no work for him there—but he declined to accept Einstein's offer to give him the money. Then Infeld suggested that he and Einstein raise the money by coauthoring a book for the general reader to be called *The Evolution of Physics*. Einstein enthusiastically agreed. Publisher Max Schuster accepted their proposal and arranged to meet the authors at Einstein's home on April 20, 1937.

Schuster took his partner, Dick Simon, along; feeling that they needed intellectual ballast to talk with Einstein, they asked writer Max Eastman to join them. Eastman had recently published an article on Marxism in *Harper's* magazine, and he and Einstein were soon discussing the subject, sitting at a small table on the porch. Eastman maintained that Marxism pretended to be scientific, but was really a religion. "Marx," he said, "declared that the world is made of matter, but proceeded to discover, mysteriously enough, that this world of matter was achieving with dialectic necessity exactly what he wanted to achieve. The whole thing was just a gigantic effort to prove that the external world is in favor of the proletarian revolution and is helping it." When Eastman added that the American philosopher John Dewey also "gets the objective facts into harmony with man's will by putting his will into the very process of determining the facts," Einstein, laughing "mischievously," looked around the table and said, "This man is wicked! He is really wicked!"

Schuster felt so much at ease with Einstein that he "even had the audacity, the nerve to tell him a few jokes, which he loved. When we got around to talking about the evolution of physics we had to deal with the speed of light. This is the first step in any book of physics. And in a rather informal and facetious way I told him that I had discovered something that's even faster than the speed of light. And he said, 'What is that?' 'The speed with which a woman arriving in Paris goes shopping.' He thought that was funny. Not only was he a good listener, but a very compassionate person, and a kind man. When we were about to leave I said, 'Is there anything we can do to help so you can confine yourself to teaching and to mathematics and physics?' And he said, 'No, I can't think of anything. Everything has been done and I am very happy.' As we were leaving he said, 'There is one thing you can do. As you drive home you will pass a gasoline station a few miles from here. I want to make sure that you give that gasoline attendant at the station a copy of my book.' He wasn't concerned with the bigshots who set the whole world at his feet, academically and scientifically. He loved the little man in the garage, the farmer."

Five days later, Albert dined with his Princeton friends, the Blackwoods, along with fellow guests Dr. and Mrs. Emil Brunner. Brunner, a visiting professor from Germany who was widely regarded as Europe's leading theologian after Karl Barth, was teaching at the nearby seminary. During dinner he and Einstein swapped stories of their fumbling attempts to master the English language. Dr. Brunner said that he and his wife spoke English at home to improve their vocabulary. One morning he came down to breakfast and told his wife, "This morning I woke up and found a spinster crawling across my bed." He had meant, of course, to say spider. Einstein loved the joke.

("No wonder," said Einstein expert Dr. Robert Schulmann. "It amuses Germans always when they hear the word 'spinster.' It sounds like 'spinne' to them, which is German for spider.")

Pursuing the book project, Einstein and Eastman met again the following June, this time in Infeld's apartment. Einstein came in looking refreshingly casual, "dressed like a student in an open shirt, an old brown-leather windbreaker, and tennis shoes with no socks."

During their conversation, Einstein revealed himself as anything but the political naif many considered him to be. On the contrary, he correctly predicted that if Hitler was victorious in his ambition to dominate Europe, America would be drawn into the war. He also predicted, with equal accuracy, an alliance between Stalin and Hitler. Of the " 'confessions' of the old Bolsheviks in the Moscow trials then in progress, he said: 'Of course they are not true. It is impossible that twenty men being caught in a conspiracy, would all react in the same way — and that in so unnatural a way as to defile themselves publicly.' "

Einstein agreed with Eastman that the Spanish civil war might have been averted had French premier Léon Blum sent the Spanish government the requested arms for defense, but added, "You must remember that Blum had a difficult situation in his own country. More than half the French army were fascists. He wasn't wise enough to do it, but he was also not strong enough to do it had he been wise." Knowing Einstein to have been "an ultra-pacifist and anti-militarist," Eastman was surprised but pleased to find him so realistic and flexible as to advocate that the democracies should stop fascism by force of arms.

Discussing Freud, Einstein said, "I had a dream once that seemed in a small way to verify one of Freud's theories. At Berlin we had a professor named Rüde whom I hated and he hated me. I heard one morning that he had died, and meeting a group of my colleagues I told them the news this way: 'They say every man does one good deed in his lifetime, and Rüde is no exception—he has died!'" That same night, Einstein told Eastman, he dreamt that he was in a lecture hall when Professor Rüde entered, looking extremely healthy and pompous. In the dream Einstein hurried to him, and shook his hand in a friendly fashion, saying, "I am so glad you are alive!" Einstein assumed that Freud's analysis of the dream would be that he was ashamed of his bitter remark about Rüde, and that the dream had relieved him of his feeling of guilt.

On a sunny afternoon later in that summer of 1937, Einstein was relaxing in his small rented cottage at Peconic, Long Island, when a yacht approached. On board were Oscar-winning actress Luise Rainer, her husband, playwright Clifford Odets, and several friends. They hoped to persuade Einstein to aid a group of recent European refugees—a cause he could never refuse. According to Odets's biographer, Margaret Brennan-Gibson, "the yacht was too large to dock at the tiny pier. Promptly, his white hair flying in the wind and a radiantly innocent smile lighting his face, Einstein arrived in order to debark the passengers in a rowboat. When it came to Luise and Odets' turn, it was evident to Odets, from the fact that he playfully pulled her hair, that Einstein found Luise attractive. She, flustered, capsized the boat and 'almost drowned the great scientist.'" Several photographs were taken of the group after that slapstick episode. When they were developed, Odets, in a jealous fury, grabbed a pair of scissors and decapitated the image of Einstein in one of them. Fortunately, other photos escaped his anger, including two showing Rainer alone with Einstein.

When Bohr stopped at Princeton on an around-the-world trip with his wife and their son, Hans, the two loving friends again engaged in intellectual combat. Many expected Bohr to finally persuade Einstein to accept quantum physics unreservedly. In fact, during their debate Bohr did most of the speaking, but he could hardly be heard beyond the first row. When Einstein responded, Bohr broke in with arguments which seemed to be supported by most of those who heard him. To Infeld, in the audience, the encounter between Einstein and Bohr was like a soccer match between Poland and West Germany, held in Warsaw. When the game was over, however, neither man had let a ball through his net.

Valentin Bargmann had a different reaction: "The discussion on quantum mechanics was not at all heated. But to the outside observer, Einstein and Bohr were talking past each other. Such a discussion, which involves the foundations of physics, would in fact need days and days, because one has to go into details, and definitions of all the concepts one wants to use. Now this was not done during the discussion I witnessed. So many things were left unsaid." Taking a larger view, John Wheeler saw the continuing debate between the two "as the greatest debate in intellectual history that I know about. In thirty years, I never heard of a debate between two greater men over a longer period of time on a deeper issue with deeper consequences for understanding this strange world of ours."

At this time, both of Einstein's sons were living in Zurich. Eduard, at twenty-seven, was in the full-time care of a male psychiatric nurse; the pleasant-featured youngster had grown into a heavy, morose schizophrenic. Michele Besso called on him at Mileva's apartment in one of the three houses in Zurich that she had bought with the Nobel Prize money, and was impressed by how well Eduard played Handel and Bach on the piano but disappointed to discover that Eduard had not left the house for a year. "Einstein's son entertained him with brilliant discourses on psychology," wrote Einstein biographers Roger Highfield and Paul Carter, "but the words came slowly from his lips, like notes from an old organ one had to play with one's fists."

Those who accepted the popular image of Einstein as a loner with few or no close emotional attachments, as the epitome of the detached intellectual, would have been surprised when less than a year after Elsa's death he invited his eldest son, Hans Albert, to be his guest on a three-month visit. Eduard couldn't join them in America even had he wanted to, as the law prevented mental patients from entering the country.

Hans Albert, now thirty-three, accepted the invitation, with an eye to the possibility of emigrating to the United States with his wife and two sons. They, in the meantime, were to remain behind in Zurich. Einstein met his son at the Manhattan pier on October 12, 1937; he fobbed off reporters' requests for interviews in his usual affable manner, and then the two men were driven away in Leon Watters's limousine.

Einstein began 1938 discouraged. He had started on his scientific quest as a skeptical empiricist, not unlike Mach, but his discoveries about gravitation had converted him into a rationalist, searching for the truth in mathematical simplicity. Now, after over twenty years of futile attempts to uncover the basic secrets of electricity, he was convinced that a new inspiration was needed: not the statistical approach of Bohr and Heisenberg, whose quantum physics he still regarded as an incomplete stopgap and, at times, derisively, as "mysticism and probability," but an entirely new approach.

Most scientists saw his stubborn pursuit of an objective reality as the outdated prejudices of an old fool. He believed they were the deluded ones, willing to call a half truth the final truth, and he pressed on almost single-handed with what would always be an irresistible challenge—to describe more and more precisely the ultimate nature of matter.

Hans Albert's visit was a success, both personally and professionally. He got on well with the father he had infuriated as a teenager, now finding him cordial, friendly, and generous. He found him surprising, too: here was a man famed as a theorist who liked nothing better than solving practical puzzles and discussing the latest inventions. Hans Albert speculated that they reminded Einstein of his happy, carefree days at the Patent Office in Bern before the horrors of what became known as the First World War. With a job offer as a researcher with the

Department of Agriculture in South Carolina, Hans Albert sent for his wife and two young sons, Bernhard and Klaus.

In the spring, Einstein wrote to Maurice Solovine: "I work earnestly always, supported by a few courageous colleagues [among them Infeld and Hoffmann]. I can still think, but my capacity for work has slackened. And then: to be dead is not so bad after all."

But he put aside thoughts of his own mortality and "our wretched times [in which] not one enlightened man is left" to enter the imaginary and controversial world of two novelists, enjoying the latest output of his friend Upton Sinclair, *The Flivver-King*, and trying to keep another novelist out of prison.

This novelist had asked Einstein to judge if his book was obscene. If so, he promised to burn it. Copies of the novel had been confiscated and the author charged with selling an obscene pamphlet. Einstein replied that he was handicapped by his unfamiliarity with the American idiom and could only sense the thrust of the writer's voice—which was definitely not pornographic. When the young man came up for trial, however, he was found guilty. One judge who knew nothing of Einstein's view said, according to the writer, "I have just read the first seven pages of the book and I think that anyone who would write that must be insane." Another said, "We don't need any more of this radical literature. The defendant deserves a prison sentence."

To see if he "deserved" prison or, perhaps, psychiatric treatment, he was sent to Bellevue's mental hospital for seven days' observation. The first three days in Bellevue were a nightmare, he told Einstein, but during the last four, "I was able to find peace for I applied Spinoza's renunciation of life to practise . . . I almost completely stifled my emotions." When the novelist was told he would be taken from Bellevue to face the judge who would determine his future, he asked for and got permission to use Einstein's supportive letters in his defense.

"Your letters interested the judge," the novelist later wrote to Einstein, thanking him for his help. "At the mention of your name his eyebrows arched and he was so angry that I had been sent to Bellevue he said, 'Ignorance! Ignorance! Ignorance!'" The case was closed, and the novelist set free.

On April 19, 1938, Einstein addressed three thousand people at a Passover Seder service in New York's Hotel Astor. Speaking in German, he welcomed the development of Palestine as a Jewish homeland and refuge, but feared the British plan to divide the country between Arabs and Jews might produce "narrow nationalism within our ranks, against which we have had to fight vigorously even without our own State. We are no longer the Jews of the Maccabee period. A return to a nation politically would mean turning away from the spiritualization of our community which we owe to the genius of our prophets. Not since Titus conquered Jerusalem have Jews suffered such oppression, but we shall survive this period, too, no matter how much sorrow and loss of life it may bring." Obviously referring to Nazi Germany, he said, "A tyranny based on anti-Semitism and maintained by terror must inevitably perish from self-generated poison."

To discredit him, his enemies in Germany spread the false rumor that he was a Communist who had attended a party congress in Russia. When it was quoted as fact in an American magazine article, Einstein told Siegmund Livingstone of the Anti-Defamation League that he was not a Communist and had never been to Russia nor to any Communist Party congress. He asked Livingstone if there was any way to stop such lies being printed without a libel suit, mentioning another one recently published, a claim by the Rosicrucian Order that he was a member of their secret occult society. Apparently a libel suit was the only answer, and Einstein had neither the time nor the inclination to go to court.

Some brave spirits in Nazi Germany still acknowledged Einstein's achievements, though Heisenberg was no longer among them. When SS leader Heinrich Himmler advised him that if he must discuss relativity in physics classes he must keep Einstein's name out of it, Heisenberg accepted this restriction. But aerodynamics expert Ludwig Prandtl of Göttingen didn't. He told Himmler that "relativity was regarded as physically correct by the overwhelming majority of physicists, [that] Einstein was a first-class physicist, and experimentalists who could not follow theoretical work should not therefore reject it as simply worthless and slander its representatives."

In the summer of 1938, Einstein returned to the cottage at Nassau Point, Peconic, on Long Island's eastern shore, and asked Leon Watters to join him. Watters drove out one Sunday morning in August with a portable phonograph and a set of records of Tchaikovsky's music, which he played at an al fresco lunch. The Buckys were also there, with Margot and Dukas. After lunch, Peter Bucky cajoled Einstein and Watters into a pipe-smoking "duel" and filmed them trying to outdo each other in smoke production.

When they stopped coughing, Einstein told Watters that he had been helping people escape from Nazi Germany to the States—most recently a cousin, Heinz Moos—by signing affidavits and each time putting two thousand dollars into an account to guarantee they would not become public charges. Although he was running out of ready money, he had recently agreed to vouch for the fiancé of a relative (Alice Kohn). He asked for Watters's assistance, assuring him that he would not be obliged to support the individual legally or morally. If a problem did arise, Einstein promised he would provide financial support. Watters agreed to help, and eventually would sign several affidavits on Einstein's behalf.

During the summer Einstein enjoyed two more books by Upton Sinclair; he wrote to him to say that he rated *Our Lady* as good as Anatole France at his best and agreed with the sentiments in *Letters to a Millionaire*. He also predicted that as a tireless fighter for truth and justice, his friend would continue to be scorned and defamed while alive and canonized after his death. It was not far from the truth.

In the late fall, the Nazis rounded up thousands of Polish Jews living in Germany and dumped them across the Polish border. Among them that bitterly cold November, huddled together in empty railroad cars or abandoned barracks, was the Grynszpan family, whose seventeen-year-old-son, Herschel, had found refuge in Paris. Distraught at the news of his family's suffering, he determined to kill the German ambassador in protest; instead, he fatally shot Ernst von Rath, a minor official representing the ambassador who, ironically, was under investigation by the Gestapo, suspected of sympathizing with the Jews.

The next day, Hitler demanded revenge for the assassination, and on the night of November 9, Nazis throughout Germany and Austria and the recently annexed Sudetenland went on a deadly rampage against any Jew they could find. Nazi propaganda minister Joseph Goebbels reported that the well-planned and coordinated pogroms, which were in fact ordered on teletype by the deputy Gestapo chief, Reinhard Heydrich, were "spontaneous anti-Jewish demonstrations."

An estimated one thousand Jews were killed or committed suicide, and Nazi gangs vandalized or destroyed 1,118 synagogues and ransacked Jewish hospitals, orphanages, and homes for the elderly. In Efurt, near Weimar, Jews were lined up and made to sing the "Horst-Wessel Song," a Nazi favorite, and to scream "Death to the Jews." Those who refused were savagely beaten with whips. Then they were all loaded into trucks and buses and taken to the Buchenwald concentration camp. After *Kristallnacht* (Crystal Night), as it was called because of the countless broken windows, more than thirty thousand Jews were imprisoned in various concentration camps or deported to Poland.

The Nazis' barbarity stunned the civilized world, as well as German insurance companies faced with the cost of replacing the broken glass. Herman Göring sympathized with the insurance brokers and complained to Heydrich, "I wish you had killed 200 Jews instead of destroying so many valuables." The Nazis solved the problem by fining Jews $400 million to pay for the damage, saying it was a penalty for the assassination.

Soon after, Einstein faced a personal tragedy: his six-year-old grandson, Klaus, died of diphtheria. He wrote to Hans Albert and his wife, saying that "the deepest sorrow that loving parents can experience has come to you," and went to South Carolina to comfort them.

Back in Princeton, although ill and overworked that winter, Einstein took the time to gather manuscripts and books from his shelves to be auctioned to aid German Jews and continued in his efforts to help them escape.

During the Christmas season, he came to the door to greet carol singers. He was sockless as usual and in shirtsleeves. The ever-vigilant Dukas followed him and put a coat over his shoulders. When they finished singing, Einstein shook hands with all of the carolers.

Meanwhile, in Germany, chemists Otto Hahn and Fritz Strassmann had just got astonishing results: by bombarding uranium with neutrons they had produced barium — an element half the mass of uranium. Hahn immediately wrote to share the extraordinary news with his former colleague, Lise Meitner. Meitner, who was Jewish, had helped to make this discovery before escaping from the Nazis, with Hahn's help, to the safety of Sweden. There, together with her nephew, physicist Otto Frisch, she now worked out that Hahn and Strassmann had done what many thought impossible — split the atom. Researchers in several countries subsequently obtained identical results and came to the same conclusion.

As soon as German physicist, Paul Harteck, heard about it, he alerted the German War Office to the "newest development in nuclear physics," which was probably capable of producing an "explosive many orders of magnitude more powerful that the conventional ones . . . That country which first makes use of it has an unsurpassable advantage over the others."

Goebbels was elated. But not Hahn, one of the few Germans Einstein would always respect. When Hahn realized the awesome possibilities, he considered throwing all his uranium into the ocean—and killing himself.

CHAPTER 30

World War II
and the Threat of Fission

1939 to 1940
———
60 to 61 years old

When future assistant Valentin Bargmann, a refugee scientist from Germany, first met Albert Einstein in 1937, he was "fully conscious that [he] was standing before one of the greatest scientists of all time." Nevertheless, Einstein was so friendly at that first meeting that after a while, said Bargmann, "I was participating in the discussions as if he was an old friend. I thought that remarkable. He had an office in Fine Hall—the institute was then on the campus of Princeton University—and Peter Bergmann [another of Einstein's assistants] and I were free to use it whenever we wanted."

They all spoke German because, although Einstein's English was fairly good, his pronunciation was miserable, according to Bargmann. "He would make mistakes, too, like 'I tink,' for 'I think.' Niels Bohr was worse because you couldn't hear him. He spoke too low, too quiet. Discussions were very open, with no pretending of authority. Einstein treated us as equals. This was the charm of Einstein, no matter who talked.

"I didn't see any aloofness or clumsiness, [which] Infeld noticed, though I, too, was aware of his loud laugh and brilliant eyes," Bargmann recalls. "But I can't understand Einstein's comment to Banesh Hoffmann when he pointed to an etching of Isaac Newton and said, 'Most great men are between the sexes.' I think Einstein maybe was trying to be funny because maybe Newton looked effeminate wearing his wig." (Some Newton biographers call him a celibate, others a latent homosexual.) Infeld said that Einstein became stranger and stranger with time, but Bargmann disagrees: "I couldn't always predict what he would say or think, but it was never a big shock or surprise."

What Bargmann found especially interesting about Einstein "was that he was enormously relaxed." But he claims that "these explanations of why he went

311

without socks, let his hair grow long, wore casual or even sloppy clothes, these explanations you read, that's not the Einstein I knew. One has to understand very often he said things as jokes and many people took these joking remarks seriously.

"I never saw him upset or agitated. This is very interesting because Einstein himself told me that as a young boy he had fits of temper, and this I couldn't imagine.

"When I knew Einstein at Princeton he wasn't teaching, so he didn't have any students. I worked with him on the unified field theory and his general theory of relativity. And if I disagreed with what he said he tried to convince me he was right."

After Mussolini adopted some of Hitler's anti-Semitic laws in Italy, Einstein's sister, Maja, who was living in Florence, accepted his invitation to join him in Princeton. Her non-Jewish husband, Paul Winteler, chose to live with his brother-in-law, Michele Besso, in Geneva, and Maja hoped to return to her husband when life in Europe was less threatening. Einstein's caring and affectionate treatment of his sister, especially when she later became fatally ill, belies the picture of him by other biographers of a man incapable of a close emotional bond with anyone.

Maja was soon enjoying weekend jaunts with Einstein in the Buckys' car. One springtime trip took them to Newark Airport in New Jersey, where a publicity man quickly joined the party. When Einstein questioned him about a radio altimeter on display, the man offered a pamphlet. "No, no," Einstein said, ignoring the proffered pamphlet, "I want *you* to explain it." His explanation got as far as, "Say the plane is going at 190 miles an hour"; then he paused for such a long time that Einstein jumped in, suggesting X should represent the sun's motion, and continued, "You say the airplane is going 190 miles an hour. When you consider C in relation to . . . " Einstein then nodded, encouraging the public relations man to finish the sentence. Completely bewildered, the publicist gasped, "Professor, you'll just have to take a pamphlet." Afterward he laughed about it, saying he knew that Einstein had been kidding by using algebra.

Another math problem from a young man awaited Einstein at home, among the mail piled on his desk. As he later explained to Gustav Bucky, it concerned "an interesting mathematical proposition which is undoubtedly correct and original. He conceived the proposition emotionally but failed in its mathematical proof. I have corrected the calculation, but in order not to compete with the young man as to the priority of the correct calculation I wrote him that the proposition is true but that my correction of his faulty calculation would be at his disposal at any time he cared to have it. In such a way he retains the unrestricted priority of the idea and [has the chance] to find the correct solution himself."

Bucky loved Einstein and marveled at his "delicacy of thought and feeling in leaving it to the young man whether he wanted to try again and to spare him the sense of defeat." Bucky surmised that the average man, had he been able and willing to help, would simply have mailed back the correct calculation and assumed he had done the right thing. "Einstein's mind is always preoccupied by his own most intricate problems, yet he pushes away his personal affairs and interrupts his subtle trains of thoughts to engage himself deeply in problems of others. He always does this without considering the man's standing or even asking who he is. I cannot see how he can spare the time to go deeply into problems presented to him in the mail and dealing with mathematics, social, legal, and political affairs, philosophical discussions . . . besides the manifold requests for help which are sometimes ridiculous or even impudent. A needy person according to Einstein has every right to occupy his valuable time. How he can crowd his work, his reading and his help for others into a single day is a mystery."

His secretary, Helen Dukas, and friend, Otto Nathan, helped, especially in writing the thousands of replies to strangers, but the thought behind them was always Einstein's. A professor who knew Nathan well recalled: "Nathan was an economist who had briefly encountered Einstein at a formal occasion in Germany. They met again in the early 1930s when Nathan was teaching economics at Princeton. It was a godsend to Einstein, who was very lonely. Everyone was being very pleasant to him but there were no old-German-socialist types around the place. Nathan, a bachelor, was filled with moral indignation against capitalism and both men commiserated with one another about Hitler and the terrible things happening in Germany. This established a strong bond of friendship on Einstein's part. Princeton did not keep Nathan, who then went to New York University. Not long after that New York University was not especially eager to keep him, and Einstein sent a tremendously glowing letter completely on his own initiative, which . . . saved that job. It was a great solace for him to have a friendship of that sort."

Perhaps because he missed his old friends so much, Einstein was eager to make new ones. Or so it seemed to Henry Abrams. The twenty-eight-year-old physician, only two years out of medical school, had few patients and little hope that things would improve. Then Dr. Conway Hiden, his next-door neighbor on Princeton's main street, asked Abrams for a favor. Hiden, who was the Einsteins' family doctor, was both a surgeon and in general practice, but now had decided to concentrate on surgery. He asked Abrams if he would take over for him. Abrams was overwhelmed and delighted.

His first visit as the Einsteins' doctor, however, was to Helen Dukas, bedridden with a cold. He found her in the front bedroom "in the highest bed I had ever seen, and shivering under a big feather-filled quilt. Suddenly she says to me in a German accent, 'Henerry, are you reddy?' With that she instantaneously pulled the quilt off. And she was lying there completely nude. Apparently that's the way doctors examine patients in Europe. I was so embarrassed! I thought she'd wear pajamas at least. And I immediately put [he chuckled] two towels over her. Then I examined her. It was just a bad cold and it cleared up very quickly. I got to love Miss Dukas. She was a lovely person and so capable. Ran everything in the house, did all the book work, and everything so perfectly."

Soon after that first visit, Dukas phoned Dr. Abrams and asked, "Would I have the time: the professor would like to sit and talk with me? So I came to the home and we sat in the small front room and talked about general subjects. And then, even though I felt I was in heaven sitting there with the professor, I didn't want to overstay or take advantage of a good thing. So after about twenty minutes I looked at my watch and said, 'Professor, would you excuse me? I have to go to the hospital to see a couple of patients.'"

From time to time, Dr. Abrams would get a call from Helen Dukas to say the professor wanted to sit and talk with him. "I was the luckiest man in the world," he recalled. "Each time, after about fifteen or twenty minutes, I'd say I had to leave to see patients. He didn't make any motion of wanting me to leave but I didn't want to wait until that would occur.

"Even though I was the family physician and close to the professor, he depended on New York doctors for his medical care. I just treated him for minor things. He was in fair health. He wasn't supposed to smoke, but he did.

"Once we sat and talked and he said, 'Of course, you know, doctors generally can't cure everything and don't know everything. Maybe that is helpful as far as population control is concerned. If doctors were to preserve every sick individual, maybe the world would be overrun with a tremendous population crisis.'"

In February 1939, an Irish-American merchant seaman whose home port was Savannah, Georgia, had typed a letter to Einstein while at sea in a storm, responding to an article by Einstein in *Collier's Magazine* (November 26, 1938): "I have just finished reading what I would like to humbly say, was the most comforting and far-reaching article for the unjustly oppressed that I ever read." He wished there were more such writings on how to fight anti-Semitism "from a man to man point of view, and not in the generally impassioned ways gathered from magazines, papers etc. . . . The tales from Nazi Germany and Italy (just to mention a couple) make you feel like a bayonet's splintering your spine . . . It's getting rough and I am in the foc'csle [*sic*] of a ship with crew quarters right up in the bow and can't write all that I w'd like to because of the sea. I am a sailor and see several pictures of you sailing your skiff."

The self-described "rough-neck sailor" had "picketed German and Italian ships and consulates with signs and shouts and more than once had to resort to fists in upholding this minority, which is not my nationality. I am Irish-American." He said he had wasted a day's pay on a disappointing *History of the Jews* by Ernest Renan—"I think you will agree if you read it"—and hoped Einstein would remedy the situation by writing a book on the same subject. He ended with, "It would be mighty nice to hear from you," signed off "Another friend," and mailed the letter to Einstein enclosed in Renan's book.

The sailor got a reply four months later when Einstein explained that he had wondered who sent the book, and had only just found the letter inside it. He was pleased with the sailor's interest in the fate of the Jews and the fight against the "political criminals ruling over unhappy countries." His hope for better times to come, he said, was founded not on statesmen but on the future existence of upright men like the sailor who longed for justice and would not compromise their ideals. Einstein said he intended to read the book and added in a postscript that he was happy to be sailing every day in his little boat.

In Germany, Max von Laue risked Hitler's wrath by sending Einstein a somewhat optimistic sixtieth-birthday greeting: "Now indeed you are safe, and beyond the reach of . . . hatred. As I know you, you have come to terms with it within and you stand *above* your fate. But, more than ever, your work is and remains beyond the reach of passion of any kind, and will endure as long as there exists a civilized community on earth."

Hitler, of course, was about to attempt to destroy many such communities. As for Einstein being beyond the reach of hatred — that would never happen.

Even now, Nazis on the lecture circuit in America were trying to discredit Einstein by branding him a swindler. Dr. Hans Winterkorn, for one, in addressing the University of Missouri's Faculty Club on "Germany—Past and Present," stated that the Weimar Republic under President Friedrich Ebert, which Einstein enthusiastically supported, had been governed by criminals. Asked for specifics, Winterkorn said that in 1919 a Berlin official had tried to get ten thousand dollars from a visiting East Prussian official under the pretext of it being used to publish a book by Einstein, who was a party to the scheme of swindling the government. Asked for his source for this story, Winterkorn said it appeared in a book, *Heimkehr*, by Gustav Winning, whom he knew to be reliable.

The charge prompted an agitated faculty member at the lecture to write to Einstein: "Doubtless much worse libels than this one have, for obvious reasons, been circulated in Germany about you. They are, of course, beneath your notice. But the story, told before a group of our Faculty here, will have serious repercussions if allowed to stand entirely uncontested. I earnestly request, therefore, that you throw what light you can upon the origin of this ridiculous fabrication."

Einstein replied on March 23, the day after he broadcast an appeal to help Jewish refugees and attended a Palestine Symphony Orchestra benefit concert: "None of the stories Mr. Winterkorn is telling in professional circles is true. Obviously they are from a Nazi emissary. Such people are your enemy. It is not true that there was corruption in the Republic of Germany. I had a very good ongoing view of the situation and believe all the government officials were honest. It would be very nice if you would tell your colleagues this."

Einstein spent much of his time now helping friends, relatives, and strangers who were escaping from Nazi-occupied territory. When his friend Felix Ehrenhaft, professor of physics at the University of Vienna, was driven out of Austria and "forced to abandon the great electromagnet whose construction had been the light of his life," he contacted Einstein for both practical and financial help. Einstein gave him both.

Freud sent Einstein a copy of his latest book before leaving Nazi-occupied Vienna for refuge in England. Einstein replied on May 4, 1939, with a fan letter:

> Your idea that Moses was a distinguished Egyptian and a member of the priestly caste has much to be said for it, also what you say about the ritual of circumcision . . . I do not know any contemporary who has presented his subject in the German language in such a masterly fashion. I have always regretted that for a nonexpert who had no experience with patients, it is hardly possible to form a judgment about the finality of conclusions in your writings. But after all this is so

with scientific achievements. One must be glad when one is able to grasp the structure of the thoughts expressed.

Einstein soon switched his thinking from Moses to Otto Hahn, whose atom-splitting discovery in Berlin just before Christmas of 1938 might help Hitler blow up the world in his attempts to dominate it.

Eugene Wigner remembered how "the news about nuclear fission had been brought to us in Princeton by Niels Bohr. We heard also that the Germans forbade the exportation of uranium from Czechoslovakia [then under Hitler's control]. And this alarmed us, as every physicist who knew about the process knew that there is a danger that a bomb by nuclear chain reaction can be created. In Chicago we received a cable from Fritz Houtermans, an Austrian who was in Switzerland in connection with German work on uranium fission. And the cable read: 'Hurry up! We are on the track!' He was fundamentally opposed to Hitler and felt it is dangerous for freedom if they succeed to make the atomic bomb before the United States. He tried to say we should hurry up and get it before they did."

Leo Szilard, a brilliant Hungarian Jew, who had worked with Einstein in the 1920s, was obsessed with the idea of such a deadly weapon in the hands of the Germans, and could talk of nothing else. He had predicted the possibility of a chain reaction for peaceful purposes almost five years before Hahn discovered the fission of uranium by neutrons. Szilard and his friend, Wigner, "feared that if the Germans got hold of the uranium deposits in the Belgian Congo they could use it to make an atomic bomb." Wigner did not take that problem as seriously as Szilard did, because he knew that the Germans had enough uranium from Czechoslovakia to make the atomic bomb. But both wondered how they could warn the Belgian government of the threat, in order to prevent the Germans from getting more. They knew that Einstein was a friend of Elizabeth, now the Belgian queen mother, and thought a warning letter from him to her might help.

"It was July [1939] and I knew Einstein was at a summer resort on Long Island," said Wigner. "I had never been there before. All we knew was that Einstein was staying in a cabin owned by a Dr. Moore at Peconic, Long Island. I drove Szilard in my car and when we reached Peconic, [we] asked people how to get to Dr. Moore's cabin. We asked several people and none knew. After half an hour we decided to give up and return to Princeton. Then we saw a boy, and instead of asking for Dr. Moore's cabin, Szilard asked, 'Where does Einstein stay?' The boy knew at once and we went to that address and found him."

Einstein didn't express any surprise at seeing the two men and invited them inside. There, said Wigner, "We told Einstein about the process which was discovered in Germany as well as the possibility of using it to set off an explosion. I don't think, as has been reported, that Einstein said, as if he had not known about the process, 'That never occurred to me.'" Most physicists, Wigner noted, realized that a chain reaction leading to an explosion was possible. "But Einstein was deeply involved in his own work," Wigner explained, "and unlikely to have been following the latest developments in physics. The papers he got, like *Nature*, he often left unread unless there was something of specific interest to him. It is not possible to read all the journals one receives: I receive eighty-three.

"Although Einstein did not foresee that nuclear energy would be released in his time, he thought it was scientifically possible. He did not know about the discovery of fission [splitting the atom], but he learned. We explained to him the fission process and he understood it in fifteen minutes. I was very impressed that he realized in such a short time the problem of the atomic bomb as a possibility. And he was very much aware of the political problem."

Szilard's account agrees with Wigner's: "[Einstein] was very quick to see the implications and perfectly willing to do anything that needed to be done. He was willing to assume the responsibility for sounding the alarm even though it was quite possible that the alarm might prove to be a false alarm. The one thing most scientists are really afraid of is to make fools of themselves. Einstein was free from such a fear and this above all is what made his position unique on this occasion."

"So," Wigner recalled, "Einstein took our warning as a possibility that there [was a] danger that the Germans [could] conquer the earth. Of course, the Germans could have conquered the earth without chain reaction."

Einstein was reluctant to write about such a matter to the Belgian queen mother; instead, he proposed sending the letter to a Belgian government official whom he knew. Wigner pointed out that it was wrong to contact a member of a foreign government without the knowledge of the U.S. State Department. They finally agreed that Einstein would send a copy of the letter with a covering note to the State Department, warning that if no reply came within two weeks the letter would be mailed to the Belgian official. Then, explained Wigner, "after [Einstein] probably asked us seven or eight questions, he dictated a letter to be sent to the Belgian ambassador."

The letter stated that it seemed "highly likely . . . that explosive bombs of unimaginable power could be constructed out of uranium . . . whose chief source was the Belgian Congo . . . It seemed necessary to take precautions to keep stocks out of the hands of potential enemies." There was little doubt who the enemy was likely to be, and if the ambassador didn't understand who the enemy was, the letter concluded: "Germany, which had offered uranium for sale after taking over a mine in Czechoslovakia, was no longer allowing exports of the material [uranium]."

According to Wigner, Einstein "didn't sign it [then] because it was in German and handwritten."

The covering letter was worded to the effect that a potential danger had arisen due to a new development in physics: would the State Department like to receive information on the subject and to warn the Belgium government? Or should Einstein himself inform the Belgians through their ambassador?

"I took the letter [to the Belgian ambassador] back to Princeton," Wigner continued, "had it translated into English and had it typed. Somebody then took it back to [Einstein]" while Wigner was on his way to California for a vacation.

The somebody was Szilard, who in the meantime wrote to Einstein that he had discussed the matter with a Roosevelt intimate, Alexander Sachs, an economist with scientific leanings. Sachs had advised Teller that it would be best to give the information to the president himself, and offered to hand-deliver Einstein's letter to Roosevelt in the White House. Unable to reach Wigner, now en route to California, Szilard, who couldn't drive, persuaded Edward Teller to chauffeur him on a second trip to Einstein on Long Island.

Szilard and Einstein together composed the new letter—addressed now to FDR. Teller, who wrote it down as they dictated, recalled Einstein saying, "Yes, yes, this would be the first time that man releases nuclear energy in a direct form rather than indirectly." Historian Richard Rhodes explains that by this, Einstein meant directly from fission, "rather than indirectly from the sun, where a different nuclear reaction produces the copious radiation that reaches the earth as sunlight."

On August 2, Szilard sent Einstein a long letter and a shorter version, each expressing their views and each addressed to Roosevelt. Einstein signed both, but said he preferred the longer one. He advised Szilard to be bold and get moving.

Sachs was a risky proposition as a go-between. The group had considered Charles Lindbergh, the aviator, but dropped him when they discovered that not only was he an isolationist but that Roosevelt hated him. The trouble with the Russian-born Sachs was, according to historian Peter Wyden, that "his sentences were endless and convoluted. His vocabulary was, to use one of his favorite words, 'fantasticated.' [But] as a gloomily 'Jeremiahesque observer' who 'adumbrated' for 'perduring' meanings, he already knew about fission from science journals and did not have to be sold on the significance of Szilard's mission." And, most important, he knew Roosevelt.

Unfortunately, the president was too busy to see him, being involved in more immediate matters of life and death. On September 1, 1939, German tanks and planes began to devastate Poland, and within a few days Britain and France joined the war against the invaders. Roosevelt declared a National Emergency on September 8 and was trying to get Congress to repeal the embargo on shipping arms abroad. Two months went by with no action from Sachs, and the delay was driving Wigner and Szilard up the wall.

Meanwhile, Einstein was having his own problems. A New Yorker writer was preparing a profile of adman Frank Finney, who promoted himself in ads in The New York Times and New York Herald Tribune as "the Einstein of Advertising." The writer wanted Einstein's reaction, hoping he would be more amused than irritated by the request, and even suggested a response: "I was as astonished as Mr. Finney would be if I should suddenly start calling myself 'The Finney of Physics.' "

Einstein's reply was fast and furious. He said Finney's use of his name, which he would have stopped had he known how to, was disgusting and tasteless, and "nothing of this kind will ever be published with my consent."

The president still seemed too busy to see Sachs, so Einstein, in an effort to force the president to get cracking, sent him two papers from the Physical Re-

view—"Neutron Production and Absorption of Uranium," by H. L. Anderson, E. Fermi, and L. Szilard, and "Instantaneous Emissions of Fast Neutrons in the Interaction of Slow Neutrons with Uranium," by Leo Szilard and Walter H. Zinn. The papers described advancements in science leading to an atomic bomb.

Even when Sachs managed to speak to Roosevelt on October 11, he found him preoccupied. Sachs then mentioned that he had paid for the trip to Washington and couldn't deduct it from his income tax and would FDR please listen. Because he feared that the president might throw the scientific memoranda into his out tray without studying them, Sachs insisted on reading every word to him. FDR listened. After Sachs told him more about what German physicists were doing, the following conversation took place:

> Roosevelt: "Alex, what you are after is to see that the Nazis don't blow us up."
> Sachs: "Precisely."
> Roosevelt: "This requires action."
> The president called "Pa" Watson [his military aide, General Edwin Watson] in, gave him relevant instructions, and then ordered, "Don't let Alex go without seeing me again." That same night Sachs returned to the White House, and Roosevelt put him in touch with Dr. Lyman Briggs, director of the Bureau of Standards.
> Some of the military [informed of the prospective enterprise] were not impressed: officers would say, "Well, this thing is so remote: what is this thing?—let's wait and see." But Watson would reply firmly, "The Boss wants it, boys. Get to work."

World War II had been raging for over a month when, on October 19, 1939, Roosevelt responded to Einstein, thanking him

> for your recent letter and important enclosure. I found this data of such importance that I have convened a board consisting of the head of the Bureau of Standards and chosen representatives of the Army and Navy to thoroughly investigate the possibilities of your suggestion regarding the element of radium. I am glad to say that Dr. Sachs will cooperate and work with the Committee and I feel this is the most practical and effective method of dealing with the subject.

He was wrong. Five months went by with no decision from the committee. Einstein had recently learned that since the outbreak of war the Nazi government had taken over the Berlin Institute of Physics. Research on uranium was being conducted there in great secrecy, under the direction of C. P. von Weizsacker, son of the German secretary of state. In Paris, Marie Curie's son-in-law, Frédéric Joliot-Curie, had almost produced a chain reaction. Sachs conveyed these warnings from Einstein to Roosevelt, hoping that the president would move the committee to action.

Three weeks later, on April 5, 1940, the President suggested a conference to tackle the subject, with Einstein selecting those who should attend. Einstein himself did not appear then or at any subsequent conference, initially pleading a heavy cold and eventually persuading others that he was no expert on the subject. Nevertheless, together with Szilard, Wigner, and their go-between Sachs, he continued to press the foot-dragging bureaucrats and thickheaded military to get moving.

Even when Einstein was asked to calculate the best way to purify uranium, vital information was kept from him because, as a foreigner, he was considered a security risk, making his work useless. Yet he was so anxious to protect the "secret calculations" that he refused to have them typed up.

All through the spring and summer of 1940, Hitler launched successful invasions: Denmark, Norway, Holland, Belgium, and Luxembourg all fell to his troops. Most of the British army escaped from the continent via Dunkirk in small boats; France surrendered to the Nazi hordes soon after the British evacuation. Now Britain stood alone.

Szilard, in despair, predicted that Hitler would win the war. Wigner resigned from the Uranium Committee, disillusioned by the delays, the pathetic funding—initially two thousand dollars—and the moronic attitude of the military. "Weapons don't win wars. The troops' morale wins wars," insisted one army representative, who mockingly compared the proposed atomic weapon to a death ray. "There's a goat at the Army's Aberdeen weapons-testing grounds, tethered to a stick," he said. "The Army has offered a big prize to anyone who can kill it with a death ray. Nobody," he concluded, "has claimed the prize."

In June 1940, Roosevelt appointed Vannevar Bush to take charge of all government science projects. Soon after this appointment, Professor Fritz Reiche arrived from Germany with another dire message from Fritz Houtermans: "Heisenberg will not be able to withstand longer the pressure from the government to go very earnestly and seriously into the making of the bomb . . . they should accelerate if they have already begun the thing."

There was a shortage of housing in Princeton in those days; because Dr. Reiche and his wife couldn't find anywhere to stay, Einstein offered them his home while he was on vacation. He told them to eat the fruit from the garden "as Nature intended" and not to bother to send any on to him. Reiche had up-to-date information on the latest scientific breakthroughs, causing Einstein to remark to him wryly: "I am so old-fashioned and stubborn that I still do not believe that the Good Lord plays dice. Had He really desired that, He would have done it fundamentally and would not, when throwing dice, have stuck to one scheme . . . Nevertheless, all appearances speak against the concept of complete accordance to law. However, I continue to search unceasingly for such a concept. When my discovery does not hold good, I blame myself and not HIM."

Einstein was back in his Princeton home on August 18, 1940, when an FBI agent called to interview him about the Dutch physicist Peter Debye, now working at Cornell University and suspected of being a Nazi spy. Einstein said that in the spring a British intelligence agent came to his home with a letter that had been removed from the mails by a British censor. The letter, addressed to Einstein, was from a chemist named Feadler (the FBI's phonetic spelling) living in Switzerland, who suggested that because Debye had been closely associated with Nazi big shot Hermann Göring, he might be in the United States for a secret purpose. Einstein told the British agent that Debye's contacts with Göring might have been to obtain funds for the Kaiser Wilhelm Institute, where he was the director of physics, and that he had informed the Cornell authorities about the letter.

Einstein went on to say that Debye wrote to him on June 15 denying the charges and explaining "that he had left Germany because of his refusal to change his citizenship from Dutch to German. Dr. Debye also advised . . . that he had to resign as director of the Kaiser Wilhelm Institute, and that since he came to the United States he had no connection with German officials and under no circumstances would he return to Germany."

Einstein had immediately replied to Debye, saying that he had received the information from abroad and "did not know whether the charges were true, but felt it his duty not to judge him on the facts but to turn the information over to an American citizen as it was of a serious nature."

The FBI report continues: "Einstein related that Debye was afraid of Goering, [and] when coming to America the past spring went through Switzerland but did not visit his old friends there which was very unusual and unlike Debye; that therefore Feadler was suspicious of Debye and requested Einstein to ascertain if he was in the United States for a secret purpose."

In talking to the FBI, Einstein did not question Debye's integrity as a scientist, but otherwise considered him to be an opportunist without a strong sense of loyalty. He said he knew Debye well enough not to trust him, especially as he had behaved very suspiciously abroad. Asked to explain, Einstein replied that as a Dutchman, Debye was expected to help his European colleagues who he knew had been persecuted since 1933, but he had never attempted to help them to find work outside Germany.

Einstein also thought Debye might make a good spy because of his organizing ability and warned that if his "motives are bad he is a very dangerous man." Einstein felt that Debye should be watched for a while to ascertain his motives. However, he pointed out that as Debye had a son staying with him in the United States, perhaps he did not intend to return to Germany.

On October 1, 1940, the world's most famous physicist at last became an American citizen. Einstein, along with Margot and Helen Dukas, took the oath of allegiance in a ceremony in Trenton, New Jersey.

In November 1940, the Uranium Commitee finally recommended the secret funding of research to see if the splitting of the uranium atom followed by a chain reaction—yet to be achieved—would provide the energy to drive submarines or cause massive destruction. This was an enormous relief to Wigner, who had recently learned that Hitler was pressuring Heisenberg, his leading theoretical physicist, into working on a superbomb.

President Roosevelt okayed the American project almost immediately. Columbia University got a forty-thousand-dollar government contract and hired Leo Szilard and Enrico Fermi to try to produce a chain reaction. At last the atomic bomb project was under way.

The FBI then began to investigate the scientists involved as possible security risks, which is why Einstein had a second meeting with an FBI agent. In contrast to his reservations about Debye, Einstein gave Szilard a rave review as a gifted idealist, with no political motivation, who was absolutely reliable. He mentioned that Szilard was already working on uranium experiments at Columbia "with an Italian by the name of Fermi, who was [also] a very trustworthy man."

Among the dozens of refugees now living in Princeton, Einstein made close friends with historian Erich Kahler and his wife, Alice, who were from Germany. They had been encouraged to settle in Princeton by another German refugee, novelist Thomas Mann, a mutual friend. Einstein first met the couple at a poetry reading in their Princeton home. A neighbor, Charles Bell, who was there, described him, said Alice Kahler, as " 'the idol of my science-loving youth, with parchment face and a corona of hair, a resigned and sphinx-like wisdom, between a Saint Bernard and an angel, as if he had lived the whole rise from brute to human, and was himself the record of the trial and achievement.' This is a bit exaggerated," she added, with a laugh. "But Einstein was the most charming friend you can imagine if he was in the mood."

He was in the mood when the Kahlers went with him to meet Suzuki Daisetsu, a Japanese philosopher, who explained Zen Buddhism to them. Einstein listened intently but made no comment. Two days later Alice Kahler met Einstein again and they discussed Eastern philosophy. " 'What I can understand of Zen Buddhism has no great meaning for me,' " he told her. " 'But I approve of Confucianism.' He thought it was more down-to-earth than other philosophies."

Alice Kahler was aware of Einstein's addiction to tobacco, which his doctor had forbidden him to use. "He loved to smoke," she said, "but his secretary, Helen Dukas, and daughter [sic], Margot, were adamant about doctor's orders. When he got tobacco as a gift he sent it with a little note to my husband, and when he came to our home he asked Erich, 'Give me a little of the stuff so I can at least smell it.' His desire was so great that he would pick up cigarette butts from the street—which was nearly tragic."

The Kahlers arranged for Ben Shahn, a painter whose works were devoted to political and social causes, to paint a realistic portrait of Einstein. He liked the work, but he hated Shahn's later attempt at "a kind of idea of the head," said Kahler. Marc Chagall offered to paint Einstein, but he wasn't interested, though he posed for many poor, unknown artists or photographers, saying that "he didn't give a damn about good or bad as long as it helped them financially," Kahler recalled. "He was such a generous human being: he wanted to help."

Now that he was a widower, Einstein occasionally received marriage proposals by mail. "I didn't think he would ever marry again. But believe me, he loved women!" Kahler said, laughing. "He was still infatuated with the ladies, but his work was more important than women."

Admirers sent Einstein gifts of cashmere sweaters, but he didn't wear them, apparently because he was allergic to wool. Knowing this, Alice Kahler "had the glorious idea to go to an army store and buy him his first sweatshirt. From then on he hardly wore anything else. Once I bought him a Swiss sweater made of cotton with a blue collar. He called back, happy as a child, and said, 'Never in my life have I owned anything as beautiful as this sweater. Even my blind cleaning woman admired me when she saw how marvelous I looked in it!' "

By late 1940, Nazi brutality in Europe was turning even the staunchest pacifists into eager advocates for armed intervention. Among them was Einstein's friend Bernard Shaw, who said, "We ought to have declared war on Germany the moment Mr. Hitler's police stole Einstein's violin." Asked if Hitler had solved the Jewish problem, Shaw replied, "He has created it." He wrote to a musician, "Can it be believed that the country which produced this great composer [Mozart], the friend of all the world, can now produce nothing better than the degenerate Adolf Hitler? If Mendelssohn were alive today he would be sharing the exile of Einstein."

Despite such world-famous supporters, Einstein still had his homegrown critics. One wrote from New York: "You may be a great scientist, but your picture in the New York Times of yesterday is another illustration of the fact that the influence of the Jew today tends to the degradation of civilized society, to social anarchy. Your attire—or lack of it—should be regarded by the authorities of Princeton as an insult to that institution. It stamps you as unfit to associate with socially decent people." This typewritten brickbat was signed "Hibernicus."

Among such unexpected letters was one that he probably never saw—from his first love, Marie Winteler. She wrote to say that she didn't want to be forgotten, that he had promised not to forget her, and would he lend her one hundred francs to ease her financial problems? She reminded him that her mother had done many favors for him and Maja, and promised more letters of a different nature if he was still interested in her. Marie reminded him that when they were together she had been a good and innocent child who understood neither herself nor life's realities. Today, she said, she led a good life guided by God. Einstein's protective secretary, Dukas, very likely kept the letter from him and simply stuck it in the files.

Three months later Marie again wrote, this time asking Einstein to send money so that she and her son could emigrate to the United States. She complained of not eating lunch for a year, and that her son was doing poorly paid manual work while more suited to an intellectual occupation. This letter, too, probably never reached Einstein.

(Neither Marie nor her son ever came to America. She died as a patient in a mental hospital in Meiringen, Switzerland, on September 24, 1957.)

What Einstein unquestionably did read was a cry for help from a relative, Brigitte Alexander-Katz. She was stranded in France with a newborn son, Didier. Her husband was a volunteer in the French army stationed in North Africa and had never seen the boy. To escape from the advancing Nazis, she had gone to the Gare de Lyon with a rucksack on her back, a bag in one hand, and her baby in the other, hoping to reach a friend's home two hundred kilometers away. Having just been released from a hospital, she was very weak. A human wall prevented her from reaching a train until a man held the boy overhead to prevent him from being squashed, and they then forced their way through to the train.

At her destination they were not allowed to leave the station and spent the night sleeping on the floor — "the baby," she wrote to Einstein, "held against my heart."

The next morning a peasant drove them in his oxcart to the home of Brigitte's friend, but the owner of the house had returned unexpectedly with a sick child and there was no room for anyone else. Brigitte had just found another place to stay when she was warned that the Germans were approaching and the village would soon be a battleground. A woman, also with a small child, then drove Brigitte and her son toward the river Loire, hoping to cross it to safety. Often stalled for hours behind other cars, they arrived just as French soldiers were about to blow up the bridge. Eventually, "a soldier had pity when he saw the poor babies and let us go." But on the other side there was nowhere to stay and nothing to eat. Their crying, as they walked the streets late at night, babies in their arms, moved a hairdresser to let them sleep in her salon. The following night they arrived at Brives, where tens of thousands of refugees slept in the streets. Again there was no food. Fortunately, Brigitte found someone to drive her to a cousin in Biarritz, where the baby was treated by a doctor.

After a week there, Brigitte and the baby, along with her cousin, were on the move again, just ahead of Nazi troops. In the unoccupied zone her cousin lodged her in a home with his old employees, who "were awfully anti-Semitic and made a very cruel life for me." Finally, Brigitte and her son reached a hotel for the poor in Lourdes. She was covered in abscesses and a doctor told her she wouldn't improve until she changed her diet. In fact, she had only enough money left to survive for another month. She concluded the letter: "I think that is all. The real aspect cannot be described. It is too sad for words."

Eventually, Brigitte managed to join her soldier husband, and Einstein heard through a mutual friend that they were trying to emigrate to Mexico. He immediately offered in his reply "to vouch for their reliability and integrity both personal and political. I have known Mrs. Brigitte Alexander-Katz, whose family is related to mine, since she was a little girl. Her husband, a very able engineer, will certainly be useful to any country which receives him. If you will send me the address of the proper Mexican authority I shall gladly send any letter of recommendation desired."

Not all the letters he received, of course, were as heart-wrenching as the one from Brigitte Alexander-Katz, even in those traumatic times. One was an unusual invitation to visit Michigan, where "for almost two years my friend and myself have [had] nightly communications with the Super-Natural. We have actually talked in a voice with what we believe to be Spirits . . . We have seen Tables and Chairs move and there was no one near them and we have had them move at our command. This is truly the most remarkable thing the world has ever known and I am sure you will agree after talking to me and seeing it for yourself." Einstein did not go to Michigan, but he wrote in a margin of the letter that if they didn't sleep well, they should take a tranquilizer.

After a brief visit with Hans Albert, Einstein joined his friend Gustav Bucky in his Manhattan workshop to discuss various projects. Unfortunately, a neighbor spread a rumor that the men were trying to produce "death rays" in two tiny rooms. Housing officials found no evidence of death-ray research, but they had the workshop demolished for failing to meet zoning laws.

As the year drew to an end, Einstein demonstrated his soft spot for eccentrics. His stuffier critics regarded this as a weakness. Marginal or controversial scholars who had had a difficult time seemed to remind him of his earlier friendships and associations, when he was an impecunious graduate student in Bern trying desperately to make a living by giving private lessons. He felt very much at home with these people.

"Cranks, of course, tried to get through to Einstein," said his assistant, Valentin Bargmann, "though Miss Dukas generally shielded him from them. But Einstein had nothing against talking to a crank. This amused him. This was the wonderful thing about Einstein. He was enormously relaxed."

Many regarded psychotherapist Dr. Wilhelm Reich as a crank, though he taught at the respected New School for Social Research, had attracted a devoted group of disciples, and published a magazine, *The Orgone Energy Bulletin*, reporting the results of his unique efforts. He claimed to have created something he named orgone energy in special accumulators the size of cigar boxes, and to have obtained remarkable results after subjecting cancerous mice to this so-called orgone energy treatment in his Maine laboratory.

On December 30, Reich asked Einstein for a meeting to discuss his discovery, which he said promised to be a breakthrough in the battle against cancer. Intrigued, Einstein offered Reich an entire afternoon.

CHAPTER 31

The Race for the Bomb

1941 to 1943

62 to 64 years old

An excited Wilhelm Reich arrived at Einstein's home on the afternoon of January 13, 1941, and was directed to his study on the second floor overlooking the back garden. Einstein was friendly and encouraging as they began a marathon conversation that lasted almost five hours, during which Reich had him look through an instrument called an orgonoscope. Exactly what Einstein saw has not been reported. Reich defined orgone as "a specific biologically effective energy which behaves in many respects differently to all that is known about electromagnetic energy." Einstein did apparently see something unusual; its obvious appeal was Reich's suggestion that it might be used "in the fight against the Fascist pestilence."

Already, according to Reich, this mysterious orgone energy promised to fight disease and might also make a spectacular war weapon.

Einstein seemed impressed. He agreed that if the temperature in an enclosed object was raised without any apparent source of energy, as Reich asserted, it would be a remarkable discovery—"a bomb." A rival, perhaps, to the incipient atomic bomb.

Having taken Reich for a fellow physicist, he was surprised to learn he was a psychiatrist. "What else do you do?" Einstein asked, and learned that Reich had been Freud's assistant for several years, and now taught experimental psychology and biopsychology at the New School and ran a publishing company. They agreed to meet again, when Reich would bring an orgone-energy accumulator for Einstein to put to the test. As Reich was leaving, he asked, "Can you understand now why everyone thinks I'm mad?"

"And how!" replied Einstein, perhaps recalling all those who had called his own ideas crazy.

Later that night, when he reached home, Reich told his wife "how exciting it was to talk to someone who knew the background of these physical phenomena, who had immediate grasp of the implications. He started to daydream of possibilities of working with Einstein at the Institute for Advanced Study . . . He hung onto this daydream for the next few weeks."

After preparing the experimental setup, Reich took it to Princeton on February 1. He claimed that Einstein could confirm that energy was created by leaving the orgone-accumulator — a metal box — on a table, and then comparing the temperature in the box with the temperature under the table. A week later, Einstein sent Reich his report. He wrote that although he had observed a difference of temperature, there was a simple explanation that had nothing to do with the orgone-accumulators. Convection current causes the air in the room to be warmer near the ceiling; this warm air is reflected by the surface of the table, while the underside of the tabletop is cooled by the colder air near the floor. Even without an orgone-accumulator, thermometers would show different temperatures if one was placed above and the other below a table. Einstein advised the bitterly disappointed Reich not to be carried away by an illusion.

Although Reich bombarded him with impassioned letters and documents supporting his experiments for months afterward, Einstein did not reply. Reich, a onetime Communist, explained the silence as "a communist-inspired conspiracy." The implication, of course, was that Einstein was one of the conspirators. He was not and never had been a Communist, however, despite his thickening FBI file, which listed peace movements he had supported that were suspected of being communist fronts. Reich's wife, Ilse, discounted the conspiracy theory, but believed that "Einstein saw the phenomena, may have had an inkling of their significance, but was unwilling to get involved in a highly controversial scientific discovery at a time when he was deeply engrossed with developing atomic energy." In fact, Einstein was right: orgone energy proved to be an illusion. He wrote to Reich once more, briefly, three years later, to deny that he had anything to do with rumors accusing Reich of being a charlatan.

Einstein doubtless had Reich as well as himself in mind when he wrote to his former assistant, Cornelius Lanczos, now teaching at Purdue University, agreeing with "your beloved Schopenhauer" that people obsessed with their own misery are "unable to be tragic but are condemned to remain stuck in tragicomedy . . . Yesterday idolized, today hated and spit upon, tomorrow forgotten, and the day after tomorrow promoted to Sainthood. The only Salvation is a sense of humor, and we will keep that as long as we draw breath."

Indeed he did—but with lapses. When Professor Friedrich Foerster paid a visit, for example, and congratulated him on finding a safe harbor, Einstein looked at him sadly and said, "Who knows for how long? I can see the dark times ahead. Barbarism is still far from being conquered."

However, fellow refugee Alice Kahler caught him in an upbeat mood for his birthday party on March 14. "He told me to come early," she said, "and he would explain the theory of relativity. I arrived half an hour early and he began to tell me, then asked, 'Did you understand?' 'Yes,' I said, 'every word.' 'But that was the old theory of Newton's, not mine,' he said. Then he started with examples of railways and so forth, a little above my brain.

"At the party everyone ate the cake, which was in the shape of a telegram by science writer William Laurence of *The New York Times*," Kahler recalled. "Everyone, except Einstein who was on a fat-free diet. When I asked if he was sorry he said, 'No, I'm not sorry at all. I can remember distinctly how it tasted. And this is what I resent about Gandhi. He never knew how these things tasted.'

"During my visit to Saranac Lake [where Einstein was on vacation] I brought him the famous Chinese Cross, one of the most complicated puzzles to be put together. He solved it in three minutes. I wouldn't have been able to do it in a thousand days. When I said so he offered to show me. And he took it apart and put it together in no time. When his son, Hans Albert, came to visit, Einstein was happy to learn his talent had been passed on to him. And he remarked, 'He did it beautifully, exactly as I do.' Einstein enjoyed solving puzzles and he had amazing ones sent from all over the world."

Alice Kahler disputes those who call Einstein's attitude to women cavalier or even callous, and laughed at the suggestion that he was a woman-hater: "Listen dear, he loved women. He was very attractive to them, and he was infatuated by the ladies. But he once said to me, 'The whole thing lasts just ten minutes and then it's all over.' [She laughed.] Yes, Einstein loved women and he wrote on my most precious photo of him and myself that he regretted I wouldn't sleep with him. I have it in my sleeping room. I would have been interested in him if I wouldn't have had a husband.

"He was such a young, innocent man when he first married. You cannot imagine how naive he was."

Kahler was also impressed by how he treated his sister. "The relationship between Einstein and Maja was very, very beautiful. They loved each other very much. Even before she became ill he read aloud to her every night. On vacation at Saranac Lake, New York, I sat in on these evening sessions. Once he was reading Herodotus and there was a not-so-interesting passage and I said, 'Maybe we should skip this part.' Einstein was absolutely appalled: 'How can you say such a thing! We might miss something important. We are not going to go through this without reading every line.'"

The Buckys were also at Saranac Lake that summer of 1941. Thomas Bucky spent eight summer vacations with Einstein from 1933 to 1941, when he went away to medical school. Thomas recalled: "He was a guest for several of those years of a group of prominent families who had a community of seven cottages at Saranac Lake. Believe it or not, someone there called him a snob, failing to realize he avoided people from shyness.

"If he had been sailing during the day, he'd come home to supper or dinner and sit around discussing any subject that came up," Bucky said. At the time, Bucky was studying to be a doctor, and recalled that he often discussed medicine with Einstein, who, although only a layman, "knew a little bit about everything."

At about nine on those summer evenings, Einstein retired to his room to read or sit in a chair for hours, quietly writing neat figures and formulas on small paper pads. He always carried those pads with him and whenever he got a free moment would write these precise, little formulas, puffing away at his pipe until the air was tinted blue.

"He could not and would not understand political maneuvering," insisted Bucky. "There was no compromising with the truth—in a true mathematician's fashion. He was absolutely uncompromising. What is right is right. If he were proved wrong he'd happily admit it, for no one, himself included, could stand in the way of or compromise truth. He argued with my father for years over some theory. Then one morning, after about two years of arguing, he came down to breakfast and said, 'You know, Dr. Bucky, I thought things over during the night. And now I see that perhaps you *are* right.'"

During that summer of 1941, Bucky saw a lot of Margot, and felt she lived in another world, completely unaware of what was going on in the professor's world or Miss Dukas's. Margot was a sweet, quiet, dainty person. Einstein especially appreciated her artistic gift for drawing, painting, and sculpture. She had a series of pets, including a parakeet named Bibo, and when Bucky went to college she took care of his dog, Chico, a wirehaired terrier.

Dukas, like Einstein, did not drive a car and depended for distant transportation on Watters or the Bucky brothers. "People seem surprised that Einstein never drove a car," said Bucky. "Why should he? My father supplied the transportation and the driver. In Germany, our chauffeur was at his disposal, and over here my brother, Peter, and I did all the driving for him. There's a photo of me driving him along Fifth Avenue in my first Ford. We went to Princeton and drove him around whenever he had to go some place, like New York. He would never ask. We took care of it. When I started dating I'd tell the girls I couldn't see them on Sundays because I was going to Princeton. One even asked, 'Why the devil are you always going to Princeton? Are you going to see Einstein?' 'Yes,' I answered, poker-faced."

In the fall of 1941, Einstein's friend and colleague, Eugene Wigner, was still desperately concerned that the Nazis would get the atomic bomb first. As Einstein did not have security clearance, Wigner doubtless expressed his fears to him in general terms. Einstein shared his concern with Upton Sinclair on October 23: "I have the feeling that most people in this country don't realize how dangerous our situation is. I must also confess that I feel the whole tragedy of the Russian situation. [In June, Hitler had invaded the Soviet Union.] The senseless destruction of so much precious work accomplished under the most difficult circumstances in only twenty years. It is very difficult indeed to keep up the idea of a deeper meaning in human history."

The arrival of an eight-year-old girl seeking help with her arithmetic homework drove such grim thoughts from his mind. Shortly after, fourteen-year-old Jane Swing and her friend, the girl's older sister, came to pick her up.

"Einstein opened the front door to us," Swing recalled. "I never saw anybody else there. When he let us in, things downstairs were all neat and tidy, but then he took us up the stairs to his study—I guess he practically lived in it—with a view of the backyard. And he invited us into this terribly messy room." She saw a long table littered with papers and books. More books were stacked on the floor. "I'm an artist," said Swing, "so I have a very untidy room, too. But I was just plain flabbergasted. I had never seen anything like that before."

At first, Swing was more impressed with the state of that room than with its occupant. "He looked like the room, though, a very untidy gentleman, hair going in every direction. He did have a tie on but that was around the wrong side of his face. He had a huge shock of gray hair and food all down the front of him. He reminded me of an untidy Mark Twain. He was highly unusual, like nothing I'd ever met up with before, with a very high voice, almost like a woman's—most unusual.

"And he asked if we'd like to have lunch and we said 'Certainly.' So he moved a whole bunch of papers from the table, opened four cans of beans with a can opener and heated them on a Sterno stove one by one, stuck a spoon in each and that was our lunch. He didn't give us anything to drink. He moved things off stools for us to sit. He was discussing mostly things with this friend's sister, about her arithmetic.

"My friend's younger sister went there practically every day. He loved this little girl and I think most of our conversation with Einstein was about her and her math problems."

Another visitor, physicist John Wheeler, remembered having called on Einstein "to explain the 'sum of histories' approach to quantum mechanics then being developed by Richard Feynman, whom I had as a graduate student. I hoped to persuade Einstein of the naturalness of quantum theory when seen in this light, connected so closely and so beautifully with the variation principle of classical mechanics. He listened to me patiently for twenty minutes. At the end he repeated that familiar remark, 'I still cannot believe that the good Lord plays dice.' Then he added in his beautifully slow, clear, well-modulated and humorous way, 'Of course, I may be wrong; but perhaps I have earned the right to make my mistakes.' Later I said to him, 'You must often be invited to other places. Are you never tempted to visit?' 'I love to travel,' he replied. 'But I hate to arrive.'"

At that time, many Americans were reluctant to become actively involved in the war in Europe. But on December 7, 1941, Japanese warplanes made an unprovoked bombing attack on Pearl Harbor, and the United States entered the war against both Japan and Germany.

Einstein and Wigner still feared that the Germans would soon win the fight with an atomic bomb, especially when Wigner calculated that, "like us, they have had almost three years since the discovery of fission to prepare a bomb. Assuming they know about plutonium . . . it would be possible for them to have six bombs by the end of 1942. On the other hand, we don't plan to have bombs in production until the last part of 1944."

He wasn't far wrong. In April 1942, General Friedrich Fromm, commander of the German Home Army, advised Albert Speer, Nazi minister of armaments, that German scientists "were on the track of a weapon which could annihilate whole cities, perhaps throw the island of England out of the fight." Speer conveyed the good news to Hitler, who appointed Göring head of the Reich Research

Council dealing with nuclear research. Scientists Otto Hahn and Werner Heisenberg briefed armament production officials on the subject. Heisenberg complained bitterly that while Americans were well funded for nuclear research, German research was being crippled through inadequate funding and the practice of drafting scientists into the armed forces. Now, he said, the Americans were probably in the lead, although Germany had been ahead a few years before.

Heisenberg advised Speer that it was scientifically possible to build an atomic bomb, but that even with massive support it would take at least two years. A cyclotron was a necessity, but it would be difficult to maintain secrecy because the only one available was in Paris. Speer suggested building a cyclotron in Germany, and Heisenberg agreed.

On June 23, Speer discussed the idea of an atom bomb with Hitler; according to Speer, the idea "quite obviously strained his intellectual capacity." Furthermore, Hitler was "plainly not delighted" when told that Heisenberg was not sure nuclear fission could be controlled and that it might transform the Third Reich "into a glowing star." But at the prospect of doing it to Britain, he exclaimed triumphantly, "That is how we will annihilate them!"

But it was not to be. In the fall of 1942, German nuclear physicists estimated it would take at least three years to produce an atom bomb. Speer expected the war to be long over by then and squashed the project, though research continued in an effort to produce nuclear-powered submarines.

Heisenberg was right about the Americans being ahead. On December 2, 1942, Enrico Fermi produced a self-sustaining chain reaction in a squash court under the football stands at the University of Chicago. He had controlled the release of nuclear energy from the atomic nucleus.

Einstein was not cleared to receive such information. As he wrote to his friend, Hans Mühsam, now living in Haifa:

> I have become a lonely old fellow. A kind of patriarchal figure who is known chiefly because he does not wear socks and is displayed on various occasions as an oddity. But in my work I am more fanatical than ever and I really entertain the hope that I have solved my old problems of the unity of the physical field. It is, however, like being in an airship in which one can cruise around in the clouds but cannot see clearly how one can return to reality, i.e. to the earth.

Other correspondents helped bring him down to a bumpy landing. One wrote to him from Good Hope, Illinois, to promote his "medicine that will cure all stomach aches. However, many people are prejudice [sic] and will not buy it. I am now convinced that your well known name should be on the bottles as an example of one who uses it. Unless you reply, I will put on the bottles a testimonial saying that you use it habitually." Einstein replied with an emphatic no, forbidding the use of his name and saying that he would never allow it to be used commercially, especially if it misled the public.

What would have misled the public was his thickening FBI file, which the authorities took seriously enough to prevent him from ever getting security clearance. FBI chief J. Edgar Hoover had targeted Einstein as a Communist suspect—

even a spy—and a threat to the United States. Agents were urged to keep track of his movements and to report his public statements. No gossip about him, however unlikely, was ignored. And though many of the invariably anonymous sources appeared to be eccentric at best, Einstein never got the chance to rebut their charges; in fact, the FBI seemed scared to approach him directly. Typical of the sources used against him was a letter from someone in Beaver Falls, Pennsylvania, who reported that at a meeting chaired by "Rabbi Wise . . . Louis Lipsky said, 'Einstein is experimenting with a ray which will help us to destroy armed opposition—aircraft, tanks and armored cars. He hopes that with it a dozen men could defeat 500. Through it 5% could rule a nation.'"

It didn't help that the FBI had recently arrested Theodore von Laue, the son of his friend Max von Laue. A German citizen and graduate student at Princeton, Theodore was suspected of spying for the Nazis a few months after the United States had joined the war against Hitler. As Theodore later recalled: "In 1942 I was arrested by the FBI because I had been sailing [as an enemy alien] without official permission along the New Jersey coast. I was interned in a camp near Camden. There were a lot of sinkings of American ships by German submarines in the area. And the FBI made a connection between my sailing on the inland waterways and the sinking of those ships. This was 1942 and things were tense. This was the atmosphere of the times, suspicion everywhere. I lived at Princeton University's graduate school and some people thought, 'Here's this German and ships are being sunk. Surely he must have some connection.' I was told—no proof, no investigation—that the janitors denounced me.

"My father, who remained in Germany, was a liberal democrat, strongly anti-Nazi. When Einstein lived at Caputh they sailed together. My father always spoke of him with admiration, and approved his leaving Germany. My father also helped a lot of German Jewish people get to the United States; physicist Rudolf Ladenburg, for example. So he kept up connections with Einstein and helped others escape the Nazis. Because my father knew something about the Nazi regime, before the war he sent me to be educated at Princeton, where he had friends."

Hearing of the arrest, Einstein wrote to U.S. Attorney General Francis Biddle on June 25, praising Max von Laue's courageous defiance of the Nazis and continued opposition to the regime—which was why, he said, von Laue wanted Theodore educated in America, away from the influence of the Nazis. He named Professors Hermann Weyl of the Institute for Advanced Study and Rudolf Ladenburg of Princeton University as character witnesses and asked Biddle to inform him of the outcome. Einstein raised the same points in a "To-Whom-It-May-Concern" letter, adding that Max von Laue had helped many colleagues who were persecuted for racial or other reasons, and that he, Einstein, had occasionally played chamber music with the son, who he was convinced opposed the Nazi regime and was loyal to the United States.

After appearing as character witnesses for Theodore at a Trenton hearing on August 5, 1942, Ladenburg and Weyl wrote urgently to the attorney general:

It may well be a matter of life and death to the father that what we were afraid to say then and are going to say now is kept in strict confidence. Max von Laue has given assistance to many men of science who became victims of Nazi persecution, sometimes helping them to escape from Germany at considerable personal risk. We have messages from him, sent through neutral countries, where he warns his son and us against German physicists who were visiting this country in 1939 and whom he suspected of having connections with the Gestapo. Nearly all his letters contain passages which by implication or, if mailed from neutral countries, even openly ridicule and defy the Nazi system. At a time when so many men of the German academic life have bowed to Nazi tyranny or Nazi successes, Max von Laue has been one of the few shining exceptions. It was an act of supreme patriotism and courage that he stayed in Germany. Our hopes for a better future in Europe after the overthrow of Hitler and his gang are based on men of von Laue's character and integrity who stayed "enduring the unendurable" (as father Laue writes in one of his letters) but never gave in. Such is the father. The son has embraced the American way of life with growing enthusiasm, not blindly but with the open mind and the discerning eye of a born historian. He wants nothing more than to prove his loyalty to this country as one of its citizens.

Internment, in itself a greater hardship to an ardent anti-Nazi like him than to most other internees, would probably mean the end of his career in this country. It would be tragic if his father's attempt to save his son's mind in its formative stage from the Nazi poison would end in his life being broken by our democracy to which it has been dedicated.

Theodore was released after four months, thanks to Einstein, Weyl, and Ladenburg. "I was treated quite well in the internment camp," Theodore von Laue recalled. What is more, the experience did not ruin his career; he became an American citizen and a noted historian.

Both the FBI and the Gestapo were keeping Einstein busy. No sooner did he hear that Theodore von Laue was free than he learned of a relative in great danger. Walter Einstein, a former German judge, had presided at a trial of Nazi thugs before 1933, when the Nazis were not yet in power. The men had been convicted. It was only a matter of time before the Nazis made him pay for it, so Einstein wrote to the chief of the Visa Division in Washington, hoping to speed the immigration of Walter and his father, Louis Einstein. (Walter did survive in the Dordogne but was unable to immigrate despite a second appeal from Einstein; but Louis was murdered by the Nazis.)

That same month he learned that his friend Paul Langevin had been arrested in France, "taken from his house by two car loads of Gestapo agents and flung into a cell of the Sante Prison, his braces taken from his trousers, and his shoelaces removed." Langevin, who was Jewish, was charged with being a pernicious influence on young people. Hundreds of students and teachers had defied the police to protest his arrest—the first public rally against the Germans since the occupation—but to no avail. Langevin's daughter, Helene, and son-in-law, Jacques Solomon, also a physicist, had been arrested by the Gestapo and accused of helping the Resistance. His Nazi captors had executed Solomon in May. Langevin was distraught over the death of Solomon and the as yet unknown fate of his own daughter.

Einstein tried to free Langevin and save Helene's life by asking scientists in neutral countries to agitate for their release. He was pessimistic, though, writing in a bitter mood to a friend: "Due to their wretched traditions the Germans are such a badly messed-up people that it will be very difficult to remedy the situation by sensible, not to speak of humane, means. I keep hoping that at the end of the war, with God's benevolent help, they will largely kill each other off." His mention of God was purely figurative, as he made clear to a contemporary correspondent, describing the Bible as "in part beautiful, in part wicked," but adding that "to take it as eternal truth seems to me also superstition which would have vanished a long time ago would its conservation not be in the interest of the privileged classes."

Langevin eventually escaped to Switzerland and his daughter survived imprisonment in Auschwitz.

In 1943, the Nazis in occupied Denmark were also about to arrest the partly Jewish Niels Bohr and ship him off to a concentration camp and probable oblivion. He escaped to neutral Sweden just ahead of the Gestapo. From there a British plane flew him to Britain, though he almost died from lack of oxygen because of a badly fitting oxygen mask. Fortunately, the pilot realized that Bohr was unconscious, rapidly lost altitude, and flew the rest of the way in oxygen-rich air just above the sea. Bohr reached America soon after his escape, and proved a valuable addition to the atom bomb project.

In June, Einstein became actively engaged in the war effort, working for the U.S. Navy's Bureau of Ordnance as a $25-a-day consultant. "I am in the Navy," he announced to friends, "but I was not required to get a Navy haircut."

An old friend, physicist George Gamow, was chosen by the navy to pick Einstein's brain every other week. "So long as the war lasts," Einstein told Gustav Bucky, "I do not want to work on anything else." He decided to skip his annual vacation and stay in Princeton to concentrate on work for the navy. Security clearance was not necessary, because the weapons he was asked to rate or improve were at an embryonic stage. He began to work on ideas for an electromagnetic device to explode a torpedo under a ship. But he declined to visit his contact, Lieutenant Stephen Brunauer, in Washington, D.C., unless it was urgent, his excuse in June being "the condition of my health," and in July because "I would be very much molested by snobbish people."

His explosives ideas apparently proved useful, because in August 1943, he wrote, "I am glad to hear that work is done concerning the torpedo." The following month, however, he was looking forward to a visit from Lieutenant Brunauer and John von Neumann, an expert on target selection, to discuss "our problem": "I do not have the feeling that much can be achieved in this matter through mathematical calculation. The reason is the same as in many other cases: you have to introduce, for the sake of simplicity, many doubtful assumptions which may essentially influence the outcome. Experiment seems to me the only reliable way of confirmation in this case."

When Thomas Bucky arrived home from Yale, where he was taking an introductory physics course, he found his father and Einstein, both enthusiastic inventors, "discussing [a] new, more effective antiaircraft defense system. They had been talking about it for hours and were so enthusiastic they told me about it. After listening for a few minutes I shot holes in their brainchild by pointing out they were violating a basic principle of elementary physics. They were flabbergasted. But after a while they laughed at themselves, and congratulated me for pointing it out."

CHAPTER 32

Einstein Goes to War

1944
---
65 years old

The war changed Einstein the crusading pacifist into an eager war worker, especially after he heard from Max Born—who had emigrated from Göttingen to Edinburgh, Scotland—and others, of the systematic killing of European Jews, and of the starvation and torture to death of some five thousand Americans in Japanese prison camps. He therefore welcomed the arrival of George Gamow every other Friday with his briefcase of confidential—but not top secret—documents outlining the navy's proposals for weapons and ways to defeat the enemy.

One suggestion was to destroy targets from above and below by sinking deepwater mines at the entrance to a Japanese naval base, with a follow-up attack by planes bombing the flight decks of Japanese aircraft carriers. Gamow, who had a reputation for saying anything for a laugh, claimed that Einstein okayed almost all such ideas with an enthusiastic, "Oh yes, very interesting, very, very ingenious."

As the two men walked from Einstein's study to lunch in the institute's cafeteria after one of their sessions, their conversation wandered from the war to astrophysics and cosmology. Once, when Gamow "mentioned [physicist] Pascual Jordan's idea of how a star can be created out of nothing . . . Einstein stopped in his tracks, and, since we were crossing a street, several cars had to stop to avoid running us down."

Einstein's navy assignment also included proposing new weapons and critiquing those in the works. Early in the year he admitted to Lieutenant Brunauer, his navy contact, that he had goofed: his recent idea for a torpedo that exploded when it was parallel to a ship wouldn't work, because the weapon would disintegrate before reaching its target. He suggested a solution to the problem—to make the head of the torpedo hollow, as a buffer between the tremendous water pressure and the explosive.

The navy was considering hiring a woman who had been Einstein's temporary secretary a decade before in California, so George Cook of Naval Intelligence was sent to Princeton to ask him if she could be trusted with classified material. Cook questioned Einstein in his home; he remembered that "after considerable probing, I was able to verify to my satisfaction that the applicant was in his opinion a loyal American of complete integrity who was a qualified secretary; but he said, 'Not for Naval Intelligence.' When I asked 'Why?' he replied, 'Because she is not intelligent.' And that is, verbatim, the way my report was submitted—whether she got the position, or not, I never learned."

Einstein took a break from war work to rebut Professor Philip Hitti, a prominent Arab-American historian. Hitti had recently testified about the Palestine problem before a congressional committee. Hitti's main points had been that the Arabs were descendants of the ancient Canaanites who held the land before the Jews, and that Jerusalem is their third holy city, toward which the early Arabs prayed. He also contended that the land was given to them by Allah as a result of a jihad, a holy war.

Einstein and his historian friend Erich Kahler believed that Hitti's presentation was "one-sided," and were working together to straighten Hitti's spin. The Einstein–Kahler writing team first identified themselves as "non-partisan Jews," and not front men for Zionists, then launched into their view of the facts:

> To the Arabs, Jerusalem is only the third holy city, to the Jews it is the first and only holy city, and Palestine is the place where their original history, their sacred history took place . . . The Jews do not resort to arguments of power or of priority. One does not get very far with historical rights. Very few peoples of the world would be entitled to their present countries if such a criterion were applied . . .
>
> We do not, and the vast majority of Jews does not, advocate the establishment of a state for the sake of national greed and self-glorification . . . In speaking up for a Jewish Palestine we want to promote the establishment of a place of refuge where persecuted human beings may find security and peace and the undisputed right to live under a law and order of their making.

In reply, Hitti emphatically restated his arguments opposing the establishment of a Jewish state.

The Einstein–Kahler writing team again responded:

> Professor Hitti says: "The Hebrews came and went. The natives remained." Now the fact is that the Israelites—we prefer to use this term because the Arabs also belong to the "Hebrew" people—the Israelites came, but they never went . . . After the Babylonian captivity . . . a true renaissance of Jewish Palestine began, leading up, on the one hand, to the elaboration of the Palestinian Talmud and, on the other, to the birth of Christianity from Judaism. If we were vindictive we could ask Professor Hitti whether he knows something of the revolt of the Maccabees and the ensuing independent kingdom of the Hashmoneans lasting nearly a century. He knows, of course . . . Jewish communities persisted in Palestine uninterruptedly throughout the ages . . .
>
> No suspicion of bias can certainly arise as to the statement of T. E. Lawrence, "Lawrence of Arabia," one of the most ardent friends the Arabs ever had: "Palestine was a decent country (in ancient times), and could so easily be made

so again. The sooner the Jews farm it all the better: their colonies are bright spots in the desert."

There is only one point in which we may agree with Professor Hitti: The Jews too have their diehards and their terrorists—although proportionally far less than other peoples. We do not shield or excuse these extremists. They are a product of the bitter experience that in our present world only threat and violence are rewarded and that fairness, sincerity and consideration get the worst of it. As far as Dr. Weizmann is concerned, however, we have to correct Professor Hitti's quotation. He never threatened the Arabs with expulsion. The passage to which Professor Hitti refers reads: "There will be complete civil and political equality of rights for all citizens without distinction of race or religion, and, in addition, the Arabs will enjoy full autonomy in their own internal affairs. But if any Arabs do not wish to remain in a Jewish state, every facility will be given to them to transfer to one of the many and vast Arab countries."

On April 14, the day the first published statement appeared, a stranger wrote to Einstein accusing him of being a nationalist. He denied the charge, saying that a comment of his in German had been mangled in translation, but that he did believe Jews needed a strong feeling of international solidarity to combat the devastating influences of their more or less hostile social environment. For that reason, he was convinced that Zionism was valuable and had saved and would continue to save many Jews from despair and a feeling of inferiority.

Although facing a heavy workload, he read Upton Sinclair's latest novel, *The Presidential Agent*, in which the art-dealer superhero, Lanny Budd, a secret agent for FDR, contacts Hitler, Göring, Stalin, Pétain, Mao, and Einstein. In an extraordinary effort to achieve authenticity in the novel, Sinclair had written "a thousand letters to persons who were eye witnesses of this or that scene, or who had access to inside information."

In a letter to Sinclair describing his reaction to the book, Einstein made no mention of the fact that four days before, on June 6, the Allies had invaded Europe. Instead, he praised the writer for giving "the American public a vivid insight into the psychological and economical background of the tragedy evolving in our generation. Only a real artist can accomplish this. For there is only the artistic way to reach a greater public and to impress people effectively. The best objective reasoning can never accomplish that. I am convinced that you have influenced political thought more effectively than nearly all of the politicians on the stage."

Despite Sinclair's exhaustive research, life inside Germany was still mostly a mystery to Americans. Sinclair, of course, had access to Western and neutral news sources, but they did not mention, for example, that Max Planck, who had played an important role in Einstein's career, had recently been bombed out of his home in Berlin-Grunewald, that his library had been looted, and that all of Einstein's letters to him had been destroyed; nor did they mention that after Planck and his wife left their home their few rescued possessions were stolen.

Einstein, meanwhile, was merely having to contend with questions about the just-published biography of him by his former stepson-in-law Dimitri Marianoff.* Speaking on Einstein's behalf, Dukas called the book unreliable and implied that the author knew far less about Einstein than he claimed. Einstein scoffed at Marianoff's report that he had lived in the same house as his subject for eight years, saying it was only for a few months at a time and in total for less than a year. Those close to Einstein agreed, though many of the incidents Marianoff recounts when he was on the spot ring true.

What was totally accurate was Marianoff's description of Einstein's mail—from weird to wonderful—and his deft and often delightful responses. For example, to a Manhattan rabbi inviting him to lecture, Einstein replied, "There are two kinds of temptations: those coming from the devil and those coming from the angels. In your case it is a temptation of the latter species. I do not know which of these temptations is more difficult to resist."

His mail also included a marriage proposal from a Manhattan widow, a request for an autographed photo from an Indian maharajah, and a letter from his friend Max Born, who was recovering from a nervous breakdown in Edinburgh. He had heard of Einstein's call for intellectuals to form an organization to prevent future wars and wrote to give his support; he also said that he had heard from reliable sources that most German scientists, including Heisenberg, were working for the Nazis, and that von Laue and Hahn were among the few exceptions.

Einstein replied on September 7, saying that he often laughed when he recalled their shared adventure some twenty-five years before, when he and Born had taken a tram to the Reichstag, naively convinced that they could talk people into becoming good democrats. Neither realized then that in such circumstances guts are much more important than brains, and he was not surprised to find that also applied to most scientists. For example, he wrote, "It was interesting to see the way in which von Laue cut himself off, step by step, from the traditions of the herd, under the influence of a very strong sense of justice."

He and Born continued to agree to disagree about quantum theory. As Einstein put it, "You believe in the God who plays dice, and I in complete law and order in a world which objectively exists, and which I, in a wildly speculative way, am trying to capture ... I firmly *believe*, but I hope that someone will discover a more realistic way, or rather a more tangible basis than it has been my lot to find. Even the great initial success of the quantum theory does not make me believe in the fundamental dice-game, although I am well aware that our younger colleagues interpret this as a consequence of senility."

Hedi Born was overjoyed to renew a relationship interrupted by the war. She told Einstein that she had become a Quaker, and that after reading his letter to her husband several times she felt as if she were standing in crystal clear air on top of Mount Everest. She, too, did not believe in a dice-playing God, but she couldn't accept that everything is predetermined, such as whether or not she would have her child inoculated against diphtheria. Hedi also repeated Einstein's

---

* Dimitri Marianoff and Palm Wayne. *Einstein: An Intimate Study of a Great Man.* New York: Doubleday, Doran & Co., 1944.

long-ago remark, "Where females are concerned, your production center is not situated in the brain," to show "how well all your shameful sayings are fixed in my memory!" She also wished she could hear him "roar with laughter once again!"

Max Born also wrote to Einstein to say that he failed to understand how Einstein could "combine an entirely mechanistic universe with the freedom of the ethical individual . . . To me a deterministic world is quite abhorrent."

In late September 1944, Einstein received heartbreaking news from Italy, sent at the request of his cousin Roberto Einstein. Roberto was the son of Albert's Uncle Jakob, Hermann's brother and business partner, the one who had once told young Albert that algebra was a merry science.

The news came from an army major stationed at Fifth Army Headquarters. He wrote:

> I had occasion to visit in this community [Troghi, a village near Florence] and interview Roberto Einstein in connection with a severe tragedy which has overtaken him. Roberto has requested that I inform you that his wife and two daughters . . . suffered death on August 3rd at the hands of the Nazi enemies. Roberto himself escaped unharmed and is presently living at his villa where he is carefully being attended by his sister-in-law and nieces. I regret that censorship does not permit me to dwell upon the tragedy which is well known to me. I sincerely regret the nature of this letter and trust that before long the war and its terrible consequences will have come to an end with the complete and final defeat of the German armies.

Two months later, Roberto himself wrote to explain that on August 3, the day before the Nazis retreated from the area, they had killed his wife, Cici, and his daughters, Nina and Luce, then burned their home while he was hiding in a nearby wood. He had heard that Maja's home was in good condition and that the library was safe. "The American Inquiry Commission for atrocities has already been here," he wrote, "and I trust that you will help me to obtain the identification and punishment of the murderers." A year later Roberto committed suicide.

Distressed by the news of his murdered relatives, Einstein was walking in the street after a heavy rainstorm. At the best of times he hardly looked where he was going. Now about to cross the street, he fell into an open storm sewer. A few moments later, Princeton photographer Alan Richards was driving along Mercer Street. He had just begun his career as a photographer in Princeton. Halfway down the block at street level, in the gutter, he saw

> two . . . outstretched [arms] and a . . . familiar mop of shaggy white hair. Acting on professional instinct, I jumped out of the car, camera in hand, and snapped a picture. Only then did I hear the croaking sounds he was making and see the strained expression on his face. I rushed over, grabbed him under the arms, and hauled out one of the greatest scientists the world has ever known . . . As I walked him the short distance to his home he implored me in a flood of German-accented emotional phrases, mingled with sobs of pain, not to publish the picture. I promised him I wouldn't and I never did. Actually, I couldn't have even

if I had wanted to. When I got home that night I discovered the picture was a blank. In my excitement I had forgotten to pull the film slide.

On December 11, Otto Stern, who had known Einstein in his Prague days, called on him at Princeton. Exactly what Stern said is not known, but he had been working on the Manhattan Project, the code name for the building of the atomic bomb, and he obviously alarmed Einstein in describing its devastating potential. The next day, with Stern's approval, Einstein wrote to Niels Bohr, suggesting that influential scientists including Bohr himself, Peter Kapitsa in the Soviet Union, Frederick Lindemann in Great Britain, and Arthur Compton in the United States, should unite to prevent "a secret technical arms race after the war . . . which will lead inevitably to preventive wars and . . . destruction even more terrible than the present destruction of life. The politicians do not appreciate the possibilities and consequently do not know the extent of the menace."

His idea, which he had already mentioned to Max Born, was for these scientists as a group to pressure political leaders into internationalizing military power—"a method that has been rejected for too long as being too adventurous . . . Don't say, at first sight, 'Impossible' but wait a day or two until you have got used to the idea." Even if there was just one chance in a thousand of it succeeding, Einstein suggested, they should discuss it further.

Bohr didn't have to wait even a second or two to think it over. This was virtually the same idea he had urged on both Churchill and FDR a few months previously. It meant, of course, sharing the atomic bomb with the Soviet Union. Churchill had turned him down flat, treating him as if he were either a naive idiot or a Soviet agent; FDR had listened patiently but later agreed with Churchill that Bohr should be watched as a possible security leak.

Bohr apparently had been persuaded that Churchill and Roosevelt knew what they were doing. When he called on Einstein at Princeton on December 22, Bohr assured him that "the responsible statesmen in America and England were fully aware" of the dangers and opportunities created by the atomic bomb.

Reassured, Einstein advised Otto Stern on December 26, 1944, that for the time being they should keep their idea to themselves.

CHAPTER 33

The Atomic Bomb

1945
66 years old

By late January, Soviet troops threatened Berlin from the east, while in the west the Allies were beating back the last desperate German counterattack, known as the Battle of the Bulge. Hundreds of thousands of Jews were being driven by their jailers from concentration camps in the east to Dachau, Belsen, Mauthausen, and other camps further west to prevent their release by the advancing Russians. Looking for influential supporters to aid their cause, knowing many would be seeking refuge in the weeks ahead, Theodore Strimling, a member of the Los Angeles Committee on Palestine, asked Einstein to approach Caltech's chairman, Robert Millikan. Einstein agreed to try.

Replying to Einstein's plea, Millikan wrote, "Thank God it looks as though the days of [Hitler's] malignant influence are about gone." He said he had been an active member of a group trying to save Jews "from persecution and extermination by Hitler and his cohorts," but that he, like many of his Jewish friends, opposed Zionists "who seem to me to be endangering the peace of the world through creating antagonism between the Moslem and Christian world." Millikan suggested that about half the local Jewish community would support his views.

Einstein clearly belonged to the other half. Moreover, his reputation for naivete and gullibility was largely undeserved. As he told Strimling, "I received an answer from Dr. Millikan in which he declined any help for your worthy cause. The explanations given for this are long but do not make the impression of sincerity. I suppose that he tries to avoid to alienate people who could give financial assistance to his institute (Oil!)"

The Institute for Advanced Study had moved to its own private campus apart from Princeton University in 1939. Einstein and the almost one hundred other members and faculty now worked in bright new offices in a building named Fuld Hall. Although Einstein and four of the original faculty members should have retired, having reached the age of sixty-five, they remained on full pay and attended faculty meetings as usual.

March 14 rolled around, but Einstein tried to ignore the fact that it was his sixty-sixth birthday. When Alan Richards arrived at Einstein's home to take an official photograph requested by Princeton University, he had a little trouble getting past the ever-vigilant Helen Dukas. Einstein had apparently forgotten to tell her that he had okayed the assignment.

Climbing the slightly rickety stairs to Einstein's study, Richards expected

> to find him waiting impatiently in a wing collar and frock coat, with all the dignity of his genius. Instead ... he was dressed in baggy slacks and an old sweater, his mustache straggly, his hair looking as if it hadn't been cut or combed for months. I was appalled. I wondered why, since he hadn't done anything about the rest of his face, he had bothered to shave ... Once when I brought him a dozen extra prints of a particularly good portrait to give to friends, he shoved the whole group aside. "I hate my pictures," he said. "Look at my face. If it weren't for this," he added, clapping his hands over his mustache in mock despair, "I'd look like a woman." On another occasion, when a young couple, at whose wedding he had been best man, brought their son—a little boy of 18 months—to meet him, the child took one look and burst into a screaming fit. The parents were speechless with embarrassment but Einstein's eyes lighted up. He smiled approvingly, patted the youngster on the top of his head, and crooned: "You're the first person in years who has told me what you really think of me."

Richards finally took the official pictures from the knees or the waist up, "but it was difficult for me to keep my eyes off those bare ankles."

On April 12, 1945, President Roosevelt died and Harry Truman took over as president. The war in Europe was almost won by then. The Allies' overwhelming airpower had paralyzed the German army. Hitler ordered his troops to fight to the last man and boy, but he didn't live to see if they obeyed his orders. On April 30, less than two days after he and longtime companion Eva Braun were married, they committed suicide—she took poison, he shot himself. A week later, on May 7, the German army surrendered unconditionally. And that was the end of the Third Reich.

Thomas Bucky, now serving in a U.S. Army intelligence unit in Germany, was aware of the worst horrors of the Nazi regime. He remembers writing to Einstein, saying, " 'The only good German is a dead one,' and 'Every German is a Nazi,' and all those terrible things one did say in those days. But Einstein replied, 'Now, look, it's pretty difficult to go against the stream when your life and livelihood are in jeopardy,' and 'many people struggled against the Nazis.' He was not almost paranoid against the Germans, as his biographer Clark said. But he did everything he could to combat Hitler."

As well as corresponding with friends like Bucky and with colleagues at home and overseas, Einstein, with his soft spot for kooks, was likely to reply to what others might consider nonsense. He was sure to answer serious questions like those from a navy ensign aboard the USS *Bougainville*, who "had quite a discussion last night with a Jesuit-educated Catholic officer who said you were an atheist and you talked with a Jesuit priest who gave you three syllogisms [sic] which you were unable to disprove; as a result of that you became a believer in a supreme intellect which governs the universe. The syllogisms were: A design demands a designer; The Universe is design; therefore there must have been a designer."

Einstein replied on July 2 in duplicate letters to both men: "I have never talked to a Jesuit priest in my life and I am astonished by the audacity to tell such lies about me. From the viewpoint of a Jesuit priest I am, of course, and always have been an atheist . . . It is always misleading to use anthropomorphical concepts in dealing with things outside the human sphere—childish analogies. We have to admire in humility the beautiful harmony of the structure of this world—as far as we can grasp it. And that is all."

At dawn on July 16, the atomic structure of the world was revealed when Einstein's famous equation, $E = mc^2$, came to life with a bang. That was the day the first atomic bomb was detonated in the southern New Mexico desert. Watching the fireball climb into the sky, project director J. Robert Oppenheimer recalled a line from Sanskrit literature: "I am become Death, the destroyer of worlds."

With Germany defeated, the Allies now concentrated on Japan. To avoid a prolonged invasion of the country and the inevitable enormous casualties on both sides, President Truman ordered the use of two atomic bombs—one to devastate Hiroshima, the other, Nagasaki.

Helen Dukas was resting at Saranac Lake, listening to the news on the radio, when, she told Einstein biographer Jamie Sayen, "they reported something . . . about the war, and then said a new kind of bomb has been dropped on Japan. And then I knew what it was because I knew about the Szilard thing in a vague way . . . As Professor Einstein came down to tea, I told him, and he said, 'Oh, Weh' [Alas, oh, my God!]."

Later that day, Einstein was more forthcoming. He told a reporter, Raymond Swing, that the world was not yet ready for the atomic age. Nevertheless, perhaps to prevent panic, he compared nuclear energy and atomic power with sunlight, explaining: "In developing atomic or nuclear energy, science did not draw upon supernatural strength, but merely imitated the action of the sun's rays. Atomic power is no more unnatural than when I sail a boat on Saranac Lake"—which he had just been doing.

Einstein even took an almost upbeat attitude to the potential effect of an atomic war, telling Swing that it would probably destroy only two-thirds of the world and would leave enough intelligent survivors and books to restore civilization. He believed that the secret of the bomb should be shared not with the United Nations or the Soviet Union, but with a world government founded by the world's three great military powers—the United States, the Soviet Union, and Great Britain. He said he was more fearful of future wars than of the possible tyranny of a world government. Whatever the decision, he did not expect the United States and Great Britain (whose scientists had joined with Americans to produce the bomb) to be able to keep their secret for very long.

Denying he was "the father of the release of atomic energy," Einstein informed Swing that he had merely suggested that the process was theoretically possible. "It became practical through the accidental discovery of chain reactions, and this was not something I could have predicted," he said. "It was discovered by Hahn in Berlin, and he himself misinterpreted what he discovered. It was Lise Meitner who provided the correct interpretation and escaped from Germany to place the information in the hands of Niels Bohr."

His final comment that day has so far proved right, at least insofar as it relates to the major powers: "It may well intimidate the human race into bringing order into its international affairs, which, without the pressure of fear, it would not do."

Two days after the bombing on August 6, Dukas wrote to her friend, Alice Kahler:

> I feel as if the atom bomb has hit me personally—and I have known of it since 1939, ever since Einstein wrote the letter to Roosevelt, who then named a committee. I nearly got a stiff arm receiving telegrams over the phone. One [reporter] even called Mrs. Marks (Born Strauss, whose family owned Macy's Department Store and the estate [on which Einstein was staying] at Saranac Lake) and another chased poor Walsh (caretaker of the estate) out of bed at 11 p.m. with a long distance call. But he said he would pick up no one at that hour. There were so many telegrams asking for a statement, starting with the *Princeton Herald* as well as the English and French press, so that Einstein did not give any statement, which is the simplest way out. One of the young people [reporters] who was a guest at the Sulzbergers from the *New York Times* [the Sulzbergers had a summer home nearby] came over late at night . . . Arthur H. Sulzberger [president of the newspaper] also called him [Einstein] constantly for a statement. But no dice either.
>
> It [the atomic bomb] was a terrible thing, and a person who was working on it told A.E. he wished they would not succeed, since Germany is finished, and one has to be happy after all that the Germans did not get it; they were very close to it indeed. [Not true. At the end of World War II, the allies captured ten leading German physicists—Heisenberg, von Laue, and Hahn among them—and kept them under guard in Farm Hall, an English country mansion. Their rooms were bugged by British intelligence operatives and their conversations revealed that not only had they abandoned the idea of working on an atomic bomb but that they were astonished at the Allies' success in producing at least two.]
>
> Churchill's adviser, Lord Cherwell, visited Einstein in Princeton this spring. Prof. Bohr likewise—I was not supposed to say anything about their being here—I myself did not know why they came—so secret was everything. I believed it was a social visit since both were old friends (Lord Cherwell was Prof. Lindemann in Oxford with whom E. always stayed) but E. took my word of honor not to tell anyone about their being here. I always heard about the bomb. One of the former assistants of E. (Leo Szilard) worked with Fermi, and he was the one who prompted E. to write the letter to Roosevelt. They were terribly afraid the Germans would succeed . . . Lise Meitner (a famous physicist who together with Otto Hahn produced the first chain reaction of the atom in Germany) deserves great praise, since she kept the secret til she was across the [German] border. She is in Stockholm . . .
>
> I am afraid of the mail—there will be quite an avalanche since it is true, as the newspapers reported, that the equations of Mass and Energy, which E. found thirty-five years ago, are the theoretical basis of the bomb. For this reason all the

telegrams. Never become famous. It is a plague. What the future will bring one cannot foresee—if mankind does not come to its senses.

Dukas would have been more despondent and wary of mankind had she known that the FBI was now going after her. Having wasted thousands of FBI man-hours investigating Einstein — and filling the files with false reports that he was a dangerous Red working on a death ray who planned to take over Hollywood — J. Edgar Hoover had opened a file on Dukas, too. Based on a tip from Mexico, he believed that Dukas might be corresponding with a man named Otto Katz in Mexico City, who was reported to be a Soviet agent.

Hoover's approach was surprisingly timid. He continued to veto suggestions to bug Einstein's phone or to interrogate him, afraid that if Einstein got wind of it, he would blow the whistle and embarrass the bureau as well as Hoover himself. Instead, he put his agents to work clipping news items from local papers with the names and descriptions of Einstein's dinner and vacation guests.

Getting nowhere with Einstein, Hoover concentrated on Dukas. He lifted a recent photo and her statistics from her naturalization certificate: born in Freiberg, Germany, on October 17, 1896, brunette, height 5 feet 2 inches, weight 101 pounds. He then put her on the National Special Censorship Watch List, so that all letters she sent out of the country and all her incoming mail from abroad were tested with chemicals for secret messages. When the results proved negative, decoding experts got to work, spending day after day filling page after page with permutations that always came out as gobbledygook.

In September, an FBI agent told Hoover that the bureau had not yet found any evidence that a suspect, Russian-born Jacob Billikopf, was a Communist spy. The FBI did know, however, that he had been involved in an argument in a Philadelphia hotel in the early 1930s in which Einstein's name had been mentioned. According to a news report quoted by the agent, Billikopf had been discussing Hitler's anti-Semitic policies with two other men. When a Dr. Wilbur Thomas, who was also in the hotel, suggested that one should not view the subject emotionally, but calmly and coolly, Billikopf sprang to his feet and recited a long list of Hitler's cruelties to the Jews, saying that it was an effrontery not to view such things with emotion. At that, another hotel guest, Dr. Friedrich Schoenmann, a professor of American civilization at the University of Berlin and an exchange professor at Harvard during World War I, attempted to soft-pedal Hitler's Jewish policy. When asked why people had left Germany, he said, "It took several weeks for your country to let Einstein in. Professor Einstein is recognized all over Europe as a Communist advocate. It was because he was a Communist rather than because he was a Jew that he was forced out of Germany."

This report did not help Hoover's case. He needed a stronger witness than a Hitler apologist to deport Einstein as a security risk.

By year's end, Hoover thought he was on to one: the Russian sympathizer Jacob Billikopf. The man who had argued so vehemently in the Philadelphia hotel over Hitler's treatment of the Jews had been corresponding with Helen Dukas. Perhaps, reasoned Hoover, Billikopf was a spy after all, and contacted Einstein through Dukas. The hunt continued.

CHAPTER 34

Toward a Jewish State

1946
___
67 years old

An FBI agent filled in the details for J. Edgar Hoover. Putting Einstein's secretary on the Watch List the previous year was paying off. Helen Dukas was, he said, in contact with "Boenheim, who is actively engaged in Communist-dominated organizations, as well as with Billikopf, a Russian-born social worker and sponsor of Russian war relief." The agent thought it unwise to attempt physical surveillance "because of the small size of the town and the outstanding name of the subject's employer," and admitted that Dukas might conceivably be innocent. (In fact, it was never established why she was in touch with the two men.) The FBI agent recommended tapping her phone anyway, in case both she and Einstein were spies, as he suspected. Afraid of repercussions if Einstein found out that his home was bugged, Hoover again denied the request. Nevertheless, he continued to take an interest in all the news reports and anonymous tips on what Einstein and Dukas were up to. These included an interview with a former inmate of an "insane asylum," who said that before World War II Einstein had invented a robot that could read the human mind, in order to send American military secrets to Germany.

None of Hoover's agents seemed able to tell him the simple truth: that Einstein sympathized with the Russian people but was a dedicated democrat who hated dictatorships of any kind.

In addition to Einstein's work on unified field theory and his massive correspondence, he was preparing to express his views about the future of Palestine and current British policy in the Middle East. The British government, which governed Palestine under a League of Nations mandate, had offered to let fifteen hundred Jewish refugees emigrate there every month, up to a total of seventy-five thousand. The Zionists and the Jewish Agency rejected the offer.

On January 11, Einstein testified before a joint Anglo-American Committee of Inquiry on the future of Palestine. Although an Arab witness was speaking as

Einstein entered the committee hearing room, many in the audience recognized him and applauded. The chairman quickly silenced them. But when Einstein was called to speak, the chairman invited the crowd to welcome him. During the applause Einstein whispered to a friend, "They ought to wait first to see what I say." He was right. His benign smile did nothing to prepare them for his message; he launched into the most wholehearted denunciation of British colonial policy that the committee had yet heard. "As a former admirer of the British system," he said, he regretted that he was now convinced that "as long as the British are ruling Palestine there will be no peace between Jews and Arabs."

Continuing in pleasant, measured tones, he condemned the British for ruling Palestine as they ruled India, where they supported big landowners who exploited the workers. He also accused the British of pursuing a policy of divide-and-conquer by encouraging clashes between Arabs and Jews, afraid that if they stopped fighting there would be no need for British rule. Furthermore, he said, the British had sabotaged the Balfour Declaration by restricting immigration and Jewish ownership of land and by tolerating corrupt Arab landowners who incited the Arab masses to attack Jewish settlers. The Jews, he said, had raised the Arabs' standard of living and this threatened Arab landowners who benefited from the continued poverty and ignorance of Arab peasants.

Einstein's supporters showed approval when he denounced the committee as a smoke screen behind which the British would continue to pursue their own selfish interests while ignoring the committee's recommendations.

A British member of the committee, Richard Crossman, M.P., got Einstein to concede that it was not "a British imperialist fiction" that Arabs might shoot Jewish refugees if they came in large numbers. Then Crossman asked him if he would like Americans to replace the British in Palestine.

Einstein replied that he was against any one power running the country. Instead, he proposed a trusteeship of several nations appointed by the United Nations until the country was ready for home rule, and that most European refugees be allowed to settle there immediately.

When asked what would happen if the Arabs resisted the immigration of the refugees, Einstein said, "They won't if they are not instigated."

He probably spoke for many Zionists in the audience until he defined his idea of home rule, which would require neither a Jewish state nor a Jewish majority. Instead, to ensure that Jews and Arabs lived in harmony, he favored government by an international entity.

At that, according to Richard Crossman, "the audience nearly jumped out of their seats."

But the damage to Zionist aspirations was quickly repaired. A few days later, Einstein agreed to sign a statement prepared by his friend, Rabbi Stephen Wise, who wrote in a covering letter to him, "Any reference to *bi-nationalism* at this time might be exceedingly hurtful in the hearing of a Committee which is not too friendly to us."

The critical part of the statement read:

A National Home I consider a territory in which the Jews have such rights that they can integrate freely within the limits of the economic absorptive possibilities and that they can purchase land without undue encroachment on the Arab peasant population. The Jews should have the right of cultural autonomy, their language should be one of the languages of the country, and a government should exist, working under strict constitutional rules that guarantee is given to both groups that no "Majorisation" of one group by the other is possible. There must be no discriminatory laws against the interest of either group.

But that was as far as Einstein would go, telling Rabbi Wise, "I am firmly convinced that a rigid demand for a 'Jewish State' will have only undesirable results for us."

Radical political journalist I. F. Stone, reporting for the liberal newspaper *PM*, was there when Einstein testified before the committee. Stone recalled that "although I was very much a Zionist and still am, I was very proud of Einstein, because to have the greatest Jewish figure of the period oppose a Jewish state as unfair to the Arabs was a very noble thing. He rose above ethnic limitations. He had concern for the Arabs, as he should have had. [Chaim] Weizmann himself wasn't for a Jewish state. It was [David] Ben-Gurion, the man Weizmann called a whirling dervish, who forced the Jewish state issue. Weizmann left the whole question unclear, because the Jews were a minority in Palestine.

"I was just on the eve of making my own trip with survivors of the Holocaust through the British blockade to Palestine and [was] very sympathetic, of course, to the Jewish and the Zionist cause. [Stone subsequently spent eight days on a ship in the Mediterranean that ran the British blockade to land hundreds of "illegal" Jewish refugees in Palestine.] I'm still not anti-Zionist, but I'm also not hostile to Arabs' aspirations and rights, too. I had gone to Palestine the previous year and fallen in love with the Jewish community there and the kibbutz movement, which seemed to be so much like Kropotkin's view of a voluntary anarchist community. Kropotkin was the radical Russian prince.

"Not a majority, but a substantial minority, was then in favor as I was of a binational state, not a Jewish state, which Einstein was also in favor of. The idea was that there would be a constitution which would recognize two languages, Hebrew and Arabic, and two different cultures in the same state. So, no matter who became the temporary or permanent majority the other would have an equal status."

Stone wrote to congratulate Einstein on his testimony. He, in turn, invited the journalist to his home for a talk, and they became lifelong friends.

Einstein's compassion for Jews and Arabs did not extend to the Germans. Despite his softening attitude in letters to Thomas Bucky, he refused to join an appeal by Professor Middledorf of Chicago University to ease the harsh surrender terms imposed on Germany. He had concluded, he said, from the opinion of the few Germans he knew who were neither Jews nor married to Jews, that anyone who signed the appeal was playing into the hands of Nazis in Germany and Nazi sympathizers in the United States. As for a statement in the appeal, that "retribution falls equally on the guilty and the innocent," he pointed out that "guilty" Germans would benefit along with the "innocent" should the appeal succeed, and estimated that guilty Germans outnumbered the innocent ten to one. He also ridiculed using the "notorious *London Economist*" as an authoritative source for conditions in Germany, saying that was like quoting Dr. Eck as an authority on the Reformation.

He even closed the door on rejoining the Prussian Academy. His reply to Arnold Sommerfeld's invitation to do so left no doubt about his feelings toward Germany: "The Germans slaughtered my Jewish brethren; I will have nothing further to do with them . . . I feel different about the few people who, insofar as it was possible, remained steadfast against Nazism. I am happy to learn you were among them."

Einstein's unyielding position was understandable. Along with his relatives murdered by the Nazis in Italy, and Louis Einstein, his cousin Lina Einstein had been murdered at Auschwitz, and his cousin Bertha Dreyfus had died at Theresienstadt.

Early in the year, Einstein resumed his lively exchange of ideas by mail with his friend and former colleague, Erwin Schrödinger, who was now working in Dublin, having left Germany because the Nazis disgusted him. Einstein, having made some progress in his work on unified field theory, mailed two unpublished papers on the subject to Schrödinger, remarking: "I am sending them to nobody else, because you are the only person known to me who is not wearing blinkers in regard to the fundamental questions in our science. The attempt depends on an idea that at first seems antiquated and unprofitable, the introduction of a non-symmetrical tensor as the only relevant field quantity . . . Pauli [who was also at the Institute for Advanced Study in Princeton] stuck out his tongue at me when I told him about it."

Schrödinger wrote on February 9 that he had studied Einstein's work for three days and was very impressed with it. Einstein was astonished at how quickly his friend had mastered the material. Schrödinger modestly replied: "You are after big game . . . You are on a lion hunt, while I am speaking of rabbits." They continued their exuberant discussion by mail, and on April 7 Einstein wrote, "This correspondence gives me great joy, because you are my closest brother and your brain runs so similarly to mine." On May 20, Einstein admitted that "inwardly I am not so certain as I have put it forward . . . We have squandered a lot of time on this thing, and the result looks like a gift from the devil's grandmother." He wondered if he could introduce probabilities into field theory instead of insisting on "the real situation" of the particles. "I have not laughed so much for a long time as over the 'gift of the devil's grandmother,'" Schrödinger replied. "You described exactly the way of the cross that I also traveled in order to end up with something that is probably even more impossible than your result."

A month later Einstein reported progress: "Thanks to the truly great skill and persistence of my assistant [Ernst] Straus, we have recently got this far." It was too far for Schrödinger. Einstein maintained that "in the new theory the transverse wave field is indeed present but as such transports no energy." Schrödinger preferred a "purely classical wave theory, in which the structure of space-time would yield gravitation, electromagnetism, and even a classical analog of the strong nuclear reactions."

Though Einstein and Schrödinger seemed to be on the same wavelength, Einstein's behavior seemed incomprehensible to politician Alan Cranston (later a U.S. senator from California). Cranston came to Princeton that spring to chair a meeting on the state of the world and American security and survival in the nuclear age. "Every now and then we had votes by a show of hands," said Cranston, "and I saw Einstein vote Yes, and then vote No for the same proposal. Which startled me. So, during the coffee break I asked him, 'Why?' And he said, 'There are so many nice people on both sides I cannot bring myself to vote against any of them.' I thought that was funny, and showed his gentle side. But he spoke with great conviction about the need to avoid nuclear war and I think he felt he bore a heavy burden in having played a part in helping bring these weapons into existence [and that] he'd have to play a part in getting them under control or abolished. Speaking of Russia, he thought it in our mutual interest to avoid a nuclear war and to tame the arms race so our economy would not be undermined. And to get research directed to purposes that would lift standards rather than consuming our best brains in a military role."

Einstein was under no illusions about how difficult it would be to prevent a nuclear war. At the same meeting he was asked: "Why is it that when the mind of man has stretched so far as to discover the structure of the atom we have been unable to devise the political means to keep the atom from destroying us?" Einstein replied, "That is simple, my friend. It is because politics is more difficult than physics."

Einstein surprised Russian writer Ilya Ehrenburg, who thought he had lost the faculty for feeling surprise. Ehrenburg recalled that

quite unexpectedly, on 14 May 1946 I was struck dumb, like a child who for the first time witnesses some extraordinary natural phenomenon: I was taken to Princeton and found myself face to face with Einstein. I spent only a few hours with him but my memory retains those long hours better than many an important event in my life. One can forget joy and trouble, but one never forgets amazement. His eyes were astonishingly young, by turns sad, alert or concentrated, then suddenly full of mischievous laughter like a boy's. He was young with the youth that years cannot subdue; he expressed it himself in this casual phrase: "I live and I feel puzzled, and all the time I try to understand." I knew that Einstein was interested in the *Black Book*, the general title of a collection of human documents: diaries, letters and statements by eyewitnesses concerning Nazi crimes against the Jewish people in the occupied territories. Einstein examined them with close attention and when he raised his eyes I read grief in them, and his lips twitched slightly. He said: "I have often said that the potentialities of knowledge are unlimited, as is the knowable. Now I think that vileness and cruelty also have no limits."

When Ehrenburg said he was going to the South to see how the Negroes lived, Einstein told him: "They live in terrible conditions. It's shameful. The actions of the legislatures in the Southern States are covered by some of the counts of the Nuremberg indictment." Soon afterward Einstein spoke to Ehrenburg of a beautiful young American girl who, defending discrimination, had asked him, " 'What would you say if your son announced he was going to marry a Negress?' He said he had replied: 'I should probably ask to meet his fiancee. But if my son announced he wanted to marry you I should certainly lose both sleep and appetite,' and his eyes lit up with a challenging gleam."

Just before Ehrenburg left, Einstein said: "The main thing now is to prevent an atomic catastrophe. It's a good thing you have come to America. I hope more Russians will come and talk to us. Mankind must prove itself more intelligent than Epimetheus who opened Pandora's box and could not shut it again." A few days later Ehrenburg "heard a familiar voice on the radio: Einstein was speaking about the deadly danger hanging over humanity, about the necessity to come to an understanding with the Russians to renounce atomic weapons, not to arm but to disarm—in this way he tried to shut the Pandora's box."

There was more bad news on the world front. The British in Palestine were forcibly removing illegal Jewish immigrants to detention camps on Cyprus. Members of the Irgun, a Jewish terrorist group, retaliated on July 22, 1946, by blowing up the British army's headquarters in Jerusalem. Ninety-two people were killed, Arabs and Jews among them. The extremists on both sides were unlikely to accept the recommendation of the Anglo-American Committee concerned with Palestine. They had echoed Einstein's proposed compromise, that the country "be neither an Arab nor a Jewish nation, [and] both Arabs and Jews be treated alike."

If Einstein needed to escape from the nightmare of a devastated world, he had merely to look through his mail from women. The shortest and sweetest note that year was from a stranger in Maine who gave her name and full address. It was: "I love you although I know I am not worthy of you."

That missive almost certainly got no reply. He did respond to a serious request from a doctor's wife whose husband had been hearing voices. She wrote:

> He feels these voices, which are not malicious but intended to be helpful, emanate from some committee which is using atomic or electronic waves for the public good . . . I realize all of this may sound fantastic to you (or does it?). I, of course, feel it is and so do the doctors. My husband is highly intelligent and has told us that if physicists said that there was no such invention he would believe it and admit that he is "crazy" . . . If your decision would influence him to the extent of cracking his defenses even a little he might be immensely more reactive to therapy . . . He may yet function constructively in society if you can help me.

Einstein was busy, but nevertheless replied immediately. He said that people suffering from hallucinations often think their experiences are caused by radio waves acting directly on the brain. Scientists, he said, had never confirmed this, and he was convinced they were right. He added:

> On the other hand it is understandable that a person suffering from these symptoms is trying to find a cause for them which conforms as well as possible with his knowledge about the outside world. Taking all this into account your husband

will probably come to the conviction that his symptoms are not created by any outside cause but by the functions of his own brain. I hope it will help him somewhat to overcome the disturbing condition.

There were two sure things about Einstein: he would never ignore a cry for help or miss a chance to make music. Dorothy Commins, the wife of Saxe Commins, the editor in chief of Random House, soon discovered Einstein's passion for music. Seeing him walking toward her on Mercer Street, she stepped off the sidewalk to let him pass. She remembers that "he touched my elbow and said in German 'Oh, no, please!' And I stepped back on the narrow curb.

"He then asked my name, and when I told him, he said, 'Now I know, you are a pianist and are living in Professor Hermann Weyl's house.' [The Comminses stayed there for the summer, while Weyl was in Zurich and before they found a permanent residence in Princeton.] Then he asked, 'Do you think we could play together?' I knew he played the violin, and my immediate reply was, 'I shall be delighted.' To my surprise, he said in German, 'Tonight?' 'Of course,' I replied. Then Professor Einstein added, 'At eight o'clock I shall be there.' From that moment I was walking on clouds.

"When my husband came home from New York with two friends I told them of my adventure. They looked at me thinking I had taken leave of my senses! Before eight o'clock, Saxe was on his way to Professor Einstein's home, where he was standing at his door holding his violin. He handed Saxe some pages of music—Bach and Corelli and a small collection of Dutch Folk Melodies arranged for violin and piano by Roentgen, the brother of the discoverer of X-rays. We played the Bach Concerto in D Minor, then turned to Corelli. Once he stopped and said in German, 'Could we play that section again, I've made so many mistakes?' In truth, he loved the piece and wanted to play it over and over."

Throughout the summer of 1946, Einstein was involved with Hollywood, as William Lanouette found during his research for a biography of Leo Szilard (*Genius in the Shadows: A Biography of Leo Szilard, The Man Behind the Bomb*). Lanouette discovered some funny exchanges between Einstein and Szilard, "when MGM wanted to do a movie about the atom bomb, *The Beginning or the End*, and needed to get Einstein's permission to portray him. Szilard became his agent, shuttling back and forth to Hollywood to rewrite some scenes so they would be accurate. This was somewhat outside the orbit of the serious arms-control work they were doing. The film opens as a pseudodocumentary in the year 2000 when people are burying remnants of atomic debris, and then it does a flashback to the making of the bomb. It's hilarious in some respects, one of the more lighthearted collaborations that Einstein and Szilard had."

In what was claimed to be a unique event, Metro-Goldwyn-Mayer's production chief, Louis B. Mayer, called a meeting of every producer in the company to ask their help in producing "the most important story ever filmed."

Einstein resisted giving his okay to being portrayed, uneasy with Hollywood's idea of "artistic truth." The script originally called for his historic meeting with Szilard and Wigner about the need to warn FDR of the danger of a Nazi atom bomb, which occurred at the Long Island vacation cottage, to take place in

Princeton. Mayer eventually agreed to make several script changes to grant Einstein's wish for "scientific" rather than "artistic" truth, but in typical Hollywood scorn for authenticity a Canadian, Hume Cronyn, played Oppenheimer, and an English actor, Godfrey Tearle, portrayed FDR. Brian Donlevy was more appropriately cast as General Groves, the military head of the Manhattan Project, and a comparative unknown, Ludwig Stossel, played Einstein — in a smoking jacket. It's doubtful if Einstein ever saw it or wanted to after reading critic Bosley Crowther's opinion that "the film is so laced with sentiment of the silliest and most theatrical nature that much of its impressiveness is marred. The sentiment derives from two romances which are fictiously twined with the military and scientific projects to develop the atomic bomb." Helen Dukas probably saw the movie and gave Einstein a laughing blow-by-blow account of the fiasco.

That same year, Time-Life made a documentary about the making of the bomb in its *March of Time* series. This too wavered toward "artistic truth," even though Einstein himself took part. But these filmmakers also avoided a trip to the far end of Long Island, using instead the porch at the back of his Princeton home to represent the historic Long Island meeting.

That year Maja suffered a stroke, and throughout the summer she remained confined to her bed. Besides his sister's paralysis, Einstein's chief concern was the fear that other nations might build atomic bombs and plunge the world into a war that could destroy civilization. He willingly headed a group to warn and inform the public. Named the Emergency Committee of Atomic Scientists, it included Harold Urey, Leo Szilard, Victor Weisskopf, Linus Pauling, and Hans Bethe—many of whom had helped build the bomb. At first, as he told his friends the Upton Sinclairs, he was merely the figurehead, "so [there is] very little trouble and work for me," but he was soon to give the committee much time and effort. Initially pleased with the response of the American public and scientists, he wondered if it would help: "We cannot deceive ourselves that the outlook is gloomy and the insidious press-campaign against Russia powerful; the policies of the government remind me of the German policies in the days of Wilhelm II [a warmongering seventeenth-century Dutch prince]."

An early response came from Ashley Montagu, an English-born Princeton University anthropologist and social biologist, who was already in the midst of planning a film on both the dangers and the possible advantages of atomic energy for the Federation of American Scientists. He phoned Einstein for advice and Dukas answered. "When I said it was for a film," Montagu recalled, "she got excited and said, 'Oh, Hollywood!' and I said, 'No, Pennsylvania.' She spoke to Einstein about it and he immediately came to the phone."

It was obviously a more appealing enterprise than the MGM and Time-Life efforts, for Einstein invited Montagu over. As Montagu recalled, "My first sight of him was very interesting. I don't know whether he was fond of dancing or not, but I have been all my life. It was very striking. There's a long corridor in the house and he was at the other end when Miss Dukas called him and said I was here. And then he seemed to glide towards me in a sort of undeliberate dance." Montagu thought it "enchanting"—as if Einstein were walking on air. "It was maybe the way someone else might whistle as they moved. He danced . . . He seemed somehow to be expressing his love of music as he moved. He was in his slippers and he cordially invited me up to his study."

Montagu was shy about meeting Einstein, having been diffident all his life about intruding on great men. "I regret that very much now because I missed many opportunities where I know it would have not been regarded as such by the likes of Havelock Ellis and H. G. Wells, feeling myself a mere worm in comparison with their enormous stature. And this is why I never wrote down anything that Einstein said to me on the occasions when we met . . .

"He asked me to read the script of the film—it took less than fifteen minutes—and didn't respond until I'd finished and then he said, 'A-one. It's just right.' The title was taken from a book published earlier that year, edited by Robert Oppenheimer, *One World or None*. Einstein liked the title.

"At the first meeting we'd got on to the question of what does one do in addition to making such films about getting people interested in seeing that nuclear energy isn't misused? And I asked, 'What do you think?' And he said, 'International law.'

"I said, 'Professor Einstein, international law exists only in textbooks on international law.'"

Einstein exclaimed that that was really an outrageous remark, then took the pipe out of his mouth and thought for several minutes. He finally said, almost mournfully, "You're quite right." Montagu then told him that of all the treaties that had ever been signed between nations every one had been broken with the exception of the one establishing the borders between Canada and the United States.

"I had read a lot of theoretical physics when relativity burst upon England," said Montagu, "so I could talk to him about . . . indeterminacy, which was a particular point of interest to us both. As a determinist, I believed that everything has an explicable cause. When I said, quoting the astronomer Henry Norris Russell, that it might be called 'the principle of limited human measurability,' he was enthusiastic about that idea. In Brownian movement, for example, where it is unpredictable where each atom would be—if we knew enough about the conditions under which the experiment was being performed we *would* be able to calculate where each atom would be, but it's because we have a limited capacity to measure this sort of thing, not that it won't eventually prove to be predictable. This was implied in his well-known remark, 'Subtle is the Lord, but malicious He is not,' which he explained meant that 'nature hides her secrets because of her essential loftiness, but not by means of a ruse.'"

Ashley Montagu often appeared on TV shows in those days in which various subjects were discussed and expert opinions sought. On one occasion, the subject was sleep and how many hours famous individuals needed. Montagu's producer knew he had spoken with Einstein before and suggested that he ask the scientist how long he usually slept. "So I called him up," Montagu recalled, "and he said, 'Seven.' And I said, 'Napoleon said he only slept three hours.' And Einstein said, 'Ah, well, he was such a big boaster!'"

The two men were soon on friendly terms, and during one of their meetings,

Montagu told Einstein a joke with a Jewish inflection which became one of his favorites. "It's about two Jewish tailors in the Bronx," said Montagu. "One of them happens to mention the name of Einstein and the other one says to him, 'Who's Einstein?'

"He says, 'You shlemiel! Who's Einstein? He's only the biggest scientist in the world.'

"'What is he the biggest scientist in the world for?'

"'Relativity'

"'What's relativity?'

"He says, 'Shlemiel! This is relativity. Supposing an old lady sits in your lap for a minute, a minute seems like an hour. But if a beautiful girl sits in your lap for an hour, an hour seems like a minute.'

"'And this is relativity?' his companion asks.

"'Yes,' he replies. 'That's relativity.'

"'And from this he earns a living?'

"Einstein laughed heartily and said it was one of the best explanations of relativity he had heard."

Einstein would also have laughed had he known the FBI was still wasting time and money trying to prove he was a dangerous Communist. Even Walter Bedell Smith, the U.S. ambassador in Moscow, got into the act, reporting the rumor in Jewish circles in Moscow that physicist Peter Kapitsa had, with official blessing, invited Einstein to emigrate to the Soviet Union, which Kapista was purported to have described as

a land of true democracy, free from selfish taint, where they together could pursue their scientific research unhampered by restrictions imposed by capitalist society. Einstein was assured that whatever . . . he might need would be immediately and completely placed at his disposal. Einstein replied in Hebrew through two Jewish members of a trade union delegation. Einstein's letter was addressed to Stalin. His two representatives were received by Molotov. In his letter Einstein expressed appreciation of the offer but before he could consider it he must ask several questions. His questions were: Why are Jewish scientists not permitted to hold prominent posts? Why are apparently unnecessary obstacles placed in the way of Jewish scientific and research workers? Why were certain Jewish professors of medical science whose outstanding contribution to medicine was well known not elected to the recently created Medical Academy? He asked other questions implying anti-Semitism . . . Molotov denied the truth of the implications. The invitation to Einstein was repeated. The Ministry for Internal Affairs immediately investigated the particular cases mentioned by Einstein. Certain high Soviet officials were discharged for anti-Semitism. The Professors named by Einstein were elected to the Academy.

Here was a heady mix of fact, fancy, and wishful thinking. The State Department took the report seriously, though, believing it indicated that the Russians were trying to recruit scientists who were knowledgeable in the field of atomic energy from the United States and other countries. As a result, the FBI took another look in the Helen Dukas file in search of incriminating evidence.

In fact, Niels Bohr and his family had been invited to move to Russia during the war (in a letter from Kapitsa dated October 28, 1943). He had shown the letter and his noncommittal reply to Kapitsa to British intelligence agents.

Bohr was in Princeton that September to receive an honorary degree. While there, he called on a former assistant, Abraham Pais, who had recently joined the Institute for Advanced Study, where he had a small office above Einstein's. Bohr said to him, "Let's go down and see Einstein." Pais still remembers "the friendliness and openness with which he greeted me, a total stranger, a youngster."

Pais had been imprisoned by the Nazis in Holland during the war and afterward worked with Bohr in Denmark. Now he sat in the corner of the room and listened as Bohr and Einstein resumed their ongoing fight over quantum mechanics. Pais was soon working with Einstein, and found "he wasn't like other truly very great men I have known. He was a man in whom the child was still very much alive. He was not juvenile, ever, in his behavior, but something like a playful lust of the child somehow lived in him forever."

When Pais related his experiences in a Gestapo prison, "Einstein was interested, as he was in all things having to do with the fate of the Jews. That sat very deep with him. He bit his lip as I told him how a friend and I had been taken prisoner by the Gestapo. We were involved with the Resistance movement regarding the fate of Jewish orphans after the war, if their parents did not come back from the camps. We had a written document about that. And it was in his pocket when we were caught and was, from the Nazi point of view, extremely incriminating. The Gestapo didn't torture me, just hit me in the face and things like that. But my friend with whom I and others shared the cell got executed. And I didn't. And that was by dint of absolute fate: I could have had the incriminating letter in my pocket."

Einstein was anticipating a visit from his old friend Maurice Solovine, whom he hoped to meet "once again in the best of all possible worlds." Its location must have been somewhere in his imagination, because he had written to Upton Sinclair a few days before to stress his concern about "the manufacturing of an anti-Russian spirit which is produced by newspapers and radio in a way of insidious hypocrisy without directly lying, but by arranging factual material to produce a false impression. May I characterize the method of leadership in fascist and democratic countries as follows: in dictatorships people are led by coercion and lies. In democracies by lies only."

But he was thankful that there were still people like Max von Laue in the world. In the same letter to Sinclair, he mentioned that his close friend and former colleague von Laue had behaved "very courageously during the Hitler-Regime, helping persecuted people and openly voicing his disapproval of governmental policy." He concluded on an upbeat note, "hoping that your courageous and unselfish fight for reason and human decency to which you gave your whole life may be successful."

CHAPTER 35

The Birth of Israel

1947 to 1948

68 to 69 years old

Einstein and Erwin Schrödinger had great fun exchanging ideas by mail across the Atlantic, hoping to advance unified field theory another step. Einstein was impressed with Schrödinger's work, writing on New Year's Day 1947, "You are a clever rascal." Schrödinger was overjoyed, replying, "No letter of nobility from emperor or king, neither the order of the garter nor the cardinal's red hat could do me greater honor than to be called a clever rascal by you in such circumstances." Schrödinger then urged Einstein to join him in Ireland, where "one can live in unbelievable peace and tranquility. This is due to the boundless lack of education and intellectual disinterest of the great majority of the population. This is naturally expressed unkindly. One can also say they are natural, simple people, who do not go in for humbug." Einstein said that he couldn't leave Princeton, where people had done so much for him and besides, it would end their correspondence, which he relished.

Suddenly, Schrödinger believed that he had discovered the key to unified field theory — the Holy Grail that had eluded Einstein for three decades. According to Schrödinger biographer Walter Moore, by "using Affine geometry Schrödinger thought he had succeeded in uniting electromagnetism and gravitation." He was overwhelmed. It was "a miracle," like "a totally unhoped for gift from God." Almost delirious with joy, he could hardly wait to share his momentous findings and quickly arranged to present them to a distinguished group at the Royal Irish Academy. He was convinced that his "Affine Field Theory," as he called it, would earn him his second Nobel Prize.

It was snowing heavily on January 27, the day he planned to make his discovery public, but Ireland's prime minister, Eamon De Valera, and some twenty others, including at least two reporters, braved the weather to hear what this great scientist had to say.

Schrödinger told them, "The nearer one approaches truth, the simpler things become. I have the honour of laying before you today the keystone of the Affine Field Theory and thereby the solution of a thirty-year-old problem . . . I beg my younger fellows of the institute to take this as a lesson: Never believe in scientific authority. Even the greatest genius can be wrong."

Much of his talk was beyond the grasp of the reporters, but before they could ask for clarification, Schrödinger was out of the lecture hall and cycling home through the falling snow. A reporter from *The Irish Press* followed him the several miles to his home, where, chain-smoking cigarettes throughout the interview, Schrödinger claimed that his work opened "up a new field in the realm of Field Physics. It is the type of thing we scientists should be doing instead of creating atomic bombs." Asked if he was confident of his new theory, Schrödinger replied, "I believe I am right. I shall look an awful fool if I am wrong."

News of the momentous scientific breakthrough was flashed around the world. *New York Times* science editor William Laurence got copies of Schrödinger's original paper and an account of his lecture and mailed them to Einstein, J. Robert Oppenheimer, and Eugene Wigner for their response.

Einstein was astonished: Schrödinger had made an empty boast. His final equations were identical to those Einstein and his assistant, Ernst Straus, had already produced, and far short of the goal of a convincing theory. True, Schrödinger had reached them by a more direct route, but his one doubtful "improvement" was to add a small cosmological constant—something that had once led Einstein astray.

Einstein sent his reply to Laurence: "Schrödinger's latest effort . . . can be judged only on the basis of its mathematical-formula qualities, but not from the point of view of 'truth' (i.e., agreement with the facts of experience). Even from this point of view I can see no special advantage over the theoretical possibilities known before, rather the opposite."

He stressed the undesirability of doing what Schrödinger had done: presenting a preliminary attempt to the public, especially in sensational terms that gave misleading ideas about the character of research: "The reader gets the impression that every five minutes there is a revolution in science, somewhat like a coup d'état in some of the smaller unstable republics. In reality one has in theoretical science a process of development to which the best brains of successive generations add by untiring labor, and so slowly lead to a deeper conception of the laws of nature. Honest reporting should do justice to this character of scientific work."

Even before reading the humiliating *New York Times* report, Schrödinger realized that he had deceived himself: he had been too eager to make his mark just one more time. Desperation at the prospect of never making another important scientific discovery, and frustration with the demeaning tasks demanded of him as a college administrator, had led him to make a fool of himself.

Schrödinger was sick from the flu when he wrote to Einstein on February 4, begging him not to be angry and giving him a rambling and confused explanation of why he had made such a colossal mistake.

He was devastated when he read Einstein's published comment, and the letter that followed on February 7, 1947, was cold comfort. Despite the euphemisms he employed, Einstein left no doubt that he had seen through what amounted to plagiarism. He suggested that they should stop writing to each other and get down to work.

Three years would pass before they resumed their correspondence.

Einstein had hardly finished with Schrödinger when his old antagonist, Sidney Hook, appeared on the scene. Hook, who was in Princeton that February for the university's bicentennial celebration, strolled to the nearby institute, saw Einstein's office door half open, and glanced in. Einstein recognized him and waved him to a chair, ready for a new round of questions.

Why, Hook asked, did Einstein still blame the entire German people for Hitler's crimes, when many Germans had been in concentration camps? Einstein replied that, of course, he excepted Hitler's victims, but that the mass of German adults had supported Hitler to the very end. He called the Germans arrogant and the American attitude toward them naive and sentimental. When Einstein said that the Russians knew the true nature of the Germans, Hook brought up a recent account of the brutality of Red Army troops in Berlin. Einstein countered that Americans always seemed eager to believe the worst about the Russians.

He smiled when Hook accused him of taking a Christian Science attitude to the Soviet Union—that there is no evil—and said he disapproved of many things the Russians did, but believed them a lesser threat to peace than Americans who called for a first-strike preventive war.

Hook pointed out that the American government did not advocate such a war and that citizens in a democracy were free to express irresponsible opinions.

Einstein seemed to have relished their spirited conversation, judging by the gleam in his eyes when they parted.

In April 1947, Einstein wrote to Maurice Solovine, recalling their joyful reunion in Princeton earlier in the year, when Solovine had stayed with him. Mentioning the recent death of Paul Langevin, Einstein called him "one of my dearest acquaintances, a true saint, and talented besides." In the same letter, he told Solovine that he had been restored to vigorous health (after suffering from liver disease) by changing to a sensible vegetarian diet. Perhaps he exaggerated, because he told Leon Watters that he was so exhausted that, rather than attend Verdi's *Requiem*, he preferred to spend his few free hours sitting quietly alone.

In the spring, Maja said she was feeling better, although the effects of her stroke were rapidly getting worse. Margot nursed her beloved aunt and kept her company, and Einstein went each evening to his sister's bedside to read to her. One of the works he read was Xenophon's *Cyrropaedia*. Xenophon was a Greek soldier, historian, and philosopher; his *Cyrropaedia* was a somewhat idealized story of ancient Persia and its founder, Cyrus the Great. Xenophon had managed to incorporate into the story his own ideas about education based on the teachings of Socrates and the Spartans, and Einstein thought it was a wonderful piece of work.

As for his own work, Einstein was still engrossed in his hunt for unifying equations. But he was not so absorbed, Einstein's assistant, Ernst Straus, noticed, as to blind himself to the needs of Tiger, the family tomcat. Seeing Tiger looking miserable because a rainstorm kept him housebound, Einstein addressed him man-to-man: "I know what's wrong, my dear, but I don't know how to turn it off." Tiger sometimes had brief visits from John Wheeler's cat. "Einstein would walk past my house going to and from the institute," said Wheeler, "and occasionally our family cat would follow him home. As soon as he arrived Einstein would phone me that our cat was at his house, he didn't want us to worry, and we could collect it when convenient."

For some time Einstein had been getting disquieting news about both Mileva and Eduard. He had kept his promise, made twenty-five years before, to care for them financially; he sent regular payments to Switzerland, and responded to Mileva's appeals for more when she faced unexpected expenses. Even so, she had been forced to sell two of the three houses she had bought with the Nobel Prize money, largely to help pay for Eduard's expensive hospital stays. In 1939, when Mileva was again in a financial bind, Einstein had taken over ownership of her house to prevent her from losing it but allowed her to retain power of attorney.

Now, in 1947, he learned that Mileva was finding it harder to care for Eduard, because she had slipped on ice and broken her leg. She still took him home for occasional visits, but when he became violent she would call for a van to drive him back. At thirty-seven he was an overweight, morose chain-smoker who heard voices and, at times, behaved irrationally. Yet some people who met him briefly in his windowless basement hospital room, including his only niece, Evelyn Einstein, recall a friendly and even charming man with a fleeting, endearing smile and a thirst for information about the world outside.

Einstein decided to sell the Zurich house and use the money to provide for Eduard's future. He wrote to Mileva's attorney friend in July, saying that "when the house has been sold and Tetel [Eduard] has a reliable guardian, and Mileva is no longer with us, I will be able to go to my grave with peace of mind." Mileva would be allowed to live in one of the apartments for the rest of her life.

The house was sold in August of 1947. Einstein expected Mileva to send him the money to ensure Eduard's future, but weeks went by and he had not heard from her.

He heard instead from an Idaho farmer who had named his son after Einstein and wanted the boy to become a scientist or of some use to humanity. Would Einstein "please write a few words of encouragement for little Albert in your spare moments?" Einstein complied, writing: "Nothing truly valuable arises from ambition or from a mere sense of duty; it stems rather from love and devotion towards men and towards objective things." In return, the grateful farmer sent Einstein a photo of his son and a sack of Idaho potatoes.

For a man who had lost contact with his daughter, Lieserl, and was not close to his sons, at least as adults, Einstein showed surprising affection for other people's children. As his neighbor Mab Cantril recalled in her unpublished memoir, "Einstein lived three doors from us and I saw him when he walked to and from the Institute for Advanced Study. Being young I often played hopscotch and, with a warm, hearty smile, Einstein would have a few jumps with me. His sense of humor may be shown through a story my father was told. A boy of five grew very fond of Einstein and frequently asked him questions as Einstein walked to work. Meanwhile the boy's mother was becoming extremely embarrassed because her son was absorbing so much of Einstein's valuable time. One day while the two were in deep conversation, Einstein suddenly burst into laughter. That night the youngster's mother asked him what was so funny. The boy replied, 'I asked Einstein if he had gone to the bathroom today.' The mother, quite shocked, asked what Einstein answered. He said, 'I'm glad to have someone to ask me a question I can answer.' "

During the summer of 1947, the sister of his old friend Michele Besso called on him and asked a question that called for diplomacy. She expressed her unspoken thought—Why are you famous and my brother isn't?—as "Why hasn't Michele made some important discovery in mathematics?" Laughing, Einstein turned Besso's failure into triumph. "This is a very good sign," he said. "Michele is a humanist, a universal spirit, too interested in too many things to become a monomaniac. Only a monomaniac gets what we commonly refer to as *results.*" When Besso was told of this conversation, he asked Einstein if he had been accurately quoted. Yes, Einstein replied, then gave a more poetic version: "A butterfly is not a mole; but that is not something any butterfly should regret."

One question that stumped him came from a ten-year-old girl in New Jersey who asked, "Would you please write to tell me what your plans are for the future?"

Whatever they were, George Wald changed them, by arriving unexpectedly. The Harvard biologist, who would win a Nobel Prize twenty years later, had an introductory letter to Einstein from Philipp Frank. While staying at the Princeton Inn, Wald recalled, he "asked a man at the desk, 'How do I get to 112 Mercer Street?' Everything in the office stopped, including a woman who had been typing. The man at the desk asked, 'Are you expected?' And I said, 'No, not yet, but I have a letter of introduction to him from a close friend.' He said, 'May I see?' and I said, 'Yes of course.' And he said, 'Would you please come back in half an hour?' When I did I was met with smiles and told, 'Everything's all clear. Go right ahead.' Outside I gave the cab driver Einstein's address and he stopped the cab, and asked, 'Are you expected?' The whole town of Princeton was protecting Einstein. It was most extraordinary.

"On my first visit I asked him to explain his friendly controversy with Niels Bohr involving the real meaning of the uncertainty principle. He did so in a fine, elementary way. The progress of science has overtaken one thing he said, which was: 'We've never seen the other face of the moon, but I'm sure it has another face.' At one point he said 'Never is a long time,' and that eventually it might be possible to get inside . . . the present limits of the uncertainty principle.

"This was our first conversation. He was very easy to converse with and rather jolly. He would say something and then lean back and laugh. He was a lovely person and being with him was a great and relaxing pleasure. Any time spent with him was precious."

FBI agents continued to monitor Einstein's activities, reporting that he sent a message to a September meeting sponsored by the Los Angeles Zionist Emergency Council and the Justice for Palestine Committee, in which he said he hoped that the United Nations would take over Palestine and give it to the Jews.

That same month he sent a birthday greeting to an old psychiatrist friend, Otto Juliusburg, now eighty, saying that neither of them grew old like everyone else: "What I mean is that we never cease to stand like curious children before the great Mystery into which we are born. That interposes a distance between us and all that is unsatisfactory in the human species . . . When in the mornings I become nauseated by the news that the *New York Times* sets before us, I always reflect that it is anyway better than Hitlerism."

In October, when he heard of Max Planck's death, Einstein wrote to his widow of the "beautiful and fruitful period" he had lived through with a man whose "gaze was directed on eternal truths, yet [who] played an active part in all that concerned humanity and the world around him." The time he spent with her "dear husband," he assured her, "will remain among the happiest memories for the rest of my life . . . I share your grief and greet you with all the former affection." Despite his bitterness towards the German people, he recognized exceptions and had felt particularly sorry for Planck, whose son, Erwin, a suspect in a plot to assassinate Hilter, had been tortured and then executed by the Nazis.

In November, the United Nations' Foreign Press Association gave the Emergency Committee of Atomic Scientists an award for "its valiant effort to make the world's nations understand the need of outlawing atomic energy as a means of war, and of developing it as an instrument of peace." Einstein accepted the award—an event that was also included in the FBI files. He was no scaremonger, however, reassuring those who feared that a chain reaction might destroy the world that if it were possible, "it would already have happened from the action of the cosmic rays, which are continually reaching the earth's surface." Nevertheless, he warned, "little civilization would survive" an atomic war.

Although little progress was made in controlling or abolishing nuclear weapons, one of Einstein's other important concerns appeared to be nearing a resolution. On November 29, the United Nations voted to support the division of Palestine between Arabs and Jews.

Despite Einstein's joy at the news that a Jewish state seemed inevitable, he was tormented by personal problems. Aside from his sister's paralysis, his own health was precarious and he was often in pain. Liver disease had compelled him to become a vegetarian. He had not heard from Mileva in months. "Everything is possible due to her taciturn and mistrusting nature," he complained to Hans Albert, and speculated that she had hidden the money from the house sale. As a last resort, he warned her attorney that if Mileva refused to send the money he would write Eduard out of his will to protect other beneficiaries.

Early in 1948, Henry Wallace announced his bid for the U.S. presidency as leader of the new Progressive Party, stressing economic and social rights for the common man and fiercely attacking Truman's foreign policy. A plant geneticist by profession, and a close friend of FDR, Wallace had been vice president during World War II (from 1941 to 1945). He attracted both a large group of devoted supporters and an equally large number of opponents who ridiculed him as an ineffectual left-wing wacko who spoke Russian, played with a boomerang, and conversed with the dead. He chose Glen Taylor, a U.S. senator from Idaho who was known as the "Singing Cowboy," as his running mate.

Einstein was reported by the National Wallace for President Committee to be a Wallace supporter, and was quoted as praising Wallace for being "clear, honest, and unassuming." Sidney Hook saw red, warning Einstein that "Wallace today is a captive of the Communist party whose devious work in other countries you are familiar with much better than most scientists. His speeches are written for him by fellow-travelers. His line is indistinguishable from that of *Pravda* and the *Daily Worker* . . . the prima-facie case against Wallace's position is so strong, that I hope you will reconsider your endorsement of it."

Einstein replied in astonishment, denying the charge:

> What I have really done was to recommend warmly Wallace's book [*Toward World Peace*] and—in one sentence—paid tribute to Wallace as a man who is above all the petty interests . . . In my opinion your views are far from objective. If you ask yourself who, since the termination of the war, has threatened his opponent to a higher degree by direct action—the Russians, the Americans, or the Americans, the Russians? The answer is, in my opinion, not doubtful and is accurately given in Wallace's book. It is, furthermore, not doubtful that the military strength of the U.S.A. is at present much greater than that of the Soviet Russia. It would therefore be sheer madness if the Russians would seek war . . . I am not blind to the serious weaknesses of the Russian system of government and I would not like to live under such a government. But it has, on the other side, great merits and it is difficult to decide whether it would have been possible for the Russians to survive by following softer methods. If . . . interested in my opinion you may read my answer to a few Russian scientists which I am enclosing.

The enclosed account showed Hook that Einstein could antagonize him *and* the Russians: four Soviet scientists had damned Einstein as a pawn of American interests whose call for world government was a cunning attempt by America to rule the world. In reply, Einstein easily demolished their arguments and clearly established that his motive was to prevent a catastrophic atomic war.

Privately he mocked them, imitating their inane responses: "On the basis of an open letter signed by Russian scientists, we may construct a parallel resolution for them: After careful consideration, and due consultation with our government, we do not know—what not to believe, what not to wish for, what not to say, and, what not to do." He was certainly disconcerted, as well, by the Communist Party's recent condemnation of Gregor Mendel's genetics as anti-Marxist. He considered Mendel's work magnificent.

Reproached by both Soviet scientists and Sidney Hook, Einstein must have felt that he was doing something right. He believed that most Americans were ignorant of Russian history—a history that largely explained Russia's almost paranoid behavior—and that American militarism and Russian belligerency were both to blame for the cold war. In his dealings with Hook, he acted some times as devil's advocate, at others like a marriage counselor trying to reconcile the contending parties rather than add to the growing animosity that threatened to turn violent.

Not knowing where Einstein was coming from, Hook replied to his letter:

> When you warned the world against the spread of Hitlerism, on what evidence did you rely? On what Hitler *said* . . . to the Nazi party, and on what Hitler *did*. I have pointed to the evidence which shows what Stalin believes—the regnant dogmas of the Communist party—and the record of his actions. Together they reveal a program of world conquest. You do not challenge the evidence. You do not present contrary evidence. Instead you ask: "Who has threatened his opponent to a higher degree by direct action—the Russians, the Americans, or the Americans, the Russians?" This is a fair question . . . I should say that the Soviet Union by her violation of the principles of the Atlantic Charter, by her violation of her agreements to permit free elections and a free press in Bulgaria, Rumania, Hungary, etc., by the coup she inspired in Czechoslovakia, by her war of nerves against Finland, and now Norway and Sweden, by her behavior in Berlin, by her sabotage of the U.N. commissions on Greece, the Balkans, and Korea, and almost all of its agencies—that by all this, the Soviet Union has been the aggressor. What actions of the U.S. do you consider as direct aggressions of equal weight against the Soviet Union?

Einstein did not reply. Born and Upton Sinclair may have persuaded him that it was futile to try to change someone whose mind was made up. Hook seemed too partisan to concede what Einstein saw as the facts: that both sides were at fault.

Max Born, who hadn't seen Einstein for almost twenty years, wrote from Magdalen College, Oxford, to say he had just seen a film about atomic energy,

> and there you were, as large as life, talking with that familiar and well-loved voice, and smiling your amiable, half-serious, half-cynical grin. I was quite moved . . . There we have been trying to puzzle things out, only to help the human race to expedite its departure from this beautiful earth! I no longer understand anything about politics: I understand neither the Americans, nor the Russians, nor any of the numerous little stinkers who are now, of all times, becoming nationalistic. Even our good Jews in Palestine have discredited their cause in this way. It is better to think about anything else.

Einstein agreed, but couldn't help himself, telling Upton Sinclair, "The day will inevitably come when you will discover that the people are deaf and dumb so that they cannot hear the voice of reason. In spite of knowing this for a long time I too have been engulfed in preaching."

Einstein briefly toyed with the idea of leaving America for another country, where "scholarship and things of the spirit were guaranteed more freedom," but when Selig Brodetsky suggested that he move to Israel, he said that at sixty-nine he was too old. At the same time, he seemed almost reconciled to the status quo when, in a letter to Lina Korcherthaler, his cousin in Uruguay, he defined life as "a curious drama in which we all appear. Good, when one does not take it too seriously. The play has neither beginning nor end and only the players change."

Henry Abrams, the Einsteins' family doctor, was back in Princeton from Greenland, where he had served during World War II. "A wonderful place to contemplate," Einstein had written in one letter to him. Now Abrams decided to specialize as an ophthalmologist. His friend, Guy Dean, then took over from him, handling the Einsteins' illnesses, while Abrams took care of their eyes.

In April 1948, Einstein accompanied his friend and colleague, Kurt Gödel, to Trenton on a strange mission. To prepare to become an American citizen, Gödel had carefully studied the Constitution. On the day before the obligatory exam, he called his economist friend, Oskar Morgenstern, in great consternation to say that he had discovered a logical flaw in the Constitution by which the United States could be transformed into a dictatorship. Morgenstern said it was a very unlikely possibility and warned him not to raise the point with the examining judge.

Gödel had chosen Morgenstern and Einstein to be his witnesses at the event. Because Gödel was likely to argue with the judge and jeopardize his chance of becoming a citizen, the two men tried to take his mind off the flaws in the Constitution by entertaining him with jokes and anecdotes on their drive to Trenton.

But Gödel couldn't help himself. When the judge began, "Formerly you held German citizenship," Gödel immediately corrected him: he was Austrian, not German.

"Anyway," said the judge, "it was under an evil dictatorship, but fortunately it's not possible in America . . ."

"On the contrary," Gödel interrupted. "I know how that can happen. And I can prove it!"

Fortunately, the judge was understanding and helped Einstein and Morgenstern to persuade the passionately logical Gödel to remain silent long enough to be sworn in as an American citizen.

At home Einstein found a letter from a young woman in Iowa who had just read *Albert Einstein: Maker of Universes*, the 1939 biography written by Gordon Garbedian, and was surprised to discover that the famous Albert Einstein was a Jew. She wrote:

> My father is a Jew hater. At dinner one night [he] said that the only good idea Hitler ever had was getting rid of the Jews. I thought again and again of that statement until . . . I realized what was behind it. My father considers himself a failure. He reasons that since his failure cannot be his . . . fault, it must be the fault of some group of people or some condition existing today. The fault then is with the Jews and the system they advocate—capitalism. And so he will continue to refuel that flame of hate inside him until it burns out any creative ability he may have . . . Since I am working my way through college, I work nights as a cashier in a popular little confectionery. One night a slightly tipsy "gentleman" in paying his bill forgot about the state tax. I asked him if he had a penny. He

grinned at me and, fishing the money from his pocket, exclaimed, "You god-
damn Jew!" . . . Certainly he was "just being funny." He knew I wasn't a Jew.
But in that instant I realized something of what a Jew must put up with day after
day. At school today a Catholic friend of mine remarked that you had recently
revised your views and professed a belief in a hereafter. If this is true, can you
tell me where I [might] find your reasons for doing so?

Einstein replied on April 18, thanking her for her

magnificent letter. If your father had been a "failure" he would not have had a
daughter like you are. The maturity of thought and expression in your letter has
astonished me very much and not less the earnestness of your striving for justice
and reason. The role reason and conscious thinking is playing in the behavior
of men is a very modest one. Traditional views and values are uncritically accepted and taken for granted nearly always and everywhere. Mostly even the
reasoning of philosophers is subconsciously and powerfully influenced by those
irrational motives. It is therefore not astonishing that the struggle against prejudices cannot have a quick success. But we can work for a slow success in expressing truth in our daily life whenever occasion arises. It is, of course, not true
that I believe in an eternal life of the individual. Such a belief is a strange
outcome of our desire not to die (instinct of self-preservation).

He signed the letter, "With my sincere respect."

In May, the FBI noted that as chairman of the Emergency Committee of
Atomic Scientists, Einstein had written to Secretary of State George C. Marshall
for permission to send an enclosed letter to Stalin about the deterioration of
relations between the United States and Russia and the drift toward war. The
letter suggested that Stalin broadcast to the American people his ideas for the
reconstruction of the postwar world and that a series of meetings of leading
scientists and citizens of the world be held to iron out existing difficulties.

Obviously Marshall sent it on, because Einstein received a response. In that
same letter, Einstein had asked Stalin to do him a favor—to help rescue, if
possible, the heroic Swedish diplomat Raoul Wallenberg. During the war, Wallenberg had been stationed in Nazi-held Budapest, Hungary, where he saved the
lives of some twenty thousand Jews by giving them passports out of Nazi territory.
He escaped several attempts by the Gestapo to arrest him, only to be taken
prisoner by Soviet troops when they entered Hungary. By then, his courageous
exploits had become legendary. His last known address was Cell 151 in Lubianka
Prison, Moscow, and no one in the West knew whether he was dead or alive.
"As an old Jew," Einstein wrote to Stalin, "I appeal to you to find and send back
to his country, Raoul Wallenberg [who], risking his own life, worked to rescue
thousands of my unhappy Jewish people." An aide replied that he had been
authorized by Stalin to say a search had been made for Wallenberg without
success. His fate is still a mystery.

The FBI apparently took no further interest in Einstein's appeal on behalf
of Wallenberg, but agents did follow up on a news account published in the
Arlington Daily. The article stated, "Einstein and ten former Nazi research brain-

trusters held a secret meeting and watched a beam of light melt a block of steel 20 by 20 inches. It was indicated that this new and secret weapon could be operated from planes and destroy entire cities." The FBI had Colonel C. C. Blakeney of Army Intelligence check it out; he concluded that "such information could have no foundation in fact."

Oblivious of the FBI's interest in his activities, Einstein was concerned over the news that the United States had suddenly changed its Palestine policy; it no longer supported partition and proposed instead to establish a United Nations trusteeship. Neither Jews nor Arabs wanted that, and soon both sides were fighting for control of the country.

When Einstein realized that the six hundred thousand Jews in Palestine were wildly outnumbered and in danger of annihilation from several Arab armies, he again renounced pacifism. On May 4, he sent a letter to a cousin in Montevideo, Uruguay, to be auctioned to support the Haganah (the military arm of the Jewish Agency in Palestine). The letter, which raised five thousand dollars, said in part:

> If we wait until the Great Powers and the United Nations fulfill their commitments to us then our Palestine brothers will be under the ground before this is accomplished. These people have done the only thing possible in the present deplorable conditions. They have taken their destiny in their own hands and fought for their rights . . . On the destiny of our Palestinians will depend, in the long run, the destiny of the remaining Jews in the world. For no one respects or bothers about those who do not fight for their rights.

President Truman was exhausted and, at times, disgusted by both Jewish and Arab advocates demanding that he change his mind again and follow their wishes. Then the Jews forced his hand.

In Tel Aviv's Municipal Museum at 4 P.M. on May 14, 1948, as fierce fighting continued throughout the country, David Ben-Gurion prepared to read the Declaration of Independence he had drafted proclaiming the rebirth of Israel to some four hundred Palestinian Jews. After the Tel Aviv Philharmonic Orchestra in an adjoining room played "Hatikvah," which became the Israeli national anthem, Ben-Gurion began speaking: "On this Sabbath eve, the fifth of Iyar, 5708, the fourteenth day of May 1948 . . ." He concluded, with a rap of his gavel, "I hereby declare the meeting adjourned. The State of Israel has come into being." A portrait of Theodor Herzl, the founder of the organized Zionist movement, was on the wall behind him as he spoke.

Ben-Gurion was awakened at one the next morning and told that President Truman had just recognized the new Jewish state.

The unexpected news astonished the American delegation at the United Nations; it made some South American delegates laugh — thinking it must be a joke — and caused crowds in Brooklyn and the Bronx to dance in the streets. The *Washington Star* expressed the sentiment of many Americans, praising "the swift and dramatic decision of the United States to take the lead among all nations" as "wise and heartening."

Einstein later called the event "the fulfillment of our dreams."

He turned from world affairs for a while to find a temporary replacement for Helen Dukas, who was going on vacation for two or three weeks. Einstein's sculptress friend, Gina Plunguian, volunteered to fill in.

Gina's daughter, Clair Gilbert, remembers how her mother "would show him letters addressed to him from children and crackpots and Indian mystics and he would say, 'How delightful!' and scribble answers on the bottom of the pages." Einstein talked to Gina "about the trivial and the cosmic and she made notes. She was there every week when she was doing his bust and when Robert Oppenheimer, then head of the institute, or others came to see him, she'd say, 'Don't mind me,' and sit in the corner and take notes of their conversations at lunch, and of Einstein's phone conversations. It was a diary of daily life in the household in a combination shorthand, German and English."

Meanwhile, as nations were born and secretaries came and went, Mileva's silence continued. As Einstein soon learned, the situation was more complicated and pitiful than he had imagined. Mileva couldn't have written even had she wanted to. She lay in a hospital, semiconscious from a stroke that had paralyzed her left side. It may have been brought on by Eduard who, home on a brief visit, had gone berserk, ransacking the apartment in search of some object that apparently did not exist.

Mileva died in the hospital on August 4, 1948. When Einstein's friend Otto Nathan and Hans Albert's wife, Frieda, went to Zurich to settle her affairs, they found eighty-five thousand francs hidden under her mattress, presumably the cash from the sale of the house. With it, Einstein was able to assure Eduard's future—most of it to be spent in the mental hospital—and hire a guardian to protect his interests.

In the fall, Einstein's research assistant, Ernst Straus, left to take on an academic position. John Kemeny was among the possible replacements. A twenty-two-year-old Hungarian refugee, Kemeny was partway through his doctoral thesis when he went to Einstein's office. The thesis Kemeny was working on about mathematical logic was of no conceivable interest to Einstein and he knew absolutely nothing about it. "So," said Kemeny, "I had to explain to him what type theory was, what set theory was ... It must have taken half an hour ... and I was feeling intensely guilty taking up his time. But he insisted ... He interrupted with a number of questions when something wasn't clear ... And then, my favorite line, I'll never forget, he said, 'That's very interesting. Now let me tell you what I'm working on,' in exactly the tone of voice as if somehow the two were of equal importance.

"To anybody who worked in math or science Einstein was the great hero you idolized. That's why the story about his insisting on me—a twenty-two-year-old nobody—telling my thesis was so embarrassing. The thesis was terribly important to me but it wasn't going to change the world.

"Then he said, 'Go home. Finish your thesis. Then come back.' " Kemeny did, and ended up working for Einstein from 1948 to 1949.

"He was wonderful at putting people at their ease," Kemeny recalled. "I can't believe it took more than a week or two before I felt fully at ease with him. There was something about his personal style, so understated, and so warm. He would always talk, certainly to any professional as if he were an equal, which of course I wasn't. But it put one totally at ease."

Kemeny found in Einstein a great enjoyment of life, even though he was bothered by not being able to smoke. The new assistant also discovered that Einstein pursued the laws of the universe with all the energy he had: "He had narrowed down his version of advancing unified field theory to one of three possibilities. They were similar but had important differences. And the year I was with him was spent trying to decide which version to publish. Which meant working some examples and getting insight as to how the versions differed from each other. He did choose one at the end and his unified field theory was published a year after."

Kemeny recalled a meeting with a very interesting visitor, a student of Niels Bohr, who tried to convince Einstein that he was completely wrong about quantum physics. "That was the hot subject of the day," Kemeny said. "I was in Einstein's office at the time and was a quiet observer of the scene. Without being able to judge who was right, the style of the two was absolutely fascinating. First of all, Einstein remained firm but totally calm throughout the entire long discussion. The physicist who had been a student and disciple of Bohr became more and more vehement. Einstein kept explaining patiently what the things were about quantum mechanics that bothered him; for example, that it does imply action at a distance. The more excited the man became, the more his arguments began, 'Yes, but Bohr said so and so,' and Einstein very carefully avoided saying anything about Bohr, and talked substance. The other continually quoted higher authority—which was Niels Bohr. I felt that Einstein clearly got the better of the argument."

Once Einstein became convinced of the rightness of something, he would be very firm on that, according to Kemeny. For example, he was absolutely convinced that there would be a unified field theory. He believed that the laws of the universe could not be fragmented into relativity on the one side and quantum mechanics on the other. On that he was absolutely firm.

Yet, Kemeny recalled, "he was not stubborn on whether he had found the right theory. But what he said was, 'If there is a theory along the kind of lines on which I developed general relativity, then I believe'—maybe he said, 'I'm sure'— 'I have found the right one.' On the other hand, I've heard him say repeatedly that it may turn out that just as general relativity requires an entirely new kind of mathematics it's possible that the correct unified field theory would require some mathematics he did not even know. 'In which case,' he said, 'I will just never find it. Somebody else will have to find it.'

"He may be right on that, but nobody yet knows whether the unified theory is correct or incorrect."

As Kemeny knew him, Einstein was extremely shy, "not person-to-person but in a group. Miss Dukas told me he was once mobbed in Atlantic City by an adoring mob, which he found a frightening experience, and since then he had shied away from large groups. But he felt very comfortable in Princeton and I've seen him in modest-sized groups being charming and relaxed."

Kemeny, aware of his own outstanding intelligence, had great respect for Einstein's mind: "I am not modest about my own brain [a brilliant mathematician and computer pioneer, Kemeny had worked on the Manhattan Project]. But in my life I have only met two human beings whose brains were clearly in order of magnitude better than mine. One was Einstein. The other was John von Neumann, the mathematician, also a fabulous person. Look, I saw Einstein at seventy when most people's mental processes slow down. And at seventy his brain was unbelievable to me. So I can only guess at what he had been at twenty-six when he did his first-rate work."

Kemeny was well aware of Einstein's craving for tobacco, often seeing him with an empty pipe in his mouth. Once Einstein begged Kemeny for a cigarette, although smoking was strictly forbidden to him for medical reasons. Einstein crumbled the cigarette into his pipe, lit it, and took just two puffs. "He was dying to have a smoke," said Kemeny.

When they took a break from work, Kemeny enjoyed Einstein's quiet sense of humor and listened to what may have been one of his favorite jokes, because Einstein repeated it more than once. "You must remember," said Kemeny, "that the moneys involved in this joke go back to another age. A man goes to have his car repaired. He's having terrible trouble with it. The mechanic looks at everything and then stops and kicks the car. After which it runs perfectly. The owner is terribly happy until he gets a bill for $25, which in the late 1940s was outrageously large. He complains bitterly and asks for an itemized bill. So the mechanic writes out the bill: 'Labor: kicking the car—25 cents. Knowing where to kick—24 dollars and 75 cents.' You see why Einstein would like that story. It's not the kicking that counts, but knowing where to kick."

After Kemeny became engaged to Jean Alexander, he took her to meet Einstein. They had hardly arrived when Kemeny was called to the phone, leaving his nineteen-year-old fiancée alone with Einstein. "While he was charming to her," said Kemeny, "she, of course, was overwhelmed. She tried to think of something to ask him." Finally, she asked if it would be a mistake for her to marry a mathematican, because she was not at all mathematically inclined. "Einstein's answer was absolutely that it was not a mistake," Kemeny recalled. "On the contrary, he worried about marriages when the couples were in fields where they might be competing.

"Jean is a highly visual person," said Kemeny, "and Einstein was sitting with the light coming in through the window behind him, coming through his hair. To this day she visibly remembers the fascination of the glitter of his hair and how his voice sort of mesmerized her. Afterward, she couldn't remember most of what he said, she was almost in an hypnotic state."

Einstein gave Jean good advice. The marriage worked and they had two children, and eventually Kemeny became president of Dartmouth College. He invited a group of students to his home and one of them asked him if Einstein had a reputation for being eccentric. Kemeny said yes, he did—"for example, the way he dressed ..." And then he stopped, because he realized that more than half of them were dressed exactly the same way Einstein always dressed. He ended by saying, "I guess Einstein was ahead of his time."

In late November 1948, Einstein wrote to Maurice Solovine that his sister, though not suffering, was sinking visibly. He still read to her every evening, most recently Aristotle's philosophical writings and a book containing Ptolemy's "odd" arguments against Aristarchus's view that the world moves around the sun. Turning to the subject of Palestine, he reproached the English for their "kind of cheap resentment which I would not have believed possible against our small Jewish tribe," but praised them for electing a socialist government, "perhaps the only ones to end outmoded capitalism without a revolution."

That fall, Linus Pauling and his wife called on Einstein; according to Pauling, their "conversation was pretty much concerned with world affairs ... It's pure nonsense to call him a dupe of the communists as some did. I have been accused of the same thing. I think he knew perfectly well who he was working for and against. I remember Senator Hennings of Missouri saying about me to an assistant secretary of state, 'Instead of Dr. Pauling following the communist line, it seems to me that the communists follow Dr. Pauling's line.' I think that applies to Einstein, too."

They were not only alike in that. As Pauling admitted, "His room was something like my own study, sort of a mess. While he was thinking he'd reach to the back of his head and twist curls of hair around his finger. My wife has emphasized that he had a well-developed sense of humor. He would break into boisterous laughter telling jokes or commenting on somebody. My wife was quite impressed with him and so was I, of course."

On December 12, a month after the Paulings' visit, agonizing stomach pains took Einstein to Brooklyn Jewish Hospital. His gallbladder was suspected as the cause of the trouble, but when surgeon Rudolph Nissen operated he found a large aneurysm of the abdominal aorta, a potentially fatal ballooning of the main artery from the heart. When Einstein recovered consciousness, he dictated answers to a reporter's written questions but refused to see the press because "my illness is not of public interest." However, when he heard a hospital official ticking off a persistent reporter seeking an interview, Einstein called out, "Don't be angry. He's got to make a living." He then invited the reporter in, shook his hand and said, "Sorry, no interview." But he thrilled his nurses by writing and signing a little jingle for each of them.

After overhearing a doctor say that the hospital was short of private rooms, Einstein insisted he was "getting much better" and asked to be moved to the ward. That way, his room could go to someone who needed it more. He was talked out of it when told he would be more trouble in the ward. Helen Dukas came to collect him a few days later, and they left by the back entrance through a gauntlet of reporters, newsreel cameramen, and almost the entire hospital staff, who were there to wish him well.

On the way home, pestered by photographers, he was snapped by one of them, sticking his tongue out at him. Einstein cut out a copy of the picture in a newspaper shortly afterward and sent it to the surgeon who had performed the recent operation, with the written wisecrack, 'To Nissen my tummy/the world my tongue!'"

Seeing a less playful photo of him taken by Philippe Halsman, a child asked, "Is that the Lord?" Another child, who saw Einstein in New York, thinner after his operation, exclaimed, "There's Gandhi!"

CHAPTER 36

The FBI Targets Einstein

1949

---

70 years old

Einstein took some persuading that Florida wasn't only for the rich before he agreed to convalesce in the sun. "We almost had to force him to join us," Thomas Bucky recalled. "We had a rented cottage at Lido Beach and he had his own rented place nearby. Then, Sarasota was quite wild and unsettled, and he enjoyed boating and walking on the white beach. Helen Dukas, Margot, and his sister were with him."

Einstein followed his surgeon's advice not to have visitors, even resisting a visit from Leon Watters, and informed Maurice Solovine that the surgery was not totally useless because it had corrected some defects. He had cut down on his letter writing recently—at least to South Africa, where a schoolgirl had confused him with Isaac Newton, dead over two hundred years, and was amazed that he was still alive. But he had been delighted by a New Year's Day message from a six-year-old boy: "I was sorry that you were sick. You are a nice man and I hope you are better soon. I learned about you in a comic book about atoms and I heard you were sick on the radio."

Such letters helped to divert him from his concern over the fate of the world, especially Israel. For months the small new nation had been battered by various Arab armies until, finally, the Israelis routed the Egyptians after a fierce six-day battle that ended on January 13. His pacifist convictions were therefore put on the spot when a member of the War Resisters International asked, "Don't you agree that Israel should recognize the legal status of conscientious objectors?" Yes, he replied, but with an important reservation: he would not presume to advise people who had overcome what seemed to be insurmountable obstacles to save their nation.

While Einstein walked on the beach, went sailing, and took long naps, the FBI's J. Edgar Hoover was galvanizing his troops. He fired off an irate order to the Newark office by airmail special delivery on March 11, 1949: "You are instructed to submit an explanation of why my instructions . . . have been disregarded . . . I will not tolerate any further delay in your failure to submit a report in this case [Helen Dukas]. You are instructed to prepare a report to reach the Bureau by March 20, 1949. You are also instructed to forward to the Bureau your recommendation as to what further investigation you intend to conduct in this case in order to prove or disprove the original allegations relating to the subject."

The answer arrived five days late: "It is not felt that there is sufficient basis for an interview with subject at this time; and in view of these limited leads [and Hoover's refusal to bug Einstein's phone], the Bureau is requested to advise of any suggested procedure and to consider authorizing discontinuance of this investigation."

When Hoover blasted the Newark office for not having sent a complete report on Dukas, he was told that they were swamped with work:

> The Agent to whom this case is assigned has been working on Loyalty cases carrying an average work load of approximately thirty-five security type cases, a major case, and assisting in the operation and maintenance of the technical equipment of this office. Because of the apparent inactivity of the subject . . . and the position in the household of Dr. Albert Einstein, matters that were considered of a more pressing nature were given preferred attention. In addition to the above, investigative leads in the case were limited by the Bureau instructions as Dr. Einstein lives on the college campus [he didn't] and it was felt that any direct inquiries concerning Helen Dukas would be brought to his attention.

Hoover took the advice to drop the case against Dukas, but reopened it when new charges came from a source in the Pentagon. The hunt was on again, though the hunters were badly handicapped by false clues. In fact, Hoover didn't have a clue.

While FBI agents scurried around following dead-end leads, preparations were afoot to celebrate Einstein's seventieth birthday, which he wanted to ignore.

The celebrations began with a biography of Einstein written for high school students, which the publisher planned to hit the bookstores on March 14. Then the institute announced that an Einstein Award had been created in his honor: a gift of fifteen thousand dollars every three years to go to individuals who made outstanding contributions to knowledge in mathematics and the physical sciences.

An institute-inspired book was also in the works to commemorate Einstein's achievements. Bohr happened to be on one of his visits to the institute and enthusiastically agreed to contribute to the book, saying he would re-create their arguments about quantum theory over the years. He got stuck, though. He invited Abraham Pais to help him out by taking down his words as he paced around the table in his office. The trouble was, said Pais, that

> Bohr never had a full sentence ready. He would often dwell on one word, coax it, implore it . . . This could go on for several minutes. At that moment the word was "Einstein." There was Bohr, almost running around the table and repeating: "Einstein . . . Einstein . . ." After a little while he walked to the window, gazed out, repeating every now and then: "Einstein . . . Einstein . . ."

At that moment the door opened very softly and Einstein tiptoed in. He indicated to me with a finger on his lips to be very quiet, his urchin smile on his face . . . Always on tiptoe he made a beeline for Bohr's tobacco pot . . .

Then Bohr, with a firm "Einstein" turned around. There they were, face to face, as if Bohr had summoned him forth. It is an understatement to say that for a moment Bohr was speechless. I myself, who had seen it coming, had felt distinctly uncanny for a moment . . . A moment later the spell was broken when Einstein explained his mission [he had come to steal Bohr's tobacco, as his doctor had only told him not to *buy* any] and soon we were all bursting with laughter.

Einstein was shown the manuscript before publication and was invited to provide comments, which would be included in the book. Stung by the critical tone—paeans of praise thickly interspersed with brickbats for his resistance to quantum theory—he complained, "This is not a jubilee book for me, but an impeachment." Responding to his critics, he restated in his comments his long-held view that quantum theory might be statistically right but that it failed to describe fully physical reality, and he provided examples of ridiculous efforts by its supporters to do so.

Louis de Broglie read the book during the time he was giving a course of lectures on wave mechanics at the Institut Henri Poincaré in Paris, and he thought Einstein's comments were brilliant. He brought up Einstein's objections to wave mechanics and Bohr's counterarguments during one lecture, and stated that he sided with Bohr's statistical interpretation. But as he continued to mull over Einstein's objections to Bohr's ideas, de Broglie's "25-year-old allegiance to the statistical interpretation seemed to vanish in thin air."

As part of the celebration for Einstein's seventieth birthday, Princeton had invited "a number of big shots to give papers on things that have grown out of relativity," said his onetime assistant, John Kemeny. "Because by this time there was an enormous literature on things that had been affected by relativity. It was held in the auditorium at Princeton that held maybe 250 people and admission was by ticket. I don't mean they were sold, but they had to control the size of the audience. Never did so many big shots try to be nice to me as then, because they thought I could get them a ticket. As Einstein's assistant I got one ticket and I had no influence over the other tickets. Once, in a quiet moment, I told Einstein about it and he said he was embarrassed by the whole business, that he felt it was an imposition because all kinds of very busy and important people would feel obliged to comment and be nice to the old man. I tried to convince him that people were dying to get in. I feel that while Einstein understood that the average person in the street idolized him without understanding what he did, he didn't understand that even the next most important physicists, with the possible exception of Bohr, also idolized him."

At last, the big day arrived. There was a hush from the audience as Einstein entered the hall. Then everyone stood to applaud him. Among the speakers were J. Robert Oppenheimer, I. I. Rabi, Eugene Wigner, and Hermann Weyl.

Soon after, according to Thomas Bucky, "Oppenheimer made fun of Einstein in a magazine article with such statements as, 'He's old. Nobody pays any attention to him anymore.' We were madder than all hell about it. But Einstein was not mad at all. He just didn't believe it and later Oppenheimer denied he had said it."

Einstein's mood was uncharacteristically subdued, however. He was distressed by his sister's suffering, his ill health, and the knowledge that some young physicists scorned him as an outmoded heretic, which reinforced his own growing feeling of failure. Deeply moved by an affectionate birthday letter from Solovine, he confided in reply that he doubted he'd ever been right about anything. He was even downbeat in decrying Solovine's idea for a book on Heraclitus, saying the subject was a stubborn and melancholy character not worth the effort.

As for the outside world, he impartially blasted America, Britain, and the Soviet Union for their aggressive posturing in the early days of the cold war. That fall, his concern for world peace took him to a meeting of the newly formed Unitarian Fellowship on the control of nuclear energy, at which the speaker was Ashley Montagu.

A few weeks later Einstein invited Montagu to tea to discuss nuclear disarmament. Sitting in Einstein's study, both men agreed that a good first step to total disarmament was to get the United States and the Soviet Union, then the only nations with nuclear weapons, to agree to disarm. Einstein, however, considered this a futile dream because of mankind's innate propensity for aggression. But there's no such thing, Montagu protested. "During my conversation with Einstein," said Montagu, "I mentioned an exchange of letters between Einstein and Freud in 1932 and later published as a pamphlet titled *Why War?* In one letter, Einstein had asked Freud if he thought wars were inevitable. And Freud had replied that 'there is no likelihood of our being able to suppress humanity's aggressive tendencies. In some happy corners of the earth, they say, where Nature brings forth abundantly whatever man desires, there flourish races whose lives go gently by, unknowing of aggression or constraint. This I can hardly credit: I would like further details about these happy folks.' Freud concluded that there was an instinct in human beings towards destructiveness and war. And he was very much afraid there was no cure. So Einstein had accepted the great authority's view that man is indeed endowed with instincts."

When Montagu first explained to Einstein that human beings have no instincts, Einstein thought that the concept was quite outrageous. He referred to what he and Freud had discussed in their exchange of letters—the instinct for aggression, which seemed to imply that wars were inevitable.

"I wasn't surprised that Einstein knew nothing about the peaceful Hopi or Zuni Indians, the Veddas of Ceylon, or the Pygmies of the Congo," said Montagu. "The evidence was available, but there was only a certain area in which Einstein read. Finally I convinced him he was wrong, that there is no instinct for aggression. This of course delighted him, being such a proponent of peace.

"I told him an instinct is an innate psychophysical disposition — to receive a stimulus in a particular manner, to react to it with a particular behavior accompanied by a particular emotion — and that human beings have no instincts. And when Einstein asked, 'What about the sex instinct?' I said, 'That's the easiest one to dispose of. Sex is a complicated drive which develops gradually and which is in no way an instinct. It is something that has to be learned. Two babies, a boy and a girl, shipwrecked on an island and who grew up together, would either commit suicide or they might by pure experiment discover how to do it. We have to learn what to do about sex. We have all the anatomical and physiological equipment but we don't know what to do with it, unless we learn, like playing the piano or dancing.' Eventually he accepted the idea.

"It confirmed what I already knew from my first encounter with him, that he had a very open mind. I soon realized that he would accept any idea, however contradictory to any he may have held, that was sufficiently explained and given scientific underpinning. And this was true of his former belief in innate depravity, 'original sin,' as it were, the belief that man is born hostile."

During their talk, Einstein also described capitalism's worst evil as "the crippling of individuals . . . Our whole educational system suffers from this evil. An exaggerated competitive attitude is inculcated into the student, who is trained to worship acquisitive success as a preparation for his future career."

When such thoughts depressed him, Dukas could snap him out of it with a selective sampling of that day's mail. She had a fairly free hand in deciding which letters he should read and answer, which might divert him, and which to put straight in the files—for the record—unseen by him.

He may have read but did not reply to a letter from a lonely widow and mother of five who wrote to him to say he reminded her so much of her late lamented husband, right down to the "large ears, hair, the prominent nose, the kind eyes."

Einstein was hardly at a loss for female companionship at Princeton—none of whom were likely to tell him he had big ears. Joanna Fanta even claimed to be his girlfriend. The vivacious daughter-in-law of Bertha Fanta, who had kept open house in Prague for young Jewish intellectuals including Einstein and Franz Kafka, Joanna worked in the map department of Princeton's library and saw Einstein almost every day. She also appeared to be his Boswell, because when she went to see him she took a notebook with her and as he spoke would record his words. She later sold the notebooks and his letters.

In June, Ella Winter, the former wife of muckraking writer Lincoln Steffens, asked Einstein to support a peace group of which Irène Joliot-Curie, the daughter of Marie Curie, was a member. To that end, Winter asked for an interview to publicize his coming on board. If Winter assumed that Einstein's disenchantment with American capitalism meant he had Communist sympathies, she learned her mistake by return of mail. Einstein declined the invitation, despite his commitment to world peace, because he was convinced that the only solution was on a supranational basis as advocated by the World Federalist Movement. His previous experience with the group she represented, he said, was that they always followed the Communist Party line.

Winter's response was fast and lively:

I have no objection whatever to working for peace with people who believe the ultimate solution is world federalism. Some of the people in the world peace movement believe in socialism; some in communism; some in Catholicism; some in feminism; some in Kant; others in Hegel or Marx. There are Protestants, Jews, Methodists, Lutherans, Negroes and agnostics. There are those who believe Russia does not have Anglo-Saxon civil liberties, like O. John Rogge, and others who believe passionately Russia has the solution to the economic problem of our day even though at present she is unable to carry such a solution into effect . . . I am not asking you to do any particular work. I am asking you, the man who told President Roosevelt about the atomic bomb, to use the enormous prestige of your name and fame now for world peace. There is no "party-line" to that; except the "line" against a small group of self-interested men in Washington and elsewhere who prefer ruling the world to the lives of millions of boys. I myself have three sons . . . A world state would be wonderful. But the first nearest next stop surely must be to fight for the right to live, to be alive to work for a world state.

Einstein praised her "very intelligent letter" but was not to be hooked. He reiterated his view that Communists were pulling the strings of the group she was helping to organize—and not pulling toward the building of a supranational state. If J. Edgar Hoover had seen those Einstein replies he would have called off the hounds.

Amid the pile of letters from people like Winter, urging him to support their causes, was a pleasant surprise: a gift from an old friend, Max Brod. Almost forty years before, he and Einstein had spent many carefree evenings of music and conversation at the Fantas' home in Prague. Now he had sent Einstein his latest novel, *Galileo in Prison*.

Einstein marveled over Brod's insight into his subject even though he had a different picture of the man. He could not imagine, for example, why Galileo had wasted his last years vainly trying to convince small-minded and superficial people, priests and politicians, of what he knew to be a scientific truth. He could not imagine himself expending such time and energy in defending relativity.

Apart from Brod, another reminder of the past came in an SOS from Violet Winteler, a sister of his first love, Marie. She was marooned in a Canadian nursing home in a desperate state, suffering from severe arthritis and apparently abandoned by the rest of her family, who had ignored her pleas for help. She asked Einstein to send her eighty or a hundred dollars to pay her debts and help her get back on her feet.

Einstein replied to the nun in Toronto who had written on Violet's behalf, saying that his sister had helped Violet financially in the past, and enclosed the requested one hundred dollars.

(Paul Winteler, Violet's brother and Maja's husband, had remained in Geneva, Switzerland, living with his brother-in-law, Michele Besso. He had hoped to be reunited with Maja, but that was not to be.)

Two visitors to Princeton that fall provided light relief: writer John Kieran and former major league baseball player Moe Berg. Kieran recalled their meeting in a conversation with his wife: "The first thing Einstein wanted to know was whether I could explain the game of baseball to him. Well that was right down my lane so I got out a little paper, took my pen and set to it. I talked as I made a few diagrams and finally I thought everything was clear. I suppose I said, 'See?' And he could *not* see. I think baseball was too simple for him to understand. He may perhaps have been searching for a mental challenge that baseball didn't supply."

The way Moe Berg told it, after Einstein had given them each a glass of tea and played his violin, he said, "Mr. Berg, you teach me baseball and I'll teach you the theory of relativity." He paused, then added: "No we must not. You will learn about relativity faster than I learn baseball."

After Einstein resumed his work at the institute, he often walked to and from the place with his friend and colleague Kurt Gödel. Somehow he interested this purest of pure mathematicians in general relativity theory. Gödel sat with Einstein one day in the math library of Fuld Hall trying to solve the gravitational field equations and came up with an intriguing conclusion: that the passing of time and the existence of change in the world was an illusion. "It seems," he said, "that one obtains an unequivocal proof for the view of those philosophers who, like Parmenides and Kant, and the modern realists, deny the objectivity of change and consider change an illusion or an appearance due to our special mode of perception."

Gödel's thesis, "An Example of a New Type of Cosmological Solution of Einstein's Field Equations of Gravity," published in 1949, embraced a "rotating universe." Matter rotating in "a vast, cosmic whirlpool" causes

> space-time projectories that loop back on themselves, returning to places they've already been. It follows from this that time is not a straight linear sequence of events, but something that bends around the universe in a curving line. You could travel from one point on the curve to another, Gödel thought, if you had a fast enough spacecraft. "By making a round trip on a rocket ship in a wide curve, it is possible in these worlds to travel into any region of past, present, and future, and back again."

Whatever the weather, Einstein chose to walk to and from the institute, frustrating many drivers who offered him a lift hoping to hear a few words of wisdom. Robert Jastrow was one of them. A fellow physicist working under Oppenheimer at the institute, Jastrow was at the wheel of his new convertible when he passed Einstein, who was also on the way to work. He stopped and offered him a lift. "Einstein said, 'No, thank you. I prefer to walk.' That was the extent of our conversation," Jastrow recalled with a laugh.

At year's end, John Kemeny suggested that Einstein "ought to get another assistant because, although I could help him in trying to work examples and to figure out if his theory might be correct, the crucial thing was to solve equations. And that takes a specialist—far from my field. I did visit a number of friends at other schools but none could help solve the equations. So he hired a young lady, Bruria Kaufman." Kaufman, a twenty-two-year-old New Yorker who had earned a Ph.D. from Columbia University at nineteen, had been John von Neumann's assistant. She later settled in Israel.

Satisfied that Einstein had the right replacement, Kemeny mentioned that he had an offer "to head up the world federalist movement nationally. I had been very active in the movement, which Einstein also supported. When I went to him for advice, he was absolutely sure this would be a terrible mistake. He said that if you're a paid head of an organization nobody will pay any attention to what you say. If you want to influence the world, make a name for yourself in your own field, and then people will listen to you on other matters as well."

Kemeny went on to become president of Dartmouth for eleven years, the first Jew to head an Ivy League college. He then returned to teaching half time and running a large computer project. He said, referring to this biography in progress, "I'm delighted about anything that can be done to keep Einstein's memory in front of people."

CHAPTER 37

The Communist
Witch-Hunt

1950 to 1951

71 to 72 years old

"To protect Einstein from cranks, Helen Dukas made an arrangement with the neighbors," recalled physics teacher Eric Rogers, "which included my wife and me. We lived next door. When the main front door was opened, Einstein remained standing behind the locked screen door. The crank would be spouting his evil or stupid ideas and any neighbor who was around was expected to come out and talk to the crank, saying such things as 'I'm sure Professor Einstein is getting very tired now,' and march the crank up the road and away. I think it always worked.

"I told him of [Wilhelm] Reich, the psychiatrist with strange ideas and Einstein spoke with him, telling me afterward, 'He's crazy, of course, but very nice.'

"Another thing I remember is how my cat would go over to Einstein's home and sit in his lap." Einstein loved and empathized with animals. So did Margot, who frequently cared for strays. Returning from a visit to friends who had a large, longhaired dog, she remarked that " 'he has so much fur you couldn't tell his front end from his rear.' To which Einstein responded, 'The main thing is that *he* knows.' "

Eric Rogers first noted Einstein's reluctance to wear socks when he was about to drive him to some celebration. It was very cold and there was snow on the ground. "Helen Dukas warned me at the front door, 'Mind about his feet. Remember he has no socks.' I asked why not and she said, 'He thinks it's undignified to let women darn them.' "

To next-door-neighbor-but-one, attorney Edward Greenbaum, Einstein was "a wonderful, marvelous character." Margot and Greenbaum's wife, both sculptors, were friends. One day Dukas phoned Greenbaum, a trustee of the Institute for Advanced Study, to say that Einstein wanted to know if he could come over to discuss the institute. Greenbaum insisted on going to Einstein's home, instead. After they'd talked for a while, Greenbaum recalled, Einstein "said he had a sort of crush on my wife, and that Margot tells him what a good sculptor she is, but 'I've never see anything she's done.' I asked, 'Why don't you come sometime to see it?' Well, like a schoolboy, he replied, 'When?' I said, 'Well, now.' It was a beautiful spring morning. He was in his slippers but he wanted to go for a walk anyway, so he came here in his slippers. My wife was in the cellar, which she calls her studio. I said, 'The doctor says he'd like to see your work.' When she answered, 'I'd love to have him come, whenever he wants,' the slippers were coming down the steps. He looked at everything and when we were walking back to his place, he said, 'Now I know.' That was typical of him."

Strangely, his need to see things for himself did not extend to the Soviet Union, about which he had no firsthand knowledge. Yet he now puzzled his admirers by having apparently fallen for Soviet propaganda. Where previously he had been evenhanded in damning both the United States and Russia for provocative actions and statements, he now appeared to have taken the side of the Communists.

The antics of U.S. Senator Joseph McCarthy may well have pushed him into the enemy camp. With his strident accusations and bullying manner, which he would soon use as chairman of a congressional committee investigating Americans suspected of being communists, McCarthy often ruined the lives of the innocent. Several of Einstein's friends—Harlow Shapley and Ashley Montagu among them—were falsely accused of being communists.

Einstein feared that the United States was becoming a fascist state, restricting civil liberties, supporting brutal dictators around the world, and mistreating black Americans at home. Where Einstein might have been expected to say, "A plague on both your houses," he chose, at least publicly, to support what he saw as the lesser of two evils. He believed that Jews under communism had a chance to survive and even to flourish. Under fascism they were doomed. Was this why he refused to make public protests about Russian atrocities even when Jews were the victims—yet condemned far less horrendous American activities?

He was dismayed when President Truman, on January 31, 1950, escalated the arms race. Truman reacted to the Soviet Union's test of a plutonium bomb by ordering increased production of various atomic weapons including the most devastating—the hydrogen bomb.

The next day, with the public alarmed by Soviet military preparations, was hardly an auspicious time for Einstein, Linus Pauling, and Thomas Mann to show what appeared to be sympathy for communists. They nevertheless joined in protesting the punishment of the attorneys who represented twelve leaders of the American Communist Party (accused of violating the Smith Act by conspiring to overthrow the government by force), asserting that in doing so they were defending the Constitution and its guarantee of a fair trial and adequate legal defense for all. The attorneys had been sentenced to jail for contempt.

Senator McCarthy didn't give a damn about fair trials. On February 9, he claimed that the U.S. State Department was riddled with card-carrying communists, and that he had a list of their names to prove it—fifty-seven in all.

A Mississippi congressman thought he had spotted another one, living in Princeton, and earned himself a *New York Post* headline: " 'Einstein Red Faker, Should Be Deported,' Rep. Rankin Screams."

The screams propelled Hoover into action. On February 15 he sent "some high lights" from the file on Professor Albert Einstein to FBI agent D. M. Ladd. Under the subhead "Organizations," the file reads: "The Bureau files reflect that Einstein is affiliated . . . with at least 33 organizations that have been cited by the Attorney General, the House Committee on Un-American Activities, or the California House Committee on Un-American Activities. He is also affiliated in one way or another with approximately 50 miscellaneous organizations which have not been cited by any of the 3 above-mentioned. He is principally a pacifist and could be considered a liberal thinker as indicated by his connections with the various organizations indicated above."

Walter Winchell, the gossip columnist and radio commentator, added to the anti-Einstein collection, sending Hoover a listener's letter that named seventeen "commie fronts" to which Einstein allegedly belonged. Winchell followed up with more "commie-front" organizations to add to the list.

Hoover still resisted confronting Einstein. In fact, the assistant commissioner of the Immigration and Naturalization Service seemed about to beat Hoover to the punch. He considered revoking Einstein's naturalization after reading in the *Tablet*, a Catholic publication, that Einstein had opposed the fascists during the Spanish civil war. American Catholics, along with Hitler and Mussolini, had generally supported the fascists. Einstein had also sponsored the Spanish Refugee Relief Campaign, classified by the House Un-American Activities Committee as a Communist front.

The biggest bombshell of all was a report linking Einstein with atomic spy Klaus Fuchs, recently sentenced by an English court to fourteen years in prison. But an FBI informant soon defused it by revealing that the two men had never met. Furthermore, although Fuchs had been released from a Canadian internment camp in 1943 after Einstein wrote to the British Home Office on his behalf, he had done so after hearing that Fuchs had been elected to the Royal Academy of Science, and long before Fuchs turned to spying.

FBI agent Ladd gave Hoover further information exonerating Einstein on March 10, 1950. According to Fuchs's father, Einstein had intervened on "my son's behalf not knowing he was a Communist." The father added that Einstein had been impressed by Fuchs's paper on nuclear energy and had considered him of value to the Allied war effort.

But Hoover's suspicions had been rearoused by a letter from Emma Rabbeis, a German woman who wrote from Berlin:

Since the announcement of the espionage affair of Dr. Klaus Fuchs I have read with growing concern all newspaper reports pertaining to Prof. Albert Einstein

... I am no informer and would not make such statements if it did not concern such weapons as atom plus hydrogen which can exterminate nations. But if an active Red such as Einstein, as I positively know, can look into the research status of such horrible weapons, one must not remain silent ... I could give very positive information that the charges against Einstein are fully justified. If you should desire it, and if it would be useful to you in preventing spies from doing any harm, I should be at your disposal with commentary.

Questioned by an army counterintelligence agent, Rabbeis told him that she had run a Berlin dress shop in the 1930s. One frequent customer was Baroness Schneider-Glend, wife of a former German consul in Japan, whose whole family were Communists. Elli, a daughter of the baroness, told Rabbeis that she was working with Einstein. Elli went to America with him, and during the voyage they caused a disturbance by refusing to stand for the German national anthem. This, said Rabbeis, proved that Einstein was a Communist. Furthermore, she believed that Einstein had had an affair with Elli and fathered her child.

No other source has confirmed Frau Rabbeis's charges, and the agent uncovered a possible motive for her attempt to smear Einstein: she had sent him a mathematical formula for winning the Berlin lottery and he had not replied to her letter. However, Hoover was inclined to believe the worst. He again encouraged his agents to keep tabs on Einstein and Dukas as a possible spy team.

Robert Oppenheimer, director of the Institute for Advanced Study, was used to being investigated. As scientific director of the Manhattan Project, he had been constantly tailed and often questioned by FBI and military counterintelligence agents. He had been cleared to continue in his top-secret job despite his links to and early sympathies with Communists. Oppenheimer had no doubt that Einstein was also a loyal American, as well as

one of the friendliest of men ... [but] he was also, in an important sense, alone. Many very great men are lonely; yet I had the impression that although he was a deep and loyal friend, the stronger human affections played a not very deep or very central part in his life taken as a whole ... I remember walking home with him on his seventy-first birthday. He said, "You know, when it's once been given to a man to do something sensible, afterward life is a little strange." ... [His] simplicity, [his] lack of clutter and [his] lack of cant, had a lot to do with his preservation throughout of a certain pure, rather Spinoza-like philosophical monism ... There was always with him a wonderful purity at once childlike and profoundly stubborn.

On May 10, Oppenheimer himself was being investigated by the FBI in California, to ensure that it was safe to allow him to receive top-secret information. There, a witness placed him at a Communist meeting in 1941. He denied ever attending a closed—or official—Communist meeting, but admitted, as he had before, that he had associated with plenty of left-wingers.

Two weeks later the FBI arrested Harry Gold for turning over to the Soviet Union atomic bomb secrets he had obtained from Klaus Fuchs.

It would be a traumatic summer. The Communist witch-hunt led by Senator Joseph McCarthy was going strong. He announced in the Senate on June 6 that the FBI had identified twenty possible agents of the Soviet Union working in the State Department, of whom three were still employed there. The State Department called the charge "absolutely false."

On June 16, soon after the Russians exploded their own atomic bomb, the FBI arrested David Greenglass for helping them to build it. He implicated his brother-in-law, Julius Rosenberg, who was soon in FBI custody. But Rosenberg's wife Ethel was not yet under suspicion.

American distrust of Communists escalated on June 25, when Communist-controlled North Korea made a lightning attack on South Korea. Three days later, North Korean troops captured the South Korean capital, Seoul. In early July, General Douglas MacArthur was appointed commander in chief of the United Nations forces defending South Korea.

And now where did Einstein stand, with the papers full of Communist spies and Communist killers? Louis Budenz gave a plausible account of Einstein's attitude. Budenz, who had been a daring union organizer, was a self-admitted former Communist and the former managing editor of the Communist *Daily Worker*. He had suggested in a recent book, *Men Without Faces*, that Einstein was being manipulated by Communists, "though occasionally he shows his independence." He elaborated in a newspaper interview: "Heading the list of those who sponsored and attended the Waldorf Astoria 'Peace' session [held in Manhattan the previous year] were Thomas Mann and Albert Einstein, neither a Communist. Mann is a novelist, long a warm defender of Moscow. The relationships of Mann and Einstein were established by what the Communists called 'remote control' while I was still part of the Red leadership. The manner of communication with Mann ran through associates of his daughter, Erika; while with Einstein, means of reaching him were set up at Princeton. In both instances, these men were persuaded to their pro-communist stands by playing on their hatred of Nazism. This I know from what I heard mentioned in Politburo meetings. No more striking illustration could be found of the way well-known men and women of unquestionable integrity are deceived and exploited by the Communists." Budenz referred to this as "Capture of the Innocents."

FBI chief Hoover suspected that Einstein was far from innocent, judging by his written reply to the army's assistant chief of staff at the Pentagon:

> Reference is made to your letter dated September 8, which forwarded a report from the European Command concerning the use of Albert Einstein's office in Berlin, Germany, until 1933 as a telegram address by agents of the Comintern. The report also reflected past activity on the part of an unnamed secretary of Einstein on behalf of the Soviet Union. There is attached for your information a blind memorandum entitled "Helen Dukas." Miss Dukas was reportedly brought to the United States as secretary by Albert Einstein in 1935 and has since been employed as secretary and housekeeper at Princeton, New Jersey, where Dr. Einstein presently lives. It is requested that European Command conduct investigation for the purpose of determining whether Helen Dukas is identical with the secretary of Albert Einstein who, prior to 1933, was active on behalf of Soviet intelligence. It is further requested that more detailed infor-

mation concerning the use of Dr. Einstein's office as a telegram address by agents of the Comintern and the part played by his secretary be furnished this Bureau ... It is pointed out in this connection that the Immigration and Naturalization Service is presently considering an investigation of Dr. Einstein for possible revocation of his citizenship."

Hoover also supplied the names of many suspects believed to have used Einstein's address, or who were aware "of the fact that the address was used in connection with Soviet activities."

Einstein did not read the *New York Daily Mirror* nor the *New York Post*, so he missed their attacks on him as a suspected Communist stooge or spy. Instead, he had to contend with a better-informed and more persistent antagonist, Sidney Hook. He even admitted to Hook that as a novice he had been involved with organizations that operated with decent arguments but dishonest intentions. In turn, Hook confided that in high school during World War I he was denounced as pro-German and almost expelled for challenging atrocity stories, spread by the Allies, in which German soldiers systematically amputated the hands of Belgian children and boiled human corpses to use the fat for industrial purposes — stories later revealed to have been propaganda.

Hook had been asked by the *Annals of the American Academy of Political and Social Science* to review Einstein's book, *Out of My Later Years*, published in 1950, and wanted him to clear up apparent inconsistencies. On one page Einstein had written that he would find it intolerable to belong to a society that consistently denied the free expression of opinion. On another page he speculated that had he been born a Russian he could have adjusted to life there. In his reply to Hook, Einstein denied any contradiction. He believed, he wrote, that the interference of the Soviet government in intellectual and artistic matters was both harmful and ridiculous. But the Russian revolution would have failed, he maintained, if those with the welfare of all the people at heart had not agreed, as a lesser evil, to a painful temporary renunciation of personal freedom.

How, Hook asked, renewing their long-standing debate, did Einstein know "the sacrifice of individual freedom is only temporary in view of the constant growth of terror?" He got no reply and didn't expect one. But he continued to be puzzled by Einstein's attitude:

> In view of the stream of revelations pouring from the press by individuals who had actually lived in the Soviet Union, I could not bring myself to believe that Einstein could be so uninformed about the true conditions of Soviet life. This was after Stalin's break with his bloody henchman Tito, after the Communist coup in Czechoslovakia, and after the Berlin blockade. Nevertheless I resolved to send him firsthand materials on the Soviet Union by individuals who had lived there, [especially] individuals who had something to say about the [often harrowing] fate of Jews and those accused of Zionism in the Soviet Union.

Einstein's response bitterly disappointed Hook: he'd try to find time to read the books but didn't expect to find anything essentially new in them. Einstein added that he was trying to form an objective view of Russian life and, never having been an enthusiast, didn't expect to be deeply disappointed by the "short-comings of this vast [Russian] empire." Hook remained mystified and disturbed "by his silence concerning the brutal treatment of Jews, especially Zionists like Einstein, in the Soviet Union and its satellites."

When asked to try to explain Einstein's silence, Hook replied: "If you're a liberal you're an antifascist. I say you also have to be an anticommunist. But these people [like Einstein] say, 'Oh, no, no!' This is a sentimental view. When it comes from Germans, especially German Jews, it's really because they feel, as Einstein told me, that the Americans don't understand the Germans and they won't punish them enough. So they're hoping the communists will punish the Germans as they should be punished. I think it's due to that subconsciously. During our second conversation, when Einstein was in a good, jovial mood, he said, 'The Germans are an arrogant people with their noses in the air.' I took him up on collective guilt, which he believed in, saying, 'How can you? There were two million Germans, non-Jews, who were in Hitler's concentration camps.' He said something like he didn't mean those Germans. But he didn't retreat from his point of view that all the Germans were responsible. But I think do-mestic terror in the Soviet Union was even greater than in Nazi Germany."

One view of Einstein's attitude is that in the 1950s "a kind of hopelessness had come over [him]. He certainly agreed with the criticism of the United States made during the so-called McCarthy experience. He thought everything was becoming terrible. This was the period when he said if he was starting again he would not become a scientist, he would become a plumber. Schopenhauer's picture used to be in one of Einstein's rooms. I think a streak of pessimism in the last years started overwhelming him. I think this affected his political judg-ment—everything is going bad, everything is hopeless in the Soviet Union and everything equally hopeless in the United States." Thomas Bucky disagrees. "I never saw him depressed. Saddened, yes. But then he would regain his cheer-fulness."

His friend and former assistant, Leopold Infeld, himself a Communist, had a different interpretation of Einstein's attitude:

His heart never bleeds and he moves through life with mild enjoyment and emotional indifference. For Einstein life is an interesting spectacle that he views with only slight interest, never torn by the tragic emotions of love or hatred. He is an objective spectator of human folly, and feelings do not impair his judgment. His interest is intellectual and when he takes sides (and he does take them!) he can be trusted more than anyone else because in his decision the "I" is not involved. The great intensity of Einstein's thought is directed outside toward the world of phenomena.

Einstein gave his own view of himself in his correspondence, writing to an inquiring stranger:

A human being is part of the whole, called by us "Universe," a part limited in time and space. He experiences himself, his thoughts, and feelings, as something separate from the rest—a kind of optical delusion of his consciousness. This delusion is a kind of prison for us, restricting us to our personal desires and to affection for a few persons nearest to us. Our task must be to free ourselves from this prison by widening our circle of compassion, to embrace all living creatures and the whole of nature in its beauty. Nobody is able to achieve this completely, but the striving for such achievement is in itself a part of the liberation and a foundation for inner security.

He offered a more specific self-assessment to his friends Max and Hedi Born:

I simply enjoy giving more than receiving in every respect, do not take myself nor the doings of the masses seriously, am not ashamed of my weaknesses and vices, and naturally take things as they come with equanimity and humor. Many people are like this, and I really cannot understand why I have been made into a kind of idol. I suppose it is just as incomprehensible as why an avalanche should be triggered off by one *particular* particle of dust, and why it should take a certain course.

Just before Christmas 1950 he answered a year-old letter from Fräulein Markwalder, the daughter of his Swiss landlady during his student days. He recalled her mother as always good and sometimes thoughtful, and claimed to have as much energy as ever. He had given up the violin, he told her, but sometimes improvised on the piano. After seventeen years in America, he said, he had not absorbed anything of the country's mentality and needed "to guard against becoming superficial in thought and feeling: it lies in the air here."

He continued to express his provocative views, telling a Minnesota high-school music director who wondered why he disapproved of marching bands that "I see in marching accompanied by rhythmic music a means for keeping the individual's body awake and his mind asleep. It was always one of the most efficient means to induce people to slaughter each other and in general to commit the most horrible crimes undisturbed by the voice of conscience."

Apparently he had cleared his own conscience, to some extent, by providing for the financial well-being of his schizophrenic son. Eduard was temporarily free from the psychiatric hospital and being cared for by Pastor Hans Freimuller in a hillside village near Zurich. Freimuller gave therapy to disturbed men through a combination of evangelism and psychoanalysis. His first impression of the forty-year-old Eduard was of a paunchy, nervous man with extraordinary, brilliant eyes "irradiated with a wondrous luminosity." They seemed to the pastor to express "goodness which looks for protection."

After weeks of almost intense piano playing in isolation, Eduard overcame his shyness and entertained the pastor's three young sons with jokes and poems. He even gave concerts for village children. Eventually, the pastor's wife found him a job addressing envelopes, which gave him a measure of independence.

A former school friend, Peter Herzog, now a teacher, lived nearby. Herzog was shocked when Eduard came to see him. The attractive young man whose intelligence had dazzled his teachers had become fearful, prematurely old, and overweight. He spoke of his fears for the future, but apparently never mentioned his father.

About that same time, U.S. Army investigators in Germany were trying to establish if Helen Dukas had been a Soviet spy while working for Einstein in Berlin. In February 1951, they found two people who had lived in the Einsteins' Berlin apartment building before the war. One, named Tetzlaff, stated that Einstein had offices in the Kaiser Wilhelm Institute for Physics and at the University of Berlin, but not in his apartment. Tetzlaff also said that Einstein and Dukas were friendly with a family named Auerbach. A former housekeeper for the Auerbach family, Marie Kulkowski, said that she had known Einstein and was still receiving food parcels from him. Lotte Schiffer, née Auerbach, was traced to London, England. Neither woman had anything incriminating to offer, however.

Stymied, the FBI's Newark office suggested "that perhaps a review of available biographies or writings concerning Einstein would reveal some of his European associates. Also, if the Bureau has available any special contacts or informants who would be in a position to approach Helen Dukas or Einstein on the pretext of writing a book or article about the influence any women or particular woman may have had on Einstein's life, this approach could be [used] to determine his Berlin associates and employees."

Hoover had "no objection" to all their suggestions, and by the end of August they had culled from Philipp Frank's 1947 biography of Einstein (*Einstein: His Life and Times*) the following names: Helen Dukas; Professor Ladenburg ("Now believed to be the Rudolph [sic] Walther Landenburg [sic], a physics professor at Princeton"); Cornelius Lanczos; Walter (Walther) Mayer; Peter Bergman (Bergmann); and Valentia (Valentin) Bargman (Bargmann). From Marianoff's biography they gathered that Marianoff had married Einstein's stepdaughter, Margot, and that he had received a lot of mail. The Newark office felt it was too risky to contact "Ladenburg in view of his close association over a great number of years with Professor Einstein." They would try to find someone else in the Princeton area with the answers.

Meanwhile, Einstein was concerned with the fate of Julius and Ethel Rosenberg. They had been sentenced to death on April 9, by Judge Irving Kaufman. He called their crime worse than murder, because

> "your conduct in putting into the hands of the Russians the A-bomb years before our best scientists predicted Russia would perfect the bomb has already caused, in my opinion, the Communist aggression in Korea, with the resultant casualties exceeding 50,000 and who knows but that millions more of innocent people may pay the price of your treason. Indeed by your betrayal you have already altered the course of history to the disadvantage of our country."

Ethel's brother, twenty-nine-year-old David Greenglass, saved his own life by testifying against her and his brother-in-law, Julius, and got a fifteen-year prison sentence.

David's wife, Ruth, wrote to Einstein saying that she and her two children would wait for her husband's release—he would be eligible for parole in four years—and asking if she could discuss their future plans with him. She explained that she and her husband had cooperated with the government, and that he had testified against his sister, not because they wanted to save their own skins, but because they realized that it was their duty as Americans and parents. What worried her was the hostility their children might face in the future, and she wondered if they should leave the country.

Einstein replied:

> Whatever a man or woman has done he or she deserves to be helped to an honest life. It seems improbable that this can be achieved in this country in the prevailing political-psychological climate. Immigration should not be too difficult for a man who has acquired skill in some technical work. [David Greenglass was a trained mechanic.] But it serves no purpose to discuss these questions now in a personal interview or otherwise.

Throughout the spring, Einstein was increasingly preoccupied with his sister's illness. Though barely able to speak, Maja was still mentally alert, and he read to her every evening until she slept. In June 1951, she broke her right arm in a fall and developed pneumonia, from which she died later in the month. Maurice Solovine was among the first to be told: "My dear sister was delivered from her horrible suffering by a gentle death four weeks ago." Einstein concluded his letter by saying, "We bear many afflictions unflinchingly but Spinoza's precarious God has made our task more difficult than our forefathers suspected." He also expressed his gratitude to the family doctor, Guy Dean, for his "precious care" and for doing "everything possible to relieve my sister's suffering." Soon after Maja was cremated, Albert wrote to a cousin: "I miss her more than one can imagine. But I am relieved that she has it behind her."

That fall, David Ben-Gurion was in the United States to raise money for Israel. En route to Philadelphia, he stopped in Princeton for a few hours with Einstein. It was mild for September, so they sat side-by-side outside on garden chairs, Ben-Gurion formally dressed with a gold emblem in his lapel given to graduates of the Israeli army's officer training school, Einstein almost formally—for him—in slacks, an open-necked white shirt, and a cashmere sweater. After being photographed shaking hands for a few press photographers, they resumed their seats and their conversation, looking, with their manes of white hair, like brothers.

Ben-Gurion later told diplomat and writer Moshe Pearlman that he was "curious to know Einstein's secret of concentrated thinking and wanted to discuss universal truth with him. Although Ben-Gurion did not have the conventional view of God, he believed," Pearlman wrote, that there must be "something infinitely superior to all we know and are capable of conceiving. Without such a being, there are certain phenomena which just cannot be explained. What is it, for example, that enables man to think? His brain is matter, just like a table. But a table does not think. The brain is part of a living organism, like my fingernail, but my fingernail cannot think. Nor can the brain think when removed from the body. But the whole of the living body taken together becomes a thinking being." When he discussed this with Einstein, "even he, with his great formula about energy and mass, agreed that there must be something behind the energy."

If Einstein explained the secret of concentrated thinking, Ben-Gurion never publicly revealed it, but he did tell a friend that he still considered Einstein to have the greatest mind of any living man. "Do you realize," he added, "that Einstein is a scientist who needs no laboratory, no equipment, no tools of any kind? He just sits in an empty room with a pencil; and a piece of paper, and his brain, thinking."

Philosopher Martin Buber, who was to teach and study at Princeton, also stopped in for a while, and he and Einstein "were delighted to discover that they both liked Ellery Queen mystery stories!" They had known each other for some forty years. It was Buber who during an early meeting had "pressed him hard" to reveal his religious belief; Einstein had replied, "What we [physicists] strive for . . . is just to draw His lines after Him." Einstein had told Hans Albert that each meeting with Buber was "a great joy."

The arrival of unexpected visitors and letters from strangers—friends and foes alike—never let up: "I am interested in physics. Would you give me a few pointers?" . . . "Just a note to let you know I think of you often and love you very much" . . . "If you believe that Israel is a place for teachers, why don't you go there in a hurry?" And all the while the FBI continued to monitor and note Einstein's activities: "A letter from Dr. Albert Einstein to the NCASP praising it and other organizations who were fighting for Willie McGee [described in the *Daily Worker* of March 27, 1951, as a Mississippi Negro victim of a rape frame-up who was seeking an appeal of a death sentence before the United States Supreme Court] whom he believed to be innocent."

In September, the New York FBI office proposed a ploy to interview an unsuspecting Dukas—by questioning her about another suspect. Hoover had waited six years to hear this suggestion. Go ahead, he said. It won't prejudice "our current investigation of Dukas. Further, such an interview might assist the Newark Office in determining the attitude and probable cooperation of Dukas . . . She is not to be informed in any way that she herself is under current investigation."

The interview was a bust. She denied that she or Einstein knew any of the "suspects" mentioned by the agents and merely confirmed what they already knew—that she had been Einstein's personal secretary since 1928.

The Dukas file was now an inch thick. Six years had passed, and Hoover was no wiser than when he started.

Einstein turned once more to his unified field theory until he realized he had gone as far as he could go. Someone else would have to provide mathematical verification. Despite years of trying, he had failed in the attempt. He mainly blamed the current crop of physicists for not having joined him in the effort, attributing their skepticism to an inability to grasp his logical and philosophical arguments.

But there was plenty of fight still left in him, and laughter. "I never read tragedy in his face," wrote his colleague Abraham Pais. "An occasional touch of sadness in him never engulfed his sense of humor."

CHAPTER 38

Conversations and Controversies

1952

73 years old

Carl Seelig, a writer with a compassionate interest in the mentally disturbed, was at work on an Einstein biography. Living in Zurich, he had easy access to several people who had known Einstein as a young man. Seelig had broken Einstein's resistance to the proposed biography by his apparent integrity and intelligence—and had kept him in a cooperative mood by sending him a regular supply of Swiss-made dried soups. Einstein enjoyed not only the taste but also the memories they evoked of idyllic times in Switzerland. Dukas was also pleased with this "manna from heaven," as the soup was just right for Einstein's restricted diet.

All previous Einstein biographers had steered clear of what had happened to Eduard after he was confined in the mental institution, as if he were dead or no longer relevant to their subject. Seelig was the first and only one to find out by asking Eduard. If anyone thought Einstein was trying to hide this aspect of his life, they were wrong, which he proved by arranging for Seelig to visit Eduard.

It is not clear what was wrong with Eduard. Had he inherited schizophrenia? Had he overreacted to the breakup of his parents' marriage and his feeling of abandonment? Was his condition the result of some overwhelming traumatic experience? His brother, Hans Albert, believed it had started with Eduard's depression after an unhappy love affair and had been exacerbated by electroshock treatment. Einstein's immediate response to Eduard's despair over unrequited love had been to tell him that women were a delightful necessity, but that it could be fatal to make them a main preoccupation. He recommended hard work to take his mind off the woman, and looking for a "plaything" rather than someone "cunning" like the woman who had rejected him. Einstein compared life to riding a bicycle—the only way to maintain your balance was to keep moving.

In Eduard's case that meant pursuing his dream of becoming a psychiatrist. Trying to make the best of things, Einstein said that through understanding and overcoming his own symptoms Eduard might be in a better position to help others.

It was now some twenty years since Eduard's breakdown, and he had spent most of them in a mental institution. But Einstein told Seelig that his son's illness was relatively mild, though severe enough because of "intense emotional inhibitions" to make it impossible for him to have a career—or, apparently, to otherwise lead a normal life.

When Seelig took Eduard to dinner in a Zurich restaurant, he seemed eager to answer questions, but blanked out when asked about his paternal grandparents. Einstein later said that this was understandable; his father had died before the boy was born and Mileva and his mother hated each other and rarely met.

What Eduard did recall was a visit to his father's Berlin apartment as a boy, when he eagerly peered through a telescope on the balcony, first at the moon, and then even more eagerly into apartments across the street. He obviously enjoyed trying to recall the past. He chain-smoked throughout the evening except when eating, and finished the meal with an ice cream sundae. Seelig and Eduard met more than once. On other occasions, Seelig took him to the theater and on walks, when the writer's dalmatian went along, too.

While Einstein continued to enjoy his biographer's soups, Eduard enjoyed Seelig's company and came to regard him as his best friend. For his part, Seelig was so moved by Eduard's "tormented and brooding . . . but also serene smile and a truthfulness that are speedily charming," that he offered to be his guardian. Einstein reluctantly declined, saying that his son already had one.

Another of Eduard's visitors was his niece, Evelyn Einstein, the adopted daughter of Hans Albert. She took Eduard on short trips outside the hospital. Evelyn, then a disgruntled teenager attending a Swiss boarding school, which she loathed, found Eduard endearing and fascinating. "We got along great guns," she said. "The two of us clicked very well. He was filled with questions. He was like a sponge. I mean, they kept him isolated from the real world. And he asked questions I found strange, like 'Well, did they decide to go into electricity on automobiles?' " (This was the early 1950s.) Eduard then explained that the only encyclopedias they let him see were from the 1920s. "The way he said things," Evelyn recalled, "I just felt he was brilliant. I wonder if he was truly schizophrenic."

Evelyn was shocked by his living conditions: "He had a private hovel, a basement room with no windows. It opened onto a corridor. It was adequately furnished and he had electric light, but it was dark and dreary and dumpy. But he was used to it, for godsakes. You get used to anything. He was an institutionalized human being when I met him. He could never have functioned outside.

"I believe they felt that the less informed he was the easier he would be to control. He told me about working in the garden, which he didn't like at all. But he said it was so important because when the Russians took over Switzerland they were going to kill everybody who didn't work. And I asked, 'Who told you that?' He said the hospital [staff] did, to get him to work in the garden."

Evelyn knew there was an account of Eduard hearing imaginary voices and being violent, but said it had never happened when she was with him.

Although Eduard bombarded Evelyn with questions about "everything," including world affairs, he never once asked her about his father, mother, or brother. And if Seelig asked Eduard's opinion of his relatives or how he felt about being incarcerated in a mental hospital, that material did not appear in the Einstein biography. (In the biography, which was published shortly after Einstein's death, Seelig neither challenged Einstein's diagnosis nor reproached him for what could be regarded as an abandonment of a son. If Eduard complained of his father's treatment, Seelig kept it to himself.)

Seelig now moved on to other sources. He contacted Maurice Solovine in Paris, seeking insight and information about the glory days of the Olympia Academy.

Having witnessed some of Einstein's troubles with Mileva, being a loving friend and his official translator into French, Solovine naturally felt cautious about talking with the biographer. He also knew that Einstein rarely discussed his private life with others. What shall I do? Solovine wrote to ask Einstein.

Tell him whatever you think best, Einstein wrote back, "and pass over whatever you wish in silence. For it is not always good to be presented to the public nude—or rather neuter." As for himself, Einstein didn't want to be involved in the book any more, even indirectly. He believed that Seelig was probing too hard and annoying his friends.

In his letter, he also referred to Solovine having expressed shocked surprise that he (Einstein) had recently used the term "miraculous" to describe the ability of scientists to understand the world. Solovine apparently feared that his friend was losing touch with reality.

Einstein reassured him that he was not turning into a mystic; on the other hand, he did not claim to know all the answers. By using the word "miraculous," he explained, he meant "inexplicable" or "unexpected"; in his view, we should expect the world to be chaotic and beyond our understanding.

Even so, Einstein believed that one thing about the world would always elude scientific probing—why it exists. Don't think I've gotten religion in my old age, he added in his reply to Solovine, but I don't want to be included among the smug atheists and positivists who think they know all the answers.

Despite the many demands on his time, Einstein continued to justify his reputation for never being too busy to help anyone who asked by sketching the answer to a geometry problem for Johanna, the fifteen-year-old daughter of screenwriter Herman Mankiewicz. She had repeatedly tried and failed to find the common external tangent of two tangential circles with radii of eight and two inches, respectively, so she wrote to ask him if he could help her out. The sketch he sent back, which resembled the side view of a steamroller, was inaccurate because his circles weren't touching. But he gave the correct equation for her to work out the answer.

While he was working on this problem, the institute's director, Robert Oppenheimer, was being grilled to determine his reliability as a consultant with access to atomic weapons secrets. As early as 1947, newspaper headlines had promised rocky days ahead for Oppenheimer. In July of that year, the Washington *Times Herald's* banner headlines had read: "U.S. ATOM SCIENTIST'S BROTHER EXPOSED AS COMMUNIST WHO WORKED ON A-BOMB . . . Frank Oppenheimer Was at Oak Ridge, Los Alamos Plants."

By 1952, Robert Oppenheimer had already been investigated for several years by counterintelligence agents. In 1949 he had appeared as a witness before the House Un-American Activities Committee. Nevertheless, no decision had been made as to his loyalty or reliability. The investigations and questions continued.

In May 1952, the questioners came to Oppenheimer's office at the Institute for Advanced Study, instead of inviting him to Washington to tell what he knew and how he thought. Knowing that his colleague was in trouble, Einstein did what he could to help, offering to speak in his support before the committee. Apparently, Oppenheimer did not take him up on the offer.

Princeton photographer Alan Richards happened to be near the institute when he saw an "old man being helplessly pushed around by a photographer and a man with a microphone rudely backing him off the sidewalk and firing questions at him." Richards told them to leave "the old man" alone, then saw it was Einstein, who "was saying that he had made a statement for Oppenheimer that morning at the Institute and that's all he would have to say." When Richards said that the reporters must have made him very angry, Einstein replied, "Oh no, I don't pay any attention to those things. You know anger dwells only in the bosoms of fools."

Soon after that encounter, Richards had arranged to photograph Einstein in his study, and noticed that he was "feverishly jotting in a notebook, not even aware that there was a cut on his face [from shaving, presumably]. The two bright streams of blood were running down his cheek. I was so alarmed that I interrupted him to suggest that I go to the medicine cabinet and get a bandage. 'Oh,' he muttered, 'it's not a matter of importance. It'll stop.' And he went on writing."

That spring, Einstein's relentless political sparring partner, Sidney Hook, came back with a flurry of punches. He was incensed by Communist charges that the United States was waging germ warfare in Korea, which the French Communist Frédéric Joliot-Curie, a Nobel Prize winner, said he had investigated and confirmed. As chairman of the American Committee for Cultural Freedom, Hook invited American Nobelists to sign a letter asking Joliot-Curie to join them in an objective scientific investigation or withdraw his statement. Einstein was the only one to refuse to sign the letter, telling Hook that although he was disappointed by Joliot-Curie's insincerity, he did not believe that "a counteraction promoted by politicians" would have the desired effect. He criticized the American scientists who had signed the letter for never having collectively protested the military abuses of science generally, an issue which he considered much more important.

Furious at having his committee's work called "a counter-action promoted by politicians," Hook struck back, saying that the group was interested in defending "those values which are essential to the preservation of a free culture," and not just in clearing the good name of the United States. He went on to say that

> we are even more concerned with the question of moral responsibility and the intellectual integrity involved in the fanning of the flames of war and hatred by such accusations as those leveled by M. Joliot-Curie . . . If it is desirable for Nobel scientists to condemn the military abuses of science, as you now seem to think, surely the action we propose at the very least can be considered a step in that direction. We, therefore, see no valid reason why, even believing as you now do, you should refuse to sign the letter to M. Joliot-Curie.

Einstein regretted that Hook had read his reply as an insult, and "subsequently realized that I had not written in that gentle mode of expression attributed to St. Francis." He conceded that it was disgraceful for Joliot-Curie, as a scientist, to make accusations that he could not be sure were true, but felt that making a fuss would be useless because it would never reach the Russian public. Instead, it would produce a fresh wave of hatred by the West for the East and in such an atmosphere a viable modus vivendi between the two would be impossible.

Einstein clung to his belief that a letter of protest signed by other Nobelists would be useless, and though he welcomed an objective investigation of the charges made by Joliot-Curie, he did not think that one should "become indignant in the face of such accusations so long as one has not accepted the obligation to forswear the first use of *bacteriological* weapons."

Later in the fall, Einstein wrote to some Japanese people who had questioned him in a letter about the atomic bomb, explaining his role in its creation. He also told them why he saw Gandhi as the greatest political genius of the time: Gandhi had shown that peaceful protests could defeat apparently unbeatable military might. Hook read a version of this letter in *The New York Times* and fired off a riposte to Einstein on November 10:

> Gandhi could have been successful only with the British or a people with the same high human values. I am afraid he would have failed utterly with the Japanese military, with the Gestapo and the SS, and the Soviet MVD. What is even worse, the new methods of "scientific" torture would have reduced him to a broken-spirited, miserable wreck of a man, stuttering back the confessions suggested by a cunning and cruel prompter manipulating the coercions of hunger, pain, and crazed desire for sleep or even death.

Einstein replied by return of mail, agreeing with Hook. He knew that the only solution to the problem was ending the arms race. He acknowledged that the methods of the Russians were evil, but thought that the first step on the road to world peace would be to neutralize and demilitarize Germany.

Finally, Hook believed, he knew Einstein's motive for refusing to criticize the Soviet Union publicly. Einstein's position was logical, Hook concluded,

> in view of his own personal experiences that he should fear the Germans more than the Russians or Japanese or Chinese. His attitude was typical of German

Jewish refugees of almost every social stratum. [Hook was a Brooklyn-born Jew.] It was not that they had a genuine sympathy for Communism—except for those who before the rise of Hitler had thrown in their lot with the Kremlin—but that their hatred of Germany and their acceptance of the doctrine of collective guilt made them willing to accept any initiative that would punish their former persecutors and their descendants. It was psychologically understandable but politically unwise.

Einstein had hardly finished explaining himself to Hook when he became embroiled in the Velikovsky affair. Historian Immanuel Velikovsky, also of Princeton, met Einstein by chance one day while walking by the lake near the institute. Velikovsky complained bitterly of the two years of "abuse and calumny" he had suffered as a result of publishing his daring history of the universe, *Worlds in Collision*, which presented his revolutionary idea that Venus had repeatedly collided with Earth and Mars about 1500 B.C.

Because Velikovsky leaned heavily on the Old Testament, the Hindu Vedas, and Roman and Greek myths as sources, scientific scholars greeted his thesis with unacademic fury, dismissing it as garbage. Astronomer Harlow Shapley, former director of the Harvard Observatory, refused to read the manuscript, saying he was not interested in Velikovsky's sensational claims because they were inconsistent with gravitational theory and violated the laws of mechanics. He added that if Velikovsky was right, "the rest of us are crazy."

Nevertheless, editor James Putnam of Macmillan had agreed to publish the book. Then he got a letter from Shapley expressing astonishment that Macmillan would venture into the "Black Arts," and threatening to sever all relations with the company if they went ahead with the book. Putnam refused to be intimidated.

Shapley, president of Science Service, which published *Science News Letter*, struck back by printing the views of five experts who denounced Velikovsky's *Worlds in Collision*, even though not one of them had read it. Soon after the denunciations appeared, the *Harvard Crimson* reported that "a surprising number of the country's reputable astronomers descended from their telescopes to denounce *Worlds in Collision*."

Yet the book took off, partly because of a sensational article about it titled "The Day the Earth Stood Still." After the book hit the best-seller list, however, Velikovsky was told to his dismay that professors at major universities were refusing to see Macmillan salesmen, and would he mind if the book was transferred to Doubleday? As Doubleday did not have a textbook division, it was immune to the threats of professors. Velikovsky reluctantly agreed, and Macmillan further ingratiated itself with the academic world by finding a sacrificial lamb; it fired James Putnam, the twenty-five-year veteran of the company who had acquired the book.

Einstein listened patiently to Velikovsky's tale of the soft underbelly of the publishing world, despite his suffering from a painful attack of phlebitis at the time—which, he wryly remarked, usually affected pregnant women. But he went easy on astronomer Harlow Shapley—perhaps because they shared similar political views—saying that his behavior could be explained but not excused. "One must give him credit," Einstein said, "that in the political arena he conducted himself courageously and independently, and just about carried his hide to the market place. Therefore it is to some extent justified if we spread the mantle of Jewish neighborly love over him, however difficult that is." It was too difficult for Velikovsky, who replied, "Too early you have thrown the mantle of Jewish compassion over Shapley . . . His being a liberal is not an excuse."

Einstein said he enjoyed his isolation—he stood virtually alone in his resistance to quantum theory—and advised Velikovsky to follow his example and not take criticism so seriously. Velikovsky responded, "Yes, there are two heretics in Princeton. Only one is glorified: the other vilified."

Einstein was intrigued, though he knew the charming and persuasive Velikovsky was an expert in few if any of the fields to which he brought his revolutionary pseudoscientific speculations. In fact, his background was medicine and psychoanalysis. For a time, as a Russian-trained physician, he had practiced medicine in Palestine. Then—like Wilhelm Reich, another man with wild ideas to whom Einstein listened patiently—he had become a psychoanalyst. He had emigrated with his wife and two daughters to New York in 1939. Every day for the next nine years he "opened and closed the library at Columbia University," he said, researching the material to produce his provocative theories.

Einstein's main objection to Velikovsky's ideas was his contention that the close encounters of the three planets had occurred when electrical and magnetic forces had overwhelmed gravity — which contradicted the laws of both dynamics and gravity.

This did not discourage Velikovsky. Obsessed with his own ideas, he later wrote to Einstein, "When, by chance, we met last week at the lake, I became aware that you are angry with me personally for my 'Worlds in Collision.' From you I had not expected this reaction." He admitted, however, that Einstein's response—that the accepted cosmological theory exactly explained the motions of the planets—had caused him to rethink, but not abandon, his ideas. He came back with the complaint that scientists excluded electromagnetic conditions in the solar system from celestial mechanics. Then, temporarily, he rested his case.

Despite the unremitting demands on his time from the Hooks and Velikovskys and scores of strangers, Einstein still kept in touch with old friends in Europe and Israel. One of these was Dr. Hans Mühsam, now seventy-six and bedridden for eleven years, paralyzed and blind in one eye. His brother, Erich, had been tortured to death in a Nazi concentration camp. In Berlin during the 1920s, Einstein and Hans Mühsam had often gone for long walks together discussing everything, but particularly Mühsam's special interest, biology.

From time to time Mühsam received letters from Einstein, some expressing the opinion that science was still clumsily primitive, and that mathematics was unable to explain the mystery of life. In the summer of 1952, Mühsam received another letter, in which Einstein wrote, "When I am busy calculating and see an insignificant insect that flies on my paper, then I feel 'Allah is great' and that for all our scientific magnificence we are miserable drops in the ocean." Yet he wrote that he still admired some enlightened individuals, among them "the few writers of antiquity [who] gradually freed themselves from superstition and uncertainty which had darkened their existence for more than five hundred years."

Admired contemporaries included Upton Sinclair, to whom Einstein wrote on November 6, saying that his latest book, A *Personal Jesus: An Essay in Biography*, showed "an honest and understanding critical attitude toward the officially accepted texts." The manuscript had been rejected dozens of times, so Sinclair had advertised for a publisher in the *New Republic*. Among the replies was one from movie actor Errol Flynn, who offered to help; as it happened, however, Sinclair was unexpectedly left a legacy that helped to finance the book's publication.

A few days after the first president of Israel, Chaim Weizmann, died on November 9, Prime Minister David Ben-Gurion asked Abba Eban, Israel's ambassador to the United States, to find out if Einstein would accept the presidency. Einstein's friend and colleague David Mitrany was with him when Eban phoned to sound him out. "His main and urgent thought," said Mitrany, "was how to spare the ambassador the embarrassment of his inevitable refusal." He did it by saying, "I know a little about nature but hardly anything about men." He then asked Eban to do what he could "to lift the siege of journalists around my house."

That was hardly in Eban's power, especially when his deputy appeared in Princeton soon after with a formal letter assuring Einstein that as Israel's president he would have complete freedom to pursue his scientific work. Einstein finally wrote in reply:

> I am deeply moved by the offer from our state of Israel, and at once saddened and ashamed because I cannot accept it. All my life I have dealt with objective matters. Hence, I lack both a natural aptitude and the experience to deal properly with people and to exercise official functions. For these reasons alone, I should be unsuited to fulfill the duties of high office, even if advancing age was not making increasing inroads on my strength. I am the more distressed over these circumstances because my relationship to the Jewish people has become my strongest human bond since I became fully aware of our precarious situation among the nations of the world. After we have lost in recent days the man who, among adverse and tragic circumstances, bore on his shoulders for many years the whole burden of leadership of our striving for independence from without, I wish from all my heart that a man be found who by his life's work and his personality may dare to assume this difficult and responsible task.

Einstein's thoughts had shifted from the creation of Israel to the birth of the universe when he arrived early for a lecture at Princeton by George Wald. Finding Wald walking up and down the street outside the lecture hall, Einstein joined him. After a brief exchange of greetings, Einstein suddenly asked, "Why do you think all the natural amino acids were left-handed?"

The disconcerted Wald, whose mind was on the lecture, "just mumbled something." And then, he later recalled, "Einstein said, 'For many years I wondered why the electron came out negative. Negative? Positive? Those are perfectly symmetrical concepts in physics. So why is the electron negative?' All I could think was. 'It won in the fight.' And he promptly said, 'That's exactly what I think about those L amino acids. They won in the fight.'

"Now the fight he was talking about was the conflict between matter and antimatter, between the negative electrons and the positive electrons, which had already been discovered. On contact they mutually annihilate each other and the masses of both. So an electron coming in contact with a positron, there is mutual annihilation and the masses of both are annihilated and turned into radiation, according to Einstein's famous formula $E = mc^2$, in which E is the energy of the radiation, m the mass that has been annihilated, c is the speed of light, three times ten to the tenth centimeters per second. So squaring that is a lot of radiation."

During their conversation, Einstein said that positive and negative were perfectly symmetrical concepts in physics, leading Wald to comment that in the big bang that started our universe, "what came into it were exactly equal amounts of matter and antimatter, of particles and antiparticles. And in the fireball of the big bang, packed to an almost unimaginable degree, they were in contact. And there must have been an enormous storm of mutual annihilation and the end of it could well have been that all the mass that went into the big bang turned into radiation and that our universe would contain no matter, but only radiation."

Some 30 years after that conversation, Wald said: "That would have been Einstein's 'fight' — how, nevertheless we find ourselves with matter in our universe, electrons and positrons . . . But there is a much more engaging thought now: that what got into the big bang involved a tiny error in symmetry. That is, for every one billion particles of antimatter there were one billion and one particles of matter. And when all the mutual annihilation was complete, one-billionth remained — and that's our present universe."

That day Einstein also spoke to Wald "quite sadly" about the unified field theory, saying "someone else was going to have to do it." Then it was time for Wald to start his lecture.

When Washington journalist I. F. Stone, who had covered Einstein's testimony before the Anglo-American Committee of Inquiry on the future of Palestine in 1946, decided to start his own newsletter, Einstein was invited to subscribe. An independent thinker with a left tilt, along with an urge to "spit in the eye" of the powerful "and do what is right," Stone attracted fifty-three hundred charter subscribers to *I. F. Stone's Weekly*. They included Bertrand Russell, Eleanor Roosevelt, and Einstein. Marilyn Monroe, perhaps nudged by her husband, playwright Arthur Miller, bought subscriptions for every member of Congress.

Stone was so thrilled to get Einstein that "we wrote to his secretary and said could we please frame the check instead of using it. She said, 'No. Everybody wants to frame his checks and it ruins his bank account. So please cash it, and I will send the check back to you after it has passed through the bank.' Which she did. I still have that original check."

Einstein recognized that Stone was a passionate antifascist who brought the instincts of a scholar and a humanitarian to the service of journalism. He invited the writer and his family—wife Esther, daughter Celia, and sons Christopher and Jeremy—to tea.

It is easy to understand his appeal for Einstein, this committed and witty whirlwind of a man who saw himself as a guerilla warrior, swooping down to a surprise attack on a stuffy bureaucracy where it least expected independent inquiry.

Christopher Stone recalled that visit: "I must have been sixteen and my brother, Jeremy, eighteen. [Their sister Celia was also a teenager.] We went to his little house in Princeton and what struck me most was the contrast in dialogic, if there is such a word; the speaking styles of Einstein and my father. My father spoke much more quickly and with a certain energy, like a chess player under a time clock, and was ready to move on as quickly as possible to the next question. Einstein had a much slower pace. Any time he didn't understand the question or what underlay it, he wanted to clarify. He took command of the pace of the exchange in a slow, professorial style that was superb, wanting to make sure he grasped everything. There was no pretense. You really felt such modesty, and as if there was an envelope around him that seemed so settled. He had such a center."

Intrigued, Christopher studied the two men. "My own dad," he later said, "was full of movement and potential movement, and set a contrast to Einstein, who settled back in a stuffed chair and put his feet up on a hassock. I remember his voice was so marvelously German, and his English full of charm. His thoughts came from within. He had no interest in moving ahead and advancing a conversation if there was any element of it that he could not understand. And I think he understood that in politics there was much that he could not understand."

The two men joked about Stone not wanting to cash Einstein's check, but the menace of McCarthy's witch-hunt for communists dominated their conversation and parallels were drawn with the situation in Hitler's Germany. "But I don't think Einstein let that vulgar, hysterical fart Joseph McCarthy upset him," said I. F. Stone. "Although he must have seen McCarthy's political adventurism as a foretaste of what had happened in Germany."

To the elder Stone, Einstein was "a very benevolent person, in a saintly sense. There are people that emanate benevolence. I had an almost mystic vision one day at Eighth Street and Sixth Avenue in New York. I ran into a deaf-and-dumb man who was a brother of the wife of my then boss, David Stern, who owned the New York Post, and this brother had been impaired that way from about one year old. Yet he became a Greek scholar and a sculptor. Very often people with serious impediments are poisoned by them. Some are not. I hadn't seen him in years and when I saw him this time I thought, 'My God! Saint Francis must have been like this!' Because the emanation of benevolence was palpable. And Einstein had a lot of that. He was more than kindly: he was loving, very modest, and impish, too. At the end he turned to my wife and said very simply, 'Now, if there are any questions the children would like to ask I'd be glad to answer them if I can.' But they were too dumbfounded to ask anything."

As Christopher remembered it, "Einstein's last shot, realizing he hadn't paid any attention to us, who were intimidated and awed, was to say [here Christopher imitated his accent], 'Do the boys haff any questions? I will answer them if I can.' It struck me then as a marvelous note of modesty. As I got older I realized he was probably in terror, wondering, 'What kind of crazy questions can teenage boys ask? Which came first, the chicken or the egg?' I'm sure he had no idea what we might come up with. I couldn't answer. I've never been so deeply moved in my life simply by the presence and atmosphere that attended some human being, as opposed to the words which I could barely understand. So I couldn't answer. I think, as tears welled up in my eyes, I thought, 'He's suggesting that I might have a question that *he* couldn't answer!' I'm sure that in my mind if that *he* could have been caps or lowercase it would have been a capital H."

His father agreed: "I loved Einstein and wrote afterwards that it was like going to tea with God, not the terrible God of the Bible, but the little child's father in heaven, very kind and wise. Yet Einstein himself was very much like a child."

A dramatically different, even menacing picture emerges from the biography collected by the FBI. A frustrated Hoover, still too timid to question Einstein, instead ordered his agents to summarize the mammoth file. When they finished, it still amounted to 1,160 pages, a weird mixture of fact and fantasy, of lies, rumors, and ravings.

CHAPTER 39

Einstein's Mercy Plea for the Rosenbergs

1953

74 years old

Einstein gave his critics fresh ammunition by urging President Truman to spare the Rosenbergs, who were facing electrocution in Sing Sing for spying, yet keeping silent about the fate of Jews in Moscow. There, nine eminent Soviet doctors—six of them Jews—were on trial for the so-called Doctors Plot. All were accused of murdering Soviet leaders by misdiagnosing their illnesses and prescribing lethal doses of medication; even of plotting to kill Stalin himself. The Jewish doctors were further charged with being members of an international Zionist spy ring. To those in the know, it was obvious that Stalin was launching a new kind of purge supported by the simmering anti-Semitism of the Russian people. A witness at the trial reported, "They put Jewish doctors on the platform and made them say dirty words about Jews as traitors, Jews as spies, about anti-Soviet activities of Jewish organizations abroad." One doctor even confessed to spying for Germany, Britain, and the United States.

I. F. Stone recalled how he attacked the Doctors Plot as anti-Semitic, not just anti-Zionist, in one of the first issues of his *Weekly:* "It's not easy to read history backwards. The parallel with Nazi Germany was not very good. A lot of Jews, liberals, and radicals still had hope in the Soviet Union. You know, it was like the French Revolution. In 1934, too, during the Great Purge Trials [in Russia], I remember reading the official transcripts and telling a colleague on the *Nation* that it was a frame-up, it didn't smell right."

Einstein received numerous letters protesting his selective attitude, including one from Worcester, Massachusetts:

> Not all of us Americans fall for your phony science, which is understood by you and Charlie Chaplin . . . Will you take up arms for my country? No, you will

not, but instead sit back and give out dictatorial commands, as in the Rosenbergs case. President Truman is intelligent enough to decide without your assistance. You are just looking for publicity. Keep to your "science."

Another, from Galveston, Texas, said:

> I was always told that the most learned man in his own field and in his profession is but the DAMNEST [sic] fool in any other field . . . Remember that we in America live in God's paradise, not you, because you are an atheist, an infidel. We do NOT want an emigrant to dictate to our good government.

From Brooklyn, a rabbi wrote:

> Your intercession in behalf of the Rosenbergs leads me to invite you to publicly condemn the forthcoming trial in Moscow of the so-called "terrorist Jewish doctors." . . . The *New York Times* for January 13th, on page one, gives the details of the latest anti-Semitic Moscow outburst, and the names of the Jewish physicians who will soon be compelled to confess to crimes they never committed, for which they will surely be executed, unless world opinion is aroused. The lives of the accused "terrorist Jewish doctors" are certainly as precious as the lives of the Rosenbergs. May I also beseech you to break your silence concerning the recent anti-Semitic trials in Prague, which ended in the hasty execution of eleven men, nine of whom were Jews, who were forced to confess to a conspiracy involving the State of Israel, the American government, the American Jewish Joint Distribution Committee [a major Jewish charitable organization]—a conspiracy which you must know never existed, and which is reminiscent of Hitler and his "Protocols."

In his reply, Einstein told the rabbi that he felt it was his duty as an American to plead for clemency for the Rosenbergs. Like his colleague Harold Urey, who had also protested the death sentences, he thought the evidence of their guilt was unconvincing, and that even if they were guilty, the punishment was too harsh. He believed it would be useless for him to publicly condemn the Soviet government for investigating crimes under the guise of justice "but based on lie and torture," as he could hardly write to Stalin, "who is probably the chief criminal," and his words would not reach the public behind the Iron Curtain. They would only fuel the hatred felt by most Americans for Russia. He did suggest that he might join a protest from the "international intelligensia," if political bias could be excluded.

In February, the FBI suddenly reactivated their investigation of Einstein because of his December 23, 1952, letter to Judge Kaufman—who had sentenced the Rosenbergs to death—which the judge had handed over to Hoover. An FBI report notes: "In part this [Einstein] letter stated . . . [that] no one would ever learn from him that he had written such a letter. Furthermore that he did not wish to challenge the jury's verdict as such although he did want to mention that for anyone who was not present in court during the trial, the guilt of the defendants was not established beyond a reasonable doubt." (To the very last, the Rosenbergs claimed they were innocent, even though they probably could have saved their lives by admitting their guilt.) The report then quoted directly from Einstein's letter:

> In any event, from all that has become known one must gain the conviction that the defendants could only have played a minor role in the transmission to a Soviet representative of the document prepared by Mr. Greenglass. This is why it would be incomprehensible if the Rosenbergs should be made to suffer a more severe punishment than Greenglass whose crime was confirmed by his own confession. It is also to be considered that not one of all the other persons who were found guilty of having betrayed information about atomic energy and who, no doubt, surrendered more important material than the incompetent Greenglass, was executed. May I finally appeal to you as a fellow human being to use the authority given you by law lest an irreparable action be taken.

Einstein's friend Dorothy Commins was his guest for tea in his study at the height of the Moscow Doctors plot controversy, but they discussed instead Debussy's music, which he found "an enigma," the movements of the planets, the spring plants-in-waiting under the deep snow outside, and his blue shirt (a gift from Alice Kahler). " 'Perhaps you could tell me,' he asked [Commins], because he knew I sewed, 'how does one put this together?' And he rolled up the cuff of his sleeve and said, 'I can't find a stitch. There isn't a machine or a hand stitch in this garment.' I said, 'I guess that's a magic garment,' and he laughed."

After Dukas brought up a tea tray and left them alone, Commins glanced around the room at the sagging bookshelves and frayed books, a photo of Gandhi and prints of Faraday and Maxwell in old-fashioned picture frames, and a small radio-phonograph. She also caught a glimpse of his sparsely furnished bedroom.

Shortly after that meeting, Commins recalled, Einstein was planning to write an article in English: "He called up my husband and asked, 'Saxe, can you help me? My syntax is all wrong.' So Saxe came over and got it all straightened out.

"My husband, then editor in chief of Random House, would take the train home from New York and get off near Einstein's house with literature Einstein wanted to read. At that time Bertrand Russell was writing and he wanted that. And a short visit would turn into a long one. They'd walk each other home, like schoolboys. Finally, Helen Dukas would call me, 'Is Professor Einstein there?' I'd say, 'Yes, he is.' She'd say, 'Send him home. He needs his dinner and it's getting cold.' And Saxe would say, 'You must go back, because dinner's waiting for you.' And then Einstein would say, 'Oh, that doesn't interest me. Let's go on walking and talking.' "

After one stimulating session with Einstein, Saxe wrote down the gist of their conversation: "We covered a good deal of ground, from moral absolutes to the limitations of science. Particularly interesting was the development of the thesis that it is necessary to put limitations on freedom, contrary to what the eighteenth-century libertarians thought. When it becomes a question between security and freedom . . . we accept limitations. If we don't, our culture imposes them. When we ask what is freedom for, we get into the problem of values and metaphysics."

Friends and enemies continued to question Einstein's values when he kept quiet about the imperiled Jewish doctors in Moscow. But where Sidney Hook, a Brooklyn rabbi, and stacks of hate mail and newspaper columns had failed to persuade him to protest against Soviet tyranny, it was probably Helen Dukas who finally succeeded. Even the FBI deduced that. (FBI document NK 100 = 29614 reads in part: "Helen Dukas . . . probably has some influence on Professor EINSTEIN.")

The proximate cause was a telegram to him from an anti-Communist magazine, the *New Leader:* "Note you support the Rosenberg clemency. In name human rights we ask you make equally forthright condemnation anti-Semitic Prague trial and imminent execution Soviet Jewish doctors."

Einstein replied in a statement to the press: "It goes without saying that the perversion of justice which manifests itself in all official trials staged by the Russian government, not only in Prague, but also the earlier ones since the second half of the '30s, deserves unconditional condemnation. Another question is what can be done from here against the course of these contemptible methods and devices." He then repeated his belief that whatever he or anyone else said publicly would never influence those in power in the Soviet Union and would only "fan the flames of mutual hatred. The most appropriate step would be, in my opinion, a kind of corporate condemnation from recognized authorities in the field of science and scholarship. The advantage of such an action would be that it would be obviously independent of politics. If such a pronouncement is to be made, it should be given to the whole press—this also to avoid the impression of an act of political propaganda."

The shrewd advice was unnecessary. Stalin's death on March 5 saved the doctors' lives. A month later, all nine were freed as innocent victims of false arrest, fake evidence, and a sham trial.

Einstein had had several good reasons to celebrate his seventy-fourth birthday—and break his diet—a few days later. The New York Baker's Union sent him a huge chocolate and marzipan cake shaped like a book on a lectern, decorated with sugar-icing roses, carnations, forget-me-nots, and violets. The banquet he attended on his birthday to raise funds for an Albert Einstein College of Medicine raised over three million dollars, almost one-third of the cost of the building. And the Moscow doctors had been freed.

At the end of March, a woman outwitted the Princeton neighbors who had volunteered to protect Einstein from pestering strangers. She arrived by taxi at midnight, and, ringing the doorbell with furious determination, she shouted out that she had an important communication for Dr. Einstein and must see him immediately. Dukas tried gentle persuasion, but the woman was adamant. She had to see Einstein at once. When Dukas said no chance, the woman became hysterical. The police were called, and they escorted her to a nearby mental asylum. That, apparently, was the last Einstein heard of her. He took no special precautions after that midnight visit and, as John Wheeler noticed, still usually walked home alone from the institute.

On April 20, a month after his own birthday, Einstein celebrated the fourth birthday of Mark Abrams, the son of his ophthalmologist. Einstein had agreed to be Mark's godfather, and Helen Dukas phoned the Abramses' home to say,

"The professor has a birthday gift he'd like to present. Could you come over?" So, Mark recalls, "my folks and uncle went with me." The uncle took a photo of Einstein "trying to figure out the Lincoln Logs — the Lego of its day — he had given me and I'm busily trying to get into a bag of these chocolate gold-foil-covered coins he also gave me, little coins made of chocolate. What I seem to remember is that Einstein had a very sweet laugh."

The following month, John Wheeler was giving a course in special and general relativity; to his great delight, Einstein invited him and his students to tea on May 16, and offered to answer their questions. A few days before the visit, Wheeler asked each student to prepare three questions he would most like to put to Einstein.

Arthur Komar, one of the eight students Wheeler took with him, had recently heard Einstein say during a lecture that the laws of physics should be simple. A student had challenged that statement by asking, "What if they aren't simple?" Einstein had replied, "Then I would not be interested in them." Asked at the same lecture why he rejected quantum mechanics, he said he could not accept the concept of a priori probability. "But," came a puzzled response, "you were the one who introduced a priori probability." "I know that," he conceded, "and have regretted it ever since. But when one is doing physics one should not let one's left hand know what one's right hand is doing." At the end of the lecture Komar had attended, Einstein had sat with a sigh of relief and announced, "This is my last examination." Now he was putting himself to the test again.

Oscar Wallace Greenberg remembered the house as being musty and having drawn shutters, but Wheeler has a brighter memory of that afternoon. "It was a friendly, agreeable meeting, and I don't recall it being musty. Why did he invite me and the students to tea? I think Einstein was just trying to be helpful. He was always very kind to me and this was the first year I gave a course in relativity, and so I suspect that might have been it."

Margot Einstein and Helen Dukas served the men tea as they sat talking around the dining room table. Wheeler remembers how "students asked questions about everything from the nature of electricity and unified field theory to the expanding universe and his position on quantum theory. Einstein responded at length and fascinatingly."

Asked about Mach's principle (that the resistance of bodies to acceleration by applied forces is determined by the effects of distant matter in the universe), Einstein admitted that he had lost his early enthusiasm for it and "perhaps there wasn't in nature anything corresponding to Mach's principle after all."

Discussing special relativity, Einstein explained that "in constructing it I knew it was not complete. So is everything that we do in our time: with one hand we believe; with the other, we doubt. I once thought temperature was a basic concept. I feel the same way about Maxwellian theory, but I am now convinced there is no cheap way out. If there are too many hypothetical elements one cannot believe one is on the right track. Then I came to logical simplicity, a desperate [man's] way to get on the right track. But one event in my life convinced me of the usefulness of logical simplicity. That was general relativity ... Present quantum theory based on special relativity is horribly complicated. For most people special relativity, electromagnetism, and gravitation are unimportant, to be added in at the end after everything else has been done. On the contrary, we have to take them into account from the beginning. Otherwise it is as if one did a classical problem and put it in the law of conservation of energy at the end."

As the visitors were about to leave, a student asked, as Wheeler remembered it: " 'Professor Einstein, what will become of this house when you are no longer living?' Einstein's face took on a humorous smile and he spoke in that beautiful, slow, slightly accented English that could have been converted immediately into printers' type: 'This house will never become a place of pilgrims come to look at the bones of a saint.' "

On June 17, according to Dorothy Commins, her husband Saxe took Illinois governor Adlai Stevenson to meet Einstein. Princeton University had awarded Stevenson an honorary degree the previous day. She recalled: "Stevenson asked, 'Can you, Professor Einstein, apply a mathematical law of probability to a prediction of the development of world conflict?' This launched Einstein into a brilliant and humane exposition of his views on the problems of world peace. When [leaving], Stevenson said, 'I also want to thank you for endorsing my candidacy during the [1952] campaign.' Professor Einstein answered, 'Do you know why I came out for you? Because I had still less confidence in the other one' [Averell Harriman, Stevenson's Democrat opponent for the presidential nomination]. Stevenson laughed uproariously."

The visit from a former presidential candidate did not divert him from his concern over the plight of the Rosenbergs. Einstein feared that the worldwide Communist demonstrations in support of the Rosenbergs only made it worse for them, by turning it into a political case and arousing political passions on both sides. To his dismay, their appeal to the U.S. Supreme Court for a stay of execution was rejected by a 5 to 4 vote. On June 19, despite continued worldwide protests—or, as he thought, partly because of them—Ethel and Julius Rosenberg were executed.

Harold Urey shared Einstein's doubts about the Rosenbergs' guilt and wrote to him from the University of Chicago's Institute for Nuclear Studies on June 25:

> It is impossible for me to express my depression over this affair ... How anyone could believe the evidence that was told in court about these people was beyond me to understand. One of our law professors spent two weeks on the case, and returned to Chicago with the flat statement that he believed the Rosenbergs were innocent. Certainly nothing in their behavior indicated that they were guilty. I believe this has damaged the United States enormously ... Your support of my position and your letter have been most heartening to me, for very few scientific people have troubled to consider the case at all. I have noticed your stand in regard to the inquisitions from legislative committees [investigating sus-

pected Communists], but I believe that your position would have been better had you based the whole matter on the first amendment to the Constitution instead of the fifth. That is, that these committees are exceeding their authority and that they have no right to question people in regard to their beliefs, and that this is a violation of freedom of the speech, the press, etc. Though I disagree with you on this particular point, I am so very glad that a man of your stature has the courage to take a stand on questions of this kind.

Einstein had already written to the liberal U.S. Supreme Court Justice William O. Douglas, who had voted for a stay of execution in the Rosenberg case: "You have struggled for the creation of a healthy public opinion in our troubled time that I cannot abstain from expressing to you my thankfulness and high appreciation." Douglas replied: "Your letter . . . written on the occasion of the Rosenberg decision has reached me. You have paid me a tribute which lightens the burdens of this dark hour—a tribute I will always cherish."

Einstein explained his view of the Rosenbergs to a correspondent in Albany, Georgia, who was puzzled by Einstein's position: "As far as I can judge no convincing proof has been brought forward in the Rosenberg case. But even if such proof had been furnished the death sentence was not justified according to the intention of the law. Those two people have been made victims of political passions."

Summing up the situation in America, Einstein told Bertrand Russell that all intellectuals, including young students, had become completely intimidated by politicians who had convinced the public that the Russians and American Communists were a threat to the country. He said that Russell was the only prominent individual to "challenge these absurdities in which the politicians have become engaged. The cruder the tales they spread, the more assured they feel of their re-election by the misguided population. This also explains why [President] Eisenhower did not dare to commute the death sentence of the two Rosenbergs, although he knew well how much their execution would injure the name of the United States abroad."

At a closed meeting of the Atomic Energy Commission, General Leslie Groves, the military head of the atomic bomb project, admitted that "the data that went out in the case of the Rosenbergs [were] of minor value." Supreme Court Justice Felix Frankfurter, one of the four justices to vote for a stay of execution, concluded that, guilty or not, the Rosenbergs were, in effect, tried for conspiracy and unfairly sentenced for treason.

It seems they were guilty but, as Einstein said, not beyond a reasonable doubt.

He partly blamed the Communist Party for the tragic outcome: for promoting the Rosenbergs as innocent victims of "American fascism." Einstein restated his position in a September 12, 1953, letter to a Los Angeles correspondent: "Unhappily, the Rosenberg case was used as an instrument by communists which circumstance had a very unfortunate influence on the course of events."

Disappointed by the failure of the American Civil Liberties Union to support the Rosenbergs — on the grounds that "the question of commutation of the death sentences . . . raised no civil liberties issues" — Einstein concluded that the ostensibly liberal organization had been taken over by reactionaries. He remedied the situation by joining I. F. Stone and others in forming the Emergency Civil Liberties Committee.

During the summer heat wave, workaholic Saxe Commins had a heart attack and was rushed to the hospital. "Although it was a hot, humid day, Einstein walked to the hospital — he didn't have a car and didn't want one — and I think the asphalt must have melted under his shoes," recalled Dorothy Commins. "When he walked into the intensive care unit and Saxe saw him he said, 'My God, Professor Einstein, how could you come on a day like this!' And Einstein said, 'Where there is love there is no question.' I shall never forget that. And that tells you how close they were. They understood each other completely. He also sent Saxe a bouquet of flowers with a message: 'Well, it needed the devil to get you to rest a bit. Heartfelt wishes.' "

For some time, Einstein's friend Max Born had been trying to persuade him that Germany was not, as Einstein asserted, "a land of mass-murderers." Born and his wife, Hedi, had returned to live in Germany after spending the war years in Edinburgh. He told Einstein that there were many good Germans who had suffered far worse under the Nazis than either Einstein or himself, especially their newfound friends, many of them, like Hedi Born herself, Quakers.

But Einstein was not impressed. He warned an American woman about to join her husband in Berlin that "you will find them [the Germans] affable, intelligent, and they will seem to agree with you, but you must not believe a one of them."

Even so, his attitude towards the Germans had mellowed somewhat. He approved, for example, of plans to restore their economy and to resume normal political and diplomatic relations — if only to avoid the disastrous consequences of the Treaty of Versailles. But he adamantly opposed rearming them.

As his anti-Nazi views were so well known, he was asked to sign a letter protesting the visit to Princeton of a mathematician who had joined the Nazi Party after the Nazis had occupied his homeland. When Einstein's assistant, Ernst Straus, handed it to him and explained the reason for the protest, Einstein replied, "You mean, they want me to sign something just to harm another human being?" and threw the letter in the wastebasket. A surprising response, until one realizes that had he signed, Einstein would have acted like a Nazi.

In Edinburgh, Max Born had made a close friend of Sir Edmund Whittaker, a brilliant mathematician. Writing about Einstein for publication, Whittaker had denigrated his importance in working out the special theory of relativity and inflated Hendrik Lorentz's role. In Born's opinion, Lorentz "probably never became a relativist at all, and only paid lip service to Einstein at times to avoid argument." Hoping to prevent Einstein from being hurt, Born warned him what to expect from Whittaker's assessment — with which, he stressed, he had vehemently disagreed.

Einstein replied on October 12: "Don't lose any sleep over your friend's book . . . I would not consider it sensible to defend the results of my work as being my own 'property,' as some old miser might defend the few coppers he had laboriously scraped together. I do not hold anything against him, nor of course against you. After all, I do not need to read the thing." If Born needed convincing that Einstein was totally indifferent to fame and glory, that reply did it.

There was nothing like a friendly letter from Einstein to raise the spirits. I. F. Stone acknowledged as much in a letter to him on October 8:

> Some days here in cold war Washington I feel like a pariah and other days like a survival [sic] from the past. Other days, when I go to press and see my inadequate little four pages, I feel like an embezzler. But today when your kind letter arrived I felt, "Gee, whizz" and "Could that be me?" and I took out the article you liked and reread it twice. My chest expanded a little and when I went out to lunch, I felt a little differently than usual. Usually I feel people think, "There's that Red bas——d," or "There's that poor Izzy Stone" but today I felt—established, recognized, even a man of some substance, who maybe doesn't write too badly, perhaps even recognized by an occasional passerby as a man whose work has been praised by Dr. Einstein.

Some would surely call Einstein a glutton for punishment: on November 8 he invited Immanuel Velikovsky—the man obsessed with his controversial cosmological theories—and his wife, Elisheva, over to talk. Greeting them with a smile, Einstein said, as he moved a heavy chair forward, "This is my Jupiter chair."

With that Velikovksy was off and running. Did Einstein know that Jupiter was the highest deity in ancient Rome, and that Zeus in Greece, Marduk in Babylonia, Amon in Egypt, and Mazda in Persia all represented the planet Jupiter? That it was worshiped by people of antiquity, its name on everyone's lips, while Apollo, the sun, the giver of life and warmth, was regarded as a secondary deity? Einstein was impressed. Did he know, Velikovsky continued, that in the *Iliad* Zeus is said to be able "to pull all the other gods and the Earth with his chain, being stronger than all of them together," and that a Byzantine scholar states in an old commentary "that this means the planet Jupiter is stronger in its pull than all the other planets combined, the Earth included?" Einstein thought it very strange that ancient people should have known this. He was obviously intrigued, because when Velikovsky rose to go, Einstein objected, "We have only just started."

Aware of his reputation as a fanatic obsessed with one idea, Velikovsky occasionally changed the subject, moving from "time" to "coincidence" and then to "accident." Einstein provided the light touches, observing that it was "an accident of unusual rarity that his chair should occupy its very position in space, but that it was no accident that we two were sitting together, because *meshugoim* [crazy people] are attracted to one another."

A few weeks later, after Velikovsky had sent Einstein a copy of his recent lecture, he and his wife were again invited to tea to discuss it—especially one point: how the sun is slightly flattened by its own rotation. Velikovsky was so absorbed in their discussion that he feared he had overstayed his welcome.

The next morning, Velikovsky was about to phone Helen Dukas to apologize, when she called him, saying that the professor would like to talk to him. Einstein said, in a clear, resonant voice that could have been mistaken for a young man's: "After our conversation last night I could not fall asleep. For the greater part of the night, I turned over in my mind the problem of the spherical form of the sun. Then before morning I put on the light and calculated the form the sun must have under the influence of rotation, and I would like to report to you."

Velikovsky was deeply impressed by Einstein's thoughtful phone call. Years later, he would remember this episode as a wonderful demonstration of "Einstein's attitude toward a scientific problem that intrigued him and, even more, his behavior toward a fellow man."

On the evening of December 8, another surprise visitor appeared on his doorstep. Albert Shadowitz called at Einstein's home and explained his plight to Helen Dukas. She was silent for what seemed like a full minute, then led the way to Einstein, who was in his study. Shadowitz wanted advice. He had worked as a physicist at the U.S. Army's Aberdeen Proving Grounds from 1941 to 1943 and for the following eight years at the Federal Telecommunications Laboratories of I.T.&T. Told that a union being formed at the plant intended to exclude Jews and blacks, he had helped to organize a rival union, the Federation of Architects, Engineers, Chemists and Technicians, which then amalgamated with the United Office and Professional Workers of America. Now Joseph McCarthy had subpoenaed Shadowitz to appear before his committee, as this union was alleged to be Communist-dominated and a possible cover for Russian espionage. What should he do?

Einstein did not ask if he was or had been a Communist, believing the First Amendment gave him the right to keep his political views to himself. However, he advised him to say up front that he was not a spy nor engaged in any other disloyal activity — if that were true — and to say nothing more, except to cite the First, not the Fifth, Amendment. He also gave Shadowitz permission to use his name in any way that would help.

At the closed hearing a few days later, Shadowitz volunteered the information that he had never been a spy. Roy Cohn, a McCarthy assistant, then began questioning him, but quit when Shadowitz refused to say if he was a Communist "following the advice of Professor Einstein." He was then told he would be expected to appear at a public hearing early in the new year.

Shadowitz wrote to Einstein: "I don't know how my wife and I could have stood up under the ordeal if it weren't for your support."

In his reply, Einstein congratulated Shadowitz for sticking to his First Amendment rights.

On December 21, Einstein attended Thomas Bucky's wedding. "It was a black-tie affair in the Terrace Room at the Plaza in Manhattan," Bucky recalled. "We hadn't invited him, knowing how much he hated formal affairs. But he turned up uninvited. He had heard about it from relatives. By his standards he was well dressed in a dark suit, shirt, collar, and tie. But he came in off the street, and over those clothes he wore a navy pea jacket and an old Russian-style hat.

A news photographer took a photo of him which appeared next day on the front page of the *New York Daily News*. People didn't realize that by his standards he was regally dressed. It was a great effort for him to attend such gatherings. They were torture for him. He hated to leave his home and lose time from his work and his thoughts. So I knew when he arrived that he really cared for me. He didn't lose much time, either. He took some of the hotel's stationery during the reception and from time to time worked on equations. I framed one of these as a memento."

CHAPTER 40

The Oppenheimer Affair

1954

---

75 years old

The press helped feed a growing public fantasy that Einstein was a kind of super–Santa Claus. A recent headline had him making another delivery: a new set of laws for the cosmos. But the headline writer got it wrong. Einstein had hit a brick wall in his latest efforts to confirm his unified field equations. He could no longer even hope for success in the foreseeable future.

This failure sapped his confidence generally. A pile of letters, as always, awaited his response. He hesitated to start on them, feeling that it was ridiculous for him to advise anyone about anything important.

George Wald caught him in this mood when he stopped in Einstein's office at the institute. He recalled that Einstein "looked at me with his face long and sad, and said, 'People keep writing to me asking, What is the meaning of life? And what am I to tell them?' That, like many things he said, struck very deep into me. I couldn't answer him."

It wasn't the only question that stumped him. Take the letter he had just opened from a thirty-nine-year-old San Francisco mother of two young sons, who chose him rather than a medical expert to ask, "Is there any chance for a person to have a baby girl? You are such a smart man and know all the answers. I'm sure if there is some possibility of Sex determination you will let me know." The letter came at an inopportune time, when Einstein was convinced that overpopulation threatened the planet and that "some practical and social activities and practices of the Catholic organizations are detrimental and even dangerous for the community as a whole, here and everywhere. I mention here only the fight against birth control at a time when overpopulation in various countries has become a serious threat to the health of people and a grave obstacle to any attempt to organize peace on this planet."

Some strangers' letters seemed written to crack him up, such as this one from a Michigan man: "An acquaintance of mine has established himself as a king in his home. He has made his wife a servant. His premise is that no great man must do such menial work. Yes, he fancies himself to be a great philosopher. Perhaps he is. I would like to show a picture of you helping Mrs. Einstein with the dishes. I would be deeply grateful if you could supply one."

Einstein's colleague and biographer, Abraham Pais, has a definite opinion about Einstein's housekeeping role: "I will give you my considered answer, but I cannot testify to it under oath. I bet you dollars for doughnuts that he never lifted a finger." In fact, he did help around the house in the early years of his first marriage, but he never claimed to be a good husband or father. He rarely met his son Hans Albert, now an engineer in California, and told the inquiring biographer Carl Seelig that he did not "exchange letters with [his schizophrenic son] Teddy," because of an "inhibition that I am not fully capable of analyzing. But it has to do with my belief that I would awaken painful feelings of various kinds if I entered into his vision in any way." He was an enigma to himself in these relationships—as irrational, perhaps, as he thought most of humanity to be.

Yet many admired his "detachment," and no one doubted that Margot and Helen Dukas loved him. So did the Buckys; the Borns; Solovine; Besso; the Commins; the Kahlers; I. F. Stone, as well as his son and wife; and countless strangers. Max Born sent him an affectionate seventy-fifth-birthday message: "There is no one in the world for whom I have more profound admiration, and to whom I am more indebted, than you." To this Hedi Born added: "Whenever the war [World War I] really got me down and I came to see you, something of your Olympian outlook on life rubbed off on me and happy once more, I went on my way."

As a witness at a public hearing of the House Un-American Activities Committee on January 8, Albert Shadowitz again refused to cooperate. The Senate cited him for contempt, and soon after he was fired from his job with the Kay Electric Company. (He eventually became a physics professor.) Some newspapers attacked Einstein for his "ingratitude" in taking refuge in the United States and then urging Shadowitz to flout its laws. Letters to Einstein were almost equally divided pro and con. One correspondent enclosed a copy of the "American Creed," which included the words, "It is my duty to love it [my country], to support its Constitution; to obey its laws; to respect its flag, and to defend it against all enemies." Einstein wrote in the margin, "This is precisely what I have done."

Grateful for Einstein's support, Shadowitz turned up at Princeton to thank him in person, this time with his wife and father, who embarrassed Shadowitz by grabbing Einsten and kissing him. But as Shadowitz's wife, Edith, explained, Einstein was "like a superman to our Jewish community because he exemplified all those things which we would like to see in Judaism. To Al's father this was the great Jew."

To celebrate Einstein's seventy-fifth birthday in March, the newly formed Emergency Civil Liberties Committee held a conference in Princeton. Norman Thomas, leader of the American Socialist Party, declined an invitation to attend because, as he told Einstein, several prominent members of the committee "have shown over the years anything but a consistent love of liberty [the implication being that they were Communists] ... I am thoroughly persuaded ... that the test of freedom in American ... is a capacity to oppose both Communism and ... McCarthyism."

Although Einsein respected Thomas, he believed, as he stressed in his reply, that America was much less endangered by its own Communists than by the hysterical hunt for them and "fellow citizens whose red tinge is weaker ... Why should America be so much more endangered than England by the English Communists? Or is one to believe that the English are politically more naive than the Americans so that they do not realize the danger they are in? No one there works with inquisitions, suspicion, oaths, etc., and still 'subversives' do not go unchecked. There, no teachers, and no university professors have been thrown out of their jobs, and the Communists there appear to have even less influence than formerly."

Einstein feared that America was becoming like the Germany of 1932, "whose democratic community, through similar means, was so deeply undermined that Hitler could quite easily deal it a deathblow ... I am firmly convinced that the same thing will happen here if those who are clear-sighted and capable of self-sacrifice do not resist."

One who did resist was news reporter Edward R. Murrow. After his denunciation of McCarthy in a television broadcast on February 24, many religious leaders and a few influential newspapers such as *The New York Times*, the *New York Herald Tribune*, and the *Washington Post* joined the anti-McCarthy faction. Even Einstein's friendly adversary, Sidney Hook, criticized McCarthy's "exaggerated and irresponsible claims and behavior." Though McCarthy still had millions of fervent admirers, before the year was out he would be censured by the U.S. Senate — a move welcomed by President Eisenhower — and his power to continue his witch-hunt would quickly evaporate.

The investigation of Robert Oppenheimer as a possible security risk had dragged on for a second year. The critical question was, could Oppenheimer still be trusted with secret information, considering his past communist associations and his ambiguous behavior and statements? Enrico Fermi, Isidor Rabi, and Hans Bethe said yes, that Oppenheimer was a brilliant man of unquestioned loyalty. But Edward Teller testified:

> In a great number of instances, I have seen Dr. Oppenheimer act—I understand that Dr. Oppenheimer acted—in a way which for me was exceedingly hard to understand, and his actions frankly appeared to me confused and complicated. To this extent I feel I would like to see vital interests of this country in hands which I understand better and therefore trust more ... If it is a question of wisdom and judgment, as demonstrated by actions since 1945, then I would say one would be wiser not to grant clearance.

The board concluded that "Oppenheimer is a loyal citizen," but had a "tendency to be coerced, or at least influenced in conduct, over a period of years." The members voted six to one not to reinstate his clearance. President Eisenhower agreed and ordered that from April 15 Oppenheimer was no longer to be entrusted with atomic secrets.

Einstein immediately phoned the Associated Press to express his admiration for Oppenheimer as a scientist and human being. The next day, while walking home for lunch with Kurt Gödel, he was involved in a near-rerun of an incident that had occurred two years before. According to one account, "he was surrounded by television cameras and newsmen. Dukas, preparing lunch, saw the journalists converging on the surprised physicist, [and] rushed outside to rescue him." As Dukas recalled, "The newsmen were coming at him and I ran down there with my dirty apron shouting, 'Professor Einstein, they are newsmen, don't talk, don't talk!' . . . [He refused to elaborate on his AP statement.] They nearly pushed him into the bushes. They came after him and I slammed the door. They were mad."

Expecting Oppenheimer's enemies to try to have him fired from his position as director of the Institute for Advanced Study, Einstein wrote to a friend, Senator Herbert Lehman, also a trustee of the institute, describing Oppenheimer as "by far the most capable director the Institute has ever had," and warning that his dismissal "would do grave harm to the reputation of the Institute." Einstein's fears were unjustified: Oppenheimer was reinstated as director.

That summer, the Comminses took novelist William Faulkner with them to have tea with Einstein. Einstein tried to encourage Faulkner to join the conversation, but he seemed to be in a catatonic trance. "What could I say to the great man," Faulkner explained to Saxe Commins afterward, "that could possibly have any significance?"

For the next few weeks, Einstein was occupied in trying to find a job for a handicapped young man with a prison record, and with Velikovsky's emotional account of how others had tried to destroy his book *Worlds in Collision*. He listened to him, sometimes far into the night, defend the book's controversial contents. Velikovsky believed that he had confirmed an Old Testament miracle by establishing that the earth had stopped rotating—or had at least slowed down—at the exact moment in the biblical account of how Joshua had commanded the sun and moon to stand still. This happened, said the quiet-spoken but impassioned Velikovsky, when a giant comet erupted from Jupiter and twice almost hit the earth during the time of Moses. Its gravitational pull caused the earth to stop or almost stop spinning, which in turn parted the Red Sea, allowing the Israelites to escape from the pursuing Egyptians in 1500 B.C. The comet also had a catastrophic effect, he went on, collapsing mountains, flooding the plains, and devastating much of the planet with hurricanes and streams of molten lava that turned vast seas to steam.

Velikovsky also speculated that deposits from the gases in the comet's tail were the manna the ancient Israelites ate in the desert. Among other Velikovsky conclusions: a comet collided with Mars and, losing its tail, ended up at the planet Venus. Mars moved close to the earth, and in 687 B.C. they nearly collided. Then came another bombshell: once, said Velikovsky, the earth had turned completely over, so that the sun rose in the west and set in the east!

Einstein's rebuttal, that Velikovsky's ideas contradicted the laws of dynamics and gravity, did not faze him for long: he came up with the suggestion that electrical and magnetic forces had overwhelmed gravity during the close encounters of the planets.

Velikovsky found a measure of personal vindication several years later, in 1962. Then, as *Mariner II* approached Venus, a former Einstein assistant, Princeton physicist Valentin Bargmann, and Columbia astronomer Lloyd Motz pointed out that Velikovsky had been the first to predict three facts about the solar system: the earth's far-reaching magnetosphere, Jupiter's radio noise, and the extremely high temperatures on Jupiter, which they rated among the most important of recent discoveries. They urged scientists to take a careful look at Velikovsky's work. *Mariner II* also showed he was right about both Jupiter's temperature and its atmosphere; its surface temperature was at least 800 degrees Fahrenheit, and its atmosphere consisted of hydrocarbons, not water or carbon dioxide as the "experts" had assumed. (These discoveries did not validate all of his theories, of course.)

While Velikovsky's enemies had driven him to a fine frenzy, Einstein's were afraid to come out of hiding. On June 12 the FBI's Hoover asked army intelligence in Germany to interview physicist Max von Laue about Einstein and Dukas. A week later, Colonel Perry, chief of the Security Division, wrote to Hoover from the Pentagon of an effort to expedite the interview, which would be forwarded immediately upon receipt. It was hardly worth the nine-month-long wait. The two officers who grilled von Laue found him willing to tell all he could remember about Helen Dukas, but that wasn't much—a fleeting memory of seeing her in Berlin at Christmas 1932, twenty-one years before. After that—a blank. He knew none of her friends, relatives, or associates. He had no idea of her politics, special interests, hobbies, or other activities. The interviewers concluded that he was telling the truth.

By now, the FBI had been investigating Einstein and Dukas for nine years. In November, anxious to bring the case to a "logical conclusion," they reviewed the latest details of the secretary-spy they suspected might be Dukas.

The scene they conjured up was Einstein's Berlin office around 1929 to 1931 (Dukas had, in fact, been his secretary for a couple of years by that time). According to informants, the office was then a hotbed of Soviet spies, having been "used as a cable drop for a Soviet Espionage Ring operating in the Far East. During this time Dr. EINSTEIN allegedly had two secretaries in his office, both of whom were Communist Sympathizers. One of these secretaries decoded the Soviet Espionage messages and in turn passed them on to a courier who was responsible for liaison between EINSTEIN'S office and Soviet officials in Moscow . . . A review of the file fails to reflect any outstanding leads in this case and the most logical remaining lead in this matter is to interview HELEN DUKAS."

The FBI report continued: "A review of [Einstein's] file reflects that one of the outstanding leads at the time is to ascertain the whereabouts of his son Edward Einstein who was last known to be residing in Switzerland." (They might have had second thoughts if they had known that their "outstanding lead" was a schizophrenic in a psychiatric hospital.) The report also noted that Einstein had recently been counseling witnesses not to testify before the House Un-American Activities Committee.

Apparently no one recommended reading a recent biography of Einstein or quizzing its author, Antonina Vallentin, who presumably would have some pertinent information about him and his children. Perhaps Einstein's cynical response to a correspondent in England justified their ignoring Vallentin: "The author of the book mentioned in your letter knows me only superficially and what she tells about me is mostly fantasy. The writing of books is mostly just a business like any other despite claims to the contrary."

Dr. Clair Gilbert, a friend of the Einstein family, was among dozens who could have put the FBI straight. She was sure that neither Einstein nor Dukas was disloyal. Gilbert had a unique view of the Dukas–Einstein relationship: "Dukas absolutely idolized Einstein and he prospered greatly under her rule. Dukas was occasionally very stern and barked at people, but he felt comfortable with her. He knew she was protecting him. He had a sly sense of humor, and in his younger days he was probably hell on wheels. But when we knew him he was relaxed and secure. Dukas did a wonderful job of being the housekeeper and screening out the world for him. He was a kind of kindly grandfatherly guy that we knew in his later years. That was one of the faces of Einstein, but not the only one. I've always thought there was a hellcat underneath."

If there was, Linus Pauling never saw it. On November 11, just after being awarded the Nobel Prize for chemistry, he called on Einstein as he had done two or three times a year since 1948. "I went a few days before the Nobel ceremony," Pauling said. "And a couple of days after the announcement I issued a strong statement supporting Oppenheimer. I'm sure I discussed it with Einstein. I think we also discussed the fact that there was no mutual trust between the Russians and Americans. Our conclusion was that the stockpiles of nuclear weapons were already getting to be of such a nature that they would force the two countries to get along together despite their mistrust."

Einstein also said to Pauling, among others, "I think I have made one mistake in my life, to have signed that letter [to FDR promoting the atomic bomb]." Pauling added: "He didn't write it, of course, he just signed it. [Szilard probably wrote most of it.] He may have edited it a little bit. Then Einstein went on to say: 'But perhaps I may be excused because we were all afraid the Germans would be getting the atomic bomb.' He didn't say that he wouldn't have signed it had he known the Germans were not far advanced in their production of an atomic bomb."

CHAPTER 41

The Last Interview

1955

76 years old

Einstein first met Marian Anderson, the African-American contralto, in the 1930s when he went backstage to congratulate her after a concert at Carnegie Hall. Some time later, when she was to perform at Princeton, she was refused a room at the local hotel. Hearing about it, Einstein invited her to stay at his home, and she accepted. When she had a return engagement at Princeton in January 1955, he again invited her to be his guest.

"Dilapidated," as he called himself, and bedridden with anemia, he made the effort to get up, descend the stairs, and welcome her. Then Margot and Helen Dukas took over. After the engagement, as Anderson was about to leave, Einstein repeated his cautious descent to see her off and say good-bye. "This, though I did not know it," she later wrote, "was really goodbye."

The FBI was not yet ready to say good-bye, however. At long last they had decided to act. A Hoover aide once again suggested interviewing Dukas, under the pretext that they were after information about others. Hoover approved, and two agents questioned her on February 23, 1955, in the Princeton home she shared with Albert and Margot Einstein.

According to the agents, Dukas "was extremely friendly and appeared quite sincere in her answers . . . did not appear to be evasive in any manner, but spoke quite freely and admitted that due to the passage of years her mind was not clear as to names, dates and places. She did not elaborate on many of her answers and the interview was conducted in a discrete [*sic*] and circumspect manner as suggested by the Bureau . . . At no time did she give any hint or indication that she was aware the investigation concerned her in any way."

Dukas told the agents she had been Einstein's private secretary since 1928, and that before then the office duties were handled by his wife, Elsa, and sometimes Einstein's eldest stepdaughter, Ilse. Elsa had handled all the correspondence, while Dukas cooked and kept house for the entire family, as well as serving as his secretary. She said that she had never been interested in politics except for opposing Hitler, "and her only interest in life has been EINSTEIN and that she follows all of his wishes . . . at no time would she have engaged in any activity without first bringing it to Dr. EINSTEIN's attention." Dukas apparently knew no one in Germany who was interested in communism "attempting to infiltrate Dr. EINSTEIN's office or the office of any other scientist traveling in Dr. EIN-STEIN's circle."

She established the whereabouts of Einstein's sons: Hans Albert was now a professor of hydraulic engineering at the University of California, Berkeley, (and not, as one FBI report said, behind the Iron Curtain) while Eduard, "a hopeless mental case," had been confined in a Zurich mental institution since 1930. Einstein himself had just recovered from a severe bout of flu, followed by anemia, that had kept him housebound for months.

Satisfied that she had told the truth, the agents concluded that if the allegations were correct and Einstein's senior secretary had been in contact with Soviet couriers, the most likely suspects were Elsa and Ilse.

The agents also reported that "additional investigation is not warranted in view of the long lapse of time since EINSTEIN's office was allegedly used by the Soviets, the lack of corroborating information, and the fact that personnel involved are scattered in many countries and in many cases are deceased. Therefore, both the DUKAS case and the EINSTEIN case are being closed in the Newark Office, and will not be reopened on the basis of this allegation, unless advised to the contrary by the Bureau."

The report's conclusion was a shrewd handling of the situation. Hoover could now claim that the investigation was a triumph; he had finally traced the likely suspects. Fortunately for his reputation, they were both dead, so they could not tell him that he had been wasting his time for the past ten years. Asked how Einstein might have responded had he known he had been the subject of an FBI investigation, his friend Thomas Bucky said, "He would have laughed."

Several others agreed—then and later. As Richard Alan Schwartz wrote in the *Nation*:

> The amount of money spent to pay people who devoted countless hours to clipping newspaper items, typing summaries of public statements and following up absurd leads must have been enormous . . . Perhaps most perplexing is the way in which history is ignored. The Depression, the rise of Nazism and World War II might never have happened as far as the FBI was concerned. That someone might have supported communist causes during the 1930s in response to an economic crisis that represented at least a temporary failure of capitalism or to the spread of fascism is not even considered.

When eighty-two-year-old Michele Besso died of a stroke on March 15, 1955, the day after Einstein's birthday, Einstein wrote a moving letter to the Besso family about one of his closest friends: "What I most admired about him as a human being is the fact that he managed to live for many years not only in peace but also in lasting harmony with a woman—an undertaking in which I twice failed rather disgracefully."

Einstein had already interrupted his work to discuss with Bertrand Russell a plan to avert the disaster they both feared was inevitable from the arms race. The plan led to the Pugwash movement, in which an international group of scientists discussed nuclear weapons and the social responsibilities of scientists. They were largely responsible for a limited test-ban treaty—an important first step.

Einstein was still weak from anemia, so this was hardly the time for I. Bernard Cohen to request an interview. But Cohen, a Harvard historian of science, knew that Isaac Newton was Einstein's hero, and "with malice aforethought in my letter to him I stressed my work on Newton. I knew he would be interested and he was. Miss Dukas told me to call when I got to Princeton on the morning of April 3. Which I did, and she said, 'How about ten?' "

Cohen appeared at Einstein's front door at exactly ten. Dukas took him to the second floor and said that the visitor had arrived. Einstein emerged from his study, his eyes shining as if he had been laughing or crying. He wore an open blue shirt, gray flannel pants, and leather slippers. He greeted Cohen with a smile, went somewhere to get his pipe (already loaded with tobacco), and led the way back to his study. It was a cold, early spring morning, and after sitting in front of a little table, he tucked a rug around his feet. He nodded to Cohen to take a chair facing him.

The visiting professor had a head full of questions but was so overcome with emotion he couldn't say a word. After a while Einstein came to the rescue, as if answering a question: "There are so many unsolved problems in physics," he said, "so much that we do not know and our theories are far from adequate." Cohen mentioned something about the theory of photons and Einstein replied, laughing, "Not a theory." He had very definite ideas as to what constituted a theory (Einstein's friend, Max Born, believed that all of his theories "were directly based on experience"), and the theory of photons failed, because it did not give a complete account of optical phenomena.

"Einstein asked about my training," said Cohen, "and how I became interested in Newton. He strongly disapproved of the vanity of scientists, including Newton, who wouldn't give credit to others. He was shocked when he learned that Newton had tried to prove, in the famous controversy over who had invented calculus, that Leibniz had been a plagiarist. And that, when what was thought to be an international committee was set up to investigate the claims, Newton anonymously directed the committee's activities. Einstein made a famous comparison apropos that controversy: that if Newton and Leibniz hadn't lived we would still have differential calculus, but if Beethoven hadn't lived we wouldn't have had the Eroica Symphony. He himself said we would have had the special relativity theory even if he hadn't lived, and he thought the person most likely to have done it was Paul Langevin. But the implication was that no one except Einstein would have done the general theory.

"He told me most emphatically that he thought the worst person to document any ideas about how discoveries are made is the discoverer. He said many people had asked him how he had come to think of this or that, but he had always found himself a very poor source of information about the genesis of his own ideas. Einstein thought that the historian is likely to have better insight into the thought processes of a scientist than the scientist himself."

Einstein also recalled with pleasure his visits to Mach, and said that Mach, Newton, Lorentz, Planck, and Maxwell were the scientists he most admired. In fact, they were the only ones Einstein ever accepted as his true precursors.

"He also discussed Velikovsky's shabby treatment by certain scientists," said Cohen. "I think he misunderstood their position and thought they were trying to suppress Velikovsky's book, and Einstein was very strongly opposed to suppressing anyone's books whether it was sense or nonsense. Shapley and others didn't want to suppress Velikovsky's book; they just didn't want it to appear in the guise of a sound textbook published by Macmillan, the leading scientific textbook publisher. Einstein once said to Velikovsky, 'Why can't you have a sense of humor about this criticism? Why are you so worked up about it?' "

Fearing that he was tiring Einstein, Cohen got up to leave several times, but Einstein pressed him to stay, saying, "There is still more to talk about." When he finally began to descend the stairs, horrified to note that they'd been talking for almost two hours, Cohen turned to thank Einstein for the conversation, missed a step, and almost fell. Einstein chuckled and said, "You must be careful here, the geometry is complicated. You see, negotiating the stairs is not really a physical problem but a problem in applied geometry."

Cohen was almost out of the door when Einstein called from his study, "Wait. Wait. I must show you my birthday present." Cohen returned and Einstein showed him what looked like a curtain rod with a cup on top and a ball attached to it by a piece of string. It was a gift from his neighbor Eric Rogers, the physics teacher, whose children Einstein helped with their homework. His eyes gleamed with delight as he demonstrated how the odd contraption illustrated the equivalence principle. He pressed the rod against the ceiling, then brought it down to the floor, and the ball popped into the cup.

A few days later Einstein and a friend visited Margot, who was in the Princeton hospital with sciatica. After leaving the hospital they discussed death, and the friend quoted James G. Frazer's view that the fear of death accounted for the birth of primitive religion. (Frazer was the author of the *Golden Bough*, a study of magic and religion.) The friend said that to him death was a mystery. "And a relief," Einstein added.

He was back in his office on April 12, when his assistant, Bruria Kaufman, saw him grimace in pain. When he seemed to have recovered, she asked if everything was comfortable. "Everything is comfortable," he replied, forcing a smile, "but I am not."

The next morning he felt well enough to receive Abba Eban, the Israeli ambassador to the United States, and Reuven Dafni, the information officer at the Israeli consulate in New York City. They were there at his invitation because he was troubled over Israel's security in a hostile world and wanted to help. Eban remembered how "Einstein opened the door to us himself, dressed in a rumpled beige sweater and equally disheveled slacks [and then asked] did I think that the media would be interested to record a talk by him to the American people and the world? I exchanged glances with Dafni, as if to say that this was the newspaperman's dream."

Einstein regarded "the birth of Israel as one of the few political acts in his lifetime which had an essentially moral quality," and wanted to alert the world to the threat to its existence. "I must challenge the conscience of the world," he said. He thought that if he spoke about Israel's successes on American television, viewers would switch channels, but not if he "boldly criticized the world powers for their attitude to Israel." His visitors enthusiastically agreed, and it was arranged for him to speak on a coast-to-coast television network as part of Israel's seventh anniversary on April 27.

Einstein then offered them coffee, and Eban expected a servant to bring it. But, to his dismay, "Einstein trotted into the kitchen, from which we soon heard the clatter of cups and pots, with an occasional piece of crockery falling to earth, as if to honor the gravity theory of our host's great predecessor, Newton." When he brought the drinks in and was informed that his talk would probably be heard by sixty million viewers, Einstein quipped, "So I shall have a chance of becoming world famous!"

Eban eagerly phoned his wife, Suzy, afterward, and when asked if anything interesting had happened, he was ready with a casual, "Well, I went to Princeton and Einstein made coffee for me." He told her that Einstein was "enchanting" but strong-willed, and might say things on live television that he, the diplomat, would prefer unsaid. Nevertheless "the overall value of his pronouncements is morally enormous."

The next afternoon Einstein was resting in his bedroom when Dukas heard him hurry to the bathroom and collapse. His doctor, Guy Dean, and two other doctors soon arrived in answer to her call. He was obviously dangerously ill, but insisted on remaining at home. Morphine eased his pain. Left alone in the house with him—Margot was still in the hospital—Dukas slept on a bed in his study next to his bedroom. She had insisted on being near him, despite his protests, and throughout the night she fed him ice cubes and mineral water to fight off dehydration.

The following day he resisted being taken to the Princeton hospital until he was told that he would otherwise be a burden to Dukas. As he was driven to the hospital, Einstein talked animatedly with an attendant, a young political economist who was a Red Cross volunteer. In the hospital he was fed intravenously and given drugs to ease his pain. Before long he was on the phone to Dukas asking for the first draft of his television speech, a copy of I. F. Stone's Weekly, and his latest notes on unified field theory.

He was checking his equations when his friend Otto Nathan arrived. Einstein put his hand on his heart and said that he now felt close to success with his theory.

Margot, only a few rooms away, was wheeled to his bedside, where he greeted her with a cheerful, "What elegance there is in your movements!" He was so changed by pain and blood loss from what was diagnosed as a ruptured aneurysm of his abdominal aorta that she didn't recognize him at first, but, she recalled, "his personality was the same as ever. He was pleased that I was a little better, joked with me, and was completely in command of himself with regard to his condition; he spoke with profound serenity—even with a touch of humor—about the doctors, and awaited his end as an imminent natural phenomenon."

Gustav Bucky phoned his son Thomas to tell him that Einstein was in the hospital. "We drove to Princeton," said Thomas Bucky, "and saw he was very ill. We discussed his condition with Dr. Dean of Princeton and Dr. Rudolf Ehrman of New York, who had been his physician in Berlin. They said he was suffering from an ailment that could be corrected by new surgical techniques and agreed to my suggestion that I talk to Dr. Frank Glenn, the chief surgeon at my hospital, the New York Hospital. He was one of the pioneers of aortic transplants and repairs. At 7:30 the next morning, half an hour before Dr. Glenn had a scheduled operation to do, I began pacing the corridors outside the operating room and intercepted him. He agreed to go to Princeton early that afternoon. After examining Einstein, Dr. Glenn said surgery was possible and suggested moving him to New York to make sure."

Einstein was in pain, but he refused to go, saying, "I do not believe in artificially prolonging life." And that was that, Thomas Bucky recalled: "I talked to him. Dad talked to him. Dr. Dean and the family tried. The answer remained no. We were talking and talking for maybe a day. All of us felt disappointed, frustrated and frantic, for he was getting more and more seriously ill and suffering more and more pain."

Einstein's son, Hans Albert, flew in from California and gave them some hope after talking with his father. "Just give me a little time," Hans Albert said. "Leave it in my hands. I'll get him to say yes tomorrow." But that tomorrow was too late.

Dr. Dean had looked in just before midnight and found him asleep. At 1:15 A.M. on April 18 his night nurse noticed he was having difficulty breathing, and went to check. Helped by another nurse, she raised his head to ease his breathing. In his drugged sleep he mumbled something in German—a language she didn't understand—then took two deep breaths, and died.

"The cause was a ruptured aneurysm of the abdominal aorta," said Thomas Bucky. "And Einstein had been right: the autopsy proved that surgery would not have been feasible." Dr. Rudolph Nissen, who had successfully performed surgery on Einstein in 1948, agreed, explaining that removing the aneurysmal sack of a seventy-six-year-old man and replacing it with a vascular graft would not have succeeded because the site was too close to branches of the kidney artery.

"As fearless as he had been all his life," wrote Margot, "so he faced death humbly and quietly. He left the world without sentimentality or regrets."

"This strange world," he had called it in a letter of condolence to Michele Besso's widow shortly before. "For those of us who believe in physics, this sepa-

ration between past, present, and future is only an illusion, however tenacious."
His heartbroken friend Alice Kahler wrote to a relative:

> The world has lost its best man, and we have lost our best friend. And it came
> so suddenly. Of course, we knew that Einstein did not feel well at all and that
> the situation was dangerous, but there was still hope we thought. We talked
> frequently to Margot who was in the hospital, too [she had been there fourteen
> days]. Monday morning Erich went to buy eggs from Jimmy the grocer and
> came home white as a sheet, could hardly walk. They told him the radio had
> announced the death of Albert Einstein. We rushed to 112 Mercer Street, where
> the Oppenheims and Dr. Nathan, Einstein's closest friend, had just arrived. Dr.
> Nathan asked me to stay in the house with Kathy, the colored maid, to answer
> the phone (messages were then transferred to the Institute for Advanced Study).
> By 12 noon the house was beleaguered by newspapermen in cars and a television
> truck from CBS. The CBS truck took shots of every person coming from or
> leaving the house and wanted to interview me, which I refused of course. Only
> after the will was read by Dr. Nathan in the hospital one heard about Einstein's
> provisions he made for being cremated and his ashes strewn to the winds, as was
> done with his sister Maja.
>
> Now I want to add that Margot has told me and she can speak freely now,
> that after the operation some years ago which the great surgeon Nissen per-
> formed, she knew one day it might happen exactly as it did since Nissen pre-
> dicted it. When he opened him that time he had a shrunk liver, the gall bladder
> was fairly intact (one had thought that the bladder was the trouble); instead an
> arteriosclerotic vein (aneurysm) of the main arteria was swollen and the surgeon
> anticipated that one day it might burst like a hemorrhage and this will be the
> end. He put a net around it to protect it and it lasted seven years. For this reason
> the doctors agree that no operation would have helped, although sometimes one
> can replace the trouble spot with a plastic insert. Margot told her stepfather that
> Nissen said that one day he might go like this—and she was making a snap with
> her fingers.

Carrying out his friend's wishes, Otto Nathan had Einstein cremated, and
scattered the ashes in the Delaware River. His brain was preserved for medical
research; some researchers hoped to find in it a clue to his genius.

Einstein left $20,000 and his house, furniture, and household goods to Mar-
got; $20,000 and all his personal clothing and personal effects to Helen Dukas—
except for his violin, which went to his only surviving grandson, Bernhard Caesar;
$15,000 to his son, Eduard; and $10,000 to his son, Hans Albert. Einstein had
appointed Nathan and Dukas as joint trustees of his estate, and had established
a trust to hold "all of my manuscripts, copyrights, publication rights, royalties
and royalty agreements, and all other literary property and rights," and any other
rights. Income from the trust was to go to Dukas until her death, and then to
Margot. Ultimately it and his papers went to the Hebrew University in Israel.

Dukas gave mementos to Einstein's friends. Abraham Pais received his last
pipe ("Its head is made of clay, its stem of reed") and the galley proof of the
second appendix to the "Generalized Theory of Gravitation," which Einstein
had wanted him to have. "I was in my thirties when that 1950 book came out,"

Pais recalled. "I read it then and reread it once every few years, always with the same thought as I turn the pages. Does the man never stop?"

Hans Albert had a distinguished career as a professor of civil engineering at the University of California at Berkeley, where he gained an international reputation. He kept a low profile so that many in the university did not know he was Albert's son. Visiting the library several times a week, he found it somewhat disconcerting to walk past the bust of his father between those of Dante and Copernicus. "Do you know what it's like to have your father a statue?" he once asked a visitor.

A movie producer who called on Hans Albert and his second wife, Elizabeth, in their modest hilltop home was not interested in statues or father–son relationships. Gazing through the window at the fabulous view below of San Francisco and the Golden Gate Bridge, the producer said, "We'd like to make a film about your father. But I need a love story."

"I can't tell you any more than there is in the books about him," Hans Albert replied.

"But I can't make a movie based on his scientific work alone," the producer pointed out. "It must have human interest. It's impossible to believe that your father didn't have any romances, apart from his two marriages."

"I don't know of any," Hans Albert said. Which was true. Nor did Elizabeth. The would-be moviemaker left disappointed.

Hans Albert died of a heart attack at Woods Hole, Massachusetts, in 1973.

Hans Albert's adopted daughter, Evelyn, lives in California, where she deprograms members of cults.

In November 1955, seven months after Albert Einstein's death, the first of his five great-grandchildren, Thomas Martin, son of Bernhard Caesar, was born in Bern. Bernhard had been brought up in California, where, he later recalled, he argued with a fellow on a trolley about a math problem: "The argument got pretty heated until at last he almost shouted, 'Who d'you think you are anyway, Einstein?' I answered, 'Yes!' and everyone on the trolley burst out laughing." Bernhard eventually moved to Switzerland to work as an experimental physicist for the Swiss army, developing armor plating for tanks. None of his children are physicists.

In 1965, Eduard died in the Swiss psychiatric hospital where he had been confined for two decades.

Margot Einstein and Helen Dukas continued to live at 112 Mercer Street. Dukas cooperated with Jamie Sayen and Abraham Pais on their biographies of Einstein and responded to other researchers by mail and telephone.

Margot continued to care for abandoned animals and to paint and sculpt. She, too, helped Sayen with his book. She died in 1986.

Helen Dukas and Otto Nathan, as executors of Einstein's estate, devoted the rest of their lives to preserving his idealized memory, collecting additional material for the Einstein Archives, and frustrating the attempts of writers and scholars to probe deeper into his life.

For more than a quarter of a century, and up to within a few days of her death in 1982, Dukas, according to Freeman Dyson, who knew her well, "conducted an enormous and worldwide correspondence. She was finding new Einstein documents in unexpected places and establishing their historical context. She was helping historians from all over the world to locate and interpret the documents already in the archives. There was nobody else who could . . . tell at a glance when it was written (if undated), what it said, and how it fitted into the story. There was nobody else who remembered even the names of all the people Einstein knew, let alone their dates, titles, jobs, and family connections." Dukas took her role as the collector of any and all Einstein-related material very seriously. Sometimes she used strong-arm tactics, snatching a book from a woman, insisting that it had only been lent to her by Einstein and was not a gift.

Otto Nathan joined Dukas in the quest to gather as much as possible of Einstein's writings. Nathan was even more aggressive than Dukas, successfully taking Hans Albert to court to prevent him from publishing extracts from his father's letters. He also threatened legal action if a physicist didn't hand over a collection of Einstein correspondence in his possession. The letters were handed over. He used such effective delaying tactics to put off the publication date of Einstein's *Collected Papers* that the first volume, *The Early Years: 1879–1902* (edited by John Stachel and translated by Anna Beck), was not published until 1987—thirty-two years after Einstein's death.

CHAPTER 42

Einstein's Legacy

After Einstein's death came the revisionists. They claimed that Einstein was a stubborn old codger who wasted the last twenty-five years of his life chasing an impossible dream. J. Robert Oppenheimer — who had once called Einstein "completely cuckoo"—said as much in 1965, on the fiftieth anniversary of the general theory of relativity. In an attempt to dispel the myth and expose the man, Oppenheimer conceded that "his early papers are paralysingly beautiful"—even though "there are many errata. Later there were none." Then, said Oppenheimer, Einstein engaged in a "noble and furious" but finally futile fight with Bohr "to prove that the quantum theory had inconsistencies." Not only that, Oppenheimer continued, Einstein also took on "an ambitious program, to combine the understanding of electricity and gravitation [unified field theory] . . . [that] left out too much that was known to physicists but had not been known much in Einstein's student days." In plain words, the man was either stubborn or stupid.

Leopold Infeld flew to his defense, as good as calling Oppenheimer a fool:

> To what errors [errata] does Oppenheimer refer? Neither I nor any other physicist with whom I have spoken understands that sentence. The work of every physicist can be divided into stages. At every stage he thinks he has completed his work on that lode of gold which he has uncovered. Then it turns out that it is only the surface vein of a much greater lode and that he must search deeper. From this point of view the work of every physicist is a step-by-step search for the truth. Newton's laws are the truth today, too, but only for small velocities. A fool might say that Newton's work was full of errors since it does not apply to high speeds approaching that of light. I know of no errors by Einstein aside from the usual printing errors and those Einstein knew well, since they drew him, in his subsequent work, closer to the truth.

Ironically, in the same year that Oppenheimer attacked Einstein's scientific reputation (1965), astrophysicists Arno Penzias and Robert Wilson "tuned in" to the background microwave radiation that pervades the universe, assumed to be the residue from the big bang. The discovery — which won them the Nobel Prize — supported Einstein's theory of creation.

Despite the scientific evidence confirming much of Einstein's work, when the first major Einstein biography, *Einstein: The Life and Times*, appeared in 1971, its author, Ronald Clark, seconded Oppenheimer's gloomy assessment: "Einstein's theory of a unified field remains unsubstantiated and current thought veers away from the universe being built in that way."

In 1976, laser experiments between the earth and moon supported Einstein's theory of general relativity. So did the work of Martin Levine and Robert Vessot of the Smithsonian Astrophysical Observatory. They shot a rocket containing an extremely accurate clock 6,200 miles into space, then checked the clock's time with an identical clock on Earth. According to Einstein's theory, the clock in space should have run 4.3 parts in 10 billion faster than the one on earth. Over the course of a year it would be about one one-hundredth of a second fast. The result confirmed Einstein's figure.

In 1983, thanks to his theory, physicists worked out the history of the universe back to a fraction of a second after the big bang. That fraction of a second would be expressed by a decimal point followed by 42 zeros and a one, or .000000000000000000000000000000000000000001, according to writer Rick Gore.

Ten years later another confirmation of general relativity came from instruments on the NASA satellite Cosmic Background Explorer (COBE). The satellite recorded hundreds of millions of measurements in space, showing that the primordial temperatures found there of just above absolute zero are uniformly distributed. Such conditions would be expected from an afterglow of the big bang—with which, according to Einstein, the universe began.

Most recently, in July 1995, news of another Einstein triumph was announced. By cooling atoms to near absolute zero, researchers in Boulder, Colorado, produced a new form of matter. Some seventy years before, Einstein and Satyendra Nath, in their Bose–Einstein condensation theory, had predicted this as a possibility.

And so it goes.

Physicist Richard Feynman believed that Einstein failed in his later work because he stopped thinking in concrete physical images and became a manipulator of equations. John Kemeny disagreed: "First of all we don't know that Einstein failed. The equations have not been solved and therefore no one knows whether they are correct. If Einstein did fail it is probably because the problem he was trying to solve was too difficult for this state of the history of physics."

In 1994, twelve leading Einstein experts announced that he was brighter than he realized, having produced his general theory of relativity in 1912, three years earlier than even he thought—confirming his views that scientists themselves are not the best judges of the evolution of their ideas.

This discovery occurred when, after three years of trying, the Einstein scholars finally deciphered the equations in a notebook Einstein kept in 1912. These equations were in a chaotic state, not unlike Einstein's study. He had started at one end, then turned the notebook over and started again; on a few pages he apparently started from both top and bottom. Some notebook pages are blank; some have been torn out. It didn't help that a front cover label was stuck on the back of the book.

What seems to have happened is that Einstein correctly formulated the theory in 1912 but rejected it as impossible because it contradicted Newtonian physics. He changed his mind three years later, in November 1915, when he delivered a series of papers to the Academy of Sciences in Berlin.

Recently there has been a resurgence of interest in trying to find a unifying force in nature. A growing number of scientists — Stephen Hawking, for one — believe that Einstein was not chasing a dream, that there may well be a unifying principle linking gravity with electromagnetism and both the strong and weak nuclear forces. This attempt at unification is the main theme in physics today, as Einstein believed, says Nobelist Chen Ning Yang.

Princeton physicist John Wheeler is among the enthusiasts: "Einstein's unified field theory has come to life in an absolutely spectacular form in the last decade in superstring theory. It's the idea that in addition to the dimensions you and I see and work with, three in space and one in time, there are additional dimensions — the geometry of which, however, is curled up into a very tight little sphere everywhere, at every point.

"Each little sphere is in a certain sense like a pipe organ pipe. Each organ pipe has air vibrating within it which gives it a musical tone, depending on its length and diameter. Each little sphere in the geometry has its own vibrations, more numerous than an organ pipe. And these vibrations describe the various fields of force that give rise to the particles we know. Period.

"This theory began with Theodor Kaluza, an associate of Einstein's, and Oscar Klein, an associate of Niels Bohr: the so-called Kaluza–Klein, a five-dimensional theory, in 1926. Nowadays, the idea is that there are additional dimensions. At the moment the favored dimensionality is four plus six. I think Einstein would have found it very sympathetic. The Kaluza–Klein idea has been generalized in a perfectly wonderful way today. So that Einstein's dream, his hope, his goal of reducing the various forces of physics to pure geometry has taken on life. It's unbelievably active today."

Asked if he shared John Wheeler's view of superstring theory, Abraham Pais replied: "That's a very complicated question. Superstring theory goes in a direction that Einstein never had dreamed of. I would not disagree, but I would not make a point of it."

"Can one say that unified field theory is not dead?"

"That you can say."

Another Einstein expert, Michio Kaku, professor of theoretical physics at City College of the City University of New York and coauthor of *Beyond Einstein*, enthusiastically agrees. He said: "Einstein was working at a tremendous handicap and that is, in the 1930s and '40s and '50s, nobody knew what the nuclear force was all about. Now we know that the nuclear force is in some sense like the missing link, linking electromagnetism to gravity. So Einstein was literally thirty years ahead of his time. He was trying to link electricity and magnetism with gravity and skipping the intermediate stage, which was the nuclear force.

"Some people therefore say that Einstein's theory was all wrong and that he was going senile towards the end of his life," Dr. Kaku said. He disagrees. He sees Einstein as a prophet who was talking about unification when Wolfgang Pauli and others were laughing behind his back. It was Pauli who made the now famous remark, "What God has torn asunder, let no one put together." According to Pauli, if God has broken electricity and magnetism from gravity, what right did Einstein have to put them back together again?

"Today, on the other hand," said Dr. Kaku, "unification is all the rage in physics. In fact, we realize unification is perhaps the dominant thread in the last two thousand years of scientific investigation.

"So we now know in one sense that Einstein was premature. He didn't understand the nuclear force because that was only worked out fairly recently. However, he was prophetic [by] talking about unification when other people were dabbling with neutrons and protons . . .

"People misinterpret Einstein's objections to quantum theory," he said. "He didn't think quantum theory was wrong, but incomplete. He wanted a unified field theory whereby matter itself comes out as a by-product of geometry."

What Einstein wanted was a super quantum theory. "At the present day," Dr. Kaku continued, "most physicists believe that the quantum theory is correct as it stands, without any higher level of understanding. And the string equations go perfectly well with the quantum theory. That's why they've generated so much excitement . . .

"Some of us are trying to take the string theory to a higher level . . . If we succeed, it would complete Einstein's quest to unify gravitation with the other forces."

But what of his impact as a human being on those who knew him personally and on the whole world? Margot Einstein summed him up as "full of harmony. He lived in his own world, and this saves people." To Henri Poincaré, Einstein had "the most original mind that I have ever known." He had "a genius superior to Newton's," according to Erwin Schrödinger. To the Max Borns, he was their dearest friend.

Niels Bohr considered Einstein's service to science and humanity "as rich and fruitful as any in the whole history of our culture" and believed that "mankind will always be indebted to Einstein for the removal of the obstacles to our outlook which were involved in the primitive notions of absolute space and time. He gave us a world picture with a unity and harmony surpassing the boldest dreams of the past." In his personal life, Bohr wrote, Einstein "was always prepared to help people in difficulties of any kind, and to him, who himself had experienced the evils of racial prejudice, the promotion of understanding among nations was a foremost endeavor."

But in recent years a new generation of revisionists have given a different view of Einstein: did he, alone, deserve all these accolades? In an effort to diminish Einstein, several authors and lecturers have been plugging the line that his first wife, Mileva Maric, was an intellectually battered woman. She, they say, deserves much of the credit for his glory as an equal partner in his triumphant scientific endeavors: something which he never acknowledged. A former *U.S. News & World Report* editor, Andrea Gabor, made this claim in her recent book, *Einstein's Wife.* It was applauded by reviewer Jill Ker Conway, a former president of Smith College, who said: "Maric seems to have been Einstein's intellectual peer, attractive at first, but too dangerous for continuing intimacy, a danger not to be passed over lightly." The review aroused a quick rebuttal from Einstein experts Robert Schulmann and Gerald Holton. The following letter from them was published in *The New York Times Book Review* on October 8, 1995:

> In Jill Ker Conway's review of Andrea Gabor's book *Einstein's Wife*, she writes that "Ms. Gabor rightly denounces Albert Einstein for concealing the major contribution of Mileva Maric, his first wife, to the formulation of the theory of relativity." This opinion goes even beyond the flights of journalistic fantasy of Ms. Gabor, which in turn was based chiefly on a nationalistic puffery of a biography of Mileva Maric, originally published in Serbia in 1969. All serious Einstein scholarship, by Abraham Pais, John Stachel and others, has shown that the scientific collaboration between the couple was slight and one-sided. The documentary evidence is that Maric encouraged and aided Einstein in the early years, when their intense passion for each other made their life at the margin of a society bearable, when he expressed and shared his groundbreaking ideas freely with her and a few friends, while developing them in isolation from the physics community. It is clear that she also helped him by looking up data and checking calculations. Einstein never acknowledged this help publicly; but neither did Maric claim more in her letters to him or others. The true collaboration which they had originally planned when they both intended careers as high-school teachers never did develop. Nor is there a shred of documentary proof of her originality as a scientist.

Holton elaborates on Mileva's importance to Einstein in his recent book, *Einstein, History and Other Passions,* in which he points out that Mileva

> left no evidence of orginality as a potential major scientist — but that, of course, is also true of most male science students. During the early, good years she and Albert longed palpably for each other's companionship. Einstein also valued her intellectually, and not just because he, an autodidact throughout his life, always needed someone who understood him, to talk with about new ideas . . . We also know [through their exchange of letters] more about Mileva's increasing demoralization or loss of self-esteem. Was it caused by the loss of Lieserl, or by her husband's rapidly increasing fame? . . . Or was it their disappointment with each other when the original vision of a cozy couple immersed in the study of science came to naught? . . .
>
> During the years when Albert and Mileva were passionately attached to each other, and particularly when she seemed in need of psychological support, he occasionally used in his letters to her such phrases as "our work" or "our investigation." Starting in 1990, a small number of writers attempted to inflate the meaning of these phrases to include the strong probability that Mileva was in fact responsible for either the physics or the mathematics in Einstein's 1905 relativity paper but was deprived of any credit for it. For a time this allegation received extensive press coverage around the world . . .

Careful analysis by established historians of physics, including John Stachel, Jurgen Renn, Robert Schulmann, and Abraham Pais, has shown that scientific collaboration between Mileva and Albert was indeed minimal and one-sided. Einstein's occasional use of the word "our" was meant chiefly to serve the emotional needs of the moment . . .

Ironically, the exaggeration of Mileva's role far beyond what she herself ever asked for or what could be proved only detracts from her real and significant place in history. She was one of the pioneers in the movement to bring women into science, even if she did not reap its benefits. At a great personal sacrifice, as it later turned out, she seems to have been essential to Albert during the onerous years of his most creative early period. She was not only an anchor of his emotional life but also a sympathetic sounding board for his highly unconventional ideas during the years when he was undergoing the quite unexpected rapid metamorphosis from eager student to scientist of the first rank.

Supposed secrets about Einstein's private life, too, still attract worldwide attention. In 1995, forty years after Einstein's death, Ludek Zakel, a sixty-three-year-old Czech physicist, claimed to be his son. Journalist Marcela Pekhackova published the evidence in *DNES*, a Prague magazine. Zakel told her that in the spring of 1932, Elsa Einstein thought she had cancer of the uterus and was afraid to consult doctors in Berlin because the Nazis were coming to power. Instead, he said, she went to Prague to seek the help of Albert's friend, the director of the Apolinare Hospital in the Czech capital, who found that Elsa didn't have a tumor. Instead, according to the account, she was pregnant. Then, on April 14, 1932, according to Zakel, both Elsa and a woman named Eva Zakel gave birth to a son in the same hospital. But Eva's child died on the day he was born. Zakel maintains that Eva was so desperate to please her husband, who wanted children, that when she learned that Albert Einstein didn't want any more, she accepted Elsa's child — later named Ludek Zakel — as her own. Elsa, presumably, returned to Berlin "cured."

To back his claim, Zakel had the following documents: a notarized birth certificate and a baptismal record affirming that he is Einstein's son, issued during the Communist era; the written, but unverified, statements of two nurses long since dead; the signed statement of his mother, a ninety-three-year-old Eva Zakel, who is no longer willing to discuss the subject. Zakel also said that at one time Margot Einstein sent a friend of hers to inform him that he was Elsa's son. In 1979, Zakel had written to a Czech newspaper, *The Evening Prague*: "I would like to thank you and the country on behalf of the Einstein family for the celebration of the anniversary of my father's 100th birthday. I am very grateful." The editorial staff of the newspaper took it as a gag or a scam to obtain a passport in order to emigrate to the Unied States. It wasn't printed.

What is the evidence that Ludek Zakel is not the son of Albert and Elsa Einstein?

Elsa's name does not appear in the records of the Apolinare Hospital (it should be remembered, however, that many records were damaged during World War II, and that her admission there might not have been recorded because she was a foreigner and a Jew). The documents could have been forged attempts to escape from Czechoslovakia during the Communist regime — the not unlikely work of a mother anxious to help her son to escape to freedom. In fact, Zakel had applied to the U.S. Embassy for citizenship many times — claiming to be the son of an American citizen — but his applications were always rejected.

Elsa did not return to Berlin from her visit to the United States with Albert until April 6, 1932. According to Zakel's scenario, she gave birth to him in Czechoslovakia eight days later. But in the preceding months she had been frequently seen and photographed in public and it is hardly likely that no one would have noticed — and that no photographs would have shown — signs of her pregnancy.

The strongest evidence against Zakel's claim is Elsa Einstein's age. She was born in 1876. In 1932 she would have been fifty-six years old, an age when it would have been a near-miracle for her to have given birth.

To Robert Schulmann, the idea "that someone would go from Berlin to Prague to check on a tumor because of the Nazis is ridiculous. And the implication that in 1932 anti-Semitism was greater in Berlin than in Prague is nonsense. After all, there were many Jewish doctors in Berlin. Almost all of Einstein's friends were Jewish doctors. Then there's the background. Einstein did not sleep with that lady [Elsa] certainly at that age. He had plenty of paramours and it's questionable whether he slept with Elsa after 1920. Zakel makes a credible case on the face of it, but most liars do if they're good at it and they believe it themselves. They might even pass a lie detector test. In such a case as this, one tries to reconstruct the facts on the basis of sound evidence — and there isn't any."

Despite the controversies that still rage over Einstein's legacy, the true measure of his impact on the world today is that his name alone symbolizes Science. Another is his enduring worldwide appeal. A pop icon on a par with Elvis Presley and Marilyn Monroe, he stares enigmatically from postcards, magazine covers, T-shirts, and larger-than-life posters. A Beverly Hills agent markets his image for television commercials. He would have hated it all.

Among his insatiable fans is a recent seventy-year-old visitor to Princeton, so eager to see inside Einstein's locked office at the Institute for Advanced Study that he hung from a second-story ledge outside the building to peek inside.

How did Einstein see himself? He had mixed feelings. In 1949, he denied that he regarded his achievements with "calm satisfaction . . . I am uncertain as to whether I was even on the right track. In me my contemporaries see both a heretic and a reactionary who has, so to speak, survived himself . . . The feeling of inadequacy comes from within. Well, it cannot be otherwise when one has a critical mind and is honest, and mood and modesty keep us in balance in spite of external influences."

Despite his self-doubts and his self-confessed failure as a husband and father, Einstein also regarded himself as blessed. As early as 1925, when the British Royal Astronomical Society had awarded him its Gold Medal, he had written in response words that would always be true for him:

> He who finds a thought that lets us penetrate even a little deeper into the eternal mystery of nature has been granted great grace. He who, in addition, experiences the recognition, sympathy, and help of the best minds of his time, has been given almost more happiness than a man can bear.

Einstein's Brain

Einstein did not want anyone to worship at his "old bones," as he sometimes put it. He got his wish: he was cremated. But not before pathologist Thomas Harvey, who performed the autopsy, removed his brain, most of which he now keeps preserved in glass jars in his Kansas home.

The preservation of Einstein's brain was supposed to have been kept secret at the time, but it made the headlines.

"I'll tell you what happened," said Helen Dukas. "There was an autopsy and the son [Hans Albert], as next of kin, of course, gave the permission. And then that Mr. Harvey, without asking, had already taken the brain away. So we couldn't do anything about it. And then he came and asked [if he could use the brain for research] and the son said, 'Under one condition. Anything published, only in scientific papers.' Well, then the journalists got hold of it. I mean *that* Professor Einstein would never have allowed."

"Harvey, who did the autopsy, was married to a librarian I knew at Princeton," explained Ashley Montagu. "He kept Einstein's brain in a jar in the basement of his house. Then he separated and divorced from his wife, and the brain was still down there. And that's how I heard of it. I was shocked to learn there was such an organ hidden in a jar down in Mrs. Harvey's basement. She said, 'I wish someone would get rid of the damn thing!'

"It was stupid from the very beginning," said Montagu, who trained as an anatomist. "It was due to ignorance as all stupidities are. Einstein might not have minded, but it wouldn't have been necessary if somebody had been wise enough to inform them that even [a] gross feature like the size has no relation whatsoever to intelligence. The human brain is about 1,350 cubic centimeters, or three and one-half pounds by weight. The largest brain on record is 2,500—almost twice the size of the normal brain—the brain of a hydrocephalic, that of an idiot. Even the most sophisticated research can't make discoveries about the normal brain through such examinations."

Twenty-four years later, in 1979, reporter Steven Levy of the *New Jersey Monthly* was curious about the fate of Einstein's brain and what, if anything, it had revealed. He traced Thomas Harvey to Wichita, Kansas, where he was the medical supervisor in a biological testing lab.

Where's the brain? Levy asked. To his amazement, in reply, Harvey picked up a cardboard box containing two glass jars. Inside one, floating in a clear solution of formaldehyde, were Einstein's cerebellum, a piece of his cerebral cortex, and a few aortic vessels. The larger jar contained several small pieces of his cerebral cortex encased in transparent blocks.

Over the years, Harvey mailed sections of Einstein's brain to researchers across America and in China, Germany, and Japan who wished to study them. Although he could have made a mint selling what was left of Einstein's brain, he resisted offers from museums and millionaires, as well as appeals from rabbis wanting to bury the remnants so that Einstein's soul may rest in peace.

Harvey lost his medical license in 1988 when he failed a three-day exam, and now (1995) at age eighty-two, he works late shifts in a Kansas plastics factory.

In a sense, Einstein got what he wanted: few of the scientists who handled his brain have treated it like a holy relic. But he would have deplored the publicity.

When questioned, Dr. Harvey said that although he had distributed sections of Einstein's brain to a number of scientists, "I have more of his brain than anybody. One researcher, Marian Diamond, says her research indicates there was a difference between his brain and the average person. Whether it shows more intelligence or not remains to be seen."

Dr. Marian Diamond, a professor of anatomy at the University of California at Berkeley, has reported finding an above-average number of glial cells, which nourish neurons, in the brain's left hemisphere, an area associated with math and language skills. Ashley Montagu is still skeptical. So is Dr. Lucy Rorke, a neuropathologist in Philadelphia. Dr. Rorke has five sets of slides containing sections of Einstein's brain and thinks "it's a rather extraordinary brain . . . Dr. Diamond [claimed] that there was an excessive number of glial cells and that this indicated that his neurons were getting extra nutrition . . . I think she didn't realize that the sections that were prepared were much thicker than one ordinarily prepares with routine sections, if one is going to try and diagnose a tumor or something like that. And her unfamiliarity with thick sections of brain led her, I think, to erroneously come to that conclusion . . .

"What is impressive about his brain is that he doesn't have the usual degenerative changes that one sees in older people, specifically in terms of the usual massive accumulations of a wear-and-tear pigment in the neurons, a substance we call lipofuscin. And it really is a beautiful, pristine brain. It looks like the brain of a young person. Of course, everybody wants to know if he had any indication of Alzheimer's plaques or tangles. There's absolutely no indication of that . . . It doesn't have any of the changes we oftentimes see in the brains of older people."

When told that Ashley Montagu thought it was nonsense to think one can discover intelligence in the brain through a physical examination, Dr. Rorke replied: "I agree completely. You can find if it's healthy, of course. But to find intelligence or genius is an impossible quest. It's impossible to predict what will happen years ahead, of course . . . What's interesting, too, is that none of his many illnesses seem to have affected the healthiness of his brain." (Dr. Harvey disagrees, saying Einstein's brain was normal for a man his age, showing the expected deterioration.)

Dr. Harvey is thinking of donating what he has left of the brain to the Hebrew University. Meanwhile, researchers are still trying to "read" slices of Einstein's brain to discover clues to his genius.

In 1993 the BBC aired a documentary called *Einstein's Brain*. While researching the film, Kevin Hull, the producer, made a shocking discovery: Henry Abrams, Einstein's ophthalmologist, had taken and kept Einstein's eyes.

Abrams recalled that during the autopsy "every doctor in the Princeton hospital came down that day as a matter of curiosity, or just to say good-bye." He happened to be in the Princeton hospital at the time and asked the hospital administrator, Jack Kauffmann (since deceased), for permission to "remove and to keep the eyes. [Kauffmann] said, 'Absolutely, yes.' The whole thing took about 20 minutes. I just needed scissors and forceps."

Abrams has kept Einstein's eyes in a New Jersey safe-deposit box for forty years. He visits them several times a year to make sure they're in good condition, checking the solution in the small glass jar in which they float. That way, he feels, he is giving Einstein immortality. According to the British *Guardian Weekly*, Abrams has recently thought of selling Einstein's eyes to ensure his grandchildren's financial future.

Soon after the article appeared, Dr. Abrams was overwhelmed with requests for interviews and eventually had his phone taken out of service for several months. In June 1995, he was back on the phone again and agreed to talk. "For forty years," he said, "I had the eyes safely put away and never gave it a thought. I think a friend or someone in the family when he was abroad mentioned something about the eyes. You know how these things can spread. The family always knew about it."

Asked if he was correctly quoted in saying, "His eyes were angelic: they gave the impression he knew everything in the world. His eyes were godlike," Abrams replied, "I didn't say that he knew everything in the world. I did say, 'When you look into his eyes you're looking into the beauties and mysteries of the world.'"

Abrams denied the report that he might sell Einstein's eyes, but when asked what will happen to them when he dies, he replied, "I'll have to ask my lawyer." His psychologist son, Mark, believes his father's motive in taking and keeping the eyes is simply an expression of veneration.

Notes

Archival Sources

The Einstein Archives are in the Hebrew University, Jerusalem. Photocopies of the documents are at Princeton University and Boston University. Princeton University Press is publishing the Einstein documents under the title *The Collected Papers of Albert Einstein*. The editors of *The Collected Papers of Albert Einstein, Volume 1: The Early Years, 1879–1902*, were John Stachel, David Cassidy, and Robert Schulmann. *Volume 5: The Swiss Years: Correspondence, 1902–1914*, was edited by Martin J. Klein, A. J. Kox, and Robert Schulmann.

The correspondence between Einstein and Upton Sinclair is located in The Upton Sinclair Manuscripts, Manuscripts Department, Lilly Library, Indiana University, Bloomington, Indiana.

I found several of my sources in the Columbia University Oral History Project collection, New York City; the Caltech Archives, Pasadena, California; and the Archives of the Burndy Library, Norwalk, Connecticut.

Richard Alan Schwartz of Florida International University in Miami kindly gave me a copy of his FBI file on Albert Einstein. The FBI provided a copy of their file on Helen Dukas, now in my possession, through the Freedom of Information Act.

The Leon L. Watters Collection is in the Rare Documents File and Box 2813 in the American Jewish Archives, Hebrew Union College, Jewish Institute of Religion, Cincinnati, Ohio.

Preface

ix *"Our life is divided betwixt folly and prudence"* Walter Moore, *Schrödinger: Life and Thought* (Cambridge, Engl.: Cambridge University Press, 1989), p. vii.

x *"Einstein's pleasure in the company of women"* Ronald W. Clark, *Einstein: The Life and Times* (New York: Avon, 1971), p. 31. *"dredging-job" in which "Einstein can do nothing right"* John Carey, review of *The Private Lives of Albert Einstein* by, Roger Highfield and Paul Carter, *Sunday Times Books*, August 29, 1993, sec. 6, p. 1.

Chapter 1: Childhood and Youth

1 *"Much too fat!"* Maja Winteler-Einstein, "Albert Einstein—A Biographical Sketch," in *The Collected Papers of Albert Einstein, Volume One: The Early Years, 1879–1902*, ed. John Stachel, et al., trans. Anna Beck (Princeton, N.J.: Princeton University Press, 1987), p. xviii. *"My parents were worried"* Banesh Hoffmann with Helen Dukas, *Albert Einstein: Creator and Rebel* (New York: Viking Press, 1972), p. 14. *"He was so good and dear"* Jette Koch to Her-

mann and Pauline Einstein, Summer 1881, Einstein Archives, Hebrew University. *Albert's response to Maja's birth* Winteler-Einstein, *Albert Einstein*, p. xviii.

2 *As Maja remembered it* Ibid., p. xv. *his parents encouraged him* Ibid., p. xix.

3 *"A sound skull is needed"* Ibid., p. xxii. *Those school days* Ibid., p. xix.

4 *At twelve, Albert* Hoffmann with Dukas, *Einstein: Creator and Rebel*, pp. 22–24. *he felt neither uncomfortable nor singled out* Philipp Frank, *Einstein: His Life and Times*, trans. George Rosen (New York: Knopf, 1947), p. 9. However, biographer Jamie Sayen quotes a statement Einstein made in 1920: "Physical assaults and insults were frequent on the way to school, though for the most part not really malicious. Even so, however, they were enough to confirm, even in a child of my age, a vivid feeling of not belonging." Jamie Sayen, *Einstein in America: The Scientist's Conscience in the Age of Hitler and Hiroshima* (New York: Crown, 1985), p. 25.

5 *Kant proposed some bizarre ideas* Will Durant, *The Story of Philosophy* (New York: Simon and Schuster, 1967), p. 199. *"where there is usually a great deal of wind"* F. Paulsen, *Immanuel Kant* (New York: Scribner, 1910), p. 82.

6 *"constantly searched for new harmonies"* Winteler-Einstein, "Albert Einstein," p. xxi. *"There, now I've got it"* Ibid. *Whether the smile* Abraham Pais, *'Subtle is the Lord' . . . : The Science and the Life of Albert Einstein* (New York: Oxford University Press, 1982), p. 38.

7 *This was too much* Hans Albert Einstein, quoted in Bela Kornitzer, *Ladies Home Journal* (April 1951): p. 136.

8 *Albert brought pen, ink, and notebook to the party* Winteler-Einstein, *Albert Einstein*, p. xxii. *Despite this traumatic failure* Hoffmann with Dukas, *Einstein: Creator and Rebel*, pp. 27, 28.

Chapter 2: First Romance

9 *"an unforgettable oasis"* Carl Seelig, *Albert Einstein: A Documentary Biography*, trans. Mervyn Savill (London: Staples Press, 1956), p. 21.

10 *"getting not-so-good grades"* Hermann Einstein to Jost Winteler, December 30, 1895, in *Collected Papers of Albert Einstein*, vol. 1, no. 14. *According to Einstein biographer* Seelig, *Albert Einstein*, pp. 17, 18.

11 *In his letters to Marie* Albert Einstein to Marie Winteler, April 21, 1896, in Stachel, et al., *Collected Papers of Albert Einstein*, vol. 1, no. 18. *"Heaven, no! It's in my blood!"* Seelig, *Albert Einstein*, p. 14. *"Fate decreed"* Ibid. *"strode energetically up and down"* Ibid., pp. 14–15.

12 *field trip to the Jura Mountains* Ibid., p. 19. *The examiner praised him* "Inspectors Report on a Music Examination," Aargau Kantonschule, March 31, 1896, in Stachel, et al., *Collected Papers of Albert Einstein*, vol. 1, no. 17. *his "beloved sweetheart"* Albert Einstein to Marie Winteler, April 21, 1896, ibid., no. 18.

13 *Then, without warning* Marie Winteler to Albert Einstein, November 25, 1896, ibid., no. 29. *"My dear, dear sweetheart!"* Marie Winteler to Albert Einstein, November 30, 1896, ibid., no. 30. *When he did write to Aarau* Albert Einstein to Pauline Winteler, May 18, 1897, ibid., no. 34.

14 *Grossmann told his own parents* Seelig, *Albert Einstein*, p. 34.

Chapter 3: To Zurich and the Polytechnic

15 *"her dreamy, ponderous nature"* Seelig, *Albert Einstein*, p. 46.

16 *She did write, in October* Mileva Maric to Albert Einstein, October 20, 1987, in Stachel, et al., *Collected Papers of Albert Einstein*, vol. 1, no. 36. *He replied four months later* Albert Einstein to Mileva Maric, February 16, 1898, ibid., no. 39. *His mother and sister teased him* Albert Einstein to Mileva Maric, March 13 or 20, 1899, ibid., no. 45. *After one weekend*

Leon L. Watters, "Comments on The Letters of Professor and Mrs. Albert Einstein to Dr. Leon L. Watters" (Leon L. Watters Collection, American Jewish Archives), p. 28.   *He never forgot his beloved "fiddle"*   Seelig, *Albert Einstein*, pp. 36, 37.

17 *Einstein's indifferent approach*   Ibid., p. 43.   *Lunchtimes at Einstein's*   Ibid., p. 35.   *a widow and her two daughters invited themselves*   Ibid., p. 15.   *"Because," he replied, "we wouldn't dream"*   Ibid., p. 36.   *He infuriated physics instructor Jean Pernet*   Clark, *Einstein: The Life and Times*, p. 39.

18 *Math professor Hermann Minkowski*   Hoffmann with Dukas, *Einstein: Creator and Rebel*, p. 85.   *His casual study habits*   Seelig, *Albert Einstein*, p. 30.   *Like Einstein, he was*   Philip Cane, *Giants of Science* (New York: Pyramid, 1961), pp. 202, 203.   *"Day and night he buried himself in books"*   Frank, *Life and Times*, p. 20.

19 *Besso believed it was Mach's influence*   Hoffmann with Dukas, *Einstein: Creator and Rebel*, p. 78.   *When Mach was a child*   Lewis S. Feuer, *Einstein and the Generations of Science* (New York: Basic Books, 1974), p. 41.

20 *With what amounted*   Hoffmann with Dukas, *Einstein: Creator and Rebel*, p. 36.   *Friedrich Adler was another friend*   Anton Reiser, *Albert Einstein: A Biographical Portrait* (New York: Albert & Charles Boni, 1930), pp. 50–51. The author, whose real name was Rudolf Kayser, was Einstein's stepson-in-law.   *Thanking Stern for those therapeutic visits*   Clark, *Einstein: The Life and Times*, p. 44.   *"Little girl, small and fine"*   Albert Einstein, August 1899, Einstein Archives.

21 *Puzzled and disappointed*   Albert Einstein to Mileva Maric, August 1889, in Stachel, et al., *Collected Papers of Albert Einstein*, vol. 1, no. 50.   *Nevertheless, Albert was not too disconsolate*   Albert Einstein to Julia Niggli, August 6, 1899, ibid., no. 51.

22 *Mileva would gladly have changed places*   Mileva Maric to Albert Einstein, after August 10 and before September 10, 1899, ibid., no. 53.   *"high castle Peace of Mind"*   Albert Einstein to Mileva Maric, September 28, 1899, ibid., no. 57.   *as if his sanity were at stake*   Ibid.   *On his return to the Paradise Hotel*   Albert Einstein to Pauline Winteler, September 11, 1899, ibid., no. 56.   *Although he believed*   Clark, *Einstein: The Life and Times*, p. 415.

23 *He once overheard a woman student*   Roger Highfield and Paul Carter, *The Private Lives of Albert Einstein* (London: Faber and Faber, 1993), pp. 39, 40.   *Afterward, a detective*   The Municipal Police Detective's Report, July 4, 1900, in Stachel, et al., *Collected Papers of Albert Einstein*, vol. 1, no. 66.   *Five students took the exam*   Adolf Hurwitz to Hermann Bleuler, President of the School Council, July 27, 1900, ibid., no. 67.   *A few days later*   Mileva Maric to Helene Kaufler, Summer 1900, ibid., no. 64.

24 *Trying to divert the blow*   Ibid.

Chapter 4: Marriage Plans

25 *he described its impact to Mileva*   Albert Einstein to Mileva Maric, July 1900, in Stachel, et al., *Collected Papers of Albert Einstein*, vol. 1, no. 68.   *"cannot gain entrance"*   Ibid. The source of this slander might have been the Wintelers, who believed that a scarlet hussy had stolen Albert from Marie; Maja, who was now part of the Winteler household and who eventually married a Winteler; or Albert's relatives and acquaintances in Zurich.   *his impassioned message*   Ibid.

26 *She admitted to being terribly afraid*   Ibid.   *Pauline soon resumed her tirades*   Ibid.   *He assured Mileva that his studies were no substitute for her presence*   Albert Einstein to Mileva Maric, August 1900, ibid., no. 71.

27 *Too preoccupied with his business*   Albert Einstein to Mileva Maric, August 6, 1900, ibid., no. 70.   *When rational arguments*   Albert Einstein to Mileva Maric, August 30 or September 6, 1900, ibid., no. 74.   *Albert won a small personal victory*   Albert Einstein to Mileva Maric, September 13, 1900, ibid., no. 75.

28 *Describing her anguish*   Mileva Maric to Helene Savic, December 11, 1900, ibid., no. 83.   *The theory was disturbing*   Max Planck, *Scientific Autobiography and Other Papers* (New

York: Philosophical Library, 1949); Max Planck, address before the Berlin Physical Society, December 14, 1900; Clark, *Einstein: The Life and Times*, pp. 66, 67.

29 *"The quantum theory was his demon"* Pais, *'Subtle is the Lord'*, p. 412. *They had taken advantage of the recent snowfalls* Mileva Maric to Helene Savic, Spring 1901, in Stachel, et al., *Collected Papers of Albert Einstein*, vol. 1, no. 87.

Chapter 5: Seeking a Position

30 *"mistrusted the unworldly, dreamy young scholar"* Reiser, *Einstein: A Biographical Portrait*, p. 65. *Einstein's medical exam* Military Service Book, March 13, 1901, in Stachel, et al., *Collected Papers of Albert Einstein*, vol. 1, no. 97. *Told of this report* Thomas Bucky, interview by the author, tape recording, September 7, 1993.

31 *"You're a clever fellow, Einstein"* Hoffmann with Dukas, *Einstein: Creator and Rebel*, p. 32. *"What we learn up to age twenty"* Charles-Noel Martin, *The Universe of Science* (New York: Hill & Wang, 1963), p. 34. *"My sweetheart has a very wicked tongue"* Mileva Maric to Helene Savic, November–December, 1901, in Stachel, et al., *Collected Papers of Albert Einstein*, vol. 1, no. 125.

32 *He concluded his appeal* Albert Einstein to Wilhelm Ostwald, March 19, 1901, ibid., no. 92. *The next month Einstein applied* Albert Einstein to Heike Kamerlingh Onnes, April 12, 1901, in Clark, *Einstein: The Life and Times*, p. 43. *"burning with desire"* Albert Einstein to Mileva Maric, March 27, 1901, in Stachel, et al., *Collected Papers of Albert Einstein*, vol. 1, no. 94. *Besso called Einstein an eagle* Seelig, *Albert Einstein*, p. 71. *Albert assured her* Albert Einstein to Mileva Maric, March 27, 1901, in Stachel, et al., *Collected Papers of Albert Einstein*, vol. 1, no. 94.

33 *"If they are roses"* Aylesa Forsee, *Albert Einstein: Theoretical Physicist* (New York: Macmillan, 1963), p. 32. *Einstein's view of Besso* Albert Einstein to Mileva Maric, April 4, 1901, in Stachel, et al., *Collected Papers of Albert Einstein*, vol. 1, no. 96. *Albert informed Mileva* Ibid. *For a while he hoped* Albert Einstein to Mileva Maric, April 10, 1901, ibid., no. 97. *Hermann demonstrated his loving concern* Hermann Einstein to Wilhelm Ostwald, April 13, 1901, in Hoffmann with Dukas, *Einstein: Creator and Rebel*, p. 33. *Albert immediately replied* Albert Einstein to Marcel Grossmann, April 14, 1901, in Seelig, *Albert Einstein*, pp. 52, 53.

34 *they rode in a horse-drawn sledge* Mileva Maric to Helene Savic, May 1901, in Stachel, et al., *Collected Papers of Albert Einstein*, vol. 1, no. 109. *"I am beside myself with joy"* Albert Einstein to Alfred Stern, May 3, 1901, ibid., no. 104.

Chapter 6: The Schoolteacher

35 *Winterthur was even better than he had expected* Albert Einstein to Mileva Maric, May 9, 1901, in Stachel, et al., *Collected Papers of Albert Einstein*, vol. 1, no. 106. *On one notable occasion* Seelig, *Albert Einstein*, p. 50. *Einstein then wrote a paper on the subject* Albert Einstein to Marcel Grossmann, September 1901, in Stachel, et al., *Collected Papers of Albert Einstein*, vol. 1, no. 122. *"I hold with Schopenhauer"* Abraham Pais, *Einstein Lived Here* (New York: Oxford University Press, 1994), p. 181. *A personal questionnaire* "Who Are You?" Questionnaire completed by Einstein, from Smithsonian Institution Libraries.

36 *"As regards science"* Albert Einstein to Marcel Grossmann, April 14, 1901, in Pais, *'Subtle is the Lord'*, p. 57. *"How delightful it was last time"* Albert Einstein to Mileva Maric, May 28, 1901, in Stachel, et al., *Collected Papers of Albert Einstein*, vol. 1, no. 111. *What caused Maja's disaffection?* Ibid.

37 *one of his letters to Mileva* Ibid. *a letter he had written to Paul Drude* Albert Einstein to Mileva Maric, July 7, 1901, ibid., no. 114. *With that decision made* Albert Einstein to Jost Winteler, July 8, 1901, ibid., no. 115. *Meanwhile, Einstein had mailed the letter to Drude*

Einstein to Maric, July 7, 1901, ibid., no. 114. *he again reassured Mileva* Ibid. *At the Paradise Hotel* Albert Einstein to Mileva Maric, July 22, 1901, ibid., no. 119.

38 *How ironic, she replied* Mileva Maric to Albert Einstein, July 31, 1901, ibid., no. 121. *Maja's attitude had changed* Ibid. *Mileva got bad news* Highfield and Carter, *Private Lives*, p. 80.

Chapter 7: Expectant Father

39 *The latter was in Schaffhausen* Clark, *Einstein: The Life and Times*, p. 44. *When a few days later* Albert Einstein to Marcel Grossmann, September 6, 1901, in Stachel, et al., *Collected Papers of Albert Einstein*, vol. 1, no. 122. *"Although such a position is not ideal"* Ibid.

40 *she mentioned "Lieserl" for the first time* Mileva Maric to Albert Einstein, November 13, 1901, ibid., no. 124. *Once, when Albert* Albert Einstein to Mileva Maric, November 28, 1901, ibid., no. 126. *In mid-December* Albert Einstein to Mileva Maric, December 12, 1901, ibid., no. 127.

41 *"unspeakably happy"* Albert Einstein to Mileva Maric, December 17, 1901, ibid., no. 128. *He seemed in an upbeat mood* Ibid. *the school's owner and head teacher, Dr. Jakob Nuesch* Seelig, *Albert Einstein*, p. 51. *By skimping on food* Einstein to Maric, December 12, 1901 in Stachel, et al., *Collected Papers of Albert Einstein*, vol. 1, no. 127. *Optimistically predicting to Mileva* Ibid. *With the approaching prospect* Einstein to Maric, December 17, 1901, ibid., no. 128.

42 *The day Einstein applied* Albert Einstein to Mileva Maric, December 19, 1901, ibid., no. 130. *After his meeting with Kleiner* Albert Einstein to Mileva Maric, December 28, 1901, ibid., no. 131. *"I wouldn't have thought it possible"* Mileva Maric to Helene Savic, Winter 1901, ibid., no. 125.

43 *As Einstein had anticipated* Albert Einstein to Conrad Habicht, February 4, 1902, ibid., no. 133. *he sent a jaunty note* Ibid. *Einstein immediately replied* Albert Einstein to Mileva Maric, February 4, 1902, ibid., no. 134.

Chapter 8: Private Lessons

44 *Private lessons* Einstein's advertisement, February 5, 1902, in Stachel, et al., *Collected Papers of Albert Einstein*, vol. 1, no. 135. *First to bite* Albert Einstein, *Letters to Solovine*, introd. Maurice Solovine, trans. Wade Baskin (New York: Philosophical Library, 1987), p. 6.

45 *"Man can do what he wills"* Pais, *Einstein Lived Here*, p. 132. *"mercifully mitigates the sense of responsibility"* Einstein, *As I See It*, p. 2. *Solovine's first and lasting impression* Maurice Solovine, interview by Otto Nathan, August 19, 1957. This interview took place two years after Einstein's death. *Einstein had withdrawn his thesis* John Stachel, interview by the author, September 28, 1995. *They kicked off with Pearson* Einstein, *Letters to Solovine*, p. 8.

46 *"There is no inductive method"* W. I. B. Beveridge, *The Art of Scientific Investigation* (New York: Vintage, 1950), p. 77. *"Is not philosophy as if written in honey?"* Ilse Rosenthal-Schneider, *Reality and Scientific Truth* (Detroit: Wayne State University Press, 1980), p. 90. *"representative of the English Enlightenment"* Frank, *Life and Times*, p. 52. *He joined Frosch at the lecture* Albert Einstein to Mileva Maric, February 8, 1902, in Stachel, et al., *Collected Papers of Albert Einstein*, vol. 1, no. 136. *the book they had both recently read* Ibid. *"a violent attack on human consciousness"* Mileva Maric to Albert Einstein, November 13, 1901, ibid., no. 124.

47 *He assured her* Albert Einstein to Mileva Maric, February 1902, ibid., no. 137. *Dismissing this chance acquaintance* Ibid. *"We strongly oppose the liaison"* Pauline Einstein to Pauline Winteler, February 20, 1902, ibid., no. 138.

48 *They indicate that the child* Highfield and Carter, *Private Lives*, p. 89. *had been frustrated by Otto Nathan* "Nathan . . . seems particularly strict in his attitude towards personal material, especially involving family matters. A case in point was a plan in the late 1950s by Einstein's son Hans Albert to publish letters in his possession from Einstein to his first wife and their children. The estate asked to see the material before publication to insure that there was no invasion of privacy. Einstein's son declined to do this and publication was blocked." John Walsh, "Waiting for the Einstein Papers," *Science* 213 (July 17, 1981): p. 309. *The author questioned Schulmann about Lieserl* Robert Schulmann, interviews by the author, tape recordings, September 30, 1988, October 4, 1988, and October 16, 1991.

Chapter 9: The Patent Office

51 *Albert came close to starving* Feuer, *Generations of Science*, p. 17. *Talmey thought he hit* Max Talmey, *The Relativity Theory Simplified and the Formative Period of Its Inventor* (New York: Falcon Press, 1932), p. 167.

52 *Easiest to handle* Frank, *Life and Times*, pp. 23, 24. *Instead of bemoaning* Ibid., p. 23. *"a worldly monastery"* Seelig, *Albert Einstein*, p. 56. *Some wit described Haller* Clark, *Einstein: The Life and Times*, p. 45. *According to a new friend* B. Kuznetsov, *Einstein*, trans. V. Talmy (Moscow: Progress Publishers, 1965), p. 40.

53 *"the discussion of the previous evening"* Einstein, *Letters to Solovine*, p. 90. *In Poincaré's groundbreaking book* Ibid. *She is last mentioned* Albert Einstein to a friend, September 4, 1903, Einstein Archives. *Albert wanted to stay* Pais, *'Subtle is the Lord'*, p. 47. *"Many years later"* Hoffmann with Dukas, *Einstein: Creator and Rebel*, p. 39. *Maja's big regret* Winteler-Einstein, "Albert Einstein," p. xvii.

Chapter 10: The Olympia Academy

54 *"Dazed" and "overwhelmed"* Hoffmann with Dukas, *Einstein: Creator and Rebel*, p. 100. *He completed a fourth research paper* Pais, *'Subtle is the Lord'*, p. 67.

55 *"Our means were frugal"* Einstein, *Letters to Solovine*, p. 8. *"when we ran our happy 'Academy'"* Albert Einstein to Maurice Solovine, November 25, 1948, ibid., p. 107. *The meetings were less than delightful* Einstein, *Letters to Solovine*, p. 13; Highfield and Carter, *Private Lives*, p. 98. *"Friends had noticed"* Peter Michelmore, *Einstein: Profile of the Man* (New York: Dodd, Mead, 1962), p. 42. *"a free-thinker"* Frank, *Life and Times*, p. 23.

56 *"did not possess to any great degree"* Ibid. *"One can really quarrel"* Paul Arthur Schilpp, ed., *Albert Einstein Philosopher-Scientist* (La Salle, Illinois, Open Court, 1995), p. 688. *One day Einstein ended their arguments* Einstein, *Letters to Solovine*, p. 8. *Some days later* Ibid., p. 11. *"When I have no special problem"* Helen Dukas with Banesh Hoffmann, *Albert Einstein: The Human Side: New Glimpses from His Archives* (Princeton, N.J.: Princeton University Press, 1979), p. 17.

57 *But they were all music lovers* Einstein, *Letters to Solovine*, pp. 11–13. *"what stamped our Academy"* Ibid., p. 11.

58 *"with pleasure and the rest of my feelings"* Albert Einstein to Conrad Habicht, April 14, 1904, in *The Collected Papers of Albert Einstein, Volume Five: The Swiss Years: Correspondence, 1902–1914*, ed. Martin J. Klein, et al., trans. Anna Beck (Princeton, N.J.: Princeton University Press, 1993), no. 18. *"There is an extraordinary similarity"* Albert Einstein to Marcel Grossmann, April 6, 1904, ibid., no. 17. *Four months later* Clark, *Einstein: The Life and Times*, p. 48. *Albert's method* David Reichinstein, *Albert Einstein: A Picture of His Life and His Conception of the World* (Prague: Stella Publishing House, 1934), p. 25.

Chapter 11: The Special Theory of Relativity

60 *a state of "psychic tension"* Alexander Moszkowski, *Einstein the Searcher: His Work Explained from Dialogues with Einstein* (Berlin: Fontane, 1921), p. 4. *One balmy day in the*

early spring   Albert Einstein, "How I Created the Theory of Relativity," *Physics Today* 35, no. 8 (August 1982): pp. 45–47.   *never discover "the true laws"*   Schilpp, *Albert Einstein: Philosopher-Scientist.*

61 *"a storm broke loose in my mind"*   Reiser, *Einstein: A Biographical Portrait.*   *"Einstein said his basic discovery"*   Banesh Hoffmann, interview by the author, tape recording, October 29, 1982.   *" 'Ideas come from God' "*   Ibid. Einstein was more circumspect about the origin of his theory when interviewed by Robert S. Shankland, a physics professor at the Case Institute of Technology, Cleveland, Ohio, in 1950, saying that in physics the solution often comes "by indirect means." Robert S. Shankland, "Conversations with Albert Einstein," *American Journal of Physics* 31 (1963): pp. 37–47.   *he greeted Besso with a casual "Thank you"*   During a talk Einstein gave at Kyoto University, December 14, 1922, quoted in Einstein, "How I Created the Theory of Relativity," p. 46.   *"The 'thing' into which Einstein had sudden insight"*   Stanley Goldberg, letter to the author, February 17, 1994. Goldberg, a physicist, is the author of *Understanding Relativity: Origins and Impact of a Scientific Revolution* (Boston: Berkhaeuser, 1984).   *"The solution came to me"*   Friedrich Herneck, *Einstein privat* (Berlin: Buchverlag der Morgen, 1978), p. 349.   *Einstein gave great credit*   Albert Einstein to Moritz Schlick, December 14, 1915, in Klein, et al., *Collected Papers of Albert Einstein*, vol. 5, p. xxiv.

62 *"Einstein was probably the first"*   George Gamow, *Thirty Years That Shook Physics: The Story of Quantum Theory* (New York: Doubleday/Anchor, 1966), p. 106.   *Holton on Michelson's and Faraday's Contributions*   Gerald Holton, interview by the author, tape recording, October 19, 1995.

63 *a bantering letter*   Seelig, *Albert Einstein*, p. 74.   *"torn from the metal"*   Albert Einstein and Leopold Infeld, *The Evolution of Physics* (New York: Simon and Schuster, 1942), p. 273.   *Because he viewed his quantum hypothesis*   I. Bernard Cohen, *Revolution in Science* (Cambridge, Mass.: Harvard University Press, 1985), p. 438.   *"Crudeness in style and slips"*   Clark, *Einstein: His Life and Times*, p. 49.   *"A precise determination of molecules"*   Albert Einstein to Jean Perrin, November 11, 1909, Einstein Archives.

64 *Experiments by French physicist Jean Perrin*   Jeremy Bernstein, *Einstein* (New York: Viking Press, 1974), p. 185.

65 *Having demolished Newton's idea*   Lincoln Barnett, *The Universe and Dr. Einstein* (New York: New American Library, 1952), p. 52.

66 *Trying to account for Michelson's failure*   Alfred M. Bork, "The Fitzgerald Contraction," *Isis*, no. 57 (1966): pp. 199–207.   *"exceedingly small"*   Arthur Eddington, *The Nature of the Physical World* (New York: Macmillan, 1931), p. 5.   *"It explains why all observers"*   Barnett, *Universe and Dr. Einstein*, p. 62.   *Eventually he agreed to visit her parents*   Albert Einstein to Mileva Maric, December 17, 1901, in Stachel, et al., *Collected Papers of Albert Einstein*, vol. 1, no. 128.

67 *He even sought some response*   Margarete von Uexküll, "Erinnerungen an Einstein," *Frankfurter Allegemeine Zeitung*, March 10, 1956.   *"Relativity includes"*   Pais, 'Subtle is the Lord', p. 15.   *"failed to take the crucial step"*   Hoffmann with Dukas, *Einstein: Creator and Rebel*, pp. 68, 78.   *"hostile [to the theory of relativity]"*   Francoise Giroud, *Marie Curie: A Life*, trans. Lydia Davis (New York: Holmes & Meier, 1986), p. 170.   *"that the relativity principle"*   Clark, *Einstein: The Life and Times*, pp. 98–99.

68 *It appeared in the November 1905 issue*   *The New Encyclopaedia Britannica*, vol. 18, *Macropaedia*, 15th ed., p. 115.   *As his sister, Maja, recalled*   Pais, 'Subtle is the Lord', pp. 149–150.

Chapter 12: "The Happiest Thought of My Life"

69 *"venerable federal ink shitter"*   Highfield and Carter, *Private Lives*, p. 119.   *arranged to meet Einstein*   Max von Laue, letter to Carl Seelig, March 13, 1952, in Clark, *Einstein: The Life and Times*, p. 657.

70 *Maja had devastating news to tell*   Highfield and Carter, *Private Lives*, pp. 23, 24.   *Asked*

once where Seelig, *Albert Einstein*, p. 154. *To his dismay* Clark, *Einstein: The Life and Times*, p. 114. *Einstein blamed his rejection on ignorance* Ibid. In the early years, even Max Born, who became an expert on the subject, called relativity "so new and revolutionary that an effort was needed to assimilate it, and not everybody was willing to do so"—including, initially, Born himself. Cohen, *Revolution in Science*, pp. 410–411.

71 *"If every gram of material"* Albert Einstein, *Ideas and Opinions*, trans. Sonja Bargmann (New York: Crown, 1982), p. 340. *"Imagine the audacity"* Hoffmann with Dukas, *Einstein: Creator and Rebel*, pp. 81, 82. *At home he dealt* G. J. Whitrow, *Einstein: The Man and His Achievement* (New York: Dover, 1967), p. 19.

72 *"the first step"* Albert Einstein, lecture at Glasgow University, June 20, 1933. *Planck's reaction* Cohen, *Revolution in Science*, p. 415. *enthusiastically supported Einstein* Barnett, *Universe and Dr. Einstein*, p. 76. *"Even my conversations with Besso"* Albert Einstein to Maurice Solovine, in Einstein, *Letters to Solovine*, p. 22.

73 *"The other candidate"* Ronald Florence, *Fritz: The Story of a Political Assassination* (New York: Dial, 1971), p. 44. *"If it is possible to obtain"* Frank, *Life and Times*, p. 75.

74 *"These expressions of our colleague Kleiner"* Pais, *'Subtle is the Lord'*, pp. 185, 186. *"The festivities ended in the Hotel National"* Seelig, *Albert Einstein*, p. 93.

75 *"Thus Einstein must be considered"* Abraham Pais, *Inward Bound* (New York: Oxford University Press, 1986), p. 248. *"had already proceeded beyond General Relativity"* Max Born, *Physics in My Generation* (Elmsford, N.Y.: Pergamon Press, 1955), p. 197. *"The more I speak with Einstein"* Friedrich Adler to Viktor Adler, October 1910, in Clark, *The Life and Times*, p. 132. *She was incensed by* Highfield and Carter, *Private Lives*, pp. 124–125. *recommended Einstein for the Nobel Prize* Pais, *'Subtle is the Lord'*, p. 506; *Nobel Foundation Calendar 1975–1976*, p. 56.

76 *Tanner's recollections of Einstein* Seelig, *Albert Einstein*, pp. 101–103. *Einstein's cavalier attitude toward math* Ibid., p. 107. *He spoke "in the same way to everybody"* Frank, *Life and Times*, p. 76.

77 *"had difficulty in following his arguments"* Carl Jung to Carl Seelig, February 25, 1953, in *C. G. Jung Letters*, eds. Gerhard Adler and Aniela Jaffé (Princeton, N.J.: Princeton University Press, 1975), pp. 108, 109. *"Although I am no mathematician"* Carl Jung to Pascual Jordan, November 10, 1934, ibid., pp. 176–178. *"After my statements about his conduct"* Alfred Kleiner to anonymous, January 18, 1911, in Pais, *'Subtle is the Lord'*, p. 193.

78 *"a warm and cheerful atmosphere"* Martin J. Klein, *Paul Ehrenfest: The Making of a Theoretical Physicist* (New York: American Elsevier, 1970), p. 303. *couldn't remember knowing any Einstein* von Uexküll, "Erinnerungen an Einstein."

Chapter 13: To Prague and Back

79 *"infinitely much paperwork"* Albert Einstein to A. Stern, February 2, 1912, in Pais, *'Subtle is the Lord'*, p. 193. *his inaugural lecture* Seelig, *Albert Einstein*, p. 121. *"the dreamy look in his eyes"* Frank, *Life and Times*, p. 79. *Einstein was ambivalent about Prague* Michelmore, *Einstein*, p. 55; Frank, *Life and Times*, pp. 80, 81; Pais, *'Subtle is the Lord'*, p. 193.

80 *Even the brown water* Michelmore, *Einstein*, pp. 54, 55. *Nohel told him* Pais, *'Subtle is the Lord,'* pp. 485, 486; Frank, *Life and Times*, pp. 82, 83. *Despite his disdain* Frank, *Life and Times*, p. 82. *Einstein was intrigued* Ibid., p. 83. *In his paper* Albert Einstein, "Über den Einfluss der Schwerkraft auf die Ausbreitung des Lichtes," *Annalen der Physik* 35, ser. 4 (1911): pp. 898–908.

81 *the theory was "so well established"* Cohen, *Revolution in Science*, p. 412. *Kraus was so disturbed* Frank, *Life and Times*, p. 173. *Princeton's Professor W. F. Magie agreed* Cohen, *Revolution in Science*, p. 413. *She ruefully remarked* Mileva Einstein-Maric to Albert Einstein, October 4, 1911, in Klein, et al., *Collected Papers of Albert Einstein*, vol. 5, no. 290.

82 *headlined in popular newspapers* Marcia Bartusiak, "At the Birth of Modern Physics,"

review of *Marie Curie: A Life*, by Susan Quinn, *Washington Post Book World*, March 19, 1995, pp. 1, 14. Bartusiak described the cause célèbre: "France's tabloid press had a heyday, and for good reason. The story involved a love nest, stolen love letters, murderous threats against Marie by Langevin's insanely jealous wife, and a duel between Langevin and the scurrilous journalist who first exposed [the] affair." Apparently there were no fatalities. *Curie was "sparklingly intelligent"* Michelmore, *Einstein*, p. 58. *"[al]though Einstein had already published"* Earl of Birkenhead, *The Professor and the Prime Minister: The Official Life of Professor F. A. Lindemann, Viscount Cherwell* (Boston: Houghton Mifflin, 1962), p. 165. *"I got on very well with Einstein"* Frederick Lindemann to his father, November 4, 1911, ibid., p. 43. *threw him a lifeline* Seelig, *Albert Einstein*, p. 130. *Einstein was waiting for him* Klein, *Paul Ehrenfest*, p. 176.

83 *"Within a few hours"* Hoffmann with Dukas, *Einstein: Creator and Rebel*, p. 96. *"Yes, we will be friends"* Klein, *Paul Ehrenfest*, pp. 176, 177. *Hans Albert said he was unaware of it* Michelmore, *Einstein*, p. 59. *"loves going to school"* Mileva Einstein to Michele Besso, March 26, 1911, Einstein Archives. *Albert confided to Elsa* Albert Einstein to Elsa Löwenthall, April 30, 1912, in Klein, et al., *Collected Papers of Albert Einstein*, vol. 5, no. 389.

84 *"I cannot tell you how sorry I am for you"* Albert Einstein to Elsa Löwenthall, May 7, 1912, ibid., no. 391. *he was in full retreat* Albert Einstein to Elsa Löwenthall, May 21, 1912, ibid., no. 399. *grounds of a mental asylum* Frank, *Life and Times*, p. 98. *he pleaded with Grossmann* Pais, *'Subtle is the Lord'*, p. 212. *One evening, after attending a colloquium* Klein, *Paul Ehrenfest*, pp. 294, 295.

85 *"A day without Einstein!"* Ehrenfest's diary, June 25, 1913, in Klein, *Paul Ehrenfest*, volume 1, p. 295. *"I occupy myself exclusively"* Albert Einstein to Arnold Sommerfeld, October 29, 1912, in Pais, *'Subtle is the Lord'*, p. 216. *Ostwald again recommended Einstein* Pais, *'Subtle is the Lord'*, pp. 503, 506; *Nobel Foundation Calendar 1975–1976*, p. 58. *Einstein wrote back twice* Albert Einstein to Elsa Löwenthall, March 14 and 23, 1913, in Klein, et al., *Collected Papers of Albert Einstein*, vol. 5, nos. 432, 434. *"the only Frenchman"* Rosalynd Pflaum, *Grand Obsession: Madame Curie and Her World* (Garden City, N.Y.: Doubleday, 1989), p. 188. *"As an older friend"* Ernst Straus to Abraham Pais, October 1979, in Pais, *'Subtle is the Lord'*, p. 239.

86 *eight of them lived in Berlin* Kuznetsov, *Einstein*, p. 194. *"There was a lot of shoptalk"* Pflaum, *Grand Obsession*, pp. 188–189. *Einstein remarked in a letter to Elsa* Klein, et al., *Collected Papers of Albert Einstein*, vol. 5, no., 554. *That fall a Dutch physicist* Pais, *'Subtle is the Lord*,' p. 506; *Nobel Foundation Calendar 1975–1976*, pp. 58–59. *"Covariant means," explains Stanley Goldberg* Stanley Goldberg, interview by the author, tape recording, May 13, 1995.

87 *"an enormous achievement"* Pais, *Inward Bound*, p. 208. *"I now have someone"* Albert Einstein to Elsa Löwenthall, October 10, 1913, in Klein, et al., *Collected Papers of Albert Einstein*, vol. 5, no. 476. *Einstein agreed to go to Berlin* Pais, *'Subtle is the Lord'*, p. 240. *"My wife howls unceasingly about Berlin"* Albert Einstein to Elsa Löwenthall, December 1913, in Highfield and Carter, *Private Lives*, pp. 164, 165. *One winter evening* Seelig, *Albert Einstein*, p. 140.

Chapter 14: The War to End All Wars

89 *German propagandists responded* Otto Nathan and Heinz Norden, eds., *Einstein on Peace* (New York: Simon and Schuster, 1961), p. 3. *produced a pro-peace Countermanifesto* Ibid., pp. 4–6.

90 *the anonymous German physicist* Albert Einstein to Paul Ehrenfest, August 19, 1914, in Klein, *Paul Ehrenfest*, pp. 300, 301. *his good friend Max von Laue* Pais, *'Subtle is the Lord'*, p. 507; *Nobel Foundation Calendar 1975–1976*, p. 59. *By December, Mileva and the boys* Albert Einstein to Mileva Maric, December 12, 1914, in Klein, et al., *Collected Papers of Albert*

Einstein, vol. 5; Highfield and Carter, *Private Lives*, p. 172. *"speaks French rather haltingly"* Romain Rolland, *Journal des Années de Guerre 1914–1919* (Paris: Albin Michel, 1952), pp. 510–511.

91 *He called a truce* Albert Einstein to Mileva Maric, November 15, 1915, in Klein, et al., *Collected Papers of Albert Einstein*, vol. 5; Highfield and Carter, *Private Lives*, p. 174. *Margot found him boiling an egg in a saucepan* Sayen, *Einstein in America*, p. 36. *"the supreme intellectual achievement"* Paul Davies, *The Edge of Infinity* (New York: Simon and Schuster, 1981), p. 176. *"Space tells matter how to move"* John Archibald Wheeler, "Einstein and Other Seekers of the Larger View" (speech, Science Policy Foundation, March 1979).

92 *"What were the seeds"* Hoffmann with Dukas, *Einstein: Creator and Rebel*, pp. 122–124. *Einstein's theory explains* Tom Siegfried, "Think of gravity as geometrical," *Miami Herald*, February 4, 1987, p. 133.

93 *Mileva had a physical and mental breakdown* Highfield and Carter, *Private Lives*, p. 177. *Valuing Zangger's friendship* Ibid., p. 156. *Besso, however, was very sympathetic* Ibid., p. 177. *Einstein rebuffed his attempts* Ibid. *Einstein's name came before* Pais, *'Subtle is the Lord'*, p. 507. *They sat in Lorentz's cheerful study* Klein, *Paul Ehrenfest*, pp. 303, 304. *his idealistic friend Friedrich Adler* Ronald Florence, *Fritz*, pp. 1–4, 199–201; Seelig, *Albert Einstein*, pp. 95–97; Frank, *Life and Times*, pp. 174, 175. *More bad news* Michelmore, *Einstein*, p. 71.

94 *"his arguments were wrong"* Frank, *Life and Times*, pp. 174, 175. *Although Einstein considered Adler* Feuer, *Generations of Science*, pp. 22–24. *"I order that unrestricted submarine war"* John Dos Passos, *Mr. Wilson's War* (Garden City, N.Y.: Doubleday, 1962), p. 194. *he collapsed in agonizing pain* Michelmore, *Einstein*, p. 71.

95 *At Elsa's insistence* Clark, *Einstein: The Life and Times*, p. 191–194; Pais, *'Subtle is the Lord'*, p. 300. *three persuasive nominations* Pais, *'Subtle is the Lord'*, p. 507. *correspondence with de Sitter* Carla Kahn and Franz Kahn, "Letters from Einstein to de Sitter on the Nature of the Universe," *Nature* 257 (October 9, 1975): pp. 451–454. *de Sitter acted as a go-between* A. Vibert Douglas, *The Life of Arthur Stanley Eddington* (London: Nelson, 1956), p. 38. *was called a masterpiece* Ibid., p. 39.

96 *Einstein's concern over Eduard's mental health* Highfield and Carter, *Private Lives*, pp. 182, 183. *Elsa worried about her reputation* Michelmore, *Einstein*, p. 79. *One inducement was Albert's promise* Pais, *'Subtle is the Lord'*, p. 300. *Charles Glover Barkla* Pais, *'Subtle is the Lord'*, p. 507; *Nobel Foundation Calendar 1975–1976*, p. 59.

97 *Born was in bed* Max Born, *My Life: Recollections of a Nobel Laureate* (New York: Scribner's, 1978), p. 184. *"Einstein was well known to be politically left-wing"* Ibid., p. 185. *"We left the palace"* Feuer, *Generations of Science*, p. 81.

98 *Einstein "lived in the midst of beautiful furniture"* Frank, *Life and Times*, p. 124. *Albert began shivering* Michelmore, *Einstein*, p. 75. *"sitting in the tub"* Sayen, *Einstein in America*, p. 86.

99 *"You won't have to go home alone"* Douglas, *Life of Arthur Stanley Eddington*, p. 40. *"the plates are still being measured"* H. A. Lorentz to Albert Einstein, October 7, 1919, in Clark, *Einstein: The Life and Times*, p. 231. *"gave a final verdict confirming Einstein"* Margaret Wilson, *Ninth Astronomer Royal: The Life of Frank Watson Dyson* (Cambridge, Engl.: Cambridge University Press, 1951), p. 193. *also waiting for the results* Pais, *'Subtle is the Lord'*, p. 508; *Nobel Foundation Calendar 1975–1976*, pp. 59–60.

Chapter 15: In the Spotlight

100 *Crouch decided to skip the meeting* Meyer Berger, *The Story of The New York Times: 1851–1951* (New York: Simon and Schuster, 1951), pp. 251–252.

101 *"In each revolution of scientific thought"* Eddington, *Physical World*, p. 353. *Heaviside was one of the few* Martin Gardner, *Fads & Fallacies in the Name of Science* (New York:

Dover, 1957), p. 80.   *"Platitudes, my boy!"*   Bernard Falk, *Five Years Dead* (London: Book Club, 1938), p. 79.   *"The supposed astronomical proofs"*   "Prof. Poor Explains Theories," *The New York Times*, November 16, 1919, sect. 3, p. 8.

102 *"seized with something like intellectual panic"*   Ibid.   *See focused on Einstein's credentials*   Quoted in an anonymous report to the FBI director, February 10, 1950, FBI file.

103 *"as a rational physicist, Einstein is a fair violinist"*   Gardner, *Fads & Fallacies*, p. 86.   *a "technical analysis"*   Anonymous report, February 10, 1950, FBI file.   *"On the subject of relativity I see red"*   Ibid.   *refuted Einstein*   Ibid.   *"At a time when electricity"*   John J. O'Neill, *Prodigal Genius: The Life of Nikola Tesla* (New York: Ives Washburn, 1944), p. 4.   *Tesla rejected Einstein's view on gravity*   Ibid.

104 *"there is no logical path"*   *The Logic of Scientific Discovery* (New York: Harper & Row, 1968).   *The consensus among intellectuals*   Blanche Patch, *Thirty Years with G. B. S.* (New York: Dodd, Mead & Co., 1951), p. 234.

105 *"the Jew lacks understanding for the truth"*   William L. Shirer, *The Nightmare Years: 1930–1940* (Boston: Little, Brown & Co., 1984), p. 185.   *Lenard and other anti-Einstein Colleagues*   Ibid.   *"I'm going to cut the throat of that dirty Jew!"*   Clark, *Einstein: The Life and Times*, p. 293.   *One admiring student*   Rosenthal-Schneider, *Reality and Scientific Truth*, p. 90.   *"whenever the opportunity offered itself"*   Ibid.

106 *"Don't you agree"*   Ibid., p. 91.   *"Why are you so famous?"*   Seelig, *Albert Einstein*, p. 80.   *"I feel now something like a whore"*   Michelmore, *Einstein*, p. 91.   *Planck's "misfortune moves me"*   Albert Einstein to Max Born, December 9, 1919, in *Born–Einstein Letters*, no. 12, p. 18.

Chapter 16: Danger Signals

107 *"all this has diminished"*   Albert Einstein to Max Born, January 27, 1920, in *Born–Einstein Letters*, comm. Max Born, trans. Irene Born (New York: Walker, 1971), p. 21.

108 *not necessarily an expression of anti-Semitism*   "Einstein (Prof.), Albert — Comments on Interplanetary Communication," *The New York Times*, February 2, 1920, p. 24.   *the true nature of the noise*   "Lectures at Berlin University on Relativity," *The New York Times*, February 17, 1920, p. 3, and February 18, 1920, p. 10.   *"wept, like other men"*   Clark, *Einstein: The Life and Times*, p. 192.   *"a maniac of ferocious genius"*   Winston Churchill, *The Gathering Storm* (Boston: Houghton Mifflin, 1948), p. 11.   *Ehrenfest was coaching Einstein by mail*   Paul Ehrenfest to Albert Einstein, March 10, 1920, in Klein, *Paul Ehrenfest*, p. 317.

109 *"The infant mortality is appalling"*   Albert Einstein to Paul Ehrenfest, April 17, 1920, ibid.   *Yet he turned down*   Reichinstein, *Albert Einstein: A Picture of His Life*, pp. 136–137.   *"pampered and overestimated"*   Albert Einstein to Paul Ehrenfest, June 6, 1920, in Klein, *Paul Ehrenfest*, p. 323.   *"Don't be impatient with me"*   Paul Ehrenfest to Albert Einstein, August 16, 1920, ibid., p. 319.   *"burning with impatience"*   Paul Ehrenfest to Albert Einstein, September 2, 1920, ibid.

110 *"an extremely sensitive lad"*   Clark, *Einstein: The Life and Times*, p. 253.   *"If you are so concerned"*   Ibid.   *"Einstein has very serious opponents"*   Joachim von Elbe, *Witness to History* (Madison, Wisc.: The Max Kade Institute for German-American studies, 1988), pp. 80–81.

111 *relativity "had never been proved"*   Irving Wallace, *The Writing of One Novel* (New York: Simon and Schuster, 1969), pp. 18–19. For his book, Wallace interviewed Nobel Prize judges.   *He was also an incipient Nazi*   Frank, *Life and Times*, p. 163.   *"had a decidedly anti-Semitic complexion"*   *The New York Times*, September 26, 1920, sect. 9, p. 14.   *an angry article*   *Berliner Tageblatt*, August 27, 1920, p. 1.

112 *"I am awfully distressed"*   Paul Ehrenfest to Albert Einstein, August 28, 1920, in Klein, *Paul Ehrenfest*, p. 321.   *"My wife and I absolutely cannot believe"*   Paul Ehrenfest to Albert Einstein, September 2, 1920, ibid., pp. 321–322.   *insults made "repeatedly and publicly"*   Albert Einstein to Paul Ehrenfest, September 10, 1920, ibid., p. 323.

113 *"We must stress"* Clark, *Einstein: The Life and Times*, p. 259. "scarcely believable filth" Max Planck to Albert Einstein, September 5, 1920, in J. L. Heilbron, *Dilemmas of an Upright Man: Max Planck as Spokesman for German Science* (Berkeley: University of California Press, 1986), p. 117. *"We should not drive away such a man"* "Einstein in Trouble," *The New York Times*, August 30, 1920, editorial, p. 8.

114 *not among the Nobel Prize winners* Pais, *'Subtle is the Lord'*, p. 508; *Nobel Foundation Calendar 1975–1976*, p. 60. *series of unusually revealing interviews* These interviews were published in book form as *Einstein the Searcher: His work explained from Dialogues with Einstein*, by Alexander Moszkowski (Berlin: Fontane, 1921). *freewheeling, wide-ranging, off-the-wall questions* In 1953, Einstein wrote to biographer Carl Seelig, saying that had he realized that every one of his casual remarks would be recorded, he would have crept further into his shell.

116 *"the gutter press will get hold of it"* Hedi Born to Albert Einstein, October 7, 1920, in *Born–Einstein Letters*, p. 39. *followed with his own warning* Max Born to Albert Einstein, October 13, 1920, ibid., pp. 39–40.

117 *"Your wife is objectively right"* Albert Einstein to Max Born, October 11, 1920, ibid., p. 40.

Chapter 17: Einstein Discovers America

118 *Einstein's New Year's resolution* Albert Einstein to Max Born, January 30, 1921, in *Born–Einstein Letters*, p. 50. *"Einstein said no"* Count Harry Kessler, *The Diaries of a Cosmopolitan — 1918–1937* (London: Weidenfeld & Nicolson, 1971), p. 137–138.

119 *Philipp Frank, his successor* Frank, *Life and Times*, pp. 170–172. *"You haven't lost anything"* Ibid., pp. 173–174. *Einstein's next stop was Vienna* Felix Ehrenhaft, "My Experiences With Einstein" (Washington, D.C.: Smithsonian Institution Libraries), p. 5.

120 *Einstein heard from Chaim Weizmann* Chaim Weizmann, *Trial and Error: The Autobiography of Chaim Weizmann* (Philadelphia: Jewish Publication Society of America, 1949), p. 266. *"making me conscious of my Jewish soul"* Albert Einstein to Kurt Blumenfeld, March 25, 1955, in Pais, *'Subtle is the Lord'*, p. 476. Einstein died on April 18, 1955. *"Falling bodies are independent"* "Einstein Arrives in U.S., Explains Relativity," *Washington Post*, April 3, 1921, pp. 1, 13.

121 *"In a certain sense"* "Einstein Sees End of Time and Space," *The New York Times*, April 4, 1921, p. 5. *"It's like the Barnum circus!"* Antonina Vallentin, *The Drama of Albert Einstein*, trans. Moura Budberg (Garden City, N.Y.: Doubleday, 1954), p. 102. *"It's like a zoological garden"* mIbid.

122 *Einstein meets with reporters at the Waldorf-Astoria* "Einstein Arrives in U.S., Explains Relativity," *The New York Times*, pp. 1, 13. *Not everyone greeted them with joyous abandon* Arthur Mann, *La Guardia: A Fighter Against His Times, 1882–1933* (New York: Lippincott, 1959), p. 119; *New York Daily News*, April 6, 1921, pp. 1–2.

123 *"Einstein really did not want to talk to reporters"* William Laurence, *Heroes for Our Times: Albert Einstein* (Harrisburg, Pa.: Stackpole, 1968). *"my brothers and comrades"* Weizmann, *Trial and Error*, pp. 274–275; "Weizmann Pleads for Palestine Aid," *The New York Times*, April 13, 1921, p. 5.

124 *"please be careful with Einstein"* Kurt Blumenfeld to Chaim Weizmann, March 15, 1921, in Pais, *'Subtle is the Lord'*, p. 315. *"Your leader, Dr. Weizmann"* Sayen, *Einstein in America*, p. 48. *"the discoverer of a theory"* Frank, *Life and Times*, p. 183. *"There I was about to enter graduate work"* I. I. Rabi, interview by the author, tape recording, October 31, 1980. *his critics were applauded* "Einstein Wrong, Brush Indicates," *The New York Times*, April 23, 1921, pp. 1, 3. *"that Palestine was full of malaria"* Vera Weizmann and David Tutaev, *The Impossible Takes Longer: The Memoirs of Vera Weizmann* (New York: Harper & Row, 1967), pp. 102–103.

125 *took no part in the bitter arguments* Bernard Richards, Columbia University Oral History Project, pp. 205, 206, 210, 308. *Churchill's party visits Palestine* Martin Gilbert, *Churchill:*

A *Life* (New York: Henry Holt, 1991), p. 434; Walter H. Thompson, *Assignment: Churchill* (New York: Farrar, Straus and Young, 1955), pp. 29, 35. *might inflame the situation* Weizmann credited Lawrence with understanding both sides of the problem "if anyone can be said to have done so, and he did his utmost to interpret the spirit of one people, and to explain the aspirations of the other, believing that close cooperation between the two peoples was to their mutual advantage. I cherish his memory on personal grounds and remember with gratitude his help in furthering the cause of the Jewish people." A. W. Lawrence, ed. *T. E. Lawrence by His Friends: Chaim Weizmann* (New York: McGraw-Hill, 1963), p. 180. *"Personally, my heart is full of sympathy"* Martin Gilbert, *Winston Churchill: The Stricken World, 1916–1922* (Boston: Houghton Mifflin, 1975), pp. 570, 584. *"with a magnificent mixture"* British politician Richard Crossmann, quoted by Abba Eban in *Abba Eban: An Autobiography* (New York: Random House, 1977), p. 61.

126 *"when you drain the marshes"* Weizmann, *Trial and Error*, p. 274. *"We are reproached"* Ibid. *Thirty Jews and ten Arabs* Gilbert, *Churchill: The Stricken World*, p. 585.

127 *"I defy anybody"* Ibid. *"The theory of relativity"* A. A. Michelson, *Studies in Optics* (Chicago: University of Chicago Press, 1927), chap. 14. *"Oh yes, the snakes flew around my head"* Dorothy Michelson Livingston, *The Master of Light* (New York: Scribner's, 1973), p. 291. *"new Columbus of science"* Sayen, *Einstein in America*, pp. 49, 50; Forsee, *Einstein: Theoretical Physicist*, p. 85. *"Subtle is the Lord"* Pais, '*Subtle is the Lord*,' p. 113.

128 *"Let's run away somewhere"* Weizmann and Tutaev, *The Impossible Takes Longer*, pp. 113–114. *an Orthodox Jew asked for directions* Frank, *Life and Times*, p. 281. *Einstein's casual conversation* Leopold Infeld, *Quest: The Making of a Scientist* (New York: Doubleday Doran, 1971), pp. 260–293; Frank, *Life and Times*, p. 77.

129 *Albert was not drawn to intellectual women* Weizmann and Tutaev, *The Impossible Takes Longer*, pp. 102–103. *Edison took a dim view* Wyn Wachhorst, *Thomas Alva Edison: The American Myth* (Cambridge, Mass.: MIT Press, 1981), p. 147. *"I wouldn't give a penny"* Ibid. *B. Lord Buckley agreed* "Einstein Sees Boston; Fails on Edison Test," *The New York Times*, May 18, 1921, p. 18. *The test drove one young man* "Holyoke Youth, Crazed by Test, Asks Police Protection," *The New York Times*, May 17, 1921, p. 19. *Einstein's response to Edison's test questions* "Einstein Sees Boston: Fails on Edison Test," p. 18.

130 *Einstein illuminates his own ideas* "Einstein Tells Why He Can't Explain," *Boston Globe*, May 18, 1921, pp. 1, 6. *"the existence of the Jewish University"* "Jewish Doctors Raise $250,000 for College," *The New York Times*, May 22, 1921, p. 21. *a surprise visit from his boyhood mentor* Talmey, *The Relativity Theory Simplified*, p. 174. *Einstein's appearance in Cleveland* "Einstein and Weizmann Raise $200,000," *Cleveland Plain Dealer*, May 26, 1921, pp. 1, 8. *"it is necessary"* Ibid. *some $200,000 was raised* Ibid.

131 *roamed the aisles of five-and-dime stores* Dmitri Marianoff and Palm Wayne, *Einstein: An Intimate Study of a Great Man*, New York: Doubleday, Doran & Co., 1944), pp. 113, 114. *Einstein at a dinner party in London* Dudley Sommer, *Haldane of Cloan: His Life and Times* (London: Allen & Unwin, 1960), p. 382. *"If your theories are sound"* Clark, *Einstein: The Life and Times*, p. 276. *Einstein placed a wreath* Ibid., p. 277. *Einstein told Lady Haldane* Ibid., p. 279.

132 *"maintained so close a friendship"* "Dr. Einstein Found America Anti-German," *The New York Times*, July 2, 1921, p. 3. *"if he had been in the country longer"* "A Genius Makes a Mistake," *The New York Times*, July 4, 1921, editorial. *"called to account for everything"* Einstein, *Ideas and Opinions*, p. 15. *described American men as henpecked* Cyril Brown, "Einstein Declares Women Rule Here," *The New York Times*, July 8, 1921, p. 9. *reactions to Einstein's indictment* "Chicago Women Resent Einstein's Opinions," *The New York Times*, July 9, 1921, p. 7. *"Bugs in the coffee"* Reiser, *Einstein: A Biographical Portrait*, p. 172.

133 *Einstein's more cautious and considered opinions of the United States* *Berliner Tageblatt*, July 7, 1921. *On Thursday afternoons* Clark, *Einstein: The Life and Times*, p. 322.

134 *Salaman began to discuss* Ibid. *"he took nothing as certain truth"* Ibid., p. 324. *"the saying of Oxenstiern"* Ibid. *enjoyed science so sensuously as Einstein* Ibid. *"I hardly understood a word"* Eugene P. Wigner, as told to Andrew Szanton, *The Recollections of Eugene*

P. Wigner (New York: Plenum Press, 1992), pp. 72–73. *"If the reviewer presented a clear picture"* Ibid., pp. 96–97.

135 *"the gift of seeing"* Born, *My Life*, p. 167. *in a desperately lonely state* Infeld, *Quest*, p. 91. *he canceled the lecture* David Cassidy, *Uncertainty: The Life and Science of Werner Heisenberg* (New York: W. H. Freeman, 1992), p. 97. *approached by a student in a quandary* Spencer Weart and Gertrud Weiss Szilard, eds., *Leo Szilard: His Version of the Facts* (Cambridge, Mass.: MIT Press, 1978), p. 11.

136 *"Einstein really liked"* William Lanouette, interview by the author, tape recording, July 18, 1995.

Chapter 18: The Nobel Prize

137 *Kessler dined at the Einsteins'* Kessler, *Diaries*, pp. 155–157.

138 *"an idealist"* Neue Rundschau, 1922. *"Now we have it!"* Gordon A. Craig, *The Germans* (New York: Putnam's, 1982), p. 143. *Palestine was just a lot of sand* Technion Journal, April 1941. *"all too typical of many assimilated German Jews"* Weizmann, *Trial and Error*, p. 289. *"one of disconcerting youth"* Charles Nordmann, "Einstein in Paris," *L'Illustration*, April 15, 1922. *"eyes wandered with amused irony"* Hilaire Cuny, in *Einstein: Such As We Knew Him*, by Louis de Broglie, et al. (Edinburgh: Peebles Press, 1979), p. 196.

139 *"He keeps taking them from me"* Vallentin, *Drama of Albert Einstein*, pp. 27–28. *"in the middle of the war"* Cuny, in de Broglie, et al., *Einstein: Such As We Knew Him*, p. 196. *"This famous mathematician isn't at all austere"* Ibid., p. 197. *"Everyone had the impression"* Ibid. *"depended greatly on their political sympathies"* Frank, *Life and Times*, p. 197.

140 *"All the students of the world"* Clark, *Einstein: The Life and Times*, p. 290. *"Those days were unforgettable"* Albert Einstein to Maurice Solovine, April 20, 1922, in Einstein, *Letters to Solovine*, p. 55. *"Why does the reflection break down"* Vallentin, *Drama of Albert Einstein*, p. 132. *Rathenau's assassination* Otto Friedrich, *Before the Deluge: A Portrait of Berlin in the 1920s* (New York: Harper & Row, 1972), pp. 58–77; Count Harry Kessler, *Walter Rathenau: His Life and Work* (New York: Harcourt, Brace & Co., 1930), pp. 376–377.

141 *"fled from Germany"* The New York Times, July 5, 1922, sect. 2, p. 1. *"the silly advertisements"* Heilbron, *Dilemmas of an Upright Man*, p. 120. *"The trouble is"* Albert Einstein to Max Planck, July 6, 1922, in Heilbron, *Dilemmas of an Upright Man*, p. 120. *"looked at entirely objectively"* Planck to Wilhelm Wien, July 9, 1922, in Heilbron, *Dilemmas of an Upright Man*, p. 120. *"My feelings for Rathenau"* Pais, *Einstein Lived Here*, p. 158.

142 *"All the shady characters of the world"* Jehuda Reinharz, *Chaim Weizmann: The Making of a Statesman* (New York: Oxford University Press, 1993), p. 392. *"However, when I was told"* Werner Heisenberg, *Physics and Beyond: Encounters and Conversations* (New York: Harper & Row, 1971), p. 44. *"bitterly ashamed to share responsibility"* Clark, *Einstein: The Life and Times*, pp. 298–299. *Japanese government officials argued* "The Man Who Beat Time," *Newsweek* (August 9, 1971): p. 67.

143 *"No living being deserves"* Janos Plesch, *Janos: The Story of a Doctor* (London: Gollanez, 1947), p. 212. *looked somewhat askance* Hoffmann with Dukas, *Einstein: Creator and Rebel*, p. 150. *He loved the country so much* Ibid. *Wallace's investigation into why Einstein had been rejected by the Nobel Prize committee* Wallace, *One Novel*, pp. 14, 18, 23–25; Irving Wallace, letter to the author, March 10, 1976.

144 *enthusiastic support* Pais, *'Subtle is the Lord'*, p. 510; *Nobel Foundation Calendar 1975–1976*, p. 60. *finally got the 1922 prize* Ibid.

145 *"greatest day of my life"* Norman Bentwich and Helen Bentwich, *Mandate Memories: 1918–1948* (London: Hogarth Press, 1965), p. 89. *"We physicists understand"* Seelig, *Albert Einstein*, p. 171. *"Collect more money"* F. H. Kisch, *Palestine Diary* (London: Gollancz, 1938), p. 31. *"whistled his relativity tune"* Hoffmann with Dukas, *Einstein: Creator and Rebel*, p. 152. *"Let's enjoy everything"* Frank, *Life and Times*, p. 201. *"You can do what you like"* Hoffmann with Dukas, *Einstein: Creator and Rebel*, p. 152.

146 *"There appeared to be no action"* Frank, *Life and Times*, p. 154. *"the sleeves of his jacket"* Michelmore, *Einstein*, p. 122. *"while it was there"* Ibid., p. 124.
147 *"What more can a man want?"* Seelig, *Albert Einstein*, p. 188. *Einstein and Tatiana interrupted during their walk* Ibid., p. 189. *the socialite who invited himself to the Einsteins'* Cuny, in de Broglie, et al., *Einstein: Such As We Knew Him*, pp. 206, 207. *"important personality"* Ibid., 207.
148 *Planck's offer to Einstein* Max Planck to Paul Ehrenfest, November 30, 1923, in Heilbron, *Dilemmas of an Upright Man*, p. 121. Einstein was staying with Ehrenfest at the time the letter was written. *"that an electron exposed to radiation"* Albert Einstein to Max and Hedi Born, April 29, 1924, in *Born–Einstein Letters*, p. 82. *"he had a strong attachment to a younger woman"* Pais, *'Subtle is the Lord'*, p. 320.
149 *"the regime of frightfulness in Russia"* Isaac Don Levine, *Eyewitness to History* (New York: Hawthorn, 1973), p. 169. *the Bolsheviks were laughable, but not as bad as they had been painted* In 1920, Einstein had used much the same words to Max Born—although, as Born later wrote, Einstein had "believed like many others that the Bolshevik revolution would mean deliverance from the principal evils of our time, militarism, bureaucratic oppression and plutocracy." Einstein was soon disillusioned; even so, "Communist writers often represented him as a supporter, or at least a precursor of their doctrine [So did the FBI!] . . . Einstein would have found that laughable." *Born–Einstein Letters*, p. 24. *the Eugenia Dickson incident* Gordon H. Garbedian, *Albert Einstein: Maker of Universes* (New York: Funk & Wagnall, 1939), p. 199.
150 *"Einstein is a man of great good nature"* Plesch, *Story of a Doctor*, p. 205. *the Lunacharsky interview and article* Kuznetsov, *Einstein*, pp. 228–230.
151 *visit from the Orthodox Jews* Esther Salaman, "A Talk With Einstein," *Listener*, September 8, 1955, pp. 370–371.
152 *Einstein's walk with Salaman* Ibid., p. 371. *"It was not the same as respect, admiration, affection"* Ibid. *"a foul smelling flower"* Albert Einstein, diary, Einstein Archives.
153 *"beautiful" though "obscure"* Ibid. *"atoms in dilute, noninteracting gas"* C. Wu, "Physics 'Holy Grail' Finally Captured," *Science News* 148 (July 15, 1995): p. 36. *the root of the problem* Albert Einstein to Mileva Einstein, October 17, 1925, in Highfield and Carter, *Private Lives*, p. 228. *Eduard was less of a worry* Highfield and Carter, *Private Lives*, pp. 230–231.
154 *"our Jews are doing a lot"* Jeremy Bernstein, *Quantum Profiles* (Princeton, N.J.: Princeton University Press, 1991), p. 160. *Hitler arrives on the scene* *Hitler's Table Talk, 1941–1944* (New York: Farrar Straus & Young, 1953), pp. 283–284, 348–349; Allan Bullock, *Hitler: A Study in Tyranny* (New York: Harper & Row, 1964), p. 121.

Chapter 19: The Uncertainty Principle

155 *"a ribald farce"* Kessler, *Diaries*, p. 281. The diary entry is dated February 15, 1926. *"I suppose your wife forgot"* Ibid. *During the animated conversation* Ibid.
156 *American physicist Dayton C. Miller* *Science* 61 (1925): p. 617. *"Experiment is the supreme judge"* Albert Einstein to Michele Besso, in *Einstein–Besso, Correspondence, 1903–1955*, ed. P. Speziali (Paris: Hermann, 1972), p. 215. *Einstein and Heisenberg discuss quantum mechanics* Heisenberg, *Physics and Beyond*, pp. 62–69. These are extracts from Heisenberg's recollection of his long conversation with Einstein. His verbatim account—presumably he wasn't taking notes—indicates an exceptional memory. *"Quantum mechanics is certainly imposing"* Albert Einstein to Max Born, December 4, 1926, in *Born–Einstein Letters*, p. 90.
157 *"I spent my most harmonious hours"* Seelig, *Albert Einstein*, p. 110. *"Everyone at the Patent Office knows"* Speziali, *Einstein–Besso, Correspondence*, p. 156. *"cheerful, sure of himself and agreeable"* Sigmund Freud to Sandor Ferenczi, January 2, 1927. *"enchanting humor"* British psychoanalyst Joan Riviere, quoted in Peter Gay, *Freud: A Life for Our Time* (New York: W. W. Norton, 1988), p. 156.
158 *"Even in the most technical discourse"* Freud's friend Fritz Wittels, quoted in Gay, *Freud:*

A Life, p. 159. "because he has had a much easier time" Sigmund Freud to Marie Bonaparte, January 11, 1927, in The Last Phase: 1919–1939, vol. 3 of The Life and Work of Sigmund Freud, by Ernest Jones, M.D. (New York: Basic Books, 1957), p. 131. "I should like very much to remain in darkness" Sayen, Einstein in America, p. 134. Freud described his son, Oliver Gay, Freud: A Life, pp. 387, 429–430. Freud and Einstein exchange correspondence Jones, The Last Phase, p. 154.

159 admitting that the general theory of relativity Marianoff and Wayne, Einstein: An Intimate Study, pp. 68, 69. "I feel so much a part of every living thing" Hedi Born to Albert Einstein, April 11, 1938, in Born–Einstein Letters, no. 74, p. 132.

160 Einstein reviews Hedi's play Ibid. "experiences, ambitions and emotions" Albert Einstein, The World As I See It (London: Franklin Watts, 1940), pp. 17–20. "I don't care" Marianoff and Wayne, Einstein: An Intimate Study, p. 135. "Why do you do this?" Ibid. "That was probably the severest shock" Kessler, Diaries, pp. 321–322. The diary entry is dated June 14, 1927.

161 "Kerr, who sat listening with his vulgar little wife" Ibid., p. 322. "There is nothing so revolutionary about my observations" Ibid. "His theoretical mind showed even in these movie expeditions" Gamow, Thirty Years That Shook Physics, p. 56.

162 his "conviction that the world could be completely divided" Werner Heisenberg, introduction to Born–Einstein Letters, comm. Max Born, trans. Irene Born (New York: Walker, 1971), p. x. "However contrasting such phenomena may at first sight appear" Ruth Moore, Niels Bohr: The Man, His Science, and the World They Changed (New York: Knopf, 1966), pp. 156, 159.

163 "One cannot bow" Ibid., p. 159. "explains why atoms and molecules keep their identity" Victor F. Weisskopf, The Privilege of Being a Physicist (New York: W. H. Freeman, 1989), p. 51. soft voice and indistinct pronunciation Harold Urey, interview by the author, tape recording, March 1, 1978. Urey, a Nobel Prize–winning chemist, told the author: "Bohr was a bad talker. He didn't speak good English and I don't think he spoke good Danish, either. He was a sort of inhibited talker. I listened to what he said, lecturing and discussing things informally, but I didn't understand it very well." "the strangest debate in the history of the understanding of the world" Fred Alan Wolf, Taking the Quantum Leap: The New Physics for Non-Scientists (New York: Harper & Row, 1989), p. 117. "the most remarkable scientific concept" Moore, Niels Bohr, p. 156. "The Lord did there" Clark, Einstein: The Life and Times, p. 342.

164 "Einstein . . . looked a bit worried" Heisenberg, Physics and Beyond, pp. 80–81. "his general kindness" Louis de Broglie, New Perspectives in Physics (Edinburgh: Peebles Press, 1962), p. 182. "he could not but agree with Bohr" Pais, 'Subtle is the Lord', p. 443. Einstein traveled with de Broglie de Broglie, New Perspectives, p. 184.

Chapter 20: The Perfect Patient

165 "lone traveler" Albert Einstein, Living Philosophies (New York: Simon and Schuster, 1931). The quotation reads in full: "I am truly a 'lone traveler' and have never belonged to my country, my home, my friends, or even my immediate family, with my whole heart." "I admire this man as no other" Albert Einstein to Johann Laub, May 19, 1909, in Pais, 'Subtle is the Lord', p. 169. "I feel an unbounded admiration for you" Albert Einstein to H. A. Lorentz, November 23, 1911, ibid. "I stand at the grave" Einstein, As I See It, p. 11. "He meant more to me personally" Einstein, Ideas and Opinions, p. 75. "I helped you work out things" Michele Besso to Albert Einstein, January 17, 1928, in Bernstein, Quantum Profiles, pp. 154–155.

166 "You could have said it a little better here" Michelmore, Einstein, p. 43. march "gaily over to Professor Einstein" Seelig, Albert Einstein, pp. 16–17. "threatened . . . by the representatives of physics" Clark, Einstein: The Life and Times, p. 347. "[Einstein] attaches no importance to outward show" Plesch, Story of a Doctor, p. 209.

167 *"what was being done for him"*   Ibid., pp. 204–205.   *Rosa then suggested her younger sister, Helen*   Pais, *Einstein Lived Here*, p. 80.   *"You have gone mad"*   Ibid.   *Helen was petrified at the prospect*   Sayen, *Einstein in America*, p. 311 n. 1.

168 *"here is an old child's corpse"*   Ibid.   *"He was like that with everyone"*   Helen Dukas, "Secretary Says Einstein Was a Lone Traveler," interview by Janet Watts, *Philadelphia Inquirer*, September 21, 1973, p. 3A.   *she "had nothing to do with his scientific work"*   Ibid.   *"she devoted her life to him"*   Bucky, interview, September 7, 1993.   *"understood her boss well"*   Ibid.

169 *"young collaborators, predominately mathematicians"*   Heisenberg, introduction to *Born–Einstein Letters*.   *Einstein's assistant Jakob Grommer*   Pais, *'Subtle is the Lord'*, p. 487.   *"intensely proud of being given such a task"*   Whitrow, *Einstein: The Man and His Achievement*, pp. 54–55.   *"Then he rested after lunch"*   Dukas, "Secretary Says Einstein Was a Lone Traveler," p. 3A.

170 *"worthy Jews basely caricatured"*   Einstein, *Ideas and Opinions*, p. 171.

171 *"whether or not the Jewish rogue has slain his Jewish father"*   Hans Haider in *Die Presse* (Vienna), quoted by Michael Z. Wise in *The Jerusalem Report* 2, no. 5 (November 21, 1991): p. 4.   *"greedy" and "inhuman" and lacked even "the moral fiber of Judas"*   Ibid.   *Freud as witness in the Halsman case*   Jones, *The Last Phase*, p. 166.

172 *"It was a suffering for him"*   Yvonne Halsman, interview by the author, tape recording, March 31, 1995.   *acknowledged his innocence*   To author-attorney Hans Ruzicka, as reported in George E. Berkley, *Vienna and Its Jews: The Tragedy of Success* (Cambridge, Mass: Abt/Madison Books, 1988), pp. 183–187.

Chapter 21: The Unified Field Theory

173 *read "with interest and pleasure"*   Albert Einstein to Michele Besso, January 5, 1929, in Speziali, *Einstein–Besso, Correspondence*.

174 *Wythe Williams interviews Einstein*   *The New York Times*, January 12, 1929, p. 1.

175 *"Einstein came to mathematics"*   *The New York Times*, February 4, 1929.   *"The relation between electricity and gravity"*   M. K. Wisehart, "A Close Look at the World's Greatest Thinker," *American* (June 1930): p. 21.   *"to achieve a formula that will account"*   Forsee, *Einstein: Theoretical Physicist*, p. 131.   *"Don't bother your mind about it"*   Edgar Ansel Mowrer, *Triumph and Turmoil: A Personal History of Our Times* (New York: Weybright and Talley, 1968), p. 199.   *"Imagine a scene in two dimensional space"*   George Sylvester Viereck, *Glimpses of the Great* (New York: Macauley, 1930), pp. 430–431.

176 *Upton Sinclair's interest in psychic phenomena*   Leon Harris, *Upton Sinclair: American Rebel* (New York: T. Y. Crowell, 1975), pp. 259–260, 261–264.   *"I . . . am convinced that the book deserves the most earnest consideration"*   Albert Einstein, preface to *Mental Radio*, by Upton Sinclair (New York: Collier Books, 1971), p. x.

177 *"human emanations"*   Vallentin, *Drama of Albert Einstein*, p. 155.   *"Even if I saw a ghost I wouldn't believe it"*   Helen Dukas, interview by the author, tape recording, September 25, 1976.   *"if you aren't confused by quantum mechanics"*   John Wheeler, quoted in John Horgan, "Quantum Philosophy," *Scientific American* 267, no. 1 (July 1992): p. 97.   *"Since it is possible now to insulate against electricity"*   "Einstein's Latest Theory," *Popular Mechanics* (April 1929): pp. 536–539.   *Since we cannot permit*   Arnold Zweig to Sigmund Freud, February 18, 1929, in *The Letters of Sigmund Freud and Arnold Zweig*, ed. Ernst L. Freud (New York: Harcourt Brace Jovanovich, 1970), pp. 4–5.   *Although I haven't much*   Sigmund Freud to Arnold Zweig, ibid., p. 5.

178 *Einstein's fiftieth birthday*   Garbedian, *Einstein: Maker of Universes*, p. 232–233.   *"Gleeful in the knowledge"*   Vallentin, *Drama of Albert Einstein*, p. 159.   *"Every pipe he lit"*   Ibid.   *"see the funny side of situations"*   Plesch, *Story of a Doctor*, p. 206.   *One American journalist*   Garbedian, *Einstein: Maker of Universes*, pp. 233–234.

179 *"If, however, the Herr Minister"* Feuer, *Generations of Science*, p. 93. *Plesch persuades the city of Berlin to give Einstein a house* Frank, *Life and Times*, pp. 221–223. *His morning paper brought news of deadly confrontations* Sefton Delmer, *Trail Sinister: An Autobiography* (London: Secker and Warburg, 1961), pp. 87–89.

180 *a "tiny glimpse" into his "ingenious theory"* Clark, *Einstein: The Life and Times*, p. 421. This account was taken from the queen's notes in her agenda book, maintained in the Royal Archives, Brussels, Belgium. *"My aim lies in smoking"* Pais, *'Subtle is the Lord'*, p. 302. *"liked beautiful women, and they in turn adored him"* Highfield and Carter, *Private Lives*, p. 207. *"leave the field clear, so to speak"* Ibid., pp. 207–208.

181 *"the Austrian interloper"* Ibid. *Wachsmann "believed that Einstein's extra-marital liaisons"* Ibid., p. 210. *Woolf's account of painting Einstein's portrait* Samuel Johnson Woolf, "Einstein at 50," *The New York Times*, August 18, 1929, Sect. 5, p. 1.

182 *Einstein "demolishes" the framed portrait* Ibid. *Einstein receives the Planck Medal* Plesch, *Story of a Doctor*, p. 210.

183 *their "conversation drifted back and forth"* Chaim Tschernowitz, "A Day with Albert Einstein," *Jewish Sentinel* 1, no. 1 (September 1934): pp. 19, 34, 44, 50. *"Why are you at a Jewish conference?"* Clark, *Einstein: The Life and Times*, p. 401. *"We never wanted Palestine for the Zionists"* Norman Rose, *Chaim Weizmann* (New York: Viking, 1986), p. 243. *Arabs' rampage in Palestine* From the March 1930 report of a British Commission of Enquiry headed by Sir Walter Shaw. The report accused Arab newspapers of inciting the riots. *"Is it not bewildering that . . . brutal massacres"* Ibid.

184 *"We must avoid leaning too much on the English"* Clark, *Einstein: The Life and Times*, pp. 402–403. *"were a pain"* Helen Dukas, interview by the author, tape recording, June 16, 1980. *"a pompous liar and hypocrite"* *Nation* (November 19, 1941). *Freud told him* John-Alexis Viereck, interview by the author, tape recording, July 29, 1995.

185 *the interview began* Viereck, *Glimpses of the Great*, pp. 432–451. The questions and answers that appear in the text are reconstructions of the original published material.

187 *fellow Jews* Ibid., pp. 450–451. *Viereck's involvement with the Nazis* Viereck, interview, July 29, 1995. After his imprisonment, Vierecks's wife, Margaret, left him, converted to Roman Catholicism, and became Bishop Fulton Sheen's personal secretary in New York City. Viereck died after a stroke in the early 1960s.

188 *Whyte meets with Einstein* Whitrow, *Einstein: The Man and His Achievement*, p. 54. Lancelot Whyte eventually became president of the British Society for the Philosophy of Science. *"It is ten o'clock"* Dudley Heathcote, unpublished manuscript (Einstein Archives, 47 539), p. 2. *"This matter affects me more deeply"* Ibid.

189 *"a policy be devised"* Ibid., pp. 4–5. *Einstein accused Brodetsky* Selig Brodetsky, *Memoirs: From Ghetto to Israel* (London: Weidenfeld & Nicolson, 1960), p. 137. *"During the greater part of my speech"* Ibid.

190 *"Should we be unable to find a way"* Albert Einstein to Chaim Weizmann, November 29, 1929, in Clark, *Einstein: The Life and Times*, pp. 402–403.

Chapter 22: On the International Lecture Circuit

191 *Asked to name the world's most popular figure* "Second in N.Y. Univ., Senior Poll of Most Admired Man in the World," *The New York Times*, June 6, 1930, p. 16. *Yet interest in Einstein's enigmatic work* "4,500 Battle in Museum to See Einstein Film; Police Quell Stampede After 8 Guards Fail," *The New York Times*, January 9, 1930, p. 1. *"Even if I saw a ghost"* Dukas, interview, June 16, 1980. *When writing samples were requested* "Expert on Writing Amazes Einstein," *The New York Times*, February 23, 1930, sect. 3, p. 3.

192 *"firm belief in physical causality"* Albert Einstein to Maurice Solovine, March 4, 1930, in Einstein, *Letters to Solovine*, p. 71.

193 *"Yes, that is a general human property"* Stanley A. Blumberg and Gwinn Owens, *Energy and Conflict: The Life and Times of Edward Teller* (New York: Putnam's, 1976), pp. 43–44. *"If*

I had been able to address your congress" Michelmore, *Einstein*, p. 146. *"the more Einstein became aware of German anti-Semitism"* Bernstein, *Einstein*, p. 172. *"Now I know there is a God in heaven!"* Yehudi Menuhin, *Unfinished Journey* (New York: Knopf, 1976), p. 96. *Einstein spoke scornfully of his heroes* "Einstein in England Offers Another Theory," *The New York Times*, June 7, 1930, pp. 1, 2.

194 *"If one gets hold of something"* Wisehart, "World's Greatest Thinker," p. 21. *"Einstein first ignored"* Robert Jastrow, "Have Astronomers Found God?" *The New York Times Magazine*, June 25, 1978, p. 19.

195 *"Eddington is the greatest authority in England"* Douglas, *Life of Arthur Stanley Eddington*, p. 104. *"To the general public [Eddington] is best known"* Ibid., pp. 103–104. *"When Eddington was asked, 'When was the world created?'"* Ibid. *"the limitless energy stored in matter"* Berger, *Story of The New York Times*, p. 354. *"putting a shadow"* Michelmore, *Einstein*, p. 146. *"Eduard [had] suffered a breakdown"* Ibid., p. 147.

196 *Eduard's obsession with an older student* Highfield and Carter, *Private Lives*, p. 234. *He did not advise Eduard* Albert Einstein to Eduard Einstein, February 5, 1930, ibid. *"How strange is the lot of us mortals!"* Einstein, *Living Philosophies*, p. 3. *"Eduard was a schizophrenic"* Dr. Elizabeth Roboz Einstein, interview by the author, tape recording, May 15, 1987.

197 *The author asked Einstein expert Robert Schulmann to elaborate* Robert Schulmann, interview by the author, tape recording, November 26, 1991. *"Tributes to Eduard were published by his friends"* In *Eduard Einstein*, by Eduard Rübel, (Bern, Switz.: Paul Haupt, 1986).

198 *Einstein at the meeting of the Committee on Intellectual Cooperation* The New York Times, July 27, 1930, pp. 1, 2. *"its blessing to the suppression of minorities"* Hoffmann with Dukas, *Einstein: Creator and Rebel*, pp. 154, 155. *"I believe in it in theory"* Rabbi F. M. Isserman, Temple Israel, St. Louis, Missouri, to Otto Nathan, February 5, 1957, Einstein Archives 50 485. *"in a weather-beaten raincoat"* Marianoff and Wayne, *Einstein: An Intimate Study*, p. 11. *"Look, Dukas"* Ibid., p. 22.

199 *"the Hitler vote is only a symptom"* "Fascists Walk Out of Berlin Council," *The New York Times*, September 19, 1930, p. 9. *"We are moving toward bad times"* Sigmund Freud to Arnold Zweig, December 7, 1930, in Freud, *Letters of Sigmund Freud and Arnold Zweig*, p. 25. *"if quantum physics is right"* John Wheeler, interview by the author, tape recording, May 30, 1988. *"going from one to another and trying to persuade them"* Bernstein, *Quantum Profiles*, p. 43.

200 *"Your people have not quite grasped"* George Bernard Shaw to John Reith, October 20, 1930, in Dan H. Laurence, ed., *Bernard Shaw: Collected Letters 1926–1950* (New York: Viking, 1988), pp. 211–212. *"learned some English since he was here last"* George Bernard Shaw to Herbert Samuel, October 23, 1930, ibid., p. 212. *"I must [talk about]"* Patch, *Thirty Years with G. B. S.*, p. 235; Michael Holroyd, "Albert Einstein Universe Maker," *The New York Times*, March 14, 1991, p. A15.

201 *"The only way of really helping the Jew in Eastern countries"* "Professor Einstein in London: Appeal for Jews of Eastern Europe," *Times* (London), October 29, 1930, p. 12. *"What does it matter?"* Patch, *Thirty Years with G.B.S.*, p. 128.

202 *Einstein's replies to the students' questions* "Einstein Tells Radicals World Is a Riddle: Says Science Cannot Tell How It All Began," *The New York Times*, November 15, 1930, p. 1. *"He is a man as free from vanity"* Weart and Szilard, *Leo Szilard: His Version of the Facts*, p. 12. *"I hate crowds and making speeches"* "Einstein Puzzled by Our Invitations," *The New York Times*, November 23, 1930, p. 13. *"Isn't it a sad commentary on commercialism?"* Ibid. *"We Jews are everywhere subject to attacks"* "Einstein Attacks British Zion Policy," *The New York Times*, December 3, 1930, p. 15.

203 *not handling "the Palestine question with objectivity"* "Einstein Says Jews Should Seek Truce," *The New York Times*, December 7, 1930, p. 11. *"these men like wolves"* "Einstein and New York Reporters," *Times* (London), December 12, 1930, p. 13. *"The professor is afraid"* "On Arrival Braves Limelight For Only 15 Minutes," *The New York Times*, December

12, 1930, pp. 1, 16. *Elsa then offered to translate* Joseph E. Persico, *Edward R. Murrow* (New York: McGraw-Hill, 1988), p. 70.

204 *Then he remembered his promise to broadcast* The New York Times, December 12, 1930, p. 16. *"to declare, before the World Disarmament Conference convenes"* Marianoff and Wayne, *Einstein: An Intimate Study*, p. 27. *"We have a very profound appreciation"* "Einstein Receives 'Keys' to the City," *The New York Times*, December 14, 1930, p. 1.

205 *the pastor "met Einstein and his wife at the door"* "Einstein Saw His Statue in Church Here; Comments That He Must Be Careful Now," *The New York Times*, December 28, 1930, p. 1. *"I might have imagined"* Vallentin, *Drama of Albert Einstein*, p. 193. *A familiar face turned up* George Sylvester Viereck, "Einstein Sees Economic Reorganization Needed to Revive Prosperity in Germany," *Washington Herald*, January 4, 1931, p. 5. *"Don't forget, Upton Sinclair was not just a spiritualist wack"* I. F. Stone, interview by the author, tape recording, August 13, 1988. *Einstein arrives in San Diego, answers reporters' questions* "Crowds Acclaim Einstein on Arrival in California," *Los Angeles Times*, January 1, 1931, pp. 1, 2.

206 *"It would take a whole library"* "Einstein Hails Age of Thinking Men at Coast Fete," *Washington Herald*, January 1, 1931, p. 3. *Coughlin pursues Einstein* Gene Coughlin, *How to Be One Yourself: A Short Cut to Membership in the Second Oldest Profession* (New York: A. S. Barnes, 1961), p. 82. *"There were tears in his eyes"* Ibid.

207 *"the irrational, the inconsistent, the droll, even the insane"* Albert Einstein, preface to *Einstein: A Biographical Portrait*, by Anton Reiser (New York: Albert & Charles Boni, 1930).

Chapter 23: Einstein in California

208 *"meeting of the leading American men of physical science"* "The Progress of Science: Einstein on His Innovations," *Times* (London), May 11, 1931, p. 7.

209 *"the whole universe is expanding"* From a speech by Harlow Shapley at a meeting of the American Association for the Advancement of Science, Cleveland, Ohio, December 31, 1930, reported in the *Washington Herald*, January 1, 1931, p. 3. *theoretical investigations "by Lemaître and Tolman"* "Prof. Einstein Begins His Work at Mt. Wilson, Hoping to Solve Problems Touching Relativity," *The New York Times*, January 3, 1931. *"The professor hopes he passed the examination"* *Los Angeles Evening Herald*, January 3, 1931, pp. 1, 2; "Einstein Calls World Misunderstandings Serious Peril: Cosmic Rays Are Mystery, Says Savant," *Washington Herald*, January 3, 1931, p. 3.

210 *the Einsteins tour First National Studio* "First National Studio Technicians Present Trick Picture of Einstein with Wife in Auto," *The New York Times*, February 4, 1931, p. 15. *Charlie Chaplin meets the Einsteins* Charles Chaplin, *My Autobiography* (New York: Simon and Schuster, 1964), p. 312.

211 *"if you happen to tell him the same joke twice"* Plesch, *Story of a Doctor*, p. 213. *"From far away I have come to you"* Bernard Jaffe, *Michelson and the Speed of Light* (New York: Doubleday, 1960), pp. 167–168.

212 *"Michelson's experiment was of considerable influence"* Ibid., pp. 100–101. *"often the scene of much conviviality"* Livingston, *Master of Light*, pp. 336–337. *"an incredibly bushy head of hair"* Ibid., p. 337.

213 *"I was amazed, too"* "Einstein Completes Unified Field Theory," *The New York Times*, January 23, 1931, p. 17. *"the farther away a galaxy is, the faster it moves"* Jastrow, "Have Astronomers Found God?" p. 20. *"The mountain is spinning!"* Forsee, *Einstein: Theoretical Physicist*, p. 126. *"I must think of everything"* "Einstein Explores Far-Away Nebulae," *The New York Times*, January 24, 1931, p. 20. *A woman reporter, Cissy Patterson* My account of Patterson's encounter with the "relatively nude" Einstein was drawn from the following sources: *Washington Herald*, pp. 1, 2; Ishbel Ross, *Ladies of the Press* (New York: Harper, 1936), pp. 220, 504; Paul Healy, *Cissy* (New York: Doubleday, 1966), p. 203; Alice Albright Hoge, *Cissy*

Patterson (New York: Random House, 1966), pp. 107–109; Ralph Martin, *Cissy: The Extraordinary Life of Eleanor Medill Patterson* (New York: Simon and Schuster, 1974), pp. 288–290.
214 *"The people are applauding you"* Seelig, *Albert Einstein*, p. 194. *watching "Einstein from the corner of my eye"* Eleanor Patterson, "Einstein Forgets Theories, Sobs at Chaplin Clowning, Master of Mysteries of Space Loses His Scientific Abstraction at First Performance of 'City Lights'," *Washington Herald*, February 10, 1931, pp. 1, 2. *incident that highlighted Elsa's innate kindness* Marianoff and Wayne, *Einstein: An Intimate Study*, pp. 83, 84.
215 *"I have just read you are coming to Pasadena"* Harris, *Upton Sinclair*, p. 262. *"There's an odd-looking old man"* Mary Craig Sinclair, *Southern Belle* (New York: Crown, 1957), pp. 339–340. *"Such was the beginning of as lovely a friendship"* Upton Sinclair, *Autobiography* (New York: Harcourt, Brace & World, 1962), pp. 259–260. *"from first to last the discoverer of relativity"* M. C. Sinclair, *Southern Belle*, p. 340. *"I was at the séance"* Dukas, interview, June 16, 1980.
216 *"When the medium went into the trance"* Ibid. *dinner at the exclusive Town House* M. C. Sinclair, *Southern Belle*, p. 340. *Mrs. Sinclair's view of Elsa* Ibid. *"America was prepared to go mad over him"* "Millionairess Offered $ to Sit Next and Violin Offered," *Outlook and Independent*, December 24, 1930. *"with the apologetic explanation"* Marianoff and Wayne, *Einstein: An Intimate Study*, p. 114. *"and if more gifts arrived"* Livingston, *Master of Light*, p. 337.
217 *"I offer it to you like a closed box"* "Einstein Drops Idea of Closed Universe," *The New York Times*, February 5, 1931, pp. 1, 17; *Times* (London), February 6, 1931. *"the native population is so thinly scattered"* Kendall Foss, "Endorses Move to Have Eastern European Jews Migrate to Peru," *The New York Times*, February 8, 1931, sect. 3, p. 4. *"I don't believe we should have two classes"* Forsee, *Einstein: Theoretical Physicist*, p. 125. *"Why does this magnificent applied science"* "Einstein Sees Lack in Applying Science," *The New York Times*, February 17, 1931, p. 6. *"my father and Einstein shared a similar socialist outlook"* David Sinclair, interview by the author, tape recording, December 3, 1982.
218 *"He always gets lost somewhere"* Marianoff and Wayne, *Einstein: An Intimate Study*, p. 184. *Lindemann "was a heavyweight of personality"* C. P. Snow, *Science and Government* (Cambridge, Mass.: Harvard University Press, 1961), p. 17. *"Essentially an amateur"* Birkenhead, *Professor and the Prime Minister*, p. 163. *"It was a great thrill"* Whitrow, *Einstein: The Man and His Achievement*, p. 58.
219 *turned "for comfort and peace to his violin"* Birkenhead, *Professor and the Prime Minister*, p. 163. *his special theory was based on the experiments of Fizean and Michelson* "The Progress of Science: Einstein and His Innovations," *Times* (London), May 11, 1931, p. 7. *Einstein's second lecture at Oxford* "Dr. Einstein at Oxford: An Expanding Universe," *Times* (London), May 18, 1931, p. 14. *"not by the pressure from behind"* "A Unified Field Theory: Electro-Magnetism and Gravitation," *Times* (London), May 25, 1931, p. 6. *walked "into a schoolboy's changing room"* John G. Griffith, *The Trusty Servant* (Winchester, Engl.: Culverlands, 1986), p. 5. *"I myself do not claim my opinion is correct"* Sayen, *Einstein in America*, p. 56.
220 *"solely prompted by pure love of justice"* Albert Einstein to Robert Millikan, August 1, 1931, Robert Millikan Papers, Caltech. *"the dark recesses of the Serbian and Croat underworld"* Churchill, *The Gathering Storm*, p. 107. *According to the two men* "Einstein Accuses Yugoslavian Rulers in Savant's Murder," *The New York Times*, May 6, 1931, pp. 1, 3. *The New York Times reported frequent protests* Ibid. *they were ridiculed for implicating "a government"* G. S. Gavrilovitch, "Murder in Yugoslavia," letter to *The New York Times*, May 11, 1931, p. 3.
221 *"For fifteen years many friends"* Luigi Criscuolo, "The Situation in Yugoslavia," letter to *The New York Times*, May 22, 1931, p. 24. *the documents on which Einstein had based his account were published* *The New York Times*, August 23, 1931, sect. 2, p. 2. *According to one source* C. L. Sulzberger, *A Long Row of Candles* (New York: Macmillan, 1969), p. 64.

Sulzberger claims that "Italians, Hungarians and Croat nationalists collaborated in arranging the assassination of Yugoslavia's King Alexander." *"a Croat terrorist who had been trained in Italy"* A. J. P. Taylor, *The Origins of the Second World War* (New York: Fawcett, 1963), p. 85.

222 *"I was much struck by the emptiness of the streets"* Bullock, *Hitler*, pp. 177–178. *"an amiable weakness of talking exclusively toward his left"* "Einstein Defines Aim of Physicists," *The New York Times*, October 5, 1931, p. 11.

223 *Mooney, his voice "choked with emotion"* Rebekah Raney to Albert Einstein, December 22, 1931, Einstein Archives 47 646. *Soviet trains didn't run on time* Laurence, *Bernard Shaw: Collected Letters*, p. 259.

224 *"the velocity of light is constant"* "California Test Supports Einstein Theory," *The New York Times*, December 31, 1931, pp. 1, 8. *"What God hath put asunder"* Clark, *Einstein: The Life and Times*, p. 405. *"Einstein's never-failing inventiveness"* Pais, *'Subtle is the Lord'*, p. 347.

Chapter 24: Weighing Options

225 *"Why is that not brought about?"* "Einstein Advocates Economic Boycott," *The New York Times*, February 28, 1932, sect. 2, p. 4.

226 *"I gave a speech, but alas!"* Nathan and Norden, *Einstein on Peace*, p. 163. *Flexner met Einstein at the Athenaeum* Abraham Flexner, *An Autobiography* (New York: Simon and Schuster, 1960), pp. 250, 251. *"book on Stalin"* Albert Einstein to Isaac Don Levine, March 15, 1932, in William R. Corson and Robert T. Crowley, *The New KGB: Engine of Soviet Power* (New York: Morrow, 1985), p. 102.

227 *"they were handing him the rope"* Mowrer, *Triumph and Turmoil*, p. 211. *"on your own terms"* Flexner, *An Autobiography*, p. 251. *"Because of the glorification of Soviet Russia"* Hoffmann with Dukas, *Einstein: Creator and Rebel*, p. 164.

228 *"I can still recall [that] when a professor"* Frank, *Life and Times*, p. 226. *the Buckys spend the day at Caputh* Bucky, interview, September 7, 1993.

229 *the families become close friends* Ibid.

230 *"a kind of thin wall of air separated Einstein from his closest friends and family"* Frida Sarsen Bucky, "You Have to Ask Forgiveness . . . Albert Einstein As I Remember Him," *Jewish Quarterly* 15, no. 4 (Winter 1967–68): p. 31. *Einstein liked coffee* Ibid. *"Einstein was absolutely conscious"* Ibid. *"You had to know how far you could go with him"* Bucky, interview, September 7, 1993. *"Aren't you chilly?"* Frank, *Life and Times*, p. 448.

231 *"simple and unpretentious as any child"* "Chance Led to the Association of Einstein With U.S. Institute," *The New York Times*, April 19, 1955, p. 1. *Antonina came to see the Einsteins* Vallentin, *Drama of Albert Einstein*, pp. 202–209.

232 *Einstein accepts Flexner's offer* Beatrice Stern, "A History of the Institute for Advanced Study, 1930–1950" (Special Collections of the Hoover Library, Western Maryland College, Westminster, Maryland, 1964), p. 139. *"My only wish is that Dr. Mayer"* Einstein to Abraham Flexner, July 30, 1932, ibid. *"because man has within himself"* Albert Einstein to Sigmund Freud, July 30, 1932, Einstein Archives.

233 *"We assume that human instincts"* Sigmund Freud to Albert Einstein, September 1932, Einstein Archives. *"Albert's will is unfathomable"* Elsa Einstein to Paul Ehrenfest, April 5, 1932, Einstein Archives. *"Our situation on this earth seems strange"* Albert Einstein, speech to the German League of Human Rights, Einstein Archives.

234 *Albert's bedroom was way down the hall* Herneck, *Einstein privat*. *"He goes to his study"* Pais, *'Subtle is the Lord'*, p. 301. *"very lovable, and gentle"* Ralph Nash, interview by the author, tape recording, August 23, 1993. *Fondiller's meeting with Einstein* William Fondiller, interview, Columbia University Oral History Project.

235 *"my dear and old friend"* Michele Besso to Albert Einstein, September 18, 1932, in

Speziali, *Einstein–Besso, Correspondence*, no. 112, p. 287. *"brought you discoveries by the armful"* Ibid. *"Someone asked me"* Ibid., p. 288.

236 *"Dear Albert Senior, Major, Maximus"* Michele Besso to Albert Einstein, October 17, 1932, ibid., no. 113, p. 289. *"How could I be angry with you"* Albert Einstein to Michele Besso, October 21, 1932, ibid., no. 114, p. 291.

237 *"Many geologists will be surprised"* "Princeton Geologist Dubious," *The New York Times*, October 18, 1932, p. 6. *"I have received a leave of absence"* "Einstein Will Head School Here, Opening Scholastic Centre," *The New York Times*, October 11, 1932, p. 1. *"Great news"* Ibid., p. 18. *"I have read with delight"* Benjamin Cardozo to Abraham Flexner, October 11, 1932, Archives of the Burndy Library, Norwalk, Connecticut. *"The Pope of Physics"* Robert Jungk, *Brighter Than a Thousand Suns* (New York: Harcourt, Brace and Company, 1958), p. 46. *"It was as if God himself"* Ed Regis, *Who Got Einstein's Office? Eccentricity and Genius at the Institute for Advanced Study* (Reading, Mass.: Addison-Wesley, 1987), p. 16. *"It's utterly untrue"* Jewish Telegraphic Agency, November 21, 1932. *fortune-tellers, mediums, and their ilk* Dukas, interview, June 16, 1980. *Einstein tops the danger list of the Woman's Patriot Corporation* FBI file.

238 *"His frequently revised theory of relativity"* Ibid. *"Never have I yet experienced from the fair sex"* Albert Einstein, "Reply to the Women of America," in Einstein, *Ideas and Opinions*, pp. 7–8. *When Geist asked* "Visa Granted: Resentment Subsides," *The New York Times*, December 6, 1932, p. 1. *"in the name of the intelligent American people"* "Einstein Resumes Packing for Voyage," *The New York Times*, December 7, 1932, p. 4.

239 *"the professor and I were quite determined"* Ibid. *"Say I am not an adherent"* George Bernard Shaw to Blanche Patch, December 6, 1932, in Laurence, *Bernard Shaw: Collected Letters*.

240 *"fists swung, chair legs were brandished"* Lowell Thomas, *History As You Heard It* (Garden City, N.Y.: Doubleday, 1957), p. 30. *Einstein was being discussed in a Berlin law court* "German Banker Charged with Using Einstein's Name in Stock Swindle," *The New York Times*, December 14, 1932, p. 8. *working at Cambridge University's Cavendish Laboratory* E. N. da Costa Andrade, *Rutherford and the Nature of the Atom* (Garden City, N.Y.: Doubleday, 1964), p. 168. *his gift of prophecy failed him now* William Manchester, *The Glory and the Dream* (New York: Bantam, 1990), p. 69. *Einstein agreed with him* Ibid.

Chapter 25: Einstein the Refugee

241 *"He's a radical and an alien Red!"* "Women Patriots Try New Ban on Einstein," *The New York Times*, January 9, 1933, sect. 1, p. 21. *Oursler questions Einstein* Fulton Oursler, *Behold This Dreamer* (Boston: Little, Brown, 1964), pp. 293–299.

242 *Leon Watters, a wealthy biochemist* Leon L. Watters, "Comments on the Letters of Professor and Mrs. Albert Einstein to Dr. Leon L. Watters" (The Leon L. Watters Collection, Jewish American Archives, Hebrew Union College, Jewish Institute of Religion, Cincinnati, Ohio, n.d.), p. 3. *When men are calling"* Ibid.

243 *"wandering about in his garden"* Ross, *Ladies of the Press*, p. 216. rolled *"madly from the fruit stands"* Ibid., p. 220. *felt the shock in his office at Caltech* Whitrow, *Einstein: The Man and His Achievement*, p. 61; "Einstein Oblivious as Earth Trembles," *The New York Times*, March 11, 1933, p. 11. *"unhindered on wholesale blackmail, robbery, violence, and murder"* Mowrer, *Triumph and Turmoil*, p. 214. *"brushed aside the Constitution"* Ibid.

244 *"If you go to Germany, Albert"* Michelmore, *Einstein*, p. 180. *Reaching Manhattan, he left* Ibid., pp. 181–182. *"Einstein has been hardly a day in New York"* Alfred Hugenberg, editorial, *Lokalanzeiger*, March 17, 1933. *Storm troopers surrounded* "Nazis Hunt Arms In Einstein Home," *The New York Times*, March 21, 1933, p. 10. *"who knew of my family's friendship with Einstein"* Thomas Bucky, interview by the author, tape recording, September 8, 1993.

245 *Planck was no Nazi* Heinrich von Ficker (on behalf of Planck) to Einstein, March 18, 1933, in Heilbron, *Dilemmas of an Upright Man*, pp. 156, 157. *von Laue immediately called a meeting* Moore, *Schrödinger*, p. 265. *"Where would we be"* Cassidy, *Uncertainty*, p. 302. *"In times like these"* Albert Einstein to Willem de Sitter, April 5, 1933, in Dukas with Hoffmann, *Einstein: The Human Side*, p. 55. *Einstein told Planck* Albert Einstein to Max Planck, April 6, 1933, in Heilbron, *Dilemmas of an Upright Man*, p. 159.

246 *"The most important example of the dangerous influence of Jewish circles"* *Völkischer Beobachter*, May 13, 1933. *"Intellectualism is dead"* Moore, *Schrödinger*, p. 270. *Planck courageously responded* Heilbron, *Dilemmas of an Upright Man*, p. 159. *"I was in Berlin"* Moore, *Schrödinger*, p. 269. *"I think the Nazis"* Birkenhead, *Professor and the Prime Minister*, p. 164.

247 *"the forced emigration of Jews"* Heilbron, *Dilemmas of an Upright Man*, p. 153. *When [Planck] extolled Jewish Germans' contributions to science* Max F. Perutz, *Is Science Necessary? Essays on Science and Scientists* (New York: Dutton, 1989), pp. 178–179. *"What the Nazi* Völkischer Beobachter *calls"* "Boycott of Jews," *Times* (London), April 3, 1933, p. 14. *lunch with Winston Churchill* Hoffmann with Dukas, *Einstein: Creator and Rebel*, p. 170.

248 *"I saw a great deal of Lindemann"* Churchill, *The Gathering Storm*, pp. 78–80, 82. *"evil monster" in Germany* Albert Einstein to Max Born, May 30, 1933, in *Born–Einstein Letters*, p. 114. *"There is no known instance"* Birkenhead, *Professor and the Prime Minister*, p. 266. *"if you bring together thirty such gentlemen"* Alan D. Beyerchen, *Scientists Under Hitler* (New Haven, Conn.: Yale University Press, 1978), p. 227 n. 39. *"the French would pray for the victims"* Richard Rhodes, *The Making of the Atomic Bomb* (New York: Simon and Schuster, 1986), p. 192. *"parting anti-Semitic accusations"* Max Born to Albert Einstein, June 2, 1933, in Cassidy, *Uncertainty*, p. 304. *"The deeper we search"* "Dr. Einstein at Oxford: History of the Atomic Theory," *Times* (London), June 14, 1933, p. 12.

249 *he took death threats* Vallentin, *Drama of Albert Einstein*, p. 231. *$5,000 bounty on his head* Clark, *Einstein: The Life and Times*, p. 494. *The well-meaning Vallentin added to the tension* Vallentin, *Drama of Albert Einstein*, p. 231. *"Safety is impossible here"* "Belgium: Einstein, Fearing Nazis, Flees to England," *Newsweek* (September 16, 1933): pp. 13–14. *"You will be astonished"* Ibid., p. 14. *"an asylum for the unfortunate"* Vallentin, *Drama of Albert Einstein*, p. 230. *"With the earth revolving"* Einstein Archives, 31 844.

250 *"Germany," said the Englishman* "Nationality of Jews: Commander Locker-Lampson's Bill," *Times* (London), July 27, 1933, p. 7. *"a mixture of the humane, the humorous, and the profound"* Jacob Epstein, *Epstein: An Autobiography* (London: Vista Books, 1963), p. 77. *"How can we save mankind"* "Dr. Einstein on Liberty, Albert Hall Speech," *Times* (London), October 4, 1933, p. 14.

251 *"Striving after knowledge"* *Truth* 37, no. 9 (September 30, 1933): p. 5. *"which are in truth"* "Professor Einstein's Political Views: Victim of Misunderstanding," *Times* (London), September 16, 1933, p. 12. *"Virtually overnight, Princeton was transformed"* Regis, *Who Got Einstein's Office?*, p. 4. *"A desk or table"* Elma Ehrlich Levinger, *Albert Einstein* (New York: Messner, 1949). *"Einstein Wears an Invisible Cloak"* *Princeton Herald*, October 20, 1993.

252 *"coming towards me a bulky gentleman"* *Princeton Alumni Weekly*, October 13, 1993. *saw him buy a vanilla ice cream cone* John A. Lampe, D. D., "How Einstein Came to Princeton," *Saturday Review* (July 7, 1956): pp. 38, 39. *Otto Nathan was among the first* Leonard B. Boudin, "Otto Nathan," *The Nation* (February 14, 1987): p. 169. *"an extraordinary sweet gentle man"* Ibid. *Carolyn Blackwood helped Elsa* James Blackwood, interview by the author, tape recording, September 1, 1994.

253 *"The whole of Princeton"* Vallentin, *Drama of Albert Einstein*, p. 235. *"puny demigods on stilts"* Albert Einstein to Queen Elizabeth (of Belgium), November 20, 1933, in Michelmore, *Einstein*, p. 196. *While shopping for herself* Freeman Dyson, *From Eros to Gaia* (New York: Pantheon, 1992), p. 301. *invitation to the White House* Regis, *Who Got Einstein's Office?*, p. 34. *"Professor Einstein has come to Princeton"* Abraham Flexner to Franklin D. Roosevelt, November 3, 1933, in Stern, "History of the Institute for Advanced Study," p.

174. *"You are a Jew faker"* Einstein Archives, 31 847. *"The Jews are just as dangerous today"* Einstein Archives, 31 848.
254 *said to resemble a church* Regis, *Who Got Einstein's Office?*, p. 55. *"I was in my first year"* Wheeler, interview, June 25, 1989.
255 *Sinclair Lewis, the prickly* Mark Schorer, *Sinclair Lewis: An American Life* (New York: McGraw-Hill, 1961), p. 578; "Einstein Eulogizes Nobel as Idealist," *The New York Times*, December 19, 1933, p. 18. *The overeager cameramen* Dukas, interview, June 16, 1980.

Chapter 26: A New Life in Princeton

256 *Leopold and Einstein correspond* Nathan F. Leopold, *Life Plus 99 Years* (New York: Popular Library, 1958), pp. 256, 257. Leopold was released from prison in 1958, married, and lived abroad as a social worker and lab technician. *"mutual love of sailing"* Sayen, *Einstein in America*, p. 66.
257 *"Einstein stopped eating and listened intently"* Watters, "Comments on the Letters," p. 9. *"a prophetic message"* E. Randol Schoenberg, "Arnold Schoenberg and Albert Einstein: Their Relationship and Views on Zionism," *Journal of the Arnold Schoenberg Institute* 10, no. 2 (November 1987): pp. 134–182. *Einstein thought Schoenberg and his music "crazy"* According to Professor Claudo Spies, "Helen Dukas mentioned [that] Einstein made some disparaging remark about Arnold Schoenberg and the Twelve-Tone idea as being 'verruckt.'" Professor Claudio Spies, letter to E. Randol Schoenberg, May 30, 1987. *"seemed more bewildered than pleased"* Dorothy R. Joralemon, "When Einstein Sat for My Mother," *50 Plus* (June 1982): pp. 43–45. *During a break in the painting* Dorothy Joralemon, interview by the author, tape recording, December 2, 1982. Joralemon is the daughter of Winifred Rieber.
258 *"I would have left him on the mountain"* Ibid. *"I didn't argue with his delusions"* Ibid. *"as irrefutable as any mathematical equation"* April 14, 1934, Einstein Archives, 31 855. *"the professor said with some glee"* Watters, "Comments on the Letters," p. 10. *"to make a day of it"* Ibid.
259 *"I was digging up the jonquils"* James Blackwood, interview by the author, tape recording, September 7, 1994. *"'On what ship will you return?'"* Ibid. *"'And besides, Jesus was a Jew'"* Ibid. *Einstein at Watters's Fifth Avenue apartment* Watters, "Comments on the Letters," p. 16.
260 *Einstein and Watters visit Untermyer's home, Greystones* Ibid., pp. 17, 18. *The boys rigged up a shortwave radio* Bucky, interview, September 7, 1993. *"thinks no more of me"* Watters, "Comments on the Letters," p. 19. *"She was part of the scenery, yes"* Blackwood, interview, September 7, 1994. *"Helen was treated extremely well"* Freeman Dyson, interview by the author, tape recording, September 6, 1994.
261 *"kindly, expressive eyes"* Watters, "Comments on the Letters," p. 20. *The Blackwoods meet Elsa aboard the* Westernland Blackwood, interview, September 7, 1994. *"had left Ilse's ashes in Europe"* Ilse's husband, Rudolf Kayser, later emigrated to the United States, where he remarried.
262 *"Please do not think"* James Blackwood, first draft of talk given to members of the Princeton Historical Society, Spring 1995, photocopy in author's possession. *the lighter side of life that summer* Bucky, interview by the author, tape recording, September 7, 1994. *"I happened to see this boat in trouble"* Harry Darlington, interview by the author, tape recording, February 11, 1988. *Watters on Einstein's sailing skills* Watters, "Comments on the Letters," pp. 21–22.
263 *"wore a raincoat but no hat"* Blackwood, interview, September 7, 1994. *The Blackwoods have dinner at the Einsteins' home* Ibid.
264 *After New Year's Day* Ibid. *Blackwood on Einstein's public and private faces* Ibid.
265 *"What about this feller"* James Blackwood, letter to the author, April 4, 1995.
266 *Einstein was also surprisingly accessible* Blackwood, interview, September 7, 1994. *"I know who you are"* Ibid. *"What Einstein meant by that"* Ibid. *"Many things which go*

under my name" Albert Einstein to George Seldes, October 13, 1954, in *The Great Quotations*, compiled by George Seldes (New York: Pocket Books, 1967), p. 1014.
267 *the Einsteins breathed a sigh of relief* Thomas Bucky, interview by the author, tape recording, April 3, 1995. *"greatly overwrought and depressed"* Watters, "Comments on the Letters," p. 24. *"I don't ask, but if I am asked I go"* Ibid. *Watters celebrated his birthday* Ibid., pp. 26–28.
268 *"There is anti-Semitism at Princeton"* Ibid. Princeton admitted to each class roughly the percentage of Jews in America. Dukas said, "We were told there was anti-Semitism before coming; not active, but snobbish. Really we never experienced anything, but we were told it was a reactionary town." Sayen, *Einstein in America*, p. 75. *Their conversation continued far into the night* Watters, "Comments on the Letters," pp. 26–27. *"When I was a very young man"* Ibid., p. 28 *"just invented a remarkable hair restorer"* November 3, 1934, Einstein Archives, 50 071.
269 *Sinclair's campaign for governor* Upton Sinclair, "Einstein As I Remember Him," *Saturday Review* (April 14, 1956): pp. 17, 18, 56. *"Where are the hundreds"* Levine, *Eyewitness to History*, p. 171. *"You can imagine that I, too, regret"* Ibid., p. 172.
270 *"I was grieved to read your statement"* Ibid., p. 173. *"During the 1920s"* Stone, interview, August 13, 1988. *"Levine had gone over to the far right"* Ibid. *if he believed in Santa Claus* Michelmore, *Einstein*, p. 202.
271 *Watters on Einstein's lecture in Pittsburgh* Watters, "Comments on the Letters," pp. 29–36.

Chapter 27: Setting In

274 *"You have more influence with him"* Watters, "Comments on the Letters," p. 12. *Einstein's stay at Watters's Manhattan apartment* Ibid., pp. 39–42.
276 *Rosso interviews Einstein* Sayen, *Einstein in America*, pp. 75, 76. *Einstein was wearing "his famous sweater"* John Oakes, interview by the author, tape recording, September 8, 1992. Some time after graduating from Princeton, Oakes became editorial page editor of *The New York Times*.
277 *"The German situation"* Birkenhead, *Professor and the Prime Minister*, p. 165. *realized that the democratic Weimar Republic* Dukas with Hoffmann, *Einsten: The Human Side*, p. 110. *the Einsteins dine at the Blackwoods' home* Blackwood, interview, September 7, 1994.
278 *"business details did not attract Einstein"* Ibid. *"We will go in there"* Marianoff and Wayne, *Einstein: An Intimate Study*, p. 200.
279 *"Elsa, very much provoked"* Ibid. *"He cooked for me all his masterpieces"* Ibid. *"were so overwhelmed by the luxury"* Mrs. Joseph Copp, interview by the author, tape recording, December 8, 1982. *Harvard was not clamoring for Einstein's presence* Harlow Shapley, *Through Rugged Ways to the Stars* (New York: Scribner's, 1969), pp. 111, 112. *"take good care of Albert"* Ibid., p. 111.
280 *"very fruitful and satisfying"* Watters, "Comments on the Letters," p. 42. *"It was a memorable summer"* Mrs. Joseph Copp, interview by the author, October 28, 1982. *"I had to pull him up"* Winifred Taylor Laubach, "The Day Albert Einstein Almost Went Down the Tube," *Yankee* (September 1974): p. 76.
281 *"no quantum-mechanical description is possible"* Bernstein, *Quantum Profiles*, p. 46. *"I was very happy"* Moore, *Schrödinger*, p. 304. *Rabbi Cohen's lengthy discussion with Einstein* Harry A. Cohen, "An Afternoon with Einstein," *Jewish Spectator* (January 1969): pp. 15–17.
282 *"did not once raise his voice"* Ibid., p. 14. *some twenty newsmen* Michelmore, *Einstein*, pp. 204–205. *Elsa felt weak and short of breath* Ibid., p. 205. *"Einstein was walking toward his home"* James Blackwood, interview by the author, September 10, 1994.
283 *"Christian pointed ahead"* Hiram Haydn, *Words and Faces* (New York: Harcourt Brace and Jovanovich, 1974), p. 164. *"leaving college to attend church"* Joseph Ceruti, *Princeton*

Alumni Weekly, October 13, 1993. *"Dear Professor Einstein, I can't have a baby"* Einstein Archives. *"Dr. Einstein: You are the prince of idiocy"* Einstein Archives, 32 018. *"the proper diet"* Einstein Archives.

284 *"Einstein has disappeared from Germany"* Elisabeth Heisenberg, *Inner Exile: Reflections of a Life with Werner Heisenberg* (Boston: Birkhauser, 1984), p. 45.

Chapter 28: Family Matters

285 *a young woman who claimed to be Einstein's illegitimate daughter* Plesch, *Story of a Doctor*, p. 211. *"with the assistance of friends"* Ibid. *"Mrs. Herrschdoerffer pretending"* Frederick Lindemann to Hermann Weyl, 1935 telegram, Einstein Archives, 51 044.

286 *The woman in question was Grete Markstein* Bial to Helen Dukas, August 12, 1936, Einstein Archives, 51 046. *"All my friends make fun of me"* Albert Einstein to Janos Plesch, 1936, Einstein Archives, 51 045. *about the obvious discrepancies* Dr. Robert Schulmann, interview by the author, tape recording, April 20, 1989.

288 *When the North Americans* Otto D. Tolischus, "Nazis Would Junk Theoretical Physics," *The New York Times*, March 9, 1936, p. 1. *von Laue had compared the Nazis' attacks* Theodore von Laue, interview by the author, tape recording, April 8, 1995. *the "weight of numbers"* Tolischus, "Nazis Would Junk Theoretical Physics," *The New York Times*, March 9, 1936, p. 19. *"the madmen in Germany"* Albert Einstein to Michele Besso, February 16, 1936, in Speziali, *Einstein–Besso, Correspondence*, no. 122, pp. 309–310.

289 *"No," Einstein replied* "Called Greatest Living Jew," *The New York Times*, March 28, 1936, p. 17. *"As in my youth"* Albert Einstein to Maja Winteler, 1936, Einstein Archives. *Infeld had known Einstein in Berlin* Banesh Hoffmann, interview by the author, tape recording, September 2, 1985.

290 *"We did all the dirty work"* Ibid. *Infeld and Einstein discussed "hundreds" of subjects* Infeld, *Quest*, pp. 285, 286. *"With Einstein there was always humor"* Thomas Bucky, interview by the author, tape recording, October 2, 1988. *"until recently I could only apprehend"* Jones, *The Last Phase*, p. 203.

291 *"I really must tell you"* Ibid., pp. 203–204. *"all the best of luck"* Albert Einstein to Leon Watters, June 24, 1936, Caltech Archives. *"You are one of the few women"* Albert Einstein to Elsbeth Grossman, September 26, 1936, in Seelig, *Albert Einstein*, p. 207.

292 *"I think you are the most considerate"* Elsa Einstein to Leon Watters, September 10, 1936, Caltech Archives. *an insider's view of the Einsteins' relationship* Bucky, interview, October 2, 1988.

293 *Ulam on Einstein and Infeld* Stanislaw Ulam, interview by the author, tape recording, November 2, 1982. Ulam became one of the fathers of the hydrogen bomb, solving what Ulam characterized as an "insoluble problem" that had stymied Edward Teller. *" 'Am I the only undergraduate' "* Henry Bach, interview by the author, tape recording, May 10, 1988. *"Are you going to cut your hair"* Ibid. *After a little boy* Sayen, *Einstein in America*, p. 69. *conversation between de Florez and Einstein* Columbia University Oral History Project collection.

294 *offer from East Coast publisher* Einstein Archives. *"Elizabeth" got straight to the point* Einstein Archives, 31872. *any of the following* The five letters from which these extracts have been taken are located in the Einstein Archives.

295 *"scientists were rated as great heretics"* Alden Hatch, *Buckminster Fuller: At Home in the Universe* (New York: Crown, 1974), p. 140. *"a chain reaction altering man's everyday world"* Robert Snyder, ed., *Buckminster Fuller: Autobiographical Monologue/Scenario* (New York: St. Martin's Press, 1980), p. 68.

296 *"felt as I had never felt of anyone before"* Ibid., p. 141. *"Young man, you amaze me"* Hatch, *Buckminster Fuller: At Home in the Universe*, p. 141–142. *"a loyal and understanding wife"* Plesch, *Story of a Doctor*, p. 220.

297 *"utterly ashen and shaken"* Banesh Hoffmann, interview by the author, tape recording,

September 6, 1985. *Hans Frank advocated "doing away with [the Jews]"* William L. Shirer, *The Rise and Fall of the Third Reich* (New York: Fawcett Crest, 1966), pp. 380, 876. *Among those he helped* Hoffmann with Dukas, *Einstein: Creator and Rebel*, pp. 152–153. *Boris Schwarz* Boris Schwarz was leader of the Indianapolis Symphony Orchestra from 1937 to 1938, and was a member of the NBC Orchestra under Arturo Toscanini from 1938 to 1939. He became an American citizen in 1943. As professor of music at Queens College of the City of New York (1941–1976), he made serious contributions to the study of Beethoven and to the history of the violin and of chamber music.

298 *A job awaited him* Hoffmann with Dukas, *Einstein: Creator and Rebel*, pp. 232–233. *"Dear Posterity!"* Dukas with Hoffmann, *Einstein: The Human Side*, p. 105.

Chapter 29: Politics at Home and Abroad

299 *"The charges against the defendants"* Sidney Hook, *Out of Step: An Unquiet Life in the 20th Century* (New York: Harper & Row, 1987), pp. 222, 223. *"I took Benjamin Stolberg with me"* Sidney Hook, interview by the author, tape recording, November 27, 1982.

300 *"in those days dedicated Communists"* Hoffman, interview, September 2, 1985. *"in a 1938 letter"* Hook, interview, November 27, 1982. *"In my presence"* Valentin Bargmann, interview by the author, tape recording, November 6, 1982.

301 *"I don't think anyone"* John Stachel, interview by the author, tape recording, November 29, 1982. *"Einstein was naive about the Soviet Union"* Sidney Hook, interview by the author, tape recording, November 29, 1982. *"I wouldn't call him naive"* Stachel, interview, November 29, 1982. *Einstein's radical friend* Otto Nathan had been an economic adviser to the Weimar Republic and later to U.S. president Herbert Hoover; he was also assistant to Treasury undersecretary Harry Dexter White, author of *The Nazi Economic System*, and coauthor of *Einstein on Peace*. He was one of the first white professors at the all-black Howard University. *"Yes, Einstein was naive"* Theodore von Laue, interview by the author, tape recording, December 12, 1994.

302 *Roosevelt opposed raising the embargo* *The Secret Diary of Harold Ickes: The Inside Struggle*, vol. 2, *1936–1939* (New York: Simon and Schuster, 1954), pp. 389, 390. *"I can assure you"* "Sympathy with Spanish Loyalists Cause Revealed by Spanish Embassy," *The New York Times*, February 5, 1937, p. 5.

303 *Meanwhile, in Nazi Germany* Beyerchen, *Scientists Under Hitler*, p. 170. *"I tried my best"* Stern, "History of the Institute for Advanced Study," p. 179. *declined Einstein's offer* Ibid., p. 180. *"Marx declared that the world is made of matter"* Max Eastman, *Einstein, Trotsky, Hemingway, Freud and Other Great Companions* (New York: Collier Books, 1962), pp. 27, 28. *Schuster "had the audacity to tell Einstein a few jokes"* Max Schuster, Columbia University Oral History Project.

304 *"I found a spinster crawling across my bed"* James Blackwood, letter to the author, April 4, 1995. *"No wonder"* Dr. Robert Schulmann, interview by the author, tape recording, April 6, 1995. *"dressed like a student"* Eastman, *Great Companions*, p. 30. *"'confessions' of the old Bolsheviks"* Ibid., p. 31. *Einstein agreed with Eastman* Ibid., p. 33.

305 *"I had a dream once"* Ibid. *"the yacht was too large"* Margaret Brennan-Gibson, *Clifford Odets* (New York: Atheneum, 1981), p. 478. *"The discussion on quantum mechanics"* Valentin Bargmann, interview by the author, tape recording, December 1, 1982. *"the greatest debate in intellectual history"* Wheeler, interview, June 25, 1989.

306 *"Einstein's son entertained him"* Highfield and Carter, *Private Lives*, p. 241. *Hans Albert speculated* Whitrow, *Einstein: The Man and His Achievements*, p. 22.

307 *"I work earnestly"* Albert Einstein to Maurice Solovine, April 10, 1938, in Einstein, *Letters to Solovine*, p. 87. *Einstein was handicapped by his unfamiliarity* Albert Einstein to a would-be writer, January 25, 1938, Einstein Archives, 52 780. Harry Hansen of *The New York World-Telegram*, and Ralph Thompson of *The New York Times*, thought well of the writer's

work. *"I have just read the first seven pages"* A would-be writer to Albert Einstein, December 27, 1938, Einstein Archives, 52 794. *"Your letters interested the judge"* A would-be writer to Albert Einstein, December 29, 1938, Einstein Archives, 52 795. *"narrow nationalism within our ranks"* The New York Times, April 20, 1938.

308 *he was not a Communist* Albert Einstein to Siegmund Livingstone, April 21, 1938, Einstein Archives. *"relativity was regarded as physically correct"* Beyerchen, *Scientists Under Hitler*, p. 163. *Einstein had been helping people escape from Nazi Germany* Watters, "Comments on the Letters," p. 46. *a tireless fighter for truth and justice* Albert Einstein to Upton Sinclair, December 19, 1938, Lilly Library, Indiana University. *Grynszpan shoots von Rath* Arthur D. Morse, *While 6 Million Died: A Chronicle of American Apathy* (New York: Ace, 1968), pp. 181, 182.

309 *"spontaneous anti-Jewish demonstrations"* *Leni Riefenstahl: A Memoir* (New York: St. Martin's Press, 1993). *"I wish you had killed 200 Jews"* Hitler adjutant Captain Weidemann, quoted in ibid. *"the deepest sorrow"* Highfield and Carter, *Private Lives*, p. 241. *During the Christmas season* Sayen, *Einstein in America*, pp. 126–127.

310 *"newest development in nuclear physics"* David Irving, *The German Atomic Bomb: The History of Nuclear Research in Nazi Germany* (New York: Da Capo Press, 1983), pp. 36, 37.

Chapter 30: World War II and the Threat of Fission

311 *"standing before one of the greatest scientists of all time"* Bargmann, interview, December 1, 1982. *"He would make mistakes, too"* Ibid. *"he was enormously relaxed"* Ibid.

312 *"No, no," Einstein said* "Einstein's Algebraic 'X's' and 'C's' Defeat Explaining Airport Aide," The New York Times, March 23, 1939, p. 25. *"an interesting mathematical proposition"* Gustav Bucky, "An Einstein Anecdote," *Jewish Frontier Magazine* (June 1939): p. 47. *"delicacy of thought"* Ibid.

313 *"Nathan was an economist"* Anonymous, interview by the author, tape recording, December 2, 1982. *Abrams on becoming the Einsteins' family doctor* Dr. Henry Abrams, interview by the author, tape recording, June 23, 1995.

314 *"I have just finished reading"* Einstein Archives, 52 8400. *"rough-neck sailor"* Ibid. *The sailor got a reply four months later* Einstein Archives, 52 841.

315 *"Now indeed you are safe"* Max von Laue to Albert Einstein, March 1939, in Dukas with Hoffmann, *Einstein: The Human Side*, pp. 104–105. *Winterkorn brands Einstein a swindler* Einstein Archives, 52 625. *"Doubtless much worse libels than this"* Ibid. *"None of the stories is true"* Einstein Archives, 52 627. *"forced to abandon the great electromagnet"* Clark, *Einstein: The Life and Times*, p. 568. *"Your idea that Moses was a distinguished Egyptian"* Jones, *The Last Phase*, p. 243.

316 *"the news about nuclear fission"* Eugene Wigner, interview by the author, tape recording, June 29, 1989. *"feared that if the Germans got hold of uranium deposits"* Ibid. *"It was July"* Ibid. *"We told Einstein about the process"* Ibid.

317 *"Einstein was very quick to see"* Leo Szilard, *The Bulletin of the Atomic Scientists* (March 1979): p. 58. *"So," Wigner recalled* Wigner, interview, June 29, 1989. *Then, explained Wigner* Ibid. *"explosive bombs of unimaginable power"* Weart and Szilard, *Leo Szilard: His Version of the Facts*, p. 83. *"I took the letter"* Wigner, interview, June 29, 1989.

318 *"Yes, yes," Teller recalled Einstein saying* Rhodes, *Making of the Atomic Bomb*, p. 307. *"his sentences were endless and convoluted"* Peter Wyden, *Day One: Before Hiroshima and After* (New York: Simon and Schuster, 1984), p. 34.

319 *"Alex, what you are after"* John Gunther, *Roosevelt in Retrospect* (New York: Pyramid, 1962), p. 316. *"for your recent letter"* Ibid. *secret research on uranium* Einstein's source for this information was chemist Peter Debye, a 1936 Nobel laureate who had been expelled from Germany when he refused to give up his Dutch citizenship and become a German citizen. Wigner, interview, June 29, 1989.

320 *Szilard predicted Hitler would win the war* Wigner, interview, June 29, 1989. *"Heisenberg will not be able to withstand the pressure"* Thomas Powers, *Heisenberg's War: The Secret History of the German Bomb* (New York: Knopf, 1993), pp. 106–107. *"I am so old-fashioned and stubborn"* Seelig, *Albert Einstein*, p. 184. *FBI agent interviews Einstein* FBI file 67CD, August 18, 1940, pp. 117–122.

321 *Einstein explains why Debye left Germany* Ibid., pp. 120–121. *Einstein did not question Debye's integrity as a scientist* Ibid. *if his "motives are bad"* Ibid., p. 122. *second meeting with an FBI agent* Ibid., November 8, 1940, pp. 129–131.

322 *Alice Kahler's recollections of Einstein* Alice Kahler, interview by the author, tape recording, April 15, 1989.

323 *"We ought to have declared war"* Michael Holroyd, *Bernard Shaw: Volume Three: 1918–1950, The Lure of Fantasy* (New York: Random House, 1991), p. 432. *"Can it be believed"* George Bernard Shaw to William J. Pickerill, October 10, 1940, in Laurence, *Bernard Shaw: Collected Letters*, p. 540. *"You may be a great scientist"* "Hibernicus" to Albert Einstein, December 16, 1940, Einstein Archives, 31 974. *First letter from Marie Winteler* Marie Winteler [Muller] to Albert Einstein, June 27, 1940, Einstein Archives, 56 354. *Three months later Marie wrote again* Marie Winteler [Muller] to Albert Einstein, September 11, 1940, Einstein Archives, 56 356. *a cry for help from a relative* Brigitte Alexander-Katz to Albert Einstein, 1940, Einstein Archives, 54 745.

324 *After a week in Biarritz* Ibid. *offered "to vouch for their reliability"* Albert Einstein to Mrs. Eryl Rudlin, March 12, 1941, The Institute for Advanced Study. *"for almost two years my friend and myself"* Einstein Archives.

325 *"Cranks tried to get through to Einstein"* Bargmann, interview, December 1, 1982.

Chapter 31: The Race for the Bomb

326 *Wilhelm Reich at Einstein's home* Ilse Ollendorff Reich, *Wilhelm Reich: A Personal Biography* (London: Elek, 1969), pp. 57–59. *"Can you understand now why everyone thinks I'm mad?"* Michel Cattier, *The Life and Work of Wilhelm Reich*, trans. Ghislaine Boulanger (New York: Avon, 1971), p. 205. *"how exciting it was"* Ollendorff Reich, *Wilhelm Reich*, pp. 57, 58.

327 *"a communist-inspired conspiracy"* Ibid., p. 57. *"Einstein saw the phenomena"* Ibid. *"your beloved Schopenhauer"* Dukas with Hoffmann, *Einstein: The Human Side*, p. 79. *"Who knows for how long?"* Seelig, *Albert Einstein*, p. 199. *"He told me to come early"* Kahler, interview, April 15, 1989.

328 *"Listen dear, he loved women"* Ibid. *"The relationship between Einstein and Maja was very beautiful"* Ibid. *"He was a guest for several years"* Bucky, interview, September 7, 1993.

329 *Bucky also saw a lot of Margot* Ibid. *"People seem surprised that Einstein never drove a car"* Ibid. *"I have the feeling"* Albert Einstein to Upton Sinclair, October 23, 1941, Lilly Library, Indiana University. *"Einstein opened the front door to us"* Jane Leonard Swing Chapman, interview by the author, tape recording, May 10, 1988.

330 *"He looked like the room"* Ibid. *"the 'sum of histories' approach to quantum mechanics"* Wheeler, interview, June 25, 1989. *"like us, they have had almost three years"* Rhodes, *Making of the Atomic Bomb*, p. 412. *"were on the track of a weapon which could annihilate whole cities"* Albert Speer, *Inside the Third Reich: Memoirs*, trans. Richard Winston and Clara Winston (New York: Macmillan, 1970), pp. 225–227.

331 *"I have become a lonely old fellow"* Reichinstein, *Albert Einstein: A Picture of His Life and World*, p. 230. *"medicine that will cure all stomach aches"* Einstein Archives, 56 065. *Einstein replied with an emphatic no* Einstein Archives, 56 066.

332 *at a meeting chaired by "Rabbi Wise"* FBI file, 1942, 100-120147-1 (84), p. 76. *"In 1942 I was arrested"* von Laue, interview, December 12, 1994.

333 *"It may well be a matter of life and death"* Rudolf Ladenburg and Hermann Weyl to

U.S. Attorney General Francis Biddle, June 25, 1942, Einstein Archives, 55 532. *"I was treated quite well"* Theodore von Laue, interview by the author, tape recording, September 27, 1994. *Walter did survive* Charles Niles, Boston University Library Special Collections, interview by the author, Oct. 15, 1995. *"taken from his house by Gestapo agents"* Robert Reid, *Marie Curie* (New York: Saturday Review Press/ E. P. Dutton, 1974), p. 318.

334 *"in part beautiful, in part wicked"* Albert Einstein to E. P. St. John, March 9, 1943, in Sayen, *Einstein in America*, p. 156. *"I am in the Navy"* Pais, 'Subtle is the Lord', p. 12. *"the condition of my health"* Albert Einstein to Lt. Stephen Brunauer, June 29, 1943, National Archives, Washington, D.C. *"I would be very much molested"* Albert Einstein to Lt. Stephen Brunauer, July 30, 1943, National Archives. *"I am glad to hear that work is done"* Albert Einstein to Lt. Stephen Brunauer, August 13, 1943, National Archives. *"I do not have the feeling that much can be achieved"* Albert Einstein to Lt. Stephen Brunauer, September 1, 1943, National Archives.

335 *"discussing their new, more effective antiaircraft defense system"* Bucky, interview, September 7, 1993.

Chapter 32: Einstein Goes to War

336 *"Oh yes, very interesting"* George Gamow, *My World Line: An Informal Autobiography* (New York: Viking Press, 1970), p. 150. *"mentioned Pascual Jordan's idea"* Ibid. *Einstein admitted that he had goofed* Albert Einstein to Commander Stephen Brunauer, January 4, 1944, Naval Sea Systems Command, National Archives.

337 *"after considerable probing"* Sayen, *Einstein in America*, pp. 150–151. *Einstein took a break from war work* Albert Einstein and Erich Kahler, "Palestine Setting of Sacred History of Jewish Race," *Princeton Herald*, April 14, 1944, pp. 1, 6, 9. *"Professor Hitti says: 'The Hebrews came and went' "* Albert Einstein and Erich Kahler, "Arabs Fare Better in Palestine Than in Arab Countries," *Princeton Herald*, April 28, 1944, pp. 1, 20.

338 *Einstein accused of being a nationalist* April 14, 1944, Einstein Archives. *Sinclair had written "a thousand letters"* Harris, *Upton Sinclair*, p. 334. *the writer had given "the American public a vivid insight"* Albert Einstein to Upton Sinclair, June 10, 1944, Lilly Library, Indiana University.

339 *"There are two kinds of temptation"* Albert Einstein to Dr. S. H. Goldenson, August 28, 1944, Einstein Archives, 55 141. *"It was interesting to see"* Albert Einstein to Max Born, September 7, 1944, in *Born–Einstein Letters*, no. 81, p. 148. *Hedi Born was overjoyed to renew the relationship* Hedi Born to Albert Einstein, October 9, 1944, ibid., no. 82, 151–154.

340 *"combine an entirely mechanistic universe"* Max Born to Albert Einstein, October 10, 1944, ibid., no. 83, p. 155. *"I had occasion to visit"* Major Milton R. Wexler to Albert Einstein, September 17, 1944, Einstein Archives, 55 048. *Two months later, Roberto himself wrote* Roberto Einstein to Albert Einstein, November 27, 1944, Einstein Archives, 55 049. *"saw two outstretched arms"* Alan Windsor Richards, *Reminiscences* (Princeton, N.J.: Harvest House Press, 1979), pp. 1, 2.

341 *Stern calls on Einstein* Wyden, *Day One*, pp. 127–128n. *"a secret technical arms race"* Albert Einstein to Niels Bohr, December 12, 1944, in Abraham Pais, *Niels Bohr's Times, in Physics, Philosophy, and Polity* (New York: Oxford University Press, 1991), p. 503. *"a method that has been rejected for too long"* Ibid. *"responsible statesmen in America and England"* Clark, *Einstein: The Life and Times*, p. 577. *Einstein advised Stern* Ibid., pp. 577, 578.

Chapter 33: The Atomic Bomb

342 *Millikan's reply to Einstein's plea* Robert Millikan to Albert Einstein, February 7, 1945, Einstein Archives, 57 222. *"I received an answer from Dr. Millikan"* Albert Einstein to Theodore Strimling, February 24, 1945, Einstein Archives, 57 223.

343 *expected "to find him waiting impatiently"* Richards, *Reminiscences*, p. 4. *"The only good German is a dead one' "* Bucky, interview, September 7, 1993. *"had quite a discussion last night"* June 10, 1945, Einstein Archives, 57 287.
344 *"I have never talked to a Jesuit priest"* July 2, 1945, Einstein Archives, 57 288. *"they reported something about the war"* Sayen, *Einstein in America*, p. 151. *Einstein talks to reporter about the A-bomb* Albert Einstein and Raymond Swing, "Einstein on the Atomic Bomb," *Atlantic Monthly* (November 1945): pp. 347–352.
345 *"I feel as if the atom bomb has hit me personally"* Helen Dukas to Alice Kahler, August 8, 1945, author's collection.
346 *Based on a tip from Mexico* January 11, 1945, Newark FBI files, no. 100-29614, pp. 1–6; Special Censorship Watch List, February 7, 1945, 28455, p. 1. *argument in a Philadelphia hotel in the early 1930s* September 4, 1945, Philadelphia FBI file no. 100-29919, pp. 1–6.

Chapter 34: Toward a Jewish State

347 *Dukas was in contact with suspected spies* January 17, 1946, FBI file no. 100-338078-26, p. 1.
348 *Einstein was called to speak* Harold A. Hinton, "Einstein Condemns Rule in Palestine: Calls Britain Unfit but Bars Jewish State and Favors UNO," *The New York Times*, January 12, 1946, p. 7. *When asked what would happen* Sayen, *Einstein in America*, pp. 234–236.
349 *that was as far as Einstein would go* Ibid., p. 236. *"although I was very much a Zionist"* I. F. Stone, interview by the author, tape recording, May 20, 1987. *Einstein refused to join an appeal to ease the surrender terms imposed on Germany* Albert Einstein to Professor Middledorf, University of Chicago, January 6, 1946, Einstein Archives.
350 *Dr. Eck on the Reformation* German theologian Johann Eck (1486–1543) was Martin Luther's most fierce and formidable opponent—and, incidentally, a greedy, sexually promiscuous alcoholic. *"The Germans slaughtered my Jewish brethren"* Albert Einstein to Arnold Sommerfeld, in Bernstein, *Einstein*, p. 169. *"I am sending them to nobody else"* Albert Einstein to Erwin Schrödinger, in Moore, *Schrödinger*, p. 424. *Schrödinger and Einstein continued their discussion by mail* Ibid., pp. 426–427.
351 *"Every now and then we had votes"* Senator Alan Cranston, interview by the author, tape recording, September 25, 1987, Washington, D.C. *"Why is it that when the mind of man"* Granville Clark, letter to the editor, *The New York Times*, April 22, 1955. *Ehrenburg recalled that "quite unexpectedly"* Ilya Ehrenburg, *Post-War Years: 1945–54* (Cleveland: World, 1967), pp. 76–77. *"They live in terrible conditions"* Ibid., p. 77.
352 *"his eyes lit up with a challenging gleam"* Einstein detested the racism he encountered in the United States. Earlier in the year, he had replied to a correspondent: "Your letter is fresh evidence of the intensity of prejudice against the Negro existing in this country. You speak with that certainty of the correctness of your own convictions characteristic for majorities—no matter what subject they deal with. A few centuries ago you would have defended the witchcraft trials in the same spirit . . . The pigmentation of the human skin is closely connected with the intensity of the rays of the sun prevailing in a given climate . . . Accordingly, the populations of Africa, anthropologically so various, all display an adaptation of pigmentation according to the intensity of the sun's rays in the climate they live in. But that these facts have anything whatever to do with inherited qualities of moral and mental character is as improbable as any other superstition. Every reasonable man must see this, if he is capable of opposing rational consideration to irrational feeling." January 24, 1946, Einstein Archives, 56 709. *"The main thing now is to prevent an atomic catastrophe"* Ehrenburg, *Post-War Years*, p. 77. *more bad news on the world front* Thomas, *History As You Heard It*, pp. 309, 311. *"I love you"* Einstein Archives, 32 006. *"He feels these voices"* Einstein Archives, 57 331. *Einstein replied immediately* Einstein Archives, 57 332.
353 *"he touched my elbow"* Dorothy Commins, "Professor Einstein in Princeton" (unpub-

lished memoir, author's collection). *"when MGM wanted to do a movie"* William Lanouette, interview by the author, tape recording, July 18, 1995.

354 *"the film is so laced with sentiment"* Bosley Crowther, "Atomic Bomb Film Starts At Capitol," *The New York Times,* February 21, 1947, p. 15. *"there is very little work for me"* Albert Einstein to the Sinclairs, June 3, 1946, Lilly Library, Indiana University. *"When I said it was for a film"* Ashley Montagu, interview by the author, tape recording, May 5, 1994. *Einstein and Montagu meet to discuss the film* Ibid.

355 *"So I called him up"* Ashley Montagu, interview by the author, tape recording, April 3, 1995.

356 *Montagu tells Einstein a relativity joke* Ibid. *"a land of true democracy"* U.S. ambassador in Moscow to FBI via State Department, July 19, 1946, FBI file no. 21278, despatch no. 240, pp. 1–2.

357 *"Let's go down and see Einstein"* Abraham Pais, interview by the author, tape recording, September 10, 1988. *Einstein "wasn't like other truly very great men I have known"* Ibid. *"Einstein was interested"* Ibid. *meet "once again in the best of all possible worlds"* Albert Einstein to Maurice Solovine, October 5, 1946, in Einstein, *Letters to Solovine,* p. 97. *"the manufacturing of an anti-Russian spirit"* Albert Einstein to Upton Sinclair, September 26, 1946, Lilly Library, Indiana University. *had behaved "very courageously"* Ibid.

Chapter 35: The Birth of Israel

358 *"one can live in unbelievable peace"* Moore, *Schrödinger,* p. 427. *by "using Affine geometry"* Walter Moore, interview by the author, tape recording, October 14, 1995. *It was snowing heavily* Moore, *Schrödinger,* p. 432.

359 *"Schrödinger's latest effort"* Ibid. *Schrödinger was sick* Ibid., p. 433.

360 *the letter that followed* Ibid. *Sidney Hook appeared on the scene* Hook, *Out of Step,* pp. 467–469. *In April 1947* Albert Einstein to Maurice Solovine, April 9, 1947, in Einstein, *Letters to Solovine,* p. 99.

361 *"I know what's wrong, my dear"* Sayen, *Einstein in America,* p. 131. *"Einstein would walk past my house"* Wheeler, interview, June 25, 1989. *Einstein deals with Mileva's financial problems* Highfield and Carter, *Private Lives,* pp. 252–253. *"Einstein lived three doors from us"* Mab Cantril's unpublished memoir is located in the Einstein Archives, 42 721.

362 *"Why hasn't Michele made some important discovery"* Niccolo Tucci, "The Great Foreigner," *New Yorker* (November 22, 1947). *"A butterfly is not a mole"* Ibid. *"Would you please write"* Einstein Archives, 46 622. *"asked a man at the desk"* George Wald, interview by the author, tape recording, May 18, 1988.

363 *"What I mean is"* Albert Einstein to Otto Juliusburger, September 29, 1947, in Dukas with Hoffmann, *Einstein: The Human Side,* p. 82. *"beautiful and fruitful period"* Clark, *Einstein: The Life and Times,* pp. 603, 604. *Einstein accepts an award* Harrison Brown, "An Early Brief Encounter," *Bulletin of the Atomic Scientists* (March 1979): p. 19. *"Everything is possible"* Highfield and Carter, *Private Lives,* p. 253.

364 *"Wallace today is a captive of the Communist party"* Hook, *Out of Step,* pp. 469–471. *"What I have really done"* Ibid., p. 471. *"On the basis of an open letter"* Michelmore, *Einstein,* p. 240.

365 *"When you warned the world"* Hook, *Out of Step,* pp. 472–473. *"and there you were"* Max Born to Albert Einstein, March 4, 1948, in *Born–Einstein Letters,* no. 85, p. 160. *"The day will inevitably come"* Albert Einstein to Upton Sinclair, March 29, 1948, Lilly Library, Indiana University. *briefly toyed with the idea of leaving America* Clark, *Einstein: The Life and Times,* p. 607. *at sixty-nine he was too old* Brodetsky, *Memoirs,* p. 289. *"a curious drama in which we all appear"* Clark, *Einstein: The Life and Times,* p. 607.

366 *"A wonderful place to contemplate"* Dr. Henry Abrams, interview, June 23, 1995. *Gödel becomes an American citizen* John L. Casti, *Searching for Certainty: What Scientists Can*

Know About the Future (New York: Morrow, 1990), pp. 373–374. *"My father is a Jew hater"* Einstein Archives, 57 705.

367 *"magnificent letter"* Einstein Archives, 57 706. *"As an old Jew"* Pais, *'Subtle is the Lord'*, p. 517. *"Einstein and ten former Nazi brain-trusters"* FBI file, 94-39617-1.

368 *"If we wait until the Great Powers"* Clark, *Einstein: The Life and Times*, p. 605. *The State of Israel comes into being* David Ben-Gurion, *Israel: A Personal History* (New York: Funk & Wagnall/Sabra, 1971), pp. 92–93. *Theodor Herzl* In 1897, Herzl wrote: "In Basle I created the Jewish State. In five years perhaps, in fifty certainly, everyone will see it." His prophecy was off by only a year. *Ben-Gurion was awakened* Moshe Pearlman, *Ben Gurion Looks Back* (New York: Shocken Books, 1970), p. 11. *astonished the American delegation* David McCullough, *Truman* (New York: Simon and Schuster, 1992), p. 618.

369 *"the fulfillment of our dreams"* Albert Einstein, in a statement he sent to the Hebrew University, Jerusalem, March 15, 1949, in Clark, *Einstein: The Life and Times*, p. 605. *"she would show him letters"* Clair Gilbert, interview by the author, tape recording, December 14, 1991. Mrs. Gilbert is the daughter of Gina Plunguian. *Mileva lay in a hospital* Highfield and Carter, *Private Lives*, p. 253. *went to Zurich to settle her affairs* Ibid., pp. 254, 255. *Kemeny becomes Einstein's research assistant* John Kemeny, interview by the author, tape recording, September 16, 1988.

372 *He read to her every evening* Albert Einstein to Maurice Solovine, April 9, 1947, in Einstein, *Letters to Solovine*, pp. 106–107. *their conversation was about world affairs* Linus Pauling, interview by the author, tape recording, March 14, 1993. *agonizing stomach pains* *New York Herald Tribune*, January 14, 1949. *On the way home* Aaron B. Lerner, *Einstein and Newton* (Minneapolis: Lerner, 1973), photo section between pp. 140, 141.

373 *"Is that the Lord?"* Sayen, *Einstein in America*, p. 293.

Chapter 36: The FBI Targets Einstein

374 *"We almost had to force him to join us"* Thomas Bucky, interview by the author, tape recording, December 23, 1994. *"I was sorry that you were sick"* Einstein Archives, 12 625. *Yes, but with an important reservation* Einstein Archives, 58 972.

375 *"You are instructed to submit an explanation"* From Director, FBI, to Newark Office, Air Mail Special Delivery, March 11, 1949. *"It is not felt that there is sufficient basis"* FBI file no. 100-29614, 78245, March 25, 1949, pp. 1–2. *"The Agent to whom this case is assigned"* FBI file no. 100-338078, April 6, 1949. *"Bohr never had a full sentence ready"* Pais, *Niels Bohr's Times*, p. 13.

376 *"This is not a jubilee book"* de Broglie, *New Perspectives*, pp. 152–153. *de Broglie's "25-year-old allegiance"* Ibid., p. 153. *Princeton had invited "a number of big shots"* John Kemeny, interview by the author, tape recording, October 23, 1987.

377 *"Oppenheimer made fun of Einstein"* Bucky, interview, December 23, 1994. *Deeply moved by an affectionate birthday letter* Albert Einstein to Maurice Solovine, March 28, 1949, in Einstein, *Letters to Solovine*, p. 111. *Einstein and Montagu discuss the human instinct for aggression* Montagu, interview, April 3, 1995. *Montagu explained that human beings have no instincts* Ibid.

378 *"the crippling of individuals"* Ibid. *lonely widow and mother of five* Einstein Archives. *Joanna Fanta claimed to be Einstein's girlfriend* Dr. Eva Short, interview by the author, tape recording, November 23, 1994. Dr. Short, a psychiatrist, is the niece of Joanna Fanta.

379 *"I have no objection whatever"* Ella Winter to Albert Einstein, June 28, 1949, Einstein Archives, 59 024. *Einstein replied to the nun* Einstein Archives, 59 022.

380 *"The first thing Einstein wanted to know"* Mrs. John Kieran, letter to the author, October 11, 1994. *"Mr. Berg, you teach me baseball"* Nicholas Dawidoff, *The Catcher Was a Spy: The Mysterious Life of Moe Berg* (New York: Pantheon, 1994), p. 250–251. *After Einstein*

resumed his work   Regis, *Who Got Einstein's Office?*, p. 63.   *A fellow physicist*   Robert Jastrow, interview by the author, tape recording, June 10, 1995.   *"ought to get another assistant"* Kemeny, interview, September 16, 1988.

Chapter 37: The Communist Witch-Hunt

382 *"To protect Einstein from cranks"*   Eric Rogers, interview by the author, tape recording, September 27, 1988.   *"'he has so much fur'"*   Sayen, *Einstein in America*, p. 131.   *"Helen Dukas warned me"*   Rogers, interview, September 27, 1988.   *"a wonderful, marvelous character"*   Edward Greenbaum, interview for the Oral History Collection, American Institute of Physics, Center for History of Physics.
384 *Rankin earned himself a headline*   Edwin R. Bayley, *Joe McCarthy and the Press* (Madison, Wisc.: University of Wisconsin Press, 1981), p. 3.   *"The Bureau files reflect that Einstein is affiliated"*   From D. M. Ladd to the Director, Subject: Professor Albert Einstein, SP 16 SKIPB, pp. 1–5.   *further information exonerating Einstein*   Ibid., March 10, 1950.   *"Since the announcement of the espionage affair"*   Translation from the German. Subject: Professor Albert Einstein, Security Matter — To the Department of State.
385 *Rabbeis informs on Einstein*   June 22, 1950, FBI files V111-12915/D-137899.   *Hoover was inclined to believe the worst*   FBI file, July 31, 1950.   *"one of the friendliest of men"*   J. Robert Oppenheimer, "On Albert Einstein," in *Einstein: A Centenary Volume*, ed. A. P. French (Cambridge, Mass.: Harvard University Press, 1979), pp. 47–48.   *It would be a traumatic summer*   Thomas, *History As You Heard It*, pp. 377–378.
386 *"though occasionally he shows his independence"*   Louis Budenz, *Men Without Faces* (New York: Harper & Brothers, 1950) pp. 211, 243.   *"Heading the list"*   *New York Daily Mirror*, August 17, 1950.   *"Reference is made to your letter"*   To Assistant Chief of Staff, G-2, Department of the Army, The Pentagon, Washington 25, D.C. — From John Edgar Hoover, pp. 1–2.
387 *In his reply to Hook*   Albert Einstein to Sidney Hook, May 16, 1950, in Hook, *Out of Step*.   *"In view of the stream of revelations"*   Ibid., pp. 477–478.   *Einstein's response bitterly disappointed Hook*   Ibid., p. 478.
388 *"his silence concerning the brutal treatment of Jews"*   Ibid.   *"If you're a liberal you're an antifascist"*   Hook, interview, November 27, 1982.   *"a kind of hopelessness had come over Einstein"*   Anonymous, interview by the author, date withheld.   *"I never saw him depressed"* Bucky, interview, December 23, 1994.   *"His heart never bleeds"*   Leopold Infeld, *Albert Einstein: His Work and Its Influence on Our World* (New York: Scribner's, 1950), p. 118.
389 *"I simply enjoy giving more than receiving"*   Albert Einstein to Max and Hedi Born, April 12, 1949, in *Born–Einstein Letters*, no. 94, p. 182.   *Just before Christmas 1950*   Seelig, *Albert Einstein*, pp. 39–40.   *"to guard against becoming superficial"*   Ibid., p. 39.   *"I see in marching"*   Einstein Archives, 61 240.   *Freimuller's impression of Eduard*   Highfield and Carter, *Private Lives*, p. 255.   *Herzog was shocked when Eduard came to see him*   Ibid.
390 *the U.S. Army investigates Dukas*   From Gustav Bard, CS Team, CIC Region VIII, February 14, 1951, D-280200, VIII-12915 1-714.   *"perhaps a review of available biographies"*   FBI files, SAC, Newark, Internal Security.   *"your conduct in putting into the hands of the Russians"* H. Montgomery Hyde, *The Atom Bomb Spies* (New York: Atheneum, 1980), p. 262.   *David's wife, Ruth*   Ruth Greenglass to Albert Einstein, Mary 1951, Einstein Archives, 41 557.
391 *"Whatever a man or woman has done"*   Albert Einstein to Ruth Greenglass, June 3, 1951, Einstein Archives, 41 559.   *"My dear sister was delivered from her horrible suffering"*   Albert Einstein to Maurice Solovine, July 30, 1951, in Einstein, *Letters to Solovine*, p. 129.   *"precious care"*   Albert Einstein to Dr. Guy Dean, June 26, 1951, Einstein Archives, 59 482.   *"I miss her"*   Albert Einstein to a cousin, June or July 1951, in Hoffmann with Dukas, *Einstein: Creator and Rebel*, p. 242.   *"curious to know Einstein's secret"*   Pearlman, *Ben Gurion Looks Back*, pp. 216–217.

392 *"Do you realize"* Robert St. John, *Ben-Gurion* (Garden City, N.Y.: Doubleday, 1959), pp. 234–235. *"were delighted to discover"* Maurice Friedman, *Martin Buber's Life and Work: The Later Years, 1945–1965* (New York: Dutton, 1983), pp. 148–149. *"A letter from Dr. Albert Einstein"* FBI file, 100-138754-835. *Willie McGee's appeal* For a fourth time the Supreme Court refused to review the case, and the thirty-two-year-old McGee was executed in the electric chair on May 8, 1951. Hodding Carter's liberal *Delta Democrat-Times* did not doubt McGee's guilt, but pointed out that in Mississippi no white man had ever been sentenced to death for rape, and that "the cause of democracy and justice is endangered . . . if a Negro continues to get one kind of punishment for rape or murder of a white victim and a white man gets another and lighter punishment for the rape or murder of a Negro victim." Quoted in "Case of Willie McGee," *The New York Times*, May 13, 1951, sect. 4, p. 2. *Go ahead, Hoover said* FBI file, SAC, Newark (65-3916) 100-338078, Subject: Espionage. *"I never read tragedy in his face"* Pais, *'Subtle is the Lord'*, p. 17.

Chapter 38: Conversations and Controversies

393 *Hans Albert believed it had started* Elizabeth Roboz Einstein, interview by the author, tape recording, August 13, 1988. Elizabeth was Hans Albert's second wife. *Einstein's immediate response* Highfield and Carter, *Private Lives*, p. 234.

394 *It was now some twenty years* Ibid., p. 256. *"tormented and brooding . . . but also serene smile"* Carl Seelig to Albert Einstein, March 22, 1952, ibid., p. 257. *"We got along great guns"* Evelyn Einstein, interview by the author, tape recording, November 13, 1994.

395 *Eduard never once asked about his father, mother, or brother* Ibid. *"and pass over whatever you wish in silence"* Albert Einstein to Maurice Solovine, March 30, 1952, in Einstein, *Letters to Solovine*, p. 131.

396 *"U.S. ATOM SCIENTIST'S BROTHER EXPOSED"* *Washington Times Herald*, July 12, 1947, p. 1. *an "old man being helplessly pushed around"* Richards, *Reminiscences*, pp. 6, 7. *"feverishly jotting in a notebook"* Ibid., p. 7. *"those values which are essential"* Hook, *Out of Step*, pp. 479–481.

397 *"become indignant in the face of such accusations"* Ibid., pp. 482–483. *"Gandhi could have been successful"* Ibid., pp. 483–484. *"in view of his own personal experiences"* Ibid., pp. 485–486.

398 *the Velikovsky affair* Ralph E. Juergens, "Minds in Chaos: A Recital of the Velikovsky Story," *American Behavioral Scientist* (September 1967): pp. 4–17. *Velikovsky was told to his dismay* Immanuel Velikovsky, *Stargazers and Gravediggers* (New York: Morrow, 1983), pp. 287–292.

399 *"When, by chance, we met last week"* Immanuel Velikovsky to Albert Einstein, August 26, 1952, Einstein Archives, 23 204. *"When I am busy calculating"* Albert Einstein to Hans Mühsam, 1940, in Clark, *Einstein: The Life and Times*, p. 612.

400 *"His main and urgent thought"* Abba Eban, *Personal Witness: Israel Through My Eyes* (New York: Putnam's, 1992), p. 228. *Wald and Einstein discuss amino acids, electrons, and the big bang* George Wald, interview, by the author, tape recording, October 9, 1985.

401 *An independent thinker* Andrew Patner, *I. F. Stone, A Portrait: Conversations with a Nonconformist* (New York: Doubleday/Anchor, 1990), p. 20. *"we wrote to his secretary"* I. F. Stone, interview by the author, tape recording, September 25, 1987.

402 *"I must have been sixteen"* Christopher Stone, interview by the author, tape recording, August 13, 1988. *"My own dad"* Ibid. *"I don't think Einstein let McCarthy bother him"* I. F. Stone, interview, September 25, 1987. *Einstein was "a very benevolent person"* Ibid.

403 *"Einstein's last shot"* Christopher Stone, interview, August 13, 1988. *"I loved Einstein"* I. F. Stone, interview, September 25, 1987.

Chapter 39: Einstein's Mercy Plea for the Rosenbergs

404 *"They put Jewish doctors on the platform"* David Shipler, *Russia: Broken Idols, Stolen Dreams* (New York: Times Books, 1983), p. 148. *"It's not easy to read history backwards"* I. F. Stone, interview, May 20, 1987. *"Not all of us Americans"* January 13, 1953, Einstein Archives, 41 575.

405 *"I was always told"* January 17, 1953, Einstein Archives, 41 603. *"Your intercession in behalf of the Rosenbergs"* January 19, 1953, Einstein Archives, 41 600. *Einstein told the rabbi* January 21, 1953, Einstein Archives, 41 601. *"In part this letter stated"* Einstein Archives, 41 547.

406 *Commins and Einstein have tea* Dorothy Commins, interview by the author, tape recording, August 13, 1988. *"He called up my husband and asked"* Ibid. *"We covered a good deal of ground"* Dorothy Commins, *What Is an Editor? Saxe Commins at Work* (Chicago: University of Chicago Press, 1978), p. 203.

407 *"Note you support the Rosenberg clemency"* FBI file, NK 100-29614, p. 3. *"It goes without saying"* Published in the *Newark Star Ledger*, January 22, 1953, ibid., p. 4.

408 *"my folks and uncle went with me"* Mark Abrams, interview by the author, tape recording, June 24, 1995. *Wheeler's students prepare questions for Einstein* Wheeler, interview, June 25, 1989. *had recently heard Einstein say* Ibid. *"It was a friendly, agreeable meeting"* Ibid.

409 *Einstein meets Adlai Stevenson* Commins, *What Is an Editor?*, p. 116. *"It is impossible for me to express"* Harold Urey to Albert Einstein, June 25, 1953, Einstein Archives, 41 556.

410 *"You have struggled for the creation"* Albert Einstein to Justice William O. Douglas, June 23, 1953, Einstein Archives, 41 576. *"Your letter . . . has reached me"* Justice William O. Douglas to Albert Einstein, June 30, 1953, Einstein Archives, 41 578. *"As far as I can judge"* August 1, 1953, Einstein Archives, 41 568. *"challenge these absurdities"* Bertrand Russell, *The Autobiography of Bertrand Russell: 1944–1969* (New York: Simon and Schuster, 1969), pp. 68–69. *"the data that went out"* Jim Miller, "The Executioner's Song," review of *The Rosenberg File*, by Ronald Radosh and Joyce Milton, *Newsweek* (September 12, 1983): p. 80. *"Unhappily, the Rosenberg Case"* Einstein Archives, 41 565. *"the question of commutation"* Walter Schneir and Miriam Schneir, *Invitation to an Inquest* (New York: Penguin Books, 1974), p. 180.

411 *"Although it was a hot, humid day"* Commins, interview, August 13, 1988. *"a land of mass-murderers"* Albert Einstein to Max Born, October 12, 1953, in *Born–Einstein Letters*, no. 103, p. 199. *Einstein was not impressed* Sayen, *Einstein in America*, p. 188. *asked to sign a protest letter* Ernst Straus, "Reminiscences," in *Albert Einstein: Historical and Cultural Perspectives: The Centennial Symposium in Jerusalem*, eds. Gerald Holton and Yehuda Elkana (Princeton, N.J.: Princeton University Press, 1982), p. 419. *"probably never became a relativist"* September 26, 1953, in *Born–Einstein Letters*, pp. 197, 198. *"Don't lose any sleep"* Albert Einstein to Max Born, October 12, 1953, ibid., p. 199.

412 *"Some days here in cold war Washington"* I. F. Stone to Albert Einstein, Einstein Archives, 61 494. *With that Velikovsky was off and running* Velikovsky, *Stargazers and Gravediggers*, p. 288. *"an accident of unusual rarity"* Ibid.

413 *"After our conversation last night' "* Ibid. *On the evening of* Sayen, *Einstein in America*, pp. 273–274. *"It was a black-tie affair"* Bucky, interview, December 23, 1994.

Chapter 40: The Oppenheimer Affair

415 *"looked at me with his face long and sad"* Wald, interview, October 9, 1985. *"Is there any chance for a person"* January 15, 1954, Einstein Archives, 32 299. *"some practical and social activities"* Responding to a reader of *The Tablet* who wondered if Einstein had been correctly quoted. Einstein Archives. *"An acquaintance of mine"* Einstein Archives, 32 298.

416 *"I will give you my considered answer"* Abraham Pais, interview by the author, tape recording, January 2, 1992. *"exchange letters with Teddy"* Highfield and Carter, *Private Lives*, p. 257. *"There is no one in the world"* Max Born to Albert Einstein, March 17, 1954, in *Born–Einstein Letters*. *As a witness* Sayen, *Einstein in America*, p. 277. *Norman Thomas declined an invitation* W. A. Swanberg, *Norman Thomas: The Last Idealist* (New York: Scribner's, 1976), p. 369.

417 *Although Einstein respected Thomas* Ibid., p. 370. *Einstein feared* A. M. Sperber, *Murrow: His Life and Times* (New York: Bantam, 1987), p. 427. *Even Einstein's friendly adversary* Hook, *Out of Step*, p. 503. *"In a great number of instances"* Peter Goodchild, *J. Robert Oppenheimer: Shatterer of Worlds* (Boston: Houghton Mifflin, 1981), p. 254. *"Oppenheimer is a loyal citizen"* Ibid., pp. 261–262.

418 *"he was surrounded by television cameras"* Sayen, *Einstein in America*, p. 286. *"'The newsmen were coming at him'"* Ibid. *Expecting Oppenheimer's enemies* Ibid., p. 289. *"What could I say"* Commins, *What Is an Editor?*, p. 217. *Velikovsky defends his book* Juergens, "Minds in Chaos," pp. 4–17.

419 *"used as a cable drop for a Soviet Espionage king"* FBI file, SAC, Newark (100-29614) (100-32986). Subject: Internal Security, November 9, 1954, pp. 1–2. *"A review of [Einstein's] file"* Ibid., p. 2.

420 *"The author of the book"* Einstein Archives, 59 419. *"Dukas absolutely idolized Einstein"* Gilbert, interview, December 14, 1991. *"I went a few days before the Nobel ceremony"* Linus Pauling, interview by the author, tape recording, February 22, 1993. *"I think I have made one mistake in my life"* Ibid.

Chapter 41: The Last Interview

421 *"This, though I did not know it"* Marian Anderson, *My Lord, What a Morning: An Autobiography* (New York: Viking, 1956), p. 267. *"was extremely friendly"* FBI file SAC, Newark (100-29614) (100-32986). Subject: Internal Security, March 9, 1955, pp. 1–8.

422 *"her only interest in life"* Ibid., p. 6. *established the whereabouts of his sons* Ibid., p. 7. *"additional investigation is not warranted"* Ibid., p. 2. *"He would have laughed"* Bucky, interview, October 2, 1988. *"The amount of money spent"* Richard Alan Schwartz, "The F.B.I. and Dr. Einstein," *Nation* 237, no. 6 (September 3–10, 1983): p. 172. *"What I most admired about him"* Albert Einstein to Michele Besso's sister and son, March 21, 1955, Pais, *'Subtle is the Lord'*, p. 302.

423 *"with malice aforethought in my letter"* I. Bernard Cohen, interview by the author, tape recording, October 9, 1985. *"There are so many unsolved problems"* Ibid. *"were directly based on experience"* Born, *My Life*, p. 167. *"Einstein asked about my training"* Cohen, interview, October 9, 1985.

424 *recalled with pleasure his visits to Mach* Ibid. *"He also discussed Velikovsky's shabby treatment"* Ibid. *"There is still more to talk about"* I. Bernard Cohen, "An Interview with Einstein," *Scientific American* 193 (1955): p. 73. *"'Wait. Wait.'"* Ibid. *Einstein and a friend discussed death* Kuznetsov, *Einstein*, p. 360. *"Everything is comfortable, but I am not"* Highfield and Carter, *Private Lives*, p. 261. *Eban and Dafni visit Einstein* Eban, *Personal Witness*, pp. 242–243.

425 *"Well, I went to Princeton"* *Abba Eban: An Autobiography*, p. 191. *As he was driven to the hospital* Infeld, *Albert Einstein*.

426 *"but his personality was the same"* Margot Einstein to Hedi Born, April 18, 1955, in *Born–Einstein Letters*, p. 234. *"We drove to Princeton"* Bucky, interview, December 23, 1994. *"I do not believe in artificially prolonging life"* Ibid. *"I talked to him"* Ibid. *"'Just give me a little time'"* Ibid. *"The cause was a ruptured aneurysm"* Ibid. *"As fearless as he had been"* Margot Einstein to Hedi Born, April 18, 1955, in *Born–Einstein Letters*, p. 234. *"This strange world"* Albert Einstein to Michele Besso's sister and son, March 21, 1955, in Pais, *'Subtle is the Lord'*, p. 302.

427 *"We are heartbroken"* Alice Kahler to Charlotte ————, April 20, 1955, author's collection. *terms of Einstein's will* Herbert E. Nass, *Wills of the Rich and Famous* (New York: Warner, 1991), pp. 221–223. The will was dated March 18, 1950. *"Its head is made of clay"* Pais, '*Subtle is the Lord*', p. 182.

428 *"Do you know what it's like"* Bela Kornitzer, "'Einstein Is My Father': An Intimate Glimpse of the World's Greatest Living Genius As Seen Through the Eyes of His Son," *Saturday Evening Post* (April 1951): p. 47. *a movie producer called on Hans Albert* Elizabeth Roboz Einstein, interview by the author, tape recording, May 15, 1982. *"The argument got pretty heated"* Kornitzer, "'Einstein Is My Father,'" p. 141.

429 *"conducted an enormous and worldwide correspondence"* Dyson, *From Eros to Gaia*, pp. 299, 300.

Chapter 42: Einstein's Legacy

430 *"completely cuckoo"* Sayen, *Einstein in America*, p. 181. *"his early papers are paralysingly beautiful"* Oppenheimer, "On Albert Einstein," p. 47. *"To what errors does Oppenheimer refer?"* Infeld, *Why I Left Canada* (Toronto: University of Toronto Press, 1978), pp. 176–177.

431 *"That fraction of a second"* Rick Gore, "The Once and Future Universe," *National Geographic* 163, no. 6 (June 1983): p. 741. *The satellite recorded "'Big Bang' theory backed by satellite data,"* *Miami Herald*, January 8, 1993, p. 3A. *"First of all"* John Kemeny, letter to the author, September 16, 1988.

432 *twelve leading Einstein experts* Chris Mihill, "Einstein Ahead of Himself," *The Guardian*, October 20, 1994. *"Einstein's unified field theory"* John Wheeler, interview by the author, tape recording, June 18, 1988. *"That's a very complicated question"* Abraham Pais, interview by the author, tape recording, December 29, 1994. *coauthor of Beyond Einstein* Dr. Michio Kaku and Jennifer Trainer, *Beyond Einstein: The Cosmic Quest for the Theory of the Universe.* (New York: Bantam, 1987). *"Einstein was working at a tremendous handicap"* Dr. Michio Kaku, interview by the author, tape recording, January 9, 1995.

433 *"full of harmony"* Margot Einstein, interview by Jamie Sayen, May 4, 1978, in Sayen, *Einstein in America*, p. 303. *To Henri Poincaré* de Broglie, et al., *Einstein: Such As We Knew Him*, pp. 196, 197. *"as rich and fruitful"* Niels Bohr, "Albert Einstein: 1879–1955," *Scientific American* 192, no. 6, (January 1955).

434 *"Maric seems to have been Einstein's intellectual peer"* Jill Ker Conway, "Having It All (or Most of It)," *The New York Times Book Review*, August 27, 1995, p. 7. *In Jill Ker Conway's review* Robert Schulmann and Gerald Holton, letter to the editor, *The New York Times Book Review*, October 8, 1995, p. 4. *"left no evidence of originality"* Gerald Holton, "Of Love, Physics and Other Passions: The Letters of Albert and Mileva," part 2, *Physics Today* (September 1994): pp. 41–42. Reprinted from Gerald Holton, *Einstein, History and Other Passions* (New York: American Institute of Physics Press, 1995).

435 *Zudek Zakel claimed to be Einstein's son* Michael Specter, "Could Physicist Be Einstein's Secret Son?" *Miami Herald*, July 23, 1995, p. 17A.

436 *the idea "that someone would go from Berlin to Prague"* Robert Schulmann, interview by the author, tape recording, September 11, 1995. *"calm satisfaction"* Albert Einstein to Maurice Solovine, March 28, 1929, in Einstein, *Letters to Solovine*, p. 111. *"He who finds a thought"* Hoffmann with Dukas, *Einstein: Creator and Rebel*, p. 253.

Appendix: Einstein's Brain

437 *"I'll tell you what happened"* Dukas, interview, June 16, 1980. *"Harvey, who did the autopsy"* Ashley Montagu, interview by the author, tape recording, November 23, 1994.

438 *When questioned* Dr. Thomas Harvey, interview by the author, tape recording, December 31, 1994.

439 *"it's a rather extraordinary brain"* Dr. Lucy Rorke, interview by the author, tape recording, December 31, 1994. *"I agree completely"* Ibid. *A scene in a 1993 BBC documentary* Jonathan Freedland, "In the Name of Science," *The Guardian Weekend*, December 17, 1994, pp. 11–15. *documentary reveals Einstein's eyes taken also* Ibid. *Soon after the article appeared* Dr. Henry Abrams, interview, June 23, 1995. *His son, Mark* Mark Abrams, interview, June 24, 1995.

Bibliography

Adler, Gerhard, and Aniela Jaffé, eds. *C. G. Jung Letters*. Princeton, N.J.: Princeton University Press, 1975.

Anderson, Marian. *My Lord, What a Morning: An Autobiography*. New York: Viking, 1956.

Barnett, Lincoln. *The Universe and Dr. Einstein*. New York: New American Library, 1954.

Barrow, John D., and Joseph Silk. *The Left Hand of Creation: The Origin and Evolution of the Expanding Universe*. New York: Basic Books, 1983.

Bayley, Edwin R. *Joe McCarthy and the Press*. Madison, Wisc.: University of Wisconsin Press, 1981.

Ben-Gurion, David. *Israel: A Personal History*. New York: Funk & Wagnall/Sabra, 1971.

Bentwich, Norman, and Helen Bentwich. *Mandate Memories: 1918–1948*. London: Hogarth Press, 1965.

Berger, Meyer. *The Story of The New York Times: 1851–1961*. New York: Simon and Schuster, 1951.

Bergmann, Peter G. *The Riddle of Gravitation*. New York: Scribner's, 1968.

Berkley, George E. *Vienna and Its Jews: The Tragedy of Success*. Cambridge, Mass.: Abt/Madison Books, 1988.

Bernstein, Jeremy. *Einstein*. New York: Viking Press, 1974.

———. *Quantum Profiles*. Princeton, N.J.: Princeton University Press, 1991.

Bernstein, Jeremy, and Gerald Feinberg. *Science and the Human Imagination*. Rutherford, N.J.: Fairleigh Dickinson University Press, 1978.

Beveridge, W. I. B. *The Art of Scientific Investigation*. New York: Vintage, 1950.

Beyerchen, Alan D. *Scientists Under Hitler*. New Haven, Conn.: Yale University Press, 1978.

Birkenhead, Earl of. *The Professor and the Prime Minister: The Official Life of Professor F. A. Lindemann, Viscount Cherwell*. Boston: Houghton Mifflin, 1962.

Blaedel, Niels. *Harmony and Unity: The Life of Niels Bohr*. Madison, Wisc.: Science Tech Publishers, 1988.

Blumberg, Stanley A., and Gwinn Owens. *Energy and Conflict: The Life and Times of Edward Teller*. New York: Putnam's, 1976.

Bohm, David. *The Special Theory of Relativity*. New York: W. A. Benjamin, 1965.

Born, Max. *My Life: Recollections of a Nobel Laureate*. New York: Scribner's, 1978.

———. *Physics in My Generation*. Elmsford, N.Y.: Pergamon Press, 1955.

Born-Einstein Letters. Commentary by Max Born. Translated by Irene Born. New York: Walker, 1971.

Boslough, John. *Stephen Hawking's Universe*. New York: Morrow, 1985.

Braunthal, Julius. *Victor and Friedrich Adler*. Vienna: Verlagder Wiener Völksbuchhaudling, 1965.

Brennan-Gibson, Margaret. *Clifford Odets*. New York: Atheneum, 1981.

Brodetsky, Selig. *Memoirs: From Ghetto to Israel*. London: Weidenfeld & Nicolson, 1960.

Bucky, Frida Sarsen. "You Have to Ask Forgiveness . . . Albert Einstein As I Remember Him." *Jewish Quarterly* 15, no. 4 (Winter 1967–68): pp. 31, 33.

Bucky, Gustav. "An Einstein Anecdote." *Jewish Frontier Magazine* (June 1939).

Budenz, Louis. *Men Without Faces*. New York: Harper & Brothers, 1950.

Bullock, Alan. *Hitler: A Study in Tyranny*. New York: Harper & Row, 1964.

Calder, Nigel. *Einstein's Universe*. Harmondsworth, Engl.: Penguin, 1979.

Calder, Ritchie. *Science in Our Lives*. New York: Signet/NAL, 1955.

Cane, Philip. *Giants of Science*. New York: Pyramid, 1961.

Capra, Fritjof. *The Tao of Physics: An Essay on Western Knowledge and Eastern Wisdom*. Cambridge, Mass.: MIT Press, 1966.

Cassidy, David C. *Uncertainty: The Life and Science of Werner Heisenberg*. New York: W. H. Freeman, 1992.

Casti, John L. *Searching for Certainty*. New York: Morrow, 1990.

Cattier, Michel. *The Life and Work of Wilhelm Reich*. Translated by Ghislaine Boulanger. New York: Avon, 1971.

Chaisson, Eric. *Relatively Speaking: Relativity, Black Holes, and the Fate of the Universe*. New York: W. W. Norton, 1988.

Chandrasekhar, S. *Eddington: The Most Distinguished Astrophysicist of His Time*. New York: Cambridge University Press, 1984.

Chaplin, Charles. *My Autobiography*. New York: Simon and Schuster, 1964.

Charon, Jean. *Cosmology: Theories of the Universe*. New York: McGraw-Hill, 1970.

Chevalier, Haakon. *Oppenheimer: The Story of a Friendship*. New York: Pocket Books, 1966.

Churchill, Winston. *The Gathering Storm*. Boston: Houghton Mifflin, 1948.

———. *Great Contemporaries*. New York: Fontana, 1965.

Clark, Ronald W. *Einstein: The Life and Times*. New York: Avon, 1971.

———. *The Life of Bertrand Russell*. New York: Knopf, 1975.

Cline, Barbara. *Men Who Made a New Physics*. New York: Signet, 1965.

Cohen, Harry A. "An Afternoon with Einstein." Jewish Spectator (January 1969): pp. 13–18.

Cohen, I. Bernard. "An Interview with Einstein." *Scientific American* 193 (1955): pp. 68–73.

———. *Revolution in Science*. Cambridge, Mass.: Harvard University Press, 1985.

Commins, Dorothy. *What Is an Editor? Saxe Commins at Work*. Chicago: University of Chicago Press, 1978.

Condon, Edward. *Reminiscences of Life In and Out of Quantum Mechanics*. New York: John Wiley & Sons, 1973.

Corson, William R., and Robert T. Crowley. *The New KGB: Engine of Soviet Power*. New York: Morrow, 1985.

Coughlin, Gene. *How to Be One Yourself: A Short Cut to Membership in the Second Oldest Profession*. New York: A. S. Barnes, 1961.

Craig, Gordon A. *The Germans*. New York: Putnam's, 1982.

Crowther, J. G. *Statesmen of Science*. Chester Springs, Pa.: Cresset Press, 1965.

Cuny, Hilaire. *Albert Einstein: The Man and His Theories*. Middlebury, Vt.: Paul S. Eriksson, 1965.

Curie, Eve. *Madame Curie*. New York: Pocket Books, 1964.

Da Costa Andrade, E. N. *Rutherford and the Nature of the Atom*. Garden City, N.Y.: Doubleday, 1964.

Davies, Paul. *The Edge of Infinity*. New York: Simon and Schuster, 1981.

Davies, Paul, and John Gribbin. *The Matter Myth*. New York: Simon and Schuster, 1992.

Dawidoff, Nicholas. *The Catcher Was a Spy: The Mysterious Life of Moe Berg*. New York: Pantheon, 1994.

de Broglie, Louis. *New Perspectives in Physics*. Edinburgh: Peebles Press, 1962.

de Broglie, Louis, et al. *Einstein: Such As We Knew Him*. Edinburgh: Peebles Press, 1979.

Delmer, Sefton. *Trail Sinister: An Autobiography*. London: Secker and Warburg, 1961.

Dilling, Elizabeth. *The Red Network*. Milwaukee, Wisc.: n.p., 1934.

Douglas, A. Vibert. *The Life of Arthur Stanley Eddington*. London: Nelson, 1956.

Dryfoos, Susan W. *Iphigene: My Life and The New York Times*. New York: Times Books, 1987.

Dukas, Helen, with Banesh Hoffmann. *Albert Einstein: The Human Side: New Glimpses from His Archives*. Princeton, N.J.: Princeton University Press, 1979.

Dyson, Freeman. *Disturbing the Universe*. New York: Harper & Row, 1979.

———. *From Eros to Gaia*. New York: Pantheon, 1992.

———. *Infinite in All Directions*. New York: Harper & Row, 1985.

Eastman, Max. *Einstein, Trotsky, Hemingway, Freud and Other Great Companions*. New York: Collier Books, 1962.

Eban, Abba. *Abba Eban: An Autobiography*. New York: Random House, 1977.
———. *Personal Witness: Israel Through My Eyes*. New York: Putnam's, 1992.
Eddington, A. S. *The Nature of the Physical World*. New York: Macmillan, 1931.
Ehrenburg, Ilya. *Post-War Years: 1945–54*. Cleveland: World, 1967.
Ehrenhaft, Felix. "My Experiences with Einstein." Washington, D.C.: Smithsonian Institution Libraries.
Einstein, Albert. "How I Created the Theory of Relativity." *Physics Today* 35, no. 8 (August 1982): pp. 45–47.
———. *Ideas and Opinions*. Translated by Sonja Bargmann. New York: Crown, 1982.
———. *Letters to Solovine*. Introduction by Maurice Solovine. Translated by Wade Baskin. New York: Philosophical Library, 1987.
———. *Living Philosophies*. New York: Simon and Schuster, 1931.
———. *Out of My Later Years*. New York: Philosophical Library, 1950.
———. Preface to *Einstein: A Biographical Portrait*, by Anton Reiser. New York: Albert & Charles Boni, 1930.
———. Preface to *Mental Radio*, by Upton Sinclair. New York: Collier Books, 1971.
———. *Relativity: The Special and General Theory: A Popular Exposition*. New York: Henry Holt, 1921.
———. *The World As I See It*. London: Franklin Watts, 1940.
Einstein, Albert, and Leopold Infeld. *The Evolution of Physics*. New York: Simon and Schuster, 1942.
Epstein, Jacob. *Epstein: An Autobiography*. London: Vista Books, 1963.
Ettinger, Elzbieta. *Rosa Luxemberg: A Life*. Boston, Mass.: Beacon, 1987.
Falk, Bernard. *Five Years Dead*. London: Book Club, 1938.
Ferris, Timothy. *Coming of Age in the Milky Way*. New York: Morrow, 1988.
———. *The Red Limit: The Search for the Edge of the Universe*. New York: Morrow, 1977.
Feuer, Lewis S. *Einstein and the Generations of Science*. New York: Basic Books, 1974.
Fitzroy, Sir Almeric. *Memoirs*. 2 vols. London: Hutchinson, n.d.
Flexner, Abraham. *An Autobiography*. New York: Simon and Schuster, 1960.
Florence, Ronald. *Fritz: The Story of a Political Assassination*. New York: Dial, 1971.
Forsee, Aylesa. *Albert Einstein: Theoretical Physicist*. New York: Macmillan, 1963.
Frank, Philipp. *Einstein: His Life and Times*. Translated by George Rosen. New York: Knopf, 1947.
French, A. P., ed. *Einstein: A Centenary Volume*. Cambridge, Mass.: Harvard University Press, 1979.
Freud, Ernst, ed. *The Letters of Sigmund Freud and Arnold Zweig*. New York: Harcourt Brace Jovanovich, 1970.
Friedman, Maurice. *Martin Buber's Life and Work: The Later Years, 1945–1965*. New York: Dutton, 1983.
Friedrich, Otto. *Before the Deluge: A Portrait of Berlin in the 1920s*. New York: Harper & Row, 1972.
Frisch, Otto. *What Little I Remember*. Cambridge, Engl.: Cambridge University Press, 1979.
Gabor, Andrea. *Einstein's Wife: Work and Marriage in the Lives of Five Great Twentieth-Century Women*. New York: Viking, 1995.
Gamow, George. *Gravity*. Garden City, New York: Doubleday/Anchor, 1962.
———. *My World Line: An Informal Autobiography*. New York: Viking Press, 1970.
———. *Thirty Years That Shook Physics: The Story of Quantum Theory*. New York: Doubleday/Anchor, 1966.
Garbedian, Gordon H. *Albert Einstein: Maker of Universes*. New York: Funk & Wagnall, 1939.
Gardner, Martin. *Fads & Fallacies in the Name of Science*. New York: Dover, 1957.
Gay, Peter. *Freud: A Life for Our Time*. New York: W. W. Norton, 1988.
Ghiselin, Brewster. *The Creative Process: A Revealing Study of Genius at Work*. New York: Mentor/NAL, 1955.
Gilbert, Martin. *Winston Churchill: The Stricken World, 1916–1922*. Boston: Houghton Mifflin, 1975.

Giroud, Francoise. *Marie Curie: A Life*. Translated by Lydia Davis. New York: Holmes & Meier, 1986.

Goldberg, Stanley. *Understanding Relativity: Origins and Impact of a Scientific Revolution*. Boston: Berkhaeuser, 1984.

Goldberg, Stanley, and Roger H. Stuewer, eds. *The Michelson Era in American Science 1870–1930*. AIP Conference Proceedings 179. New York: American Institute of Physics, 1988.

Goodchild, Peter. *J. Robert Oppenheimer: Shatterer of Worlds*. Boston: Houghton Mifflin, 1981.

Griffith, John G. *The Trusty Servant*. Winchester, Engl.: Culverlands, 1986.

Grunberger, Richard. *A Social History of Germany, 1933 to 1945*. New York: Holt, Rinehart & Winston, 1971.

Gunther, John. *Roosevelt in Retrospect*. New York: Pyramid, 1962.

Hahn, Otto. *My Life*. London: Herder & Herder, 1970.

Haldane, Viscount. *An Autobiography*. New York: Doubleday, 1929.

Halsman, Philippe. *Sight and Insight*. New York: Doubleday, 1972.

Hardy, Alister, Robert Harvie, and Arthur Koestler. *The Challenge of Chance*. New York: Random House, 1973.

Harris, Leon. *Upton Sinclair: American Rebel*. New York: T. Y. Crowell, 1975.

Harrod, Roy F. *The Prof*. London: Macmillan, 1959.

Hatch, Alden. *Buckminster Fuller: At Home in the Universe*. New York: Crown, 1974.

Hawking, Stephen W. *A Brief History of Time*. New York: Bantam, 1988.

Haydn, Hiram. *Words and Faces*. New York: Harcourt Brace and Jovanovich, 1974.

Hazen, Robert, and James Trefil. *Science Matters*. New York: Doubleday, 1990.

Healy, Paul. *Cissy*. New York: Doubleday, 1966.

Heilbron, J. L. *The Dilemmas of an Upright Man: Max Planck as Spokesman for German Science*. Berkeley: University of California Press, 1986.

Heisenberg, Elisabeth. *Inner Exile: Reflections of a Life with Werner Heisenberg*. Boston: Birkhauser, 1984.

Heisenberg, Werner. Introduction to *Born–Einstein Letters*. Commentary by Max Born. Translated by Irene Born. New York: Walker, 1971.

———. *Physics and Beyond: Encounters and Conversations*. New York: Harper & Row, 1971.

———. *Physics and Philosophy*. New York: Harper Torch, 1958.

Herneck, Friedrich. *Einstein privat*. Berlin: Buchverlag der Morgen, 1978.

Hey, Tony, and Patrick Walters. *The Quantum Universe*. New York: Cambridge University Press, 1986.

Highfield, Roger, and Paul Carter. *The Private Lives of Albert Einstein*. London: Faber and Faber, 1993.

Hoffmann, Banesh, with Helen Dukas. *Albert Einstein: Creator and Rebel*. New York: Viking, 1972.

Hoge, Alice Albright. *Cissy Patterson*. New York: Random House, 1966.

Holroyd, Michael. *Bernard Shaw: Volume Three, 1918–1950, The Lure of Fantasy*. New York: Random House, 1991.

Holton, Gerald. *Thematic Origins of Scientific Thought: Kepler to Einstein*. Cambridge, Mass.: Harvard University Press, 1988.

———. *Einstein, History and Other Passions*. New York: American Institute of Physics Press, 1995.

Hook, Sidney. *Out of Step: An Unquiet Life in the 20th Century*. New York: Harper & Row, 1987.

Hubble, Edwin Powell. *The Problem of the Expanding Universe*. Washington, D.C.: Smithsonian Institution Press, 1943.

Hyde, H. Montgomery. *The Atom Bomb Spies*. New York: Atheneum, 1980.

Ickes, Harold. *The Secret Diary of Harold Ickes: The Inside Struggle*. Vol. 2: 1936–1939. New York: Simon and Schuster, 1954.

Infeld, Leopold. *Albert Einstein: His Work and Its Influence on Our World*. New York: Scribner's, 1950.

————. *Quest: The Making of a Scientist.* New York: Doubleday Doran, 1971.

————. *Why I Left America: Reflections on Science and Politics.* Edited by Lewis Pyenson. Translated by Helen Infeld. Montreal: McGill-Queen's University Press, 1978.

————. *Why I Left Canada.* Toronto: University of Toronto Press, 1978.

Irving, David. *The German Atomic Bomb: The History of Nuclear Research in Nazi Germany.* New York: Da Capo Press, 1983.

Jackman, Jarrell, and Carla Borden, eds. *The Muses Flee Hitler.* Washington, D.C.: Smithsonian Institution Press, 1983.

Jaffe, Bernard. *Michelson and the Speed of Light.* New York: Doubleday, 1960.

Jastrow, Robert. *God and the Astronomers.* New York: W. W. Norton, 1978.

————. "Have Astronomers Found God?" *The New York Times Magazine,* June 25, 1978, pp. 18–20, 22, 24, 26, 29.

Johnson, Niel M. *George Sylvester Viereck.* Champaign, Ill.: University of Illinois Press, 1972.

Jones, Ernest, M.D. *The Last Phase: 1919–1939.* Vol. 3 of *The Life and Work of Sigmund Freud.* New York: Basic Books, 1957.

Juergens, Ralph E. "Minds in Chaos: A Recital of the Velikovsky Story." *American Behavioral Scientist* (September 1967): pp. 4–17.

Jungk, Robert. *Brighter Than a Thousand Suns.* New York: Harcourt, Brace and Company, 1958.

Kaku, Dr. Michio, and Jennifer Trainer. *Beyond Einstein: The Cosmic Quest for the Theory of the Universe.* New York: Bantam, 1987.

Kessler, Count Harry. *The Diaries of a Cosmopolitan—1918–1937.* London: Weidenfeld & Nicolson, 1971.

————. *Walter Rathenau: His Life and Work.* New York: Harcourt, Brace & Co., 1930.

Kirsten, C., and H. J. Treder, eds. *Albert Einstein in Berlin: 1913–1933.* Berlin: Acadamie Verlag, 1979.

Kisch, F. H. *Palestine Diary.* London: Gollancz, 1938.

Klein, Martin J. *Paul Ehrenfest: The Making of a Theoretical Physicist.* New York: American Elsevier, 1970.

Klein, Martin J., A. J. Kox, and Robert Schulmann. *The Collected Papers of Albert Einstein, Volume Five: The Swiss Years: Correspondence, 1902–1914.* Translated by Anna Beck. Princeton, N.J.: Princeton University Press, 1993.

Kornitzer, Bela. " 'Einstein Is My Father': An Intimate Glimpse of the World's Greatest Living Genius As Seen Through the Eyes of His Son." *Saturday Evening Post* (April 1951).

Kuznetsov, B. *Einstein.* Translated by V. Talmy. Moscow: Progress Publishers, 1965.

Lampe, John A., D. D. "How Einstein Came to Princeton. *Saturday Review* (July 7, 1956): pp. 38–39.

Lanouette, William, and Bela Szilard. *Genius in the Shadows: A Biography of Leo Szilard.* New York: Scribner's, 1992.

Lash, Joseph P. *Eleanor and Franklin.* New York: Signet/NAL, 1971.

Laquer, Walter. *Stalin: The Glasnost Revelations.* New York: Scribner's, 1990.

Laurence, Dan H., ed. *Bernard Shaw: Collected Letters 1926–1950.* New York: Viking, 1988.

Laurence, William. *Heroes for Our Times: Albert Einstein.* Harrisburg, Pa.: Stackpole, 1968.

Lawrence, A. W., ed. *T. E. Lawrence by His Friends: Chaim Weizmann.* New York: McGraw-Hill, 1963.

Leopold, Nathan. *Life Plus 99 Years.* New York: Popular Library, 1958.

Lerner, Aaron B. *Einstein and Newton.* Minneapolis: Lerner, 1973.

Levine, Isaac Don. *Eyewitness to History.* New York: Hawthorn, 1973.

Levinger, Elma Ehrlich. *Albert Einstein.* New York: Messner, 1949.

Lightman, Alan, and Roberta Brower. *Origins: The Lives & Worlds of Modern Cosmologists.* Cambridge, Mass.: Harvard University Press, 1990.

Livingston, Dorothy Michelson. *The Master of Light.* New York: Scribner's, 1973.

McCullough, David. *Truman.* New York: Simon and Schuster, 1992.

MacDonald, D. K. C. *Faraday, Maxwell and Kelvin*. Garden City, N.Y.: Doubleday, 1964.
Manchester, William. *The Glory and the Dream*. New York: Bantam, 1990.
Mann, Arthur. *La Guardia: A Fighter Against His Times, 1882–1933*. New York: Lippincott, 1959.
Mann, W. Edward. *Orgone, Reich & Eros*. New York: Simon and Schuster, 1973.
Marianoff, Dimitri, and Palm Wayne. *Einstein: An Intimate Study of a Great Man*. New York: Doubleday, Doran & Co., 1944.
Martin, Charles-Noel. *The Universe of Science*. New York: Hill & Wang, 1963.
Martin, Ralph. *Cissy: The Extraordinary Life of Eleanor Medill Patterson*. New York: Simon and Schuster, 1974.
Mason, A. T. *Brandeis: A Free Man's Life*. New York: Viking, 1946.
Massey, Sir Harrie. *The New Age in Physics*. New York: Basic Books, 1950.
Maurice, Frederick. *Haldane 1915 to 1928*. Westport, Conn.: Greenwood, 1970.
Mencken, H. L. *The Diary of H. L. Mencken*. Edited by Charles A. Fecher. New York: Knopf, 1989.
Menuhin, Yehudi. *Unfinished Journey*. New York: Knopf, 1976.
Michelmore, Peter. *Einstein: Profile of the Man*. New York: Dodd, Mead, 1962.
Michelson, A. A. *Studies in Optics*. Chicago: University of Chicago Press, 1927.
Miller, Arthur I. *Albert Einstein's Special Theory of Relativity*. Reading, Mass.: Addison-Wesley, 1981.
Moore, Ruth. *Niels Bohr: The Man, His Science, and the World They Changed*. New York: Knopf, 1966.
Moore, Walter. *Schrödinger: Life and Thought*. Cambridge, Engl.: Cambridge University Press, 1989.
Morris, Richard. *The Edges of Science*. New York: Prentice Hall Press, 1990.
Morse, Arthur D. *While 6 Million Died: A Chronicle of American Apathy*. New York: Ace, 1968.
Moszkowski, Alex. *Einstein the Searcher: His Work Explained from Dialogues with Einstein*. Berlin: Fontane, 1921.
Mowrer, Edgar Ansel. *Triumph and Turmoil: A Personal History of Our Times*. New York: Weybright and Talley, 1968.
Nathan, Otto, and Hans Norden, eds. *Einstein on Peace*. New York: Simon and Schuster, 1961.
Ollendorf Reich, Ilse. *Wilhelm Reich: A Personal Biography*. London: Elek, 1969.
O'Neill, John J. *Prodigal Genius: The Life of Nikola Tesla*. New York: Ives Washburn, 1944.
Oppenheimer, J. Robert. "On Albert Einstein." In *Einstein: A Centenary Volume*, edited by A. P. French. Cambridge, Mass.: Harvard University Press, 1979.
Oursler, Fulton. *Behold This Dreamer*. Boston: Little, Brown, 1964.
Overbye, Dennis. *Lonely Hearts of the Cosmos: The Scientific Quest for the Secret of the Universe*. New York: HarperCollins, 1991.
Pagels, Heinz. *Perfect Symmetry: The Search for the Beginning of Time*. New York: Simon and Schuster, 1985.
Pais, Abraham. *Einstein Lived Here*. New York: Oxford University Press, 1994.
———. *Inward Bound*. New York: Oxford University Press, 1986.
———. *Niels Bohr's Times, in Physics, Philosophy, and Polity*. New York: Oxford University Press, 1991.
———. *'Subtle is the Lord . . .': The Science and the Life of Albert Einstein*. New York: Oxford University Press, 1982.
Parry, Albert. *Peter Kapitsa on Life and Science*. New York: Macmillan, 1968.
Patch, Blanche. *Thirty Years with G. B. S.* New York: Dodd, Mead & Co., 1951.
Patner, Andrew. *I. F. Stone, A Portrait: Conversations with a Nonconformist*. New York: Doubleday/Anchor, 1990.
Paulsen, F. *Immanuel Kant*. New York: Scribner, 1910.
Pearlman, Moshe. *Ben Gurion Looks Back*. New York: Schocken, 1970.
Persico, Joseph E. *Edward R. Murrow*. New York: McGraw-Hill, 1988.

Perutz, Max. *Is Science Necessary? Essays on Science and Scientists.* New York: Dutton, 1989.

Peukert, Detlev. *Inside Nazi Germany.* New Haven, Conn.: Yale University Press, 1987.

Pflaum, Rosalynd. *Grand Obsession: Madame Curie and Her World.* Garden City, N.Y.: Doubleday, 1989.

Planck, Max. *Scientific Autobiography and Other Papers.* New York: Philosophical Library, 1949.

Plesch, Janos. *Janos: The Story of a Doctor.* London: Gollancz, 1947.

Powers, Thomas. *Heisenberg's War: The Secret History of the German Bomb.* New York: Knopf, 1993.

Pyke, Magnus. *The Boundaries of Science.* Harmondsworth, Engl.: Pelican, 1963.

Rabi, I. I., et al. *Oppenheimer.* New York: Scribner's, 1969.

Radosh, Ronald, and Joyce Milton. *The Rosenberg File.* New York: Holt, Rinehart & Winston, 1983.

Rapoport, Victor Yakov. *The Doctors Plot of 1953.* Cambridge, Mass.: Harvard University Press, 1991.

Regis, Ed. *Who Got Einstein's Office? Eccentricity and Genius at the Institute for Advanced Study.* Reading, Mass.: Addison-Wesley, 1987.

Reich, Wilhelm. *Passion of Youth.* New York: Farrar, Straus & Giroux, 1988.

Reichinstein, David. *Albert Einstein: A Picture of His Life and His Conception of the World.* Prague: Stella Publishing House, 1934.

Reid, Robert. *Marie Curie.* New York: Saturday Review Press/E. P. Dutton, 1974.

Reinharz, Jehuda. *Chaim Weizmann: The Making of a Statesman.* New York: Oxford University Press, 1993.

Reiser, Anton. *Albert Einstein: A Biographical Portrait.* New York: Albert & Charles Boni, 1930.

Rhodes, Richard. *The Making of the Atomic Bomb.* New York: Simon and Schuster, 1986.

Richards, Alan Windsor. *Reminiscences.* Princeton, N.J.: Harvest House Press, 1979.

Riefenstahl, Leni. *Leni Riefenstahl: A Memoir.* New York: St. Martin's Press, 1993.

Robert, Marthe. *The Psychoanalytic Revolution: Sigmund Freud's Life and Achievement.* New York: Harcourt, Brace & World, 1966.

Rolland, Romain. *Journal des Années de Guerre 1914–1919.* Paris: Albin Michel, 1952.

———. *Memoires.* Paris: Albin Michel, 1956.

Rose, Norman. *Chaim Weizmann.* New York: Viking, 1986.

Rosenthal-Schneider, Ilse. *Reality and Scientific Truth.* Detroit: Wayne State University Press, 1980.

Ross, Ishbel. *Ladies of the Press.* New York: Harper, 1936.

Rowan-Robinson, Michael. *The Cosmological Distance Ladder: Distance and Time in the Universe.* New York: W. H. Freeman, 1985.

Rübel, Eduard. *Eduard Einstein.* Bern, Switz.: Paul Haupt, 1986.

Russell, Bertrand. *The Autobiography of Bertrand Russell: 1944–1969.* New York: Simon and Schuster, 1969.

———. *The Scientific Outlook.* New York: W. W. Norton, 1931.

Sachar, Howard Morley. *The Course of Modern Jewish History.* New York: Delta, 1958.

Safranski, Rudiger. *Schopenhauer.* Cambridge, Mass.: Harvard University Press, 1987.

St. John, Robert. *Ben-Gurion.* Garden City, N.Y.: Doubleday, 1959.

Sayen, Jamie. *Einstein in America: The Scientist's Conscience in the Age of Hitler and Hiroshima.* New York: Crown, 1985.

Schilpp, Paul Arthur, ed. *Albert Einstein: Philosopher-Scientist.* La Salle, Ill.: Open Court, 1995.

Schneir, Walter, and Miriam Schneir. *Invitation to an Inquest.* New York: Penguin, 1974.

Schoenberg, E. Randol. "Arnold Schoenberg and Albert Einstein: Their Relationship and Views on Zionism." *Journal of the Arnold Schoenberg Institute* 10, no. 2 (November 1987): pp. 134–182.

Schonberg, Harold C. *The Lives of the Great Composers.* New York: W. W. Norton, 1970.

Schulmann, Robert, and Jurgen Renn, eds. *Albert Einstein/Mileva Maric—Love Letters.* Princeton, N.J.: Princeton University Press, 1992.

Schwartz, Joseph. *The Creative Moment.* New York: HarperCollins, 1992.
Schwartz, Richard Alan. "The F.B.I. and Dr. Einstein." *Nation* 237, no. 6 (September 3–10, 1983): 168–173.
Schwinger, Julian. *Einstein's Legacy: The Unity of Space and Time.* New York: Scientific American Books, 1986.
Seelig, Carl. *Albert Einstein: A Documentary Biography.* Translated by Mervyn Savill. London: Staples Press, 1956.
Seldes, George, comp. *The Great Quotations.* New York: Pocket Books, 1967.
Serafini, Anthony. *Linus Pauling: A Man and His Science.* New York: Paragon House, 1989.
Shapley, Harlow. *Through Rugged Ways to the Stars.* New York: Scribner's, 1969.
Shipler, David. *Russia: Broken Idols, Stolen Dreams.* New York: Times Books, 1983.
Shirer, William L. *The Nightmare Years: 1930–1940.* Boston: Little, Brown & Co., 1984.
———. *The Rise and Fall of the Third Reich: A History of Nazi Germany.* New York: Fawcett Crest, 1966.
Sinclair, Mary Craig. *Southern Belle.* New York: Crown, 1957.
Sinclair, Upton. *Autobiography.* New York: Harcourt, Brace & World, 1962.
———. "Einstein As I Remember Him." *Saturday Review* (April 14, 1956).
———. *Mental Radio.* New York: Collier Books, 1971.
Smith, Gene. *The Ends of Greatness: Haig, Pétain, Rathenau, and Eden, Victims of History.* New York: Crown, 1990.
Snow, C. P. *Science and Government.* Cambridge, Mass.: Harvard University Press, 1961.
———. *Variety of Men.* New York: Scribner's, 1967.
Snyder, Robert, ed. *Buckminster Fuller: Autobiographical Monologue/Scenario.* New York: St. Martin's Press, 1980.
Sommer, Dudley. *Haldane of Cloan: His Life and Times.* London: Allen & Unwin, 1960.
Speer, Albert. *Inside the Third Reich: Memoirs.* Translated by Richard Winston and Clara Winston. New York: Macmillan, 1970.
Sperber, A. M. *Murrow: His Life and Times.* New York: Bantam, 1987.
Speziali, P., ed. *Einstein–Besso, Correspondence, 1903–1955.* Paris: Hermann, 1972.
Stachel, John, et al., eds. *The Collected Papers of Albert Einstein, Volume One: The Early Years, 1879–1902.* Translated by Anna Beck. Princeton, N.J.: Princeton University Press, 1987.
Stern, Beatrice M. "A History of the Institute for Advanced Study, 1930–1950." Special Collections of the Hoover Library, Western Maryland College, Westminster, Maryland, 1964.
Stern, Fritz. *Dreams and Delusions: The Drama of German History.* New York: Knopf, 1989.
Stone, I. F. *The Haunted Fifties.* New York: Random House, 1963.
Straus, Ernst. "Reminiscences." In *Einstein: Historical and Cultural Perspectives: The Centennial Symposium in Jerusalem,* edited by Gerald Holton and Yehuda Elkana. Princeton, N.J.: Princeton University Press, 1982.
Sullivan, Walter. *Black Holes: The Edge of Space, the End of Time.* Garden City, N.Y.: Doubleday, 1979.
Sulzberger, C. L. *A Long Row of Candles.* New York: Macmillan, 1969.
Swanberg, W. A. *Norman Thomas: The Last Idealist.* New York: Scribner's, 1976.
Szilard, Leo. *Recollections.* Cambridge, Engl.: Cambridge University Press, 1968.
Talmey, Max. *The Relativity Theory Simplified and the Formative Period of Its Inventor.* New York: Falcon Press, 1932.
Taton, Rene. *Science in the Twentieth Century.* New York: Basic Books, 1964.
Taylor, A. J. P. *The Origins of the Second World War.* New York: Fawcett, 1963.
Taylor, S. J. *Stalin's Apologist: Walter Duranty, The New York Times's Man in Moscow.* New York: Oxford University Press, 1990.
Teller, Edward. *Better a Shield Than a Sword.* New York: Free Press, 1987.
Thomas, Lowell. *History As You Heard It.* Garden City, N.Y.: Doubleday, 1957.
Thomson, J. J. *Recollections and Reflections.* London: G. Bell & Sons, 1936.
Trbuhovic-Gjuric, Desanka. *Das tragicsche Leben der Mileva Einstein-Maric.* Bern, Switz.: Paul Haupt, 1983.

Tschernowitz, Chaim. "A Day with Albert Einstein." *Jewish Sentinel* 1, no. 1 (September 1934): pp. 19, 34, 44, 50.

Ulam, Stanislaw. *Adventures of a Mathematician.* New York: Scribner's, 1976.

Vallentin, Antonina. *The Drama of Albert Einstein.* Translated by Moura Budberg. Garden City, N.Y.: Doubleday, 1954.

Velikovsky, Immanuel. *Stargazers and Gravediggers.* New York: Morrow, 1983.

Viereck, George Sylvester. *Glimpses of the Great.* New York: Macauley, 1930.

von Elbe, Joachim. *Witness to History.* Madison, Wisc.: The Max Kade Institute for German-American Studies, 1988.

Wachhorst, Wyn. *Thomas Alva Edison: The American Myth.* Cambridge, Mass.: MIT Press, 1981.

Walker, Mark. *Nazi Science: Myth, Truth, and the German Atomic Bomb.* New York: Plenum Press, 1995.

Wallace, Irving. *The Writing of One Novel.* New York: Simon and Schuster, 1969.

Watters, Leon L. "Comments on the Letters of Professor and Mrs. Albert Einstein to Dr. Leon L. Watters." The Leon L. Watters Collection, Jewish American Archives, Hebrew Union College, Jewish Institute of Religion, Cincinnati, Ohio, n.d.

Weart, Spencer, and Gertrud Weiss Szilard, eds. *Leo Szilard: His Version of the Facts.* Cambridge, Mass.: MIT Press, 1978.

Weinberg, Steven. *The First Three Minutes: A Modern View of the Origin of the Universe.* New York: Basic Books, 1977.

Weisberg, Robert. *The Myth of Genius.* New York: W. H. Freeman, 1986.

Weisskopf, Victor F. *The Privilege of Being a Physicist.* New York: W. H. Freeman, 1989.

Weizmann, Chaim. *Trial and Error: The Autobiography of Chaim Weizmann.* Philadelphia: Jewish Publication Society of America, 1949.

Weizmann, Vera, and David Tutaev. *The Impossible Takes Longer: The Memoirs of Vera Weizmann.* New York: Harper & Row, 1967.

Wheeler, James Archibald. *A Journey Into Gravity and Spacetime.* New York: Scientific American Library, 1990.

Whitrow, G. J., ed. *Einstein: The Man and His Achievement.* New York: Dover, 1973.

Whyte, Lancelot. *Focus and Diversions.* New York: Braziller, 1963.

Wigner, Eugene P., as told to Andrew Szanton. *The Recollections of Eugene P. Wigner.* New York: Plenum Press, 1992.

Will, Clifford. *Was Einstein Right?* New York: Basic Books, 1986.

Wilson, Margaret. *Ninth Astronomer Royal: The Life of Frank Watson Dyson.* Cambridge, Engl.: Cambridge University Press, 1951.

Winteler-Einstein, Maja. "Albert Einstein—A Biographical Sketch." Preface to *The Collected Papers of Albert Einstein, Volume One: The Early Years, 1879–1902.* Edited by John Stachel, et al. Translated by Anna Beck. Princeton, N.J.: Princeton University Press, 1987.

Wisehart, M. K. "A Close Look at the World's Greatest Thinker." *American* (June 1930).

Wolf, Fred Alan. *Taking the Quantum Leap: The New Physics for Non-Scientists.* New York: Harper & Row, 1989.

Wyden, Peter. *Day One: Before Hiroshima and After.* New York: Simon and Schuster, 1984.

Zukav, Gary. *The Dancing Wu Li Master: An Overview of Physics.* New York: Bantam, 1980.

Index

Abrams, Henry, 313–314, 366, 439
Abrams, Mark, 407–408, 439
Academy of Sciences, Berlin, 152, 432
Adams, Walter, 209, 211, 217
Adler, Friedrich (Fritz), 20, 51, 73–74, 75, 93–94
Adler, Viktor, 20, 73, 94
Alexander, James, 254
Alexander, Jean, 371
Alexander I, king of Yugoslavia, 220–221
Alexander-Katz, Brigitte, 323–324
American Civil Liberties Union, 410–411
American Committee for Cultural Freedom, 396–397
American Mathematical Society, 273
Anderson, H. L., 319
Anderson, Marian, 421
Anglo-American Committee of Inquiry, 347–348, 352, 401
Annalen der Physik, 28, 54, 58, 62, 63–64, 67–68, 81, 92
Anti-Defamation League, 308
Anti-Relativity League, 111
anti-Semitism
  AE's view of, 20, 109, 199, 248, 270
  German, 105, 109, 110–111, 138, 141–142, 199, 231, 247, 248, 307
  in Soviet Union, 270, 404
arms race, 341, 351, 352, 354, 365, 377–378, 397
Arrhenius, Svante August, 99
atomic bomb. *See also* arms race
  AE's response to use of, 344–345
  concern over German capability, 316, 317, 329, 330–331
  and FDR, 318, 319, 320, 321, 341, 345, 353–354, 420
  first detonated, 344
  idea of sharing with Soviet Union, 341, 344
  Soviets' first explosion, 386
  use against Japan, 344

Atomic Energy Commission, 410
Auerbach family, 390
Augsburg, Anita, 17

Bach, Henry, 293
Bad Nauheim (conference), 112–113
Baldwin, Roger, 149, 270
Balfour Declaration, 125, 183, 184, 348
Bamberger, Louis, 226, 232
Bargmann, Valentin, 305, 311–312, 325, 390, 419
Barkla, Charles, 96
Barnard Medal, 155
Barnett, Lincoln, 66
Barrois, Mrs. Georges, 282
Barthou, Jean-Louis, 221
Battelli, Prof., 33
Beck, Anna, 429
Beginning or the End, The, 353
Bell, Charles, 322
Ben-Gurion, David, 349, 368, 391–392, 400
Bentwich, Helen, 145
Bentwich, Norman, 145
Berg, Moe, 379–380
Bergmann, Hugo, 81
Bergmann, Peter, 311
Bergson, Henri, 139
Berlin University, 85, 87–88, 89, 97, 133, 134–136, 148, 196
Bern University, 70, 73
Besso, Anna Winteler, 32, 70
Besso, Michele
  AE's view of, 32, 33, 362
  death of, 422
  family connections with AE, 70, 96, 165–166, 235, 312, 379
  meets Paul Ehrenfest, 85
  relationship with AE, 19, 32–33, 51, 57, 58, 60, 61, 72, 75, 154, 157, 165–166, 235–236, 288, 362, 416
Bethe, Hans, 354, 417

Biddle, Francis, 332
Bierberback, Ludwig, 105
big bang theory, 431
Billikopf, Jacob, 346, 347
Billings, Warren, 219, 220, 223
Blackwood, Andrew, 252, 261, 263, 277–278
Blackwood, Andrew Jr., 266
Blackwood, Carolyn, 252, 259, 261, 262, 263
Blackwood, James, 259, 260, 261, 263–266, 277, 278, 282
Blakeney, C. C., 368
Bloomberg, Andy, 280
Blumenfeld, Kurt, 120, 124
Bogdan, Dana, 238
Bohr, Niels
    AE introduces complementarity first, 75
    AE's view of, 87, 110
    and arms race, 341
    debate with AE over quantum theory, 164, 199–200, 305, 357, 362, 370
    escapes Nazis to Sweden, 334
    and Heisenberg, 161–164, 168
    meets AE, 110
    as speaker, 311, 375–376
    and splitting of atom, 316, 345
    and uncertainty principle, 162
    viewed by I. I. Rabi, 124
    view of AE, 110, 433
    view of quantum theory, 161–164, 177
    visits Princeton, 305, 316, 357
    wins Barnard Medal, 155
Boltzmann, Ludwig, 28, 35
bomb. See atomic bomb
Bonaparte, Marie, 158
book burning, 246
Born, Hedi
    and AE's Moszkowski interview, 116, 117
    dispute with Elsa Einstein, 118
    relationship with AE, 148, 159–160, 339–340, 389, 416, 433
Born, Max
    AE disagrees with view of quantum theory, 148, 156, 162, 339
    and AE talk to radical students in Berlin, 97
    and anti-Semitism, 248

    correspondence with AE, 156, 248, 279, 336, 339, 341, 365
    defends AE at conference, 113
    emigrates to Scotland, 336
    meets AE, 75
    relationship with AE, 116, 117, 118, 297, 339, 389, 411–412, 416, 433
    returns to Germany, 365, 411
    view of AE, 75, 112, 135
    view of general theory of relativity, 91
    view of theories, 423
Bose, Satyendra Nath, 153, 431
Bose-Einstein statistics, 153, 431
Bragg, Henry, 84
Bragg, Lawrence, 84
Brandeis, Louis, 120, 124–125
Breit, Gregory, 254
Briggs, Lyman, 319
Brillouin, Marcel, 144
Brod, Max, 81, 379
Brodetsky, Selig, 189–190, 365
Brown, Robert, 64
Brownian motion, 64, 69
Brunauer, Stephen, 334, 336
Brunner, Emil, 304
Brush, Charles Francis, 124
Buber, Martin, 392
Buckley, B. Lord, 129
Bucky, Frida, 229–230
Bucky, Gustav
    and AE in New York, 267
    and AE's last illness, 426
    as inventor, 228, 296, 324–325, 335
    and Nazis, 245
    as physician, 30, 230
    view of AE, 312–313
Bucky, Thomas
    and AE, 228, 229, 262, 296, 328–329, 335
    and AE's last illness, 426
    AE's letters to, 343, 349
    and Nazis, 244–245, 343
    in U.S. Army, 343
    view of AE, 228, 229, 230, 290, 328–329, 376, 388, 422
    view of AE and Elsa, 292
    view of AE's physical condition, 30
    view of Helen Dukas, 168
    wedding of, 413–414

Bucky family
  and AE, 228–230, 260, 267, 270, 289,
    308, 328, 416
  leaves Germany, 258–259
  and Nazis, 244–245
Budenz, Louis, 386
Burckhardt, Heinrich, 63
Bureau of Ordnance, 334
Burgess, Starling, 293, 295
Burkman, Alexander, 270
Bush, Vannevar, 320
Butler, Nicholas Murray, 204
Byland, Hans, 11–12, 166
Byland, Willy, 166

Cahen, Louis, 39, 43
California Institute of Technology
  AE returns to, 223–226, 241–243
  AE's first visit, 202, 208–217
  hopes to recruit AE, 233
Campbell, William, 211
Cantril, Mab, 361
Cardozo, Benjamin, 237
Carter, Paul, 180, 305
Cartmel, William, 287–288
Ceruti, Joseph, 283
Chadwick, James, 240
Chagall, Marc, 322
Chancellor, Sir John, 184
Chaplin, Charlie, 210–211, 214, 215,
    217
Chavan, Lucien, 52, 56, 74
Chen Ning Yang, 432
Chinese Cross, 328
Christ Church College, Oxford, 218–
    219, 227, 233, 248–249, 255
Churchill, Winston, 108, 125, 126, 184,
    225, 247, 250, 341, 345
City College of New York, 124
City Lights, 214
Clark, Ronald, 431
Clemenceau, Georges, 184
Cohen, Harry, 281–282
Cohen, I. Bernard, 72, 423–424
Cohen, Morris, 124
Cohn, Roy, 413
College de France, 138, 139
Columbia University, 124, 321
Commins, Dorothy, 353, 406, 409, 411, 416

Commins, Saxe, 353, 406, 411, 416
Communists. See also Soviet Union
  AE's view of, 149, 251, 300, 378–379,
    397–398
  AE viewed as, 237, 241, 244, 251, 307–
    308, 327, 383, 386–387
  in Germany, 109, 179, 201, 227, 240,
    243
  and McCarthy, 383, 384, 385–386,
    402, 413, 417
Compton, Arthur, 147, 341
Comte, Auguste, 26
Conway, Jill Ker, 434
Cook, George, 337
Copernicus, 160–161
Copp, Mrs. Joseph, 279, 280
cosmology, 194–195, 209, 217, 219
Cottingham, E. T., 99
Cotton Club, 267
Coughlin, Gene, 206–207
Cranston, Alan, 351
Croatia, 220
Crossman, Richard, 348
Crouch, Henry, 100
Curie, Marie, 74, 81, 139
  AE's view of, 116, 152, 264
  friendship with Einsteins, 85, 86
  and Langevin, 82
  and League of Nations, 140, 198
  and radium discovery, 71

Dafni, Reuven, 424–425
Daisetsu, Suzuki, 322
Dalen, Nils Gustav, 85
Darboux, Gaston, 103
Darlington, Harry, 262
Darrow, Clarence, 243
Dean, Guy, 366, 391, 425, 426
de Broglie, Louis, 82, 164, 169, 376
de Broglie, M., 82
Debs, Eugene V., 223
Debye, Peter, 320–321
de Donder, Theophile, 144
de Florez, Luis, 293–294
de Haas, Wander Johannes, 91, 109
Democritus, 192
Dernburg, Bernhard, 137
Derzbacher, Julius. See Koch, Julius
de Sitter, Willem, 95, 98, 217, 237, 245

De Valera, Eamon, 358
Diamond, Marian, 438
Dickson, Eugenia, 149–151
Dingle, Herbert, 242
disarmament. *See* arms race
Dobbie, Gen., 184
Doctors Plot, 404–407
Dostoyevsky, Fyodor, 108
Douglas, Allie Vibert, 195
Douglas, William O., 410
Dreyfus, Bertha, 350
Drude, Paul, 37, 68
Dukas, Helen
  and AE's brain, 437
  and AE's correspondence, 268–269,
    283, 294, 313
  as AE's personal guardian, 168, 169–
    170, 258, 266, 276, 282, 290, 309,
    322, 325, 343, 372, 393, 406, 407,
    408, 413, 418, 421, 425
  and AE's possibly illegitimate daughter,
    286
  after AE's death, 427, 428–429
  arrival in Princeton, 251, 253
  becomes American citizen, 278, 321
  defends AE against Marianoff's book,
    339
  descriptions of, 167, 313–314
  hiring of, 167–168, 169, 172
  interviewed by FBI, 392, 421–422
  investigated by FBI, 346, 347, 356, 375,
    386, 390, 419, 420
  and Otto Nathan, 313
  persuades AE to speak out against
    Soviets, 407
  relationship with AE, 168, 172, 260–
    261, 416, 420
  travels with AE, 205, 210, 246, 251
  and use of atomic bomb against Japan,
    344, 345–346
  view of AE, 53, 177, 237
Dukas, Rosa, 167
Dyson, Frank, 99
Dyson, Freeman, 253, 260, 429

$E = mc^2$, 71, 344
earthquake, 242
Eban, Abba, 400, 424–425
Ebert, Friedrich, 97, 107, 315

Eddington, Arthur, 64, 131, 194, 200
  and relativity, 95, 98–99, 100, 129–
    130, 195, 256
  and solar eclipse findings, 98–99, 100,
    129–130
Edison, Thomas, 129–130
Ehrat, Jakob, 17, 20, 23, 27, 34, 42
Ehrenburg, Ilya, 351–352
Ehrenfest, Paul
  AE visits in Leiden, 77, 93, 108, 109–
    110, 147, 179
  at Fifth Solvay Conference, 163, 164
  in Prague, 82–83
  response to attacks on AE, 111–112
  suicide of, 250
  view of AE and Lorentz, 78
Ehrenfest, Tatiana, 147
Ehrenhaft, Felix, 93, 96, 119–120, 144,
  315
Ehrman, Rudolf, 426
Eigner, Eugene, 376
Einstein, Abraham (paternal grandfather),
  2
Einstein, Albert
  and American politics, 219–220, 223,
    225, 383, 417
  appearance of, 76, 166, 274, 322, 323
  applies for jobs after college, 31–32,
    33, 37, 39
  and arms race, 341, 351, 352, 354, 365,
    377–378, 397
  attempts to discredit, 244–245, 249,
    284, 287–288, 307–308, 315
  attitude toward Soviet Union, 269–270,
    347, 360, 364–365, 383, 387–389,
    397–398, 404–406, 407
  awarded Nobel Prize, 63, 143–144,
    145, 146
  becomes American citizen, 278–279,
    321
  becomes German citizen, 97
  becomes involved with Elsa, 83–84,
    85, 86, 87, 89, 90
  becomes Swiss citizen, 23, 28, 30, 31
  and behavior of light, 26, 58–59, 63,
    65
  birth of, 1
  and birth of son Eduard, 76
  and birth of son Hans Albert, 58

brain preserved, 427, 437–439
and Brownian motion, 64, 69
builds summer home at Caputh, 179
at Caltech, 208–217, 223–226, 241–243
characterizations of, 7, 8, 16, 17, 31, 37, 57, 60, 82, 87, 98, 149–152, 150, 166, 225, 228, 258, 290, 314–315, 325, 343–344, 377, 382–383, 388
childhood of, 1–4
at Christ Church College, Oxford, 219, 233, 248–249, 255
in college, 12–14, 17–20, 22–23
and daughter Lieserl, 40, 43, 47–48, 49, 50, 51, 53, 55, 286, 287, 361
death of, 426
and death of Elsa, 296–297
and death of father, 53, 54
decision to leave Germany, 231–232, 237, 239–240
as determinist, 156, 185
and doctoral thesis, 28, 29, 40, 41, 42, 45, 63
efforts to help Jews, 80, 81, 104, 120–131, 177, 193, 200–201, 245, 309, 315
and $E = mc^2$, 71, 344
and Emergency Committee of Atomic Scientists, 354, 363, 367
end of marriage to Mileva, 89, 90, 93, 96, 97–98
estate of, 427
and FBI, 321, 331–332, 347, 367–368, 374–375, 419–420
as friend, 36, 77–78, 165
as fund-raiser, 120–131, 200–201, 218, 244, 257
and general theory of relativity, 71–72, 91–93, 101–104, 409, 423, 431–432
and German politics, 97, 107, 199, 201–202, 220, 227, 228
gives Mileva his Nobel Prize money, 96, 146
and gravitation, 72, 81, 85–86, 91–93, 305
health of, 94–95, 97, 152, 167–168, 169, 172, 268, 360, 363, 366, 372, 425
interviewed by Alexander Moszkowski, 114–116, 118

interviewed by George Sylvester Viereck, 184–187
as inventor, 228, 296, 324–325, 335
as Jew, 4–5, 185–186, 231, 268
joins Institute for Advanced Study at Princeton, 249–250, 251–252, 253, 254
last interview of, 423–424
and League of Nations, 140, 142, 145–146, 149, 198, 232
lectures in Leiden, 77, 93, 108, 109–110, 147, 179
legacy of, 430–436
letter to FDR regarding atomic bomb, 318–319, 345, 353–354, 420
letter to Stalin, 367
and Marie Winteler, 9, 10, 11, 12–13, 21, 22
marries Elsa, 98
marries Mileva, 54
and math, 5, 76
meeting with Carl Jung, 76–77
meeting with FDR, 253, 256–257
meeting with Sigmund Freud, 157–158
meets Mileva, 13
as musician, 2–3, 6, 11, 12, 16–17, 35, 40, 80–81, 187, 353
navy work during World War II, 334, 336–337
and Newton, 59, 131, 166, 211, 246, 327, 374
offered presidency of Israel, 400
and Olympia Academy, 54–59
opposition to, 110–113, 135, 237–238, 253–254, 323
as pacifist, 11, 90–91, 104–105, 118, 154, 225, 227, 239, 244, 374
papers smuggled out of Europe, 244, 259, 261, 263
papers submitted to *Annalen der Physik*, 28, 58, 62, 63–64, 67–68, 92
philosophy of life, 196, 388–389
in Prague, 77, 79–81, 84
proposed for Nobel Prize, 75, 85, 86, 90, 93, 95, 96, 99, 114
pursuit of objective reality, 27, 281, 306

Einstein, Albert (*continued*)
and quantum theory, 28–29, 63, 81, 84, 148–149, 153, 154, 156, 162, 164, 199–200, 330, 339, 376, 433
and radical students, 97, 201–202
recruited to Berlin University, 85, 87–88
recruited to University of Zurich, 73–74
rejected for military service, 6, 30
relationships with his children, 83, 90, 95–96, 166, 183, 197, 361, 395, 416
relationships with his professors, 13–14, 17–18, 23, 31
relationships with press, 120–121, 122, 123, 130, 132, 139, 184, 203–204, 205, 206, 208–209, 289
relationships with women, 49, 148, 159–160, 180–181, 322, 328, 352, 378
relationship with Elsa, 96, 180, 181, 264, 268, 280, 291, 292
relationship with Mileva, 15–16, 21, 22, 23–24, 34, 51, 53, 55, 56, 66, 69, 75, 83, 85, 87
relationship with sister Maja, 3, 26, 36–37, 38, 230, 312, 328, 360, 391
relativity theory criticized, 81, 100–104
and religion, 4–5, 266, 320, 334
and Rosenberg case, 390, 405–406, 409–411
and sailing, 229, 262–263, 280
and security clearance, 320, 331
sees Mileva for last time, 247
sees son Eduard for last time, 247
self-observations by, 20, 21, 274–275, 289, 388–389, 436
smoking of, 242, 274, 279, 308, 322, 371
as social activist, 193, 217, 219–220, 223, 225
socialist views of, 173, 217, 225, 229, 289, 300
and special theory of relativity, 60–62, 64–68, 70–71, 81, 408, 423
and splitting of atom, 271, 316–317, 344–345
as student, 3–7, 8, 9–10, 12–14, 17–20, 22–23
and Swiss Patent Office job, 33, 42, 52, 58, 74
as target of Nazis, 105, 110–111, 138, 141, 142, 146, 231, 243, 249, 288
as teacher, 7, 13, 34, 35, 39–40, 41–42, 43, 44–45, 73, 76, 133–135
testifies before Anglo-American Committee of Inquiry, 347–348
and unified field theory, 153, 173–177, 192–193, 198, 292–293, 350, 370, 392, 401, 415
viewed as Communist, 237, 241, 244, 251, 307–308, 327, 383, 386–387
viewed by Friedrich Adler, 20
viewed by Valentin Bargmann, 311
viewed by Hans Byland, 11–12
viewed by Thomas Bucky, 228, 229, 290
viewed by Harry Cohen, 282
viewed by Mrs. Copp, 280
viewed by Louis de Broglie, 164
viewed by Abba Eban, 425
viewed by Jakob Ehrat, 20
viewed by Marcel Grossman, 14, 20
viewed by Hiram Haydn, 283
viewed by Banesh Hoffmann, 290
viewed by Roy Harrod, 218
viewed by John Kemeny, 369–372
viewed by Frederick Lindemann, 82
viewed by A. V. Lunacharsky, 151
viewed by Susanne Markwalder, 17
viewed by I. I. Rabi, 124
viewed by Romain Rolland, 90–91
viewed by I. F. Stone, 402
viewed by Eugene Wigner, 134–135
viewed by others, 20, 56–57, 128, 155
views on age of earth, 236
views on America and Americans, 131–133, 216, 227, 289, 365, 383, 417
views on anti-Semitism, 20, 109, 199, 248, 268, 270, 367
views on birth of Israel, 425
views on capitalism, 378
views on Christianity, 4, 5, 259, 277–278
views on Communism, 149, 251, 378–379, 397–398
views on cosmology, 194–195, 209, 217, 219
views on discrimination, 281–282
views on eastern religion, 322
views on Freud, 158, 186, 305

views on Germans, 91, 248, 276–277,
  281, 343, 349–350, 360, 388, 411
views on Hitler, 204, 242, 247, 266,
  270, 277
views on Japanese, 143, 225–226
views on knowledge, 129–130
views on Mussolini, 242
views on Nazis, 246–247, 251, 288
views on Palestine, 183–184, 188–190,
  202–203, 307, 337–338, 347–349,
  352
views on parapsychology, 46, 191–192,
  215–216, 237
views on philosophy, 44–45, 46
views on physicians, 230, 313–314
views on Prague, 79–80, 82
visits to England, 218–219, 227, 233,
  248–249, 255
visits to United States, 120–131, 202–
  226
visit to Palestine, 144–145
and women, 49, 116, 132, 148, 152,
  159–160, 180–181, 322, 328, 352,
  378
and world peace, 227, 233–234
and Zionists, 81, 120, 124–125, 183,
  185, 217, 337, 338, 342
and Zurich Polytechnic, 7, 8, 12–13,
  31, 82, 84–86
Einstein, Bernhard Caesar (grandson),
  427
Einstein, Eduard (son)
  in AE's FBI file, 419, 422
  AE's last visit with, 247
  AE's relationship with, 106, 158, 183,
    235–236, 363, 369, 395
  Besso as surrogate father, 96, 165–166,
    235, 306
  birth of, 76
  breakdown of, 96, 195–198
  care of, 306, 361, 389, 427
  and Carl Seelig, 197–198, 393–394,
    395
  death of, 428
  descriptions of, 96, 106, 146, 153–154,
    197, 306, 389
  and niece Evelyn, 394–395
Einstein, Elizabeth Roboz (daughter-in-
  law), 48, 196

Einstein, Elsa Löwenthall (second wife)
  arrival in Princeton, 251, 252–253
  becomes involved with AE, 83–84, 85,
    86, 87, 89, 90
  and Blackwoods, 259, 261–262, 264
  in California, 210–211, 214–215
  death of, 296–297
  and decision to leave Germany, 231–
    232, 239
  descriptions of, 151, 214, 228, 296–297
  and Eugenia Dickson, 149–150
  health of, 267, 271, 282, 291
  and Ilse's illness and death, 259, 261–
    262
  and Moszkowski interview, 117, 118
  relationship with AE, 96, 180, 181,
    264, 268, 280, 291, 292
  as translator, 120, 122, 127
  view of AE, 129, 274
Einstein, Evelyn (granddaughter), 361,
  394–395, 428
Einstein, Frieda Knecht (daughter-in-law),
  153, 160, 369
Einstein, Hans Albert (son)
  after AE's death, 427, 429
  birth of, 58
  and brother Eduard, 247, 393
  at Caputh, 179, 180
  career of, 428
  and Curies, 86
  death of, 428
  and father's papers, 48, 429
  and FBI, 422
  marriage to Frieda Knecht, 153, 160,
    369
  relationship with AE, 90, 146, 196,
    235, 306, 324, 328, 363, 416, 426
  view of Mileva, 55
Einstein, Hermann (father)
  and birth of AE, 1
  as businessman, 2, 6, 7, 27, 32, 36
  death of, 53, 54
  health of, 51
  opposes AE's intended marriage, 27
  writes to Wilhelm Ostwald, 33
Einstein, Hindel (paternal grandmother),
  2
Einstein, Ilse (step-daughter), 87, 90, 110,
  223, 244, 246, 422

Einstein, Ilse (*continued*)
  illness and death of, 259, 261
  marries Rudolf Kayser, 148
Einstein, Jakob (uncle), 2, 4, 8
Einstein, Klaus (grandson), 309
Einstein, Lina (cousin), 350
Einstein, Louis (German relative), 333
Einstein, Maja (sister)
  and AE's intended marriage, 23, 26, 40
  childhood remembrances of AE, 1, 2,
    3, 6, 7–8
  death of, 391
  and death of father and mother, 53,
    107, 108
  health of, 354, 360, 391
  relationship with AE, 3, 26, 36–37, 38,
    230, 312, 328, 360, 391
  as wife of Paul Winteler, 32, 70, 109,
    312, 379
Einstein, Margot (step-daughter), 87, 90,
    98, 246, 263, 264, 291, 322, 329, 408,
    416, 421, 424, 425–426
  after AE's death, 427, 428
  becomes American citizen, 278, 321
  death of, 428
  and Dimitri Marianoff, 159, 198, 223,
    234, 244, 245, 267
  view of AE, 433
Einstein, Mileva Maric (first wife)
  AE's family opposes marriage, 23, 24,
    25–27, 29, 38, 49–50
  background of, 15–16
  birth of son Eduard, 75, 76
  birth of son Hans Albert, 58
  and daughter Lieserl, 36, 40, 43, 47–
    48, 50, 51, 53, 55, 287
  death of, 197, 369
  descriptions of, 55–56
  fails final college exam, 23, 38
  leaves AE, 89, 97
  marries AE, 54
  meets AE, 13
  mental illness in family, 15, 27, 40, 96,
    197
  as partner in AE's work, 33, 38, 55,
    434–435
  relationship with AE, 21, 22, 23–24,
    34, 51, 53, 55, 66, 75, 83, 85, 87
  sees AE for last time, 247

  and son Eduard, 195, 196, 197, 361,
    363
  suffers breakdown, 93
Einstein, Pauline (mother)
  and birth of AE, 1
  death of, 107
  and marriage of AE and Mileva, 53
  opposes AE's intended marriage, 24,
    25–27, 29, 42, 47
  passion for music, 2
Einstein, Roberto (cousin), 340
Einstein, Thomas Martin (great grand-
    son), 428
Einstein, Walter (German relative), 333
Einstein Award, 375
Eisenhower, Dwight D., 410, 417–418
Eisenstein, Sergey, 216
Elizabeth, Queen of Belgium, 179–180,
    249, 316, 317
Ellis, Havelock, 355
Emden, Robert, 144
Emergency Civil Liberties Committee,
    411, 416
Emergency Committee of Atomic Scien-
    tists, 354, 363, 367
Epstein, Jacob, 250
Epstein, Paul, 209

Falconer, Bruce, 123
Fanta, Bertha, 81, 378
Fanta, Joanna, 378
Faraday, Michael, 18, 59, 61, 62, 72, 185,
    193–194, 406
Federal Bureau of Investigation (FBI)
  interviews Helen Dukas, 392, 421–422
  investigates AE, 331–332, 356, 362–
    363, 367–368, 374–375, 384–385,
    403, 405, 419–420
  investigates Helen Dukas, 346, 347, 390
Fermi, Enrico, 319, 321, 331, 345, 417
Ferrière, Gustave, 76
Feynman, Richard, 124, 330, 431
Field, Richard, 237
Finney, Frank, 318
First National Studio, 210
Fisch, Adolf, 76
Fishbein, Morris, 296
Fitzgerald, George Francis, 62, 66
Fizeau, Armand, 219

Flammarion, Camille, 115
Flattery, Martin, 258
Flexner, Abraham, 226, 227, 230–231, 231, 237, 249–250, 251, 253, 254
Flynn, Errol, 400
Foerster, Friedrich, 327
Fondiller, William, 234
Ford, Henry, 110, 184
Forel, Auguste, 40, 46
Forster, Aime, 70
Fosdick, Harry, 205
Frank, Philipp, 55, 79, 84, 94, 98, 119, 128, 139, 228, 362, 390
Frankfurter, Felix, 410
Franz Joseph, Emperor, 79, 83
Frazer, James G., 424
Freikorps, 107–108
Freimuller, Hans, 389
Freud, Oliver, 158
Freud, Sigmund, 184, 326
    correspondence with AE, 177, 199, 232–233, 290–291, 315, 377
    meeting with AE, 157–158
Freundlich, Erwin, 98
Freundlich, Kathe, 108
Friedmann, Alexander, 194, 219
Frisch, Otto, 309
Fromm, Friedrich, 330
Frosch, Hans, 46–47
Frothingham, Mrs. Randolph, 237–238
Fruendlich, Erwin, 98
Fuchs, Klaus, 384–385
Fuld, Mrs. Felix, 226
Fuller, Buckminster, 294–296
Furth, R. H., 168

Gabor, Andrea, 434
Gabor, Dennis, 134
Galileo, 288, 379
Gamow, George, 62, 161, 334, 336
Gandhi, Mahatma, 154, 328, 397, 406
Garbedian, Gordon, 366
Gasser, Prof., 34
Gauss, Christian, 283
Gehrcke, Ernst, 105, 111
Geiser, Carl, 18
Geist, Raymond, 238
general theory of relativity, 71–72, 91–93, 100–104, 409, 423, 431–432

George, David Lloyd, 247–248
German League of Human Rights, 233–234
Germany. See also Hitler, Adolf; Nazis
    AE decides to leave, 231–232, 237, 239–240
    AE's papers smuggled out, 244, 259, 261, 263
    anti-Semitism in, 105, 109, 110–111, 138, 141–142, 199, 231, 247, 248
    Communists in, 109, 179, 201, 227, 240, 243
    concern over atomic bomb capability, 316, 317, 329, 330–331
Gestapo, 243, 309. See also Nazis
Gilbert, Clair, 369, 420
Gillette, George Francis, 103
Glenn, Frank, 426
Gödel, Kurt, 254, 266–267, 366, 380, 418
Goebbels, Joseph, 246, 309, 310
Gold, Harry, 385
Goldberg, Stanley, 61, 86
Goldman, Henry, 275
Goldschmidt, Rudolph, 248
Gore, Rick, 431
Göring, Hermann, 147, 309, 321, 330
Goudsmit, Samuel, 164
Gray, Mrs. James, 241
Green, Paul, 267
Greenbaum, Edward, 382–383
Greenglass, David, 386, 390
Greenglass, Ruth, 390-391
Grommer, Jakob, 169
Grossmann, Marcel
    as colleague of AE at Zurich Polytechnic, 84, 85, 86
    death of, 291–292
    early view of AE, 14, 20
    offers AE post at Zurich Polytechnic, 82
    relationship with AE, 23, 33, 36, 39, 51, 58
Groves, Leslie, 354, 410
Grynszpan, Herschel, 308
Guillaume, Charles Édouard, 114
Gutenburg, Beno, 242

Haas, Arthur, 95
Haber, Fritz, 91, 148, 247

Habicht, Conrad, 39, 43, 45, 47, 51, 56, 57, 63, 69, 70
Habicht, Paul, 59, 70
Hadamard, Jacques, 144
Haenisch, Konrad, 111
Hahn, Otto, 248, 309, 310, 316, 331, 339, 345
Haider, Hans, 172
Haldane, Viscount, 131
Haller, Friedrich, 33, 47, 52, 58, 74
Halsman, Liouba, 170, 171
Halsman, Philippe, 170–172, 222, 373
Halsman, Yvonne, 172
Hamilton, Ian, 131
Harding, Warren G., 125
Harriman, Averell, 409
Harrod, Roy, 218
Harteck, Paul, 310
Hartmann, Gustave, 123
Harvard University, 279
Harvey, Thomas, 437–439
Hasenohrl, F., 105
Hauptmann, Gerhart, 160, 192
Haydn, Hiram, 283
Hearst, William Randolph, 214
Heathcote, Dudley, 188–189
Heaviside, Oliver, 101
Hebrew University, 130, 131, 145, 427
Hedin, Sven, 143
Hegel, Georg Wilhelm Friedrich, 44
Heisenberg, Werner
    AE disagrees with, 148–149, 156, 169, 272
    and Bohr, 161–164, 168
    as friend and supporter of AE, 284
    ideas on quantum theory, 156, 161–164
    meets AE, 148–149
    misses meeting AE, 135, 142
    and Nazis, 288, 308, 339, 345
    and nuclear research, 331
    pressured by Hitler, 321
    and uncertainty principle, 153, 154, 162
Helm, George, 64
Helmholtz, Hermann, 19
Hermann, Jakob, 8
Herneck, Friedrich, 180
Herrschdoerffer, Mrs., 286–287

Hertz, Heinrich, 19, 155
Herzburger, Max, 134
Herzog, Albin, 8
Herzog, Peter, 389
Heydrich, Reinhard, 309
Hibben, John Grier, 127
Hiden, Conway, 313
Highfield, Roger, 180, 305
Hilferding, Rudolf, 179
Himmler, Heinrich, 308
Hiroshima, 344
Hitler, Adolf
    AE's view of, 204, 242, 266, 270, 277
    and atomic bomb, 316, 329, 330–331
    death of, 343
    invades Europe, 320
    invades Poland, 318
    invades Soviet Union, 329
    and pacifism, 249, 323, 336
    and Planck, 245, 247
    rise of, 108, 141, 147, 148, 154, 185, 199, 221, 225, 242
    takes power, 243
Hitti, Philip, 337–338
Hoffmann, Banesh, 53, 61, 67, 71, 91–92, 254, 289–290, 307, 311
Hollywood, 210–211, 353–354
Holton, Gerald, 62, 434
Hook, Sidney, 299–301, 360, 364, 365, 387, 388, 396–398, 406, 417
Hoover, J. Edgar, 330–331, 346, 347, 374–375, 379, 384, 385, 386–387, 390, 403, 422
Hopf, Ludwig, 75, 76, 80, 248
House Un-American Activities Committee, 396, 416, 420
Houston, William, 277
Houtermans, Fritz, 316, 320
Hubble, Edwin, 194, 209, 211, 213, 217, 219, 236
Hull, Kevin, 439
Hume, David, 45, 46, 57, 61
Hupka, Josef, 171

illegitimate daughter stories. See "Lieserl"
Immigration and Naturalization Service, 384
Infeld, Leopold, 128, 135, 256, 289–290, 293, 307, 311, 388, 430

Institute for Advanced Study
  AE arrives at, 254–255
  new building for, 342–343
  Oppenheimer as director, 418
  recruits AE, 227, 249–250
international law, 355

Jaffe, Bernard, 212
James, William, 19
Japan, 225, 344
Jastrow, Robert, 194, 380
Jaumann, Gustav, 77
Jeans, James, 131, 236, 250
"Jewish science," 105, 288
Jews, AE's efforts to help, 80, 81, 104,
  120–131, 177, 193, 200–201, 245,
  309, 315. See also anti-Semitism; Pales-
  tine; Zionists
Joliot-Curie, Frédéric, 319, 396, 397
Joliot-Curie, Irène, 378
Jordan, Pascual, 154, 336
Julius, Willem, 114
Juliusburg, Otto, 363
Jung, Carl, 76–77
Jung, Prof., 33

Kafka, Franz, 81, 378
Kahler, Alice, 322, 327–328, 406, 416, 427
Kahler, Erich, 322, 337–338, 416
Kaiser Wilhelm Institute, 133, 136, 169,
  321
Kaku, Michio, 432–433
Kaluza, Theodor, 432
Kaluza-Klein theory, 432
Kant, Immanuel, 5, 7, 105, 115–116
Kapitsa, Peter, 341, 356–357
Kapp, Wolfgang, 108
Katzenellenbogen, Estella, 181
Katzenstein, Moritz, 160, 167
Kauffmann, Jack, 439
Kaufler, Helene, 23–24. See also Savic,
  Helene Kaufler
Kaufman, Bruria, 380–381, 424
Kaufman, Judge, 405
Kaufmann, Edgar, 271
Kayser, Rudolf, 148, 207, 223, 244
Kemeny, John, 254, 369–372, 376, 380,
  381, 431
Kennedy, Roy, 224

Kepler, Johannes, 246
Kessler, Harry, 118–119, 137, 140, 141,
  155, 161
Kieran, John, 379
Kilbreth, Mary, 237
King's College, 131
Kirchhoff, Gustav, 19, 26
Kirov, Sergey, 269
Kisch, Frederick, 145
Klein, Oscar, 432
Kleiner, Alfred
  and AE's Ph.D. thesis, 28, 40, 41, 42, 63
  view of AE, 70, 73–74, 77
Knecht, Frieda. See Einstein, Frieda
  Knecht (daughter-in-law)
Koch, Caesar, 8, 179
Koch, Jette, 1
Koch, Julie, 21, 27
Koch, Julius, 1, 2
Kohn, Alice, 308
Kollros, Louis, 23
Komar, Arthur, 408
Konenkov, S., 275
Korea, 386
Kottler, Friedrich, 86–87
Kovarik, Alois, 236
Kowalevski, Gerhard, 79
Kraus, Oscar, 81
Kristallnacht, 309
Kropotkin, Prince Petr A., 349
Kuhnwald, Gottfried, 248

Ladd, D. M., 384
Ladenburg, Ralph, 332–333, 390
Laemmle, Carl, 210
La Guardia, Fiorello, 122–123, 251, 276
Lampa, Anton, 83
Lampe, John, 252
Lanczos, Cornelius, 169, 173, 181, 192,
  327
Langevin, Paul
  and AE's trip to France, 136, 138, 140
  AE's view of, 85, 86, 360, 423
  arrested by Nazis, 333–334
  death of, 360
  in Leiden, 109
  and Marie Curie, 82
  proposes AE for Nobel Prize, 144
  at Solvay Conference, 81

Langevin, Paul (*continued*)
  and special theory of relativity, 423
  view of AE's move to U.S., 237
Langford, Samuel, 101
Lanouette, William, 136, 353
Laski, Harold, 131
Laub, Jakob, 72–73
Laurence, William, 123, 328, 359
Lawrence, T. E., 125, 337
League for the Rights of Man, 220
League of Nations, 140, 142, 145–146,
  149, 198, 232
League of the New Fatherland, 90
Lebach, Margarette, 180
Lefshetz, S., 271, 273
Lehman, Irving, 291
Leibniz, Gottfried Wilhelm von, 423
Leibus, Rudolph, 105
Lemaître, Georges, 194, 209, 219
Lemm, Franz, 178
Lenard, Philipp
  anti-Semitism of, 105, 110–111, 116,
    141–142, 246, 288
  as enemy of AE, 105, 110–111, 113,
    143, 144, 246
  and Nobel Prize for AE, 143, 144
  scientific work of, 37, 73, 105, 110
Lenard (Philipp) Institute, 284
Lenz, Wilhelm, 288
Leopold, Nathan, 256
Lessing, Theodore, 249
Levi-Civita, Tullio, 103
Levin, Schmarya, 124–125
Levine, Isaac Don, 149, 226–227, 269–270
Levine, Martin, 431
Levy, Steven, 438
Lewis, Sinclair, 255
"Lieserl," 36, 40, 43, 47–48, 49, 50, 51,
  53, 55, 286–287, 361
Lindbergh, Charles, 191, 253, 318
Lindemann, Frederick, 82, 131, 218, 246,
  247, 248, 255, 277, 285, 341, 345
Lipsky, Louis, 332
Livingston, Dorothy Michaelson, 212
Livingstone, Siegmund, 308
Locarno Treaty, 154
Locker-Lampson, Oliver, 250, 255
Lodge, Oliver, 102, 195
Loeb, Richard, 256

Lorcherthaler, Lina, 365
Lorentz, Hendrik
  AE cites as source for special theory of
    relativity, 62, 66
  AE's view of, 77–78, 81, 165, 424
  AE visits in Leiden, 93
  death of, 165
  and Eddington photographs, 99
  as important source, 256
  and Nobel Prize for AE, 96, 114
  retires, 148
  at Solvay conferences, 81, 163
  viewed by others, 78, 82, 411
Lorentz transformation, 66
Lossros, Louis, 84
Löwenthall, Elsa, 83–84. *See also* Ein-
  stein, Elsa Löwenthall (second wife)
Ludendorff, Erich, 242
Ludwig, Emil, 186, 188
Lunacharsky, A. V., 150–151
Lynch, Arthur, 103, 239

MacArthur, Douglas, 386
Mach, Ernst, 19, 20, 45, 61, 64, 80, 104,
  424
Mach's principle, 19, 408
Magie, W. F., 81
Manhattan Project, 331, 341, 354, 371
Mankiewicz, Herman, 395
Mankiewicz, Johanna, 395
Mann, Heinrich, 220, 221
Mann, Thomas, 171, 322, 383, 386
Marianoff, Dimitri
  and biography of AE, 159, 207, 265,
    338–339
  and Margot Einstein, 159, 198, 223,
    234, 244, 245, 267
Maric, Mileva. *See* Einstein, Mileva
  Maric (first wife)
Maric, Zorka, 15, 27, 40, 96, 197
Markstein, Grete, 286–287
Markstein, Helene, 286
Markwalder, Susanne, 16–17, 389
Marshall, George C., 367
Maxwell, James, 18, 19, 59, 62, 67, 152,
  185, 193–194, 406, 424
Mayer, Louis B., 269, 353–354
Mayer, Walther, 181, 192, 205, 210, 223,
  232, 237, 246, 250, 251, 254, 293

McCarthy, Joseph, 383, 384, 385–386, 402, 413, 417
McGee, Willie, 392
McPherson, Aimee Semple, 269
Meinhardt, Willy, 166, 167
Meitner, Lise, 133, 148, 309, 345
Mellon Institute, 273
Mendel, Elsa, 180
Mental Radio, 176
Menuhin, Yehudi, 193
Messersmith, George, 238
Metro-Goldwyn-Mayer, 353–354
Meyer, Edgar, 99, 144
Meyer, Stefan, 144
Michelmore, Peter, 55
Michelson, Albert, 62, 66, 102, 104, 126–127, 211, 212, 213, 219
Mie, Gustav, 113
Miklas, Wilhelm, 171
Mill, John Stuart, 45–46
Miller, Dayton C., 103, 127, 156
Millikan, Robert
    with AE at Caltech, 209, 211, 212, 215, 217, 225
    and cosmic rays, 209
    defends AE, 132
    invites AE to Caltech, 164, 233
    and Jewish appeals, 342
    meets AE, 126
    at Mellon Institute, 273
    and photoelectric effect, 126
    uneasiness with AE's politics, 219–220, 223, 225
    view of AE and Nobel Prize, 144
Milyukov, Pavel, 150
Minkowski, Hermann, 18, 31, 72, 256
Mitrany, David, 101, 400
Modern Times, 217
Montagu, Ashley, 354–356, 377–378, 383, 437, 438, 439
Mooney, Tom, 219, 220, 223
Moore, Walter, 358
Moos, Heinz, 308
Morgenstern, Oskar, 366
Morgenthau, Henry, 275
Moszkowski, Alexander, 66, 114–116, 118
Motz, Lloyd, 419
Mount Wilson Observatory, 202, 209, 213, 225

Mowrer, Edgar Ansel, 175–176
Muhlberg, Fritz, 12
Mühsam, Hans, 331, 399
Muller, Wilhelm, 105
Murrow, Edward R., 417
Mussolini, Benito, 242, 312

Nagasaki, 344
Nathan, Otto, 48, 252, 313, 369, 425, 427
    after AE's death, 427, 428–429
Naunyn, Bernhard, 86, 90, 144
Nazis. *See also* Germany
    and anti-Semitism, 154, 231, 244, 245, 288
    and atomic bomb, 316, 329, 330–331
    attempts to discredit AE, 244–245, 249, 284, 288, 307–308, 315
    and Communists, 109, 179, 227, 240, 243
    and *Kristallnacht*, 309
    and pacifism, 249, 323, 336
    and Polish Jews, 308–309
    and Stalin, 227
    treatment of Einstein's property, 244, 245, 274
Nernst, Walther, 81, 85, 86, 91, 111, 113, 133, 148, 155
Neusch, Jakob, 39, 41–42
Newton, Isaac
    AE pays tribute to, 131, 166
    AE's view of, 152, 185, 311, 423, 424, 425
    compared with AE, 59, 211, 246, 327, 374
New York Times
    and AE, 174, 204, 205, 237, 288
    AE meets Adolph Ochs, 123
    AE's letters to, 220
    reports on Eddington's solar eclipse findings, 100–101
Nicolai, Georg Friedrich, 89
Nicolson, Harold, 227
Niggli, Julia, 20, 21, 22
Nissen, Rudolph, 372, 426
Nobel, Alfred, 255
Nobel Prize
    AE finally wins, 63, 126, 143–144, 145
    AE first proposed for, 75
    AE gives prize money to Mileva, 96, 146

Nobel Prize (*continued*)
    AE nominated for, 85, 86, 90, 93, 95,
        96, 99, 114
    and Philipp Lenard, 143, 144
Nohel, Emil, 80
Nordmann, Charles, 138, 139, 140
Nordström, Gunnar, 144
Nottingham University, 193–194
nuclear disarmament. *See* arms race
Nuesch, Jakob, 41, 43

Oakes, John, 276
O'Brien, John, 251, 276
Ochs, Adolph, 123
Odets, Clifford, 305
Olympia Academy, 54–59, 68, 187
Onnes, Heike Kamerlingh, 32, 86, 114
Oppenheimer, J. Robert
    as director of Institute for Advanced
        Study, 369, 385, 395–396, 418
    investigation of, 385, 395–396, 417–
        418, 420
    and Manhattan Project, 344, 385
    and movie about atomic bomb, 354,
        355
    and Schrödinger, 359
    viewed by I. I. Rabi, 124
    view of AE, 376–377, 430
orgone, 326–327
Ornstein, Leonard Solomon, 114
Oseen, Carl Wilhelm, 144
Ostoja, Roman, 176, 215
Ostwald, Wilhelm, 31–32, 33, 64, 74, 75,
    85, 86
Ott, Emil, 10, 11
Oursler, Fulton, 241

Pacelli, Eugenio, 140
pacifism
    and AE, 11, 90–91, 104–105, 118,
        154, 225, 239, 244, 374
    in face of Nazi aggression, 249, 323, 336
Page, Leigh, 287–288
Painlevé, Paul, 103, 139, 171
Pais, Abraham, 29, 67, 75, 148, 165, 169,
    357, 375, 392, 416, 427, 428, 432, 435
Palestine. *See also* Jews, AE's efforts to
    help; Zionists

AE on fund-raising tour for Hebrew
    University, 120–131
AE support for binational state, 348–
    349, 352
AE's view of, 183–184, 188–190, 202–
    293, 307, 337–338, 347–349, 352
AE visits, 144–145
Arab attack, 183–184
Truman recognizes Jewish state, 368
parapsychology, 191–192, 215–216, 237
Patch, Blanche, 239
Patent Office. *See* Swiss Patent Office
Patterson, Cissy, 213–214
Pauli, Wolfgang, 75, 135, 162, 192, 224,
    350
Pauling, Linus, 354, 372, 383, 420
Pearl Harbor, 330
Pearlman, Moshe, 391
Pearson, Karl, 45
Pekhackova, Marcela, 435
Penzias, Arno, 430
Pernet, Jean, 17–18
Perrin, Jean, 64
Peterson, Carl, 251–252
Pflaum, Rosalynd, 86
Picard, Émile, 103
Pick, Georg, 80
Pirandello, Luigi, 279
Planck, Erwin, 363
Planck, Max
    AE extends his quantum theory, 63, 87
    AE's view of, 76, 86, 424
    approves German aggression in WWI,
        89
    attacked by Nazis, 338
    as colleague of AE in Berlin, 110, 112,
        113, 141, 148, 196
    at conference on quantum theory, 81
    death of, 363
    defended by Heisenberg, 288
    as editorial director of *Annalen der Phy-
        sik,* 58
    meets AE, 75
    offers AE post at Berlin University, 85,
        87–88
    and origin of quantum theory, 28–29
    personal tragedies of, 106, 363
    proposes AE for 1919 Nobel prize, 99

proposes AE for Prussian Academy of Sciences, 88
response to Nazi treatment of Jewish colleagues, 113, 141, 245, 246, 247, 248
speaks with Hitler, 245, 247
stays in Germany, 284
view of AE's work on gravitation, 72, 85
wins 1918 Nobel Prize, 99
writes to AE about special relativity paper, 67, 69
Planck Medal, 182–183
Plesch, Janos, 107, 150, 166, 178, 179, 182, 183, 188, 198–199, 211, 285, 286, 296
Plunguian, Gina, 369
Podolsky, Boris, 281
Poincaré, Henri, 53, 64, 67, 81, 82, 103, 288, 433
Poincaré Institute, 187
Poor, Charles Lane, 101–102, 121
Poulton, Edward, 144
Pour le Mérité medal, 155
Prandtl, Ludwig, 308
Pringsheim, Ernst, 85
Prussian Academy of Sciences, 88, 155, 165, 173–174, 233, 237, 245, 246, 284, 288, 350
Pupin, Michael, 124, 234
Putnam, James, 398

quantum theory
AE disagrees with Bohr, 164, 199–200, 305, 357, 362, 370, 433
and Born, 148, 156, 162, 339
Copenhagen interpretation, 161–164
and Heisenberg, 156, 161–164
and Planck, 28–29, 63, 81, 87

Rabbeis, Emma, 384–385
Rabi, I. I., 124, 376, 417
Raditch, Stefan, 220, 221
Rainer, Luise, 305
Raney, Rebekah, 223
Rathenau, Walter
assassination of, 140, 141
as foreign minister, 118, 137–138
Rayleigh, Lord, 131
Red Menace. See Communists

Regis, Ed, 237, 251
Reich, Ilse, 327
Reich, Wilhelm, 325, 326–327, 382, 399
Reiche, Fritz, 320
Reiman, Otto, 191–192
Reiser, Anton, 207
Reith, John, 200
relativity. See general theory of relativity; special theory of relativity
Renan, Ernest, 314
Renn, Jurgen, 435
Rhodes, Richard, 318
Ricci, Curbastro Gregorio, 103
Richards, Alan, 340–341, 343, 396
Rieber, Winifred, 257–258
Righi, Augusto, 33
Ripley, Robert, 281
Riverside Church, 205
Robertson, Howard, 270, 271, 273
Roboz, Elizabeth. See Einstein, Elizabeth Roboz (daughter-in-law)
Rogers, Eric, 382, 424
Rogge, O. John, 379
Rolland, Romain, 90–91
Rolph, James, 219
Roosevelt, Eleanor, 256, 401
Roosevelt, Franklin
AE meets, 253, 256–257
and atomic bomb, 318, 319, 320, 321, 341, 345, 353–354, 420
death of, 343
viewed by AE as socialist, 289
Rorke, Lucy, 438, 439
Rosen, Nathan, 267, 281
Rosenberg case, 386, 390, 404, 405–406, 409–411
Rosenfeld, Kurt, 260
Rosenfeld, Leon, 199
Rosenthal-Schneider, Ilse, 105
Rosso, Henry, 276
Rothschild, Lord, 131, 200
Royal Astronomical Society, 95, 155, 194, 436
Royal Society, 155
Rubel, Eduard, 195
Rubens, Otto, 111, 113
Rüde, Prof., 305
Russell, Bertrand, 401, 406, 410, 423
Russell, Henry Norris, 355

Russia. *See* Soviet Union
Rust, Bernhard, 245
Rutherford, Ernest, 81, 87, 240, 250

Sachs, Alexander, 318, 319
St. John, Charles Edward, 95, 121, 209, 211
Salaman, Esther, 133–134, 151, 152
Salter, Wallace, 239
Samuel, Herbert, 125, 126, 144, 189
Savic, Helene Kaufler, 28, 40, 69, 85, 87, 93
Sayen, Jamie, 344, 428
Scandinavian Society of Science, 146
Schaefer, Clemens, 85
Schiffer, Lotte, 390
Schilpp, Paul, 131
Schlaefke, Fritz, 229
Schlesinger, Frank, 236
Schmid, Anna, 20–21, 75
Schneider-Glend, Baroness, 385
Schoenberg, Arnold, 257
Schoenmann, Friedrich, 346
Schonberg, Harold, 257
Schopenhauer, Arthur, 35, 44–45, 196, 255, 327, 388
Schrödinger, Erwin, 133, 169, 246, 281, 350–351, 433
and unified field theory, 350, 358–360
Schulmann, Robert
views on AE's illegitimate daughter, 286–287
views on Eduard, 197–198
views on Lieserl issue, 48–50, 286–287
views on Ludek Zakel's claim, 436
views on Mileva as AE's partner in science, 434, 435
Schwartz, Paul, 244
Schwartz, Richard Alan, 422
Schwarz, Paul, 203
Scopes, John, 243
Scottsboro boys, 219, 223
See, Thomas, 102–103, 241
Seeley, Evelyn, 242
Seelig, Carl, 1, 15, 74, 77, 197–198, 393–394, 416
Seibert, Barbara, 209
Seldes, George, 266
Serbia, 220

Shadowitz, Albert, 413, 416
Shahn, Ben, 322
Shapley, Harlow, 208–209, 244, 279, 383, 398, 399, 424
Shaw, George Bernard
introduces AE at fund-raiser, 200–201
meets with AE, 223
view of AE, 104, 131, 160, 239
view of Nazis, 323
Silberstein, Ludwig, 287–288
Sinclair, David, 217
Sinclair, Mary Craig, 176
Sinclair, Upton
AE writes preface to *Mental Radio*, 176
books of, 307, 308, 338, 400
correspondence with AE, 176–177, 205
friendship with AE, 217, 241, 329, 354, 357, 365
meets AE, 215–216
runs for governor of CA, 269
view of Viereck, 184
Smith, Robert, 289
Smith, Walter Bedell, 356
Society of German Natural Scientists and Physicians, 112–113
Solomon, Jacques, 333
Solovine, Maurice
approached by AE biographer, 395
and book about Democritus, 192
hears AE speak in Paris, 139, 187
living in Paris, 68, 69, 120, 187
meets AE, 44
and Olympia Academy, 55, 395
relationship with AE, 51, 53, 56, 57, 68, 69, 109, 140, 187, 357, 360, 377, 395, 416
as translator of AE into French, 69, 234, 236
view of AE, 45, 395
Solvay, Ernest, 81
Solvay Conference, 81–82, 163–164, 199
Sommerfeld, Arnold, 75, 81, 112, 144, 350
Soviet Union
AE's attitude toward, 227, 269–270, 347, 360, 364–365, 383, 387–389, 397–398, 404–406, 407
and anti-Semitism, 270, 404

Doctors Plot, 404–407
explodes atomic bomb, 386
Hitler invades, 329
and idea of sharing atomic bomb, 341, 344
Spang, Harry, 206–207
Spanish Civil War, 302
Spear, Nathaniel, 271
special theory of relativity, 60–62, 64–68, 70–72, 81, 408, 423
Speer, Albert, 330–331
Spinoza, Baruch, 22, 186, 192, 245, 307
splitting of atom, 271, 316–317, 344–345. *See also* atomic bomb
Stachel, John, 48, 301, 429, 435
Stalin, Josef
    AE writes to, 297, 367
    death of, 407
    and Doctors Plot, 404–407
    and Isaac Don Levine, 226–227, 269–270
    and Moscow Treason Trials, 299–302
Stark, Johannes, 70, 99, 105, 284, 288
Steinhardt, Alice, 98
Stern, Alfred, 17, 20, 31, 34, 157
Stern, David, 402
Stern, Otto, 341
Stevenson, Adlai, 409
Stevenson, Mrs. J. Ross, 252
Stone, Christopher, 402, 403
Stone, I. F., 205, 270, 349, 401–403, 404, 411, 412, 416
Strassmann, Fritz, 309
Stráus, Ernst, 350, 359, 360, 369, 411
Strimling, Theodore, 342
Sufflay, Milan, 220–221
Swedenborg, Emanuel, 115–116
Swing, Jane, 329–330
Swing, Raymond, 344
Swiss Patent Office, 33, 42, 52, 58, 69, 74, 157
Swope, Mrs. Gerard, 238
Szilard, Leo, 135–136, 187, 202, 248, 317, 318, 319, 320, 321, 322, 344, 345, 353, 354, 420

Talmey, Max, 4, 5, 51, 130
Talmud, Max. *See* Talmey, Max
Tanner, Hans, 76

Taylor, A. J. P., 221
Taylor, Glen, 364
Teller, Edward, 193, 318, 417
Tesla, Nikola, 103–104
Thalberg, Irving, 269
Thomas, Norman, 416–417
Thomas, Wilbur, 346
Thompson, Dorothy, 225
Thomson, Joseph, 101, 195
Thorndike, Edward, 224
Tiley, Mort, 280
Time-Life, 354
Tolischus, Otto D., 288
Tolman, Richard, 202, 208, 209, 211, 217
Trotsky, Leon, 179, 297, 299–300
Truman, Harry
    and atomic bomb, 344, 383
    becomes president, 343
    recognizes Jewish state, 368
    and Rosenberg case, 404
Tschernowitz, Chaim, 183

Ulam, Stanislaw, 293
uncertainty principle, 153, 154, 162
unified field theory
    AE lectures on, 192–193, 219, 254–255, 292–293
    AE's discouragement over, 392, 401, 415
    AE works on, 153, 173–177, 192, 198, 350, 370
    and Schrödinger, 150, 158–160
United Nations, 363, 368
Universal Studios, 210
University of Geneva, 74
University of Prague, 77, 79
University of Zurich, 28, 63, 70, 73–74
Untermyer, Samuel, 121, 213, 260
Urey, Harold, 354, 405, 409–410
U.S. Navy, 334, 336–337

Vallentin, Antonina, 177, 196, 231, 249, 265, 420
Van Anda, Carr, 174, 204
van der Waals, Johannes Diderik, 75
Veblen, Oswald, 232, 254
Velikovsky, Immanuel, 398–399, 412–413, 418–419, 424
Veltichko, Dimitrov, 221

Vessot, Robert, 431
Viereck, George Sylvester, 184–187, 205
Viereck, John Alexis, 184–185
Voigt, Woldemar, 40
von Ficker, Heinrich, 245
von Hevesy, George, 87
von Hindenburg, Paul, 199, 227, 235, 239–240, 242
von Laue, Max
  and AE, 84, 141
  captured by Nazis, 345
  as colleague of AE in Berlin, 133, 134, 148, 196, 284
  courage praised by AE, 332, 357
  defends AE, 111, 113, 245, 288
  interviewed by FBI, 419
  meets AE, 69–70
  and Nazis, 245, 284, 314–315, 339
  and Nobel Prize for AE, 135
  as Planck's assistant, 69–70
  son arrested as spy, 332
  view of AE's work, 175
  wins Nobel Prize, 90
von Laue, Theodore, 301–302, 332–333
von Neumann, John, 254, 334, 371, 381
von Rath, Ernst, 308–309
von Seeckt, Hans, 231
von Stürgkh, Karl, 93
von Uexküll, Margarete, 23, 67, 78
von Waldeyer-Hartz, Wilhelm, 114
von Weizsacker, C. P., 319

Wachsmann, Konrad, 181
Wagner, Ernst, 144
Wald, George, 362, 400–401, 415
Waldow, Herta, 180–181
Walker, Jimmy, 204
Wallace, Henry, 363–364
Wallace, Irving, 143
Wallace, Sylvia, 143–144
Wallace-Greenberg, Oscar, 408
Wallenberg, Raoul, 367
Walter, Bruno, 193
Warburg, Emil, 95, 96, 99, 114, 144
Warburg, Max, 141
Warner Brothers Studios, 210
War Resisters International, 238, 374
Watson, Edwin, 319
Watt, Robert, 241

Watters, Leon
  accompanies AE to Mellon Institute, 271–273
  AE visits in New York, 257, 259–260, 267–268, 274–276
  as diarist, 261
  drives AE, 258, 308
  meets AE, 242
  remarries, 291
  view of AE, 262–263
  visits AE in Old Lyme, Conn., 280
Weber, Heinrich, 8, 18, 31
Weber, Prof., 36
Webster, A. G., 124
Weimar Republic, 107, 227, 243, 252, 277, 315
Weiss, Pierre, 95, 109
Weisskopf, Victor, 354
Weizmann, Chaim, 137, 138, 145, 183, 184, 218, 349
  in aftermath of Rathenau assassination, 142
  death of, 400
  on U.S. fund-raising tour, 120, 122, 123, 124, 125–126, 130
Weizmann, Vera, 120, 128, 129
Wells, H. G., 104, 200, 201, 295
Wertheimer, Max, 97
Weyl, Hermann, 285–286, 332, 333, 353, 376
Weyland, Paul, 105, 110, 111, 113, 116
Wharton, James, 148
Wheeler, John, 91, 163, 254, 330, 361, 407, 408, 409, 432
Whitehead, Alfred Lord, 101, 131
Whittaker, Edmund, 411–412
Whyte, Lancelot Law, 169, 187–188
Wiedemann, Eilhard, 58
Wien, Wilhelm, 72, 85, 86, 96
Wigner, Eugene, 134, 193, 316, 319, 320, 321, 329, 330, 353, 359
Williams, Wythe, 174
Wilson, Robert, 430
Wilson, Woodrow, 94
Winchell, Walter, 384
Winning, Gustav, 315
Winteler, Anna, 32, 70
Winteler, Jost, 9–10, 13
Winteler, Julius, 70

Winteler, Marie, 9, 10, 11
  as AE's first love, 12–13, 21, 22, 36–37
  in later life, 48–49, 323
Winteler, Paul, 32, 70, 109, 312, 379
Winteler, Pauline, 9, 22
Winteler, Violet, 379
Winter, Ella, 378–379
Winterkorn, Hans, 315
Winternitz, Moritz, 80–81
Wirth, Joseph, 140, 141
Wise, Stephen, 253, 275, 348–349
Wise, Stephen, Mrs., 275
Wohlwend, Hans, 35
Wolff, Theodor, 141
Woman's Patriot Corporation, 237, 241, 242–243
Woolf, Samuel Johnson, 181–182
World Federalist Movement, 378, 381

World War I, 89–99
Wyden, Peter, 318

Xenophon, 360

Yugoslavia, 220–221

Zakel, Ludek, 435–436
Zangger, Heinrich, 69, 87, 93, 96, 166
Zeeman, Pieter, 114
Zen Buddhism, 322
Zinn, Walter H., 319
Zionists, 81, 120, 123, 124–125, 183, 185, 217, 337, 338, 342, 349
Zurich Polytechnic, 7, 8, 12–13, 16–18, 22–24, 31, 82, 84–86
Zweig, Arnold, 177
Zwicky, Fritz, 175, 236